Red Dot Design Yearbook 2015/2016

Edited by Peter Zec

reddot award
product design

About this book

"Working" presents the current best products relating to the workplace. All of the products in this book are of outstanding design quality and have been successful in one of the world's largest and most renowned design competitions, the Red Dot Design Award. This book documents the results of the current competition in the field of "Working", also presenting its most important players – the design team of the year, the designers of the best products and the jury members.

Über dieses Buch

„Working" stellt die aktuell besten Produkte rund um das Themengebiet „Arbeitsplatz" vor. Alle Produkte in diesem Buch sind von herausragender gestalterischer Qualität, ausgezeichnet in einem der größten und renommiertesten Designwettbewerbe der Welt, dem Red Dot Design Award. Dieses Buch dokumentiert die Ergebnisse des aktuellen Wettbewerbs im Bereich „Working" und stellt zudem seine wichtigsten Akteure vor – das Designteam des Jahres, die Designer der besten Produkte und die Jurymitglieder.

Contents
Inhalt

Professor Dr Peter Zec
Preface of the editor
Vorwort des Herausgebers

Dear Readers,

With the Red Dot Design Yearbook 2015/2016, you are holding a very special book in your hands. In the three volumes, "Living", "Doing" and "Working", it presents to you the winners of this year's Red Dot Award: Product Design, which make up the current state of the art in all areas of product design. At the same time this compendium is a piece of living history, because this year marks the 60th anniversary of the Red Dot Award.

Since 1955, we have set out every year in search of the best that the world of industrial design has to offer. What began 60 years ago as a small industrial show in Essen with a selection of around 100 products, most of which were German, has developed consistently over the past six decades. The competition has grown, ventured out into the world and become cosmopolitan.

A visible sign of this new, international gearing appeared 15 years ago, when the name of the competition was changed to the "Red Dot Design Award" and a new logo, the Red Dot, was introduced. That logo has become synonymous with good design – and the Red Dot Award is now the largest international competition for product design. In this anniversary year alone, 1,994 companies and designers from 56 countries entered 4,928 products in the competition for a Red Dot.

Yet through all the changes, one thing has stayed the same: The heart of the competition is and will always be its independent and fair jury. The following pages will show you this year's selection by our design experts.

I wish you an inspiring read.

Sincerely, Peter Zec

Liebe Leserin, lieber Leser,

mit dem Red Dot Design Yearbook 2015/2016 halten Sie ein besonderes Buch in den Händen: Es präsentiert Ihnen in den drei Bänden „Living", „Doing" und „Working" die Gewinner des diesjährigen Red Dot Award: Product Design und damit den aktuellen State of the Art in allen Bereichen des Produktdesigns. Zugleich ist dieses Kompendium jedoch auch ein Stück Zeitgeschichte, denn der Red Dot Award feiert dieses Jahr sein 60-jähriges Bestehen.

Seit 1955 machen wir uns jedes Jahr aufs Neue auf die Suche nach dem Besten, das die Welt des Industriedesigns zu bieten hat. Was vor 60 Jahren als kleine Industrieschau in Essen mit der Auswahl von rund 100 vornehmlich deutschen Produkten seinen Anfang nahm, hat sich im Laufe der vergangenen sechs Jahrzehnte beständig weiterentwickelt. Der Wettbewerb ist gewachsen, in die Welt hinausgegangen und weltoffen geworden.

Als sichtbares Zeichen für diese neue, internationale Ausrichtung bekam er vor 15 Jahren mit der Änderung zu „Red Dot Design Award" nicht nur einen neuen Namen, sondern auch ein neues Logo, den Red Dot. Dieses Logo ist heute ein Synonym für gutes Design – und der Red Dot Award mittlerweile der größte internationale Wettbewerb für Produktdesign. Alleine in diesem Jubiläumsjahr bewarben sich 1.994 Unternehmen und Designer aus 56 Ländern mit 4.928 Produkten um eine Auszeichnung mit dem Red Dot.

Eine Sache gibt es jedoch, die bei allen Veränderungen gleich geblieben ist: Das Herzstück des Wettbewerbs ist und bleibt seine unabhängige und faire Jury. Welche Auswahl unsere Designexperten in diesem Jahr getroffen haben, das sehen Sie auf den folgenden Seiten.

Ich wünsche Ihnen eine inspirierende Lektüre.

Ihr Peter Zec

The title "Red Dot: Design Team of the Year" is bestowed on a design team that has garnered attention through its outstanding overall design achievements. This year, the title goes to Robert Sachon and the Bosch Home Appliances Design Team. This award is the only one of its kind in the world and is extremely highly regarded even outside of the design scene.

Mit der Auszeichnung „Red Dot: Design Team of the Year" wird ein Designteam geehrt, das durch seine herausragende gestalterische Gesamtleistung auf sich aufmerksam gemacht hat. In diesem Jahr geht sie an Robert Sachon und das Bosch Home Appliances Design Team. Diese Würdigung ist einzigartig auf der Welt und genießt über die Designszene hinaus höchstes Ansehen.

In recognition of its feat, the Red Dot: Design Team of the Year receives the "Radius" trophy. This sculpture was designed and crafted by the Weinstadt-Schnaidt based designer, Simon Peter Eiber.

Als Anerkennung erhält das Red Dot: Design Team of the Year den Wanderpokal „Radius". Die Skulptur wurde entworfen und angefertigt von dem Designer Simon Peter Eiber aus Weinstadt-Schnaidt.

2015	Robert Sachon & Bosch Home Appliances Design Team
2014	Veryday
2013	Lenovo Design & User Experience Team
2012	Michael Mauer & Style Porsche
2011	The Grohe Design Team led by Paul Flowers
2010	Stephan Niehaus & Hilti Design Team
2009	Susan Perkins & Tupperware World Wide Design Team
2008	Michael Laude & Bose Design Team
2007	Chris Bangle & Design Team BMW Group
2006	LG Corporate Design Center
2005	Adidas Design Team
2004	Pininfarina Design Team
2003	Nokia Design Team
2002	Apple Industrial Design Team
2001	Festo Design Team
2000	Sony Design Team
1999	Audi Design Team
1998	Philips Design Team
1997	Michele De Lucchi Design Team
1996	Bill Moggridge & Ideo Design Team
1995	Herbert Schultes & Siemens Design Team
1994	Bruno Sacco & Mercedes-Benz Design Team
1993	Hartmut Esslinger & Frogdesign
1992	Alexander Neumeister & Neumeister Design
1991	Reiner Moll & Partner & Moll Design
1990	Slany Design Team
1989	Braun Design Team
1988	Leybold AG Design Team

Red Dot: Design Team of the Year 2015
Robert Sachon & Bosch Home Appliances Design Team
Technology and Design for a better Quality of Life

For more than 125 years, the name Bosch has stood for pioneering technology and outstanding quality. And for more than 80 years, home appliances made by Bosch have lived up to this name. With the Bosch brand, consumers worldwide associate efficient functionality, reliable quality and internationally acclaimed design. In the Red Dot Award: Product Design alone, the design team led by Robert Sachon has received more than two hundred Red Dots and several Red Dot: Best of the Bests in the past 10 years. To honour this achievement, the design team of Robert Bosch Hausgeräte GmbH is being recognised this year for its consistently high and groundbreaking design performance with the honorary title "Red Dot: Design Team of the Year".

Seit mehr als 125 Jahren steht der Name Bosch für wegweisende Technik und herausragende Qualität. Diesem Anspruch sind auch die Hausgeräte von Bosch seit über 80 Jahren verpflichtet. Mit der Marke Bosch verbinden Konsumenten weltweit effiziente Funktionalität, verlässliche Qualität und ein international ausgezeichnetes Design. Allein im Red Dot Award: Product Design erhielt das Designteam unter der Leitung von Robert Sachon in den vergangenen 10 Jahren mehr als zweihundert Auszeichnungen mit dem Red Dot und darüber hinaus mehrere Auszeichnungen mit dem Red Dot: Best of the Best. In diesem Jahr wird das Designteam der Robert Bosch Hausgeräte GmbH für seine kontinuierlich hohen und wegweisenden Gestaltungsleistungen mit dem Ehrentitel „Red Dot: Design Team of the Year" ausgezeichnet.

Quality of Life

Robert Sachon, Global Design Director of the Bosch brand, summarises the brand values of the company as follows: "Bosch has a long tradition spanning more than 125 years, and has always stood for technical perfection and superior quality." Bosch home appliances always put people first, in keeping with the principles of the company's founder Robert Bosch: technology and quality, credibility and trust, along with a sense of responsibility for people's well-being.

As early as 1932, Robert Bosch wrote that "progress in the development of technology serves, in the broad sense of the term, to afford the largest service to humanity – technology that is intended and able to provide all of humanity with enhanced opportunities and happiness in life." In the wake of the global economic crisis, Robert Bosch anticipated a notion, or value, that was not to enter the public consciousness until the early 1970s: the concept of quality of life.

Bosch develops premium-quality modern home appliances that facilitate people's daily lives and that help them attain a better quality of life. Nevertheless, only few home appliances have a place in the "design hall of fame". Indeed, as a category, home appliances have been largely neglected in the history of design, despite the fact that these products have, due to their technical functionality and everyday use, a high impact on the quality of life.

The Refrigerator

Home appliances from Bosch first saw the light of day at the Leipzig Spring Fair in 1933. The fair visitors must have been impressed when Bosch presented its first refrigerator: a metal barrel with a 60-litre net capacity. At the time, refrigerators were still regarded as "machines" given the obtrusive metal cylinder that gave them more of a tool character. The round form, which we today associate more with a washing machine, served the technical purpose of saving energy and retaining the cold optimally inside the appliance. A refrigerator in the shape of a cabinet, more suitable for daily use, was introduced on the market three years later. Resting on high feet, this newer model looked more like a piece of furniture than a technical machine. In 1949, Bosch then presented the refrigerator that, with its subtle, sober and highly industrialised form, was to become a design classic like no other home appliance.

In 1996, Bosch brought out a new edition of the 1949 design classic, thereby turning the model dating from the time of Germany's post-war economic miracle into a historical icon. Nevertheless, this new edition, introduced in a world that was rapidly changing and becoming faster with the rise of globalisation and the Internet, was designed to remain familiar and recognisable – in other words, to be "cool and calm" at once. That said, the new edition of the classic was much more than a nostalgic reminder of a bygone age. Equipped with modern technology, the Bosch Classic Edition awakened new desires and quickly advanced to become a longtime bestseller on its own terms.

Die Lebensqualität

„Bosch hat eine lange Tradition von mehr als 125 Jahren und steht seit jeher für die Themen ‚Technische Perfektion' und ‚Überlegene Qualität'", fasst Robert Sachon, Global Design Director der Marke Bosch, die Markenwerte des Unternehmens zusammen. Bosch Hausgeräte stellen immer den Nutzen für den Menschen in den Mittelpunkt, getreu den Grundsätzen des Firmengründers Robert Bosch: Technik und Qualität, Glaubwürdigkeit und Vertrauen sowie Verantwortungsbewusstsein zum Wohle der Menschen.

Bereits 1932 schrieb Robert Bosch, „dass die Fortschritte in der Entwicklung der Technik im vollem Umfange des Wortes dazu dienen, der Menschheit die größten Dienste zu leisten. Der Technik, die dazu bestimmt und in der Lage ist, der gesamten Menschheit ein Höchstmaß an Lebensmöglichkeit und Lebensglück zu verschaffen." Unter dem Eindruck der Weltwirtschaftskrise nimmt Robert Bosch einen Gedanken vorweg, der erst zu Beginn der 1970er Jahre in die öffentliche Wahrnehmung treten soll: den Begriff der Lebensqualität.

Bosch entwickelt moderne Haushaltsgeräte von höchster Qualität, die den Alltag der Menschen erleichtern und ihnen zu mehr Lebensqualität verhelfen. Indes ist nur wenigen Hausgeräten ein Platz im Olymp des Designs vorbehalten; wie überhaupt den Hausgeräten nicht immer der gebührende Platz in der Geschichte des Designs eingeräumt wird, obgleich sie aufgrund ihrer technischen Funktionalität und ihres alltäglichen Gebrauchs einen hohen Einfluss auf die Lebensqualität haben, weil sie die Hausarbeit enorm erleichtern und das Leben insgesamt verändern.

Der Kühlschrank

Die Geburtsstunde der Hausgeräte von Bosch schlägt auf der Leipziger Frühjahrsmesse 1933. Die Messebesucher dürften nicht schlecht gestaunt haben, als Bosch seinen ersten Kühlschrank vorstellt: eine Trommel mit 60 Litern Nutzinhalt. Zur damaligen Zeit ist es durchaus üblich, von einer Kältemaschine zu sprechen, da der Metallzylinder unmissverständlich seinen Werkzeugcharakter zum Ausdruck bringt. Die runde Form, die man heute eher mit einer Waschmaschine assoziiert, hat technische Gründe. Es geht darum, Energie zu sparen und die Kälte optimal im Gerät zu halten. Die für den Alltagsgebrauch besser geeignete Schrankform kommt drei Jahre später auf den Markt und erinnert mit ihren hohen Füßen mehr an ein Möbelstück als an eine technische Maschine. 1949 stellt Bosch dann den Kühlschrank vor, der es mit seiner zurückhaltenden, sachlichen und industriell geprägten Form wie kaum ein anderes Hausgerät zum Designklassiker schafft.

1996 bringt Bosch eine Re-Edition der Designikone von 1949 heraus und setzt dem Modell aus der Zeit des deutschen Wirtschaftswunders damit ein Denkmal. In der Zeit der Globalisierung und der sich anbahnenden Veränderung durch das Internet ist die Wiederauflage des Klassikers so etwas wie die eiserne Ration an Vertrautheit in einer sich rasant beschleunigenden Welt: ein Kälte- und Ruhepol zugleich. Doch die erneute Auflage des Klassikers ist weit mehr als die nostalgische Erinnerung an eine verloren gegangene Zeit. Ausgestattet mit moderner Technik weckt die Bosch Classic Edition schnell neue Begehrlichkeiten und avanciert wiederum selbst zum Best- und Longseller.

An evolutionary design approach. In 1933, Bosch unveils a surprise in the form of the first Bosch-manufactured refrigerator. It is compact and round. The appliance is launched on the market in 1949, and its typical shape becomes the archetype for refrigerators and goes on to become an iconic design of the Bosch brand. In 2014, Bosch presents a modern reinterpretation of the design classic in the form of the CoolClassic.
Ein evolutionärer Designansatz. 1933 überrascht Bosch mit dem ersten Kühlschrank aus eigener Produktion. Er ist kompakt und er ist rund. 1949 kommt das Gerät auf den Markt, dessen typische Form den Archetyp des Kühlschranks prägte und der zur Designikone der Marke Bosch avancieren sollte. 2014 stellt Bosch mit dem CoolClassic eine moderne Neuinterpretation des Designklassikers vor.

In 2014, Bosch introduced yet another model, the CoolClassic. In contrast to the 1996 edition, which sought the greatest possible loyalty to the original at the level of form, the CoolClassic aimed for a new interpretation and evolution of the refrigerator. While its rounded corners, sturdy metal housing and large Bosch lettering evoke the original model, the new CoolClassic has a charm of its own. Moreover, behind the façade, touchpad controls and a digital display allow for optimal control.

The CoolClassic is exemplary of a design principle that has been applied to many models of leading car brands. The principle pursues two objectives: the right combination of tradition and innovation, of identity and difference; and an evolutionary approach to design that makes the appliances unmistakably identifiable as "a Bosch" while at the same time underscoring their autonomy. Overall, both the Bosch Classic Edition and the CoolClassic are representative of a certain lifestyle that renders them suitable for use as stand-alone appliances in lofts and living areas as well as in offices, agencies and art studios. Their ultimate design message is that comfort and enjoyment already begin before opening the refrigerator door.

In addition, with its "ColorGlass Edition" of reduced, purist and handle-free models featuring coloured glass fronts, the Bosch brand has once again proved to be a design pioneer, particularly in response to the trend of open-concept kitchen and living spaces that call for a more homelike look of appliances.

2014 kommt mit dem CoolClassic dann eine moderne Neuinterpretation auf den Markt. Im Unterschied zur Re-Edition von 1996, die sich formal um eine möglichst große Originaltreue bemüht, geht es beim Re-Design des neuen CoolClassic um eine Überarbeitung und Weiterentwicklung des Kühlschranks. Die abgerundeten Ecken, das robuste Metallgehäuse und der große Bosch-Schriftzug erinnern zwar noch an das Original, der neue CoolClassic versprüht aber seinen eigenen Charme. Hinter den Fronten verbirgt sich moderne Elektronik, die durch berührungsempfindliche Tasten gesteuert wird und digital ablesbar ist.

Hier kommt ein gestalterisches Prinzip zum Ausdruck, das sich auch bei vielen Modellen erfolgreicher Automarken findet: das gelungene Zusammenspiel von Tradition und Innovation, von Identität und Differenz – ein evolutionärer Ansatz in der Gestaltung, der die Hausgeräte eindeutig der Marke Bosch zuordnet und zugleich ihre Eigenständigkeit unterstreicht. Beiden Geräten, der Bosch Classic Edition wie dem CoolClassic ist der Lifestyle-Charakter gemein, den sie als Solitär im Loft und Wohnbereich, aber auch in Büros, Agenturen und Ateliers unterstreichen. Hier beginnt der Genuss eben schon vor dem Öffnen der Kühlschranktür.

Daneben gelingt es der Marke mit ihrer „ColorGlass Edition" – puristisch reduzierten, grifflosen Modellen mit farbigen Glasfronten – gestalterisch nach vorne zu blicken und der Öffnung der Küche zum Wohnraum und damit einhergehenden wohnlicheren Gerätekonzepten als Trendsetter voranzuschreiten.

The Stove and the Oven

As with the refrigerator, the history of the kitchen stove was initially shaped by engineers. Similar to the "cooling machine", stoves were at first called "cooking machines". And as with refrigerator design, the initial aim of stove design was to emulate furniture and to achieve technical feasibility. The electric stove was not marketed until the 1930s, like the refrigerator, and required the concentration of heat within a narrow space accommodating three or four hotplates of different diameters. In retrospect, the evolution of the cast-iron hotplates of the early years to the modern hobs made of ceramic glass is exemplary of a technological optimisation process during which the energy efficiency of the appliances was continuously improved.

In parallel to the technological development, the design of the control elements and the usage of the appliances evolved. After all, the stove and oven continue to be two of the most used appliances in the kitchen – so much so that Robert Sachon seeks inspiration for his work while cooking, possibly explaining his special focus on these two appliances. As a result, he often prefers to work with them on weekends and enjoys "dealing with appliances that allow one to engage directly and creatively with the product." As head of brand design, Sachon is concerned with the question of which design elements lend themselves to being transferred to other products or categories.

The Series 8 built-in appliances, distinguished with a Red Dot: Best of the Best award, are characterised by a simple, sleek design that prioritises the compatibility of different appliances as well as an easy, intuitive operating concept. The defining element is the flush-surface, centralised control ring with which the user can control the wide range of functions with a simple turn. Moreover, trends occurring in other areas are of relevance for the design of home appliances. For example, open-living concepts have repercussions for the design and integration of home appliances. "The current trend is to design concepts that fit well into the new living environments," explains Robert Sachon. "Integration as a whole is an important topic."

The Kitchen Machine

In 1952, Bosch introduced the first food processor on the market. With a name that says it all – the Neuzeit I (Modern Era I) – this product symbolised the beginning of a new era, since it decreased the number of actions and movements required for preparing food. Neuzeit I was able to stir, knead, cut, chop, mash, grate and grind.

In today's busy world, these small home and kitchen devices are indispensable helpers. As with all other equipment categories, here too the main objective is to improve the usefulness and convenience for the user with good design and innovative technology. Robert Sachon summarises the Bosch corporate philosophy as follows: "We simply try to give the appliances the best that there is."

Der Herd und der Ofen

Die Geschichte des Küchenherds ruht wie die Entwicklung des Kühlschranks zunächst in den Händen von Ingenieuren. Analog zur Kältemaschine spricht man zu Beginn der Entwicklung noch von einer Kochmaschine. Und ebenso wie die Form des Kühlschranks orientiert sich auch die Gestaltung des Herdes zunächst an dem Vorbild des Mobiliars und der technischen Machbarkeit. Der Elektroherd setzt sich parallel zum Kühlschrank erst in den 1930er Jahren durch, schließlich musste zunächst die Wärme auf einen engen Raum konzentriert werden, der drei oder vier Herdplatten mit unterschiedlichen Durchmessern Platz bot. Die Entwicklung von den gusseisernen Herdplatten der Anfangsjahre bis zu den modernen Kochfeldern aus Glaskeramik liest sich rückblickend wie ein technischer Optimierungsprozess, bei dem die Energieeffizienz der Geräte laufend verbessert wurde.

Parallel zur technischen Entwicklung hat sich aber auch die Gestaltung der Bedienelemente und der Gebrauch der Geräte verändert. Herd und Backofen sind nach wie vor zwei der am häufigsten genutzten Geräte in der Küche. Und auch Robert Sachon holt sich nicht zuletzt beim Kochen Inspiration für seine Arbeit. Daher gilt beiden Geräten sein besonderes Augenmerk. Gerade an den Wochenenden nimmt er sich Zeit dafür, „weil man es mit Geräten zu tun hat, die es einem unmittelbar erlauben, kreativ mit dem Produkt umzugehen", sagt Sachon. Als Leiter des Markendesigns beschäftigt ihn dabei auch die Frage, „welche gestalterischen Elemente sich eignen, um sie auf andere Produkte oder Kategorien zu übertragen".

Das mit dem Red Dot: Best of the Best ausgezeichnete Einbauprogramm der Serie 8 zeichnet sich durch eine einfache, klare Linienführung aus, die die perfekte Kombinierbarkeit unterschiedlicher Geräte in den Vordergrund stellt, sowie durch ein einfaches, intuitives Bedienkonzept. Prägendes Element ist der flächenbündig integrierte, zentrale Bedienring, mit dem der Nutzer mit einem Dreh die Vielzahl der Gerätefunktionen im Griff hat. Auch Trends aus anderen Bereichen sind für das Hausgeräte-Design von Bedeutung. Offene Wohnkonzepte wirken sich beispielsweise auch auf das Design und die Integration von Hausgeräten aus. „Der Trend geht aktuell zu Gestaltungskonzepten, die sich gut in die neuen Wohnwelten einfügen", erklärt Robert Sachon. „Integration überhaupt ist ein wichtiges Thema."

Die Küchenmaschine

1952 bringt Bosch die erste Küchenmaschine auf den Markt: die „Neuzeit I". Ihr Name ist Programm. Sie steht für den Beginn einer neuen Zeit, da sie den Hausfrauen viele Handgriffe abnimmt und die Zubereitung des Essens erleichtert. Die „Neuzeit I" kann rühren, kneten, hacken, schnitzeln, pürieren, mahlen und reiben.

Heute sind im modernen Alltag die kleinen Haus- und Küchengeräte unverzichtbare Helfer. Wie bei allen anderen Gerätekategorien gilt auch hier die Prämisse, den Nutzen und den Komfort für den Anwender mit guter Gestaltung und innovativer Technik zu verbessern. „Wir versuchen einfach, den Geräten das Beste mitzugeben", bringt Robert Sachon den Anspruch von Bosch auf den Punkt.

A new design language. The Series 8 range of built-in appliances awarded the Red Dot: Best of the Best sets new standards in the design of user interfaces. The defining element is the flush-surface, centralised control ring with which the user can control the wide range of functions with a simple turn.

Eine neue Designsprache. Das mit dem Red Dot: Best of the Best ausgezeichnete Einbauprogramm der Serie 8 setzt neue Maßstäbe in der Gestaltung von User Interfaces. Prägendes Element ist der flächenbündig integrierte, zentrale Bedienring, mit dem der Nutzer mit einem Dreh die Vielzahl der Gerätefunktionen im Griff hat.

Easy to use and featuring high-quality materials and consistent design: the Styline series of small home appliances from Bosch is not only modestly elegant but also uncompromisingly useful.

Hochwertige Materialien, einfache Bedienbarkeit und gestalterisch in einer Linie: Die Styline-Serie der kleinen Hausgeräte von Bosch präsentiert sich nicht nur zurückhaltend elegant, sondern ist auch kompromisslos nützlich.

A good example is the food processor MUM, which received a Red Dot: Best of the Best award as early as 2011. Similar to its predecessor Neuzeit I, its trademark is also versatility and energy-efficient performance. And, with an emphasis on high technological performance, the MUM exhibits the evolutionary design approach unique to Bosch. From one product generation to the next, these appliances affirm and promote the brand identity. Whether large or small home appliances: all are created with the holistic, evolutionary design approach of the Bosch brand, and all follow the uniform design language and uniform design principles derived from the brand values.

The Washing Machine

The first washing machine from Bosch was put into series in 1958, followed in 1960 by the first fully automatic washing machine that allowed clothes to be laundered in a single wash, rinse and spin cycle. In 1972, a new unit was introduced that combined washer and dryer in one. Since then, the washing programme options are being continually refined and expanded. All of these appliances contribute to making the days of hard manual labour a thing of the past. Moreover, the technology and design of the appliances have not only laid the foundation for a better quality of life in the household but also for social change. To the extent that the chore of washing increasingly became a minor matter, women liberated themselves from their role as housewife, and working women were no longer the exception to the rule.

Ein gelungenes Beispiel ist die Küchenmaschine MUM, die bereits 2011 mit dem Red Dot: Best of the Best ausgezeichnet wird. Analog zu ihrem historischen Vorgänger, der „Neuzeit I", ist auch ihr Markenzeichen die Vielseitigkeit und die energieeffiziente Leistung. Im Einklang mit der hohen technischen Leistung spiegelt sich in der Formensprache der Küchenmaschine MUM aber auch der evolutionäre Gestaltungsansatz mustergültig wider. Von Generation zu Generation kommt im Produktdesign auch die Anbindung an die Marke Bosch zum Ausdruck. Ob große oder kleine Hausgeräte – allen wird der ganzheitliche, evolutionäre Designansatz der Marke Bosch zuteil, folgen sie doch sämtlich einer den Markenwerten entspringenden, einheitlichen Designsprache und einheitlichen Gestaltungsprinzipien.

Die Waschmaschine

Die erste Waschmaschine aus dem Hause Bosch geht 1958 in Serie. Es ist noch kein Waschvollautomat. Dieser kommt 1960 auf den Markt und ermöglicht es, die Wäsche erstmals in einem einzigen Durchlauf waschen, spülen und schleudern zu lassen. 1972 werden die Waschmaschine und der Trockner dann eins. Seither wird das Geräteprogramm immer weiter verfeinert und ausgebaut. All diese Geräte sorgen dafür, dass die Zeiten der mühevollen Handarbeit langsam zu Ende gehen. Die Technik und die Gestaltung der Geräte legen nicht nur den Grundstein für mehr Lebensqualität im Haushalt, sondern auch für eine gesellschaftliche Veränderung. Das Thema Waschen wird mehr und mehr zur Nebensache. Die Frauen emanzipieren sich von ihrer Rolle als Hausfrau. Berufstätige Frauen sind keine Ausnahme mehr.

Although women's liberation is usually discussed from the standpoint of a social and political movement, the topic merits further examination from the perspective of home appliance design. Indeed, the changing effect of technology and design extends beyond the material manifestation of the products. On the one hand, design concepts in the home appliance sector are always oriented towards long-term societal trends. Yet on the other hand, the designed products themselves have an impact on society, in particular through their technical function and their user-oriented utility. Here, at the latest, it becomes clear that quality of life is far more than the sum of the products that we own. The importance and impact of design go beyond what is materialised in a product, becoming discernible only in the changing conditions of use and their impact.

The Dishwasher

In 1964, the first dishwasher from Bosch went into serial production. Thanks to technological developments, dishwashers have since become very quiet. So much so that the remaining time of the programme cycles is now explicitly indicated and of late even projected onto the floor so that the user knows when the machine has finished the washing and drying process.

"It's possible to tolerate certain sounds, but it's fair to say that we're probably better off without them," says Robert Sachon. "Essentially, a common priority of all product categories is to ensure that appliances are quiet, so as not to disturb users. At the same time, we have to generate and design the sounds that consumers are supposed to hear which is not always easy." This shows that the team led by Robert Sachon values not only the quality of the product and its exterior design, but also the quality of the experience, from the operation of and interaction with the appliance to the sound design and haptics. Finally, Bosch wants design perfection to be experienceable with all the senses.

The Internet of Things – Design 4.0

Bosch undoubtedly invests heavily in the development and design of new technologies. And Bosch wouldn't be Bosch if it didn't capitalise on emerging technological opportunities to achieve more convenience and ease of use in home appliances.

"Under the slogan the 'Internet of Things', visions have been presented at trade fairs for years," explains Robert Sachon. "With the digitisation and networking of appliances, these visions are now becoming reality and taking on concrete forms," says the chief designer. With the "Home Connect" app, Bosch's digitally connected home appliances can be controlled using a smartphone or tablet PC. Among these is the Series 8 oven, distinguished with a Red Dot: Best of the Best. "'Home Connect' is a good example of how we generate real added value for our customers with new technologies," says Jörg Gieselmann, Executive Vice President Corporate Brand Bosch at BSH Hausgeräte GmbH. He adds: "With the app, setting up the appliance, making or changing basic settings or operating it when away from home becomes child's play. In this way we're creating wholly new possibilities and freedoms in the household."

Während die Emanzipation der Frauen häufig unter dem Gesichtspunkt einer sozialen und politischen Bewegung verhandelt wird, kann es sich durchaus lohnen, das Thema auch aus der Perspektive des Hausgeräte-Designs zu beleuchten. Über die materialisierte Form der Produkte hinaus zeigt sich die verändernde Wirkung von Technik und Design. Auf der einen Seite orientieren sich Gestaltungskonzepte im Hausgerätebereich immer auch an den langfristigen gesellschaftlichen Trends. Auf der anderen Seite nehmen die gestalteten Produkte wiederum selbst Einfluss auf die Gesellschaft, insbesondere über ihre technische Funktion und ihren am Nutzer orientierten Gebrauch. Spätestens hier wird deutlich, dass Lebensqualität weit mehr ist als die Summe der Produkte, die wir besitzen. Die Bedeutung und die Wirkung des Designs gehen über das hinaus, was sich in einem Produkt materialisiert. Sie werden erst in den sich verändernden Bedingungen der Nutzung und des Gebrauchs und deren Wirkung ablesbar.

Der Geschirrspüler

1964 geht der erste Geschirrspüler von Bosch in Serie. Aufgrund der technischen Entwicklung sind die Geschirrspüler inzwischen so leise und geräuscharm geworden, dass die Gestalter über zusätzliche Projektionsflächen die Restlaufzeit der Geräte auf den Boden projizieren, damit der Benutzer weiß, wann das Gerät den Spül- und Trocknungsvorgang beendet hat.

„Man kann mit bestimmten Geräuschen gut leben, man kann ohne sie vielleicht noch etwas besser leben", sagt Robert Sachon. „Es zieht sich im Grunde durch alle Produktkategorien hindurch, dass die Geräte immer angenehm leise arbeiten, damit man eben nicht gestört wird. Und die Geräusche, die der Konsument wahrnehmen soll, versuchen wir bewusst zu gestalten, was nicht immer einfach ist", so Sachon. Hier zeigt sich, dass das Team um Robert Sachon stets versucht, über die Produktqualität hinaus eine Erlebnisqualität zu vermitteln, bei der neben der äußeren Gestaltung auch die Bedienung und Interaktion mit den Geräten, das Sounddesign und die Haptik eine große Rolle spielen. Schließlich will Bosch im Design Perfektion mit allen Sinnen erfahrbar machen.

Das Internet der Dinge – Design 4.0

Bosch gehört ohne Zweifel zu den Unternehmen, die viel in die Entwicklung und Gestaltung neuer Technologien investieren. Und Bosch wäre nicht Bosch, wenn das Unternehmen die sich bietenden technischen Möglichkeiten nicht auch für eine einfachere Handhabung und leichtere Bedienbarkeit der Hausgeräte nutzen würde.

„Unter dem Schlagwort ‚Internet der Dinge' gibt es ja bereits seit Jahren Visionen, die beispielsweise auf Messen vorgestellt werden", erläutert Robert Sachon. „Mit der Digitalisierung und der Vernetzung der Geräte werden diese Visionen heute Realität und nehmen ganz konkrete Formen an", so der Chefdesigner. Mit der „Home Connect"-App können vernetzte Hausgeräte von Bosch mithilfe eines Smartphones oder Tablets gesteuert werden, zum Beispiel der Serie 8 Backofen, der mit dem Red Dot: Best of the Best ausgezeichnet wurde. „Home Connect ist ein gutes Beispiel dafür, wie wir mit den neuen Technologien echte Mehrwerte für unsere Kunden

With the effortless implementation of digitally networked home appliances, Bosch is once again setting a new standard in the design and use of such appliances. And, as in the early twentieth century, the kitchen is today again becoming a new focal point of communication, even if under the changed information and communication environment of the twenty-first century. Of course, these developments also affect the daily work and professional identity of designers. As Robert Sachon explains, "In the past our team was very heavily influenced by industrial designers, while nowadays we are focusing more on the topic of user interface design." Yet this is only to be expected since the perceptions and communication patterns of consumers are evolving. Essentially, it means that technological changes always lead to changes in the design of the future.

generieren", erklärt Jörg Gieselmann, Executive Vice President Corporate Brand Bosch der BSH Hausgeräte GmbH. „Die Verbraucher können mithilfe der App kinderleicht und einfach ihr Gerät in Betrieb nehmen, Grundeinstellungen vornehmen und verändern oder es von unterwegs aus bedienen. So schaffen wir ganz neue Möglichkeiten und Freiheiten im Haushalt."

Mit der mühelosen Benutzung digital vernetzter Hausgeräte setzt Bosch abermals einen neuen Standard in der Gestaltung und im Gebrauch von Hausgeräten. Und wie bereits zu Beginn des 20. Jahrhunderts wird die Küche heute wieder zu einem neuen Mittelpunkt der Kommunikation, wenn auch unter den veränderten Informations- und Kommunikationsbedingungen des 21. Jahrhunderts. Natürlich beeinflussen diese Entwicklungen auch die tägliche Arbeit und das Selbstverständnis der Designer, wie Robert Sachon erläutert: „Wurde unser Team früher sehr stark von Industriedesignern geprägt, so verstärken wir uns heute beim Thema ‚User Interface Design'." Das kann auch nicht anders sein, denn auch die Wahrnehmung und die Kommunikation der Konsumenten entwickeln sich weiter. Insofern ergeben sich grundsätzlich aus den technologischen Veränderungen immer auch Veränderungen für das Design der Zukunft.

The Internet of Things: with the "Home Connect" app, Bosch's digitally connected home appliances can be controlled using a smartphone or tablet PC, true to the Bosch philosophy "Invented for Life".
Das Internet der Dinge: Mit der „Home Connect"-App können vernetzte Hausgeräte von Bosch mithilfe eines Smartphones oder Tablets gesteuert werden, getreu der Bosch-Philosophie „Technik fürs Leben".

Values

When Robert Bosch founded his company in 1886 – a workshop for precision mechanics and electrical engineering in Stuttgart – housework still meant hard work. Hobs were heated with an open fire. Dishes and laundry had to be laboriously cleaned by hand. Food had to be salted or boiled. Housecleaning and the weekly washing day certainly consumed a lot of time and energy.

Robert Bosch recognised this problem and developed technical solutions which others hadn't even imagined. Some 47 years after the founding of the company, in 1933, Robert Bosch moved into the production of home appliances, a division dedicated to making people's everyday lives easier. And still during the Great Depression, he spearheaded a massive overhaul of the Bosch Group, by then globally active in the field of automotive and industrial technology, in order to modernise and diversify its activities.

From the outset it was one of Robert Bosch's principles to produce the best possible quality at all times. As early as 1918, he wrote that "it has always been an unbearable thought that someone might prove, upon examining one of my products, that my performance is inferior in some way. That's why I've always made a point of releasing only products that have passed all quality tests, in other words, that were the best of the best." This maxim essentially guides the corporate philosophy to this day: technology and quality, credibility and trust, along with a sense of responsibility for people's well-being.

And in 1921 he wrote: "I've always acted according to the principle that I would rather lose money than trust. The integrity of my promises, the belief in the value of my products and in my word have always meant more to me than temporary gain." To this day, Robert Bosch and the example he set have a significant impact on the company. The founder is still the reference point that he has always been, and which he will still be tomorrow.

Throughout the years, the design team of Robert Bosch Hausgeräte GmbH has succeeded in finding a modern and appropriate interpretation of the values which its founder embodied and which manifest in the company's brand to this day. It has set pioneering standards in home appliance design and made a significant contribution to improving the quality of life. To acknowledge the design performance of the entire team, which has garnered attention over the years with its high-quality products and consistently outstanding design, the honorary title "Red Dot: Design Team of the Year" for 2015 is bestowed on Robert Sachon and the Bosch Home Appliances Design Team.

Die Werte

Als Robert Bosch im Jahr 1886 sein Unternehmen gründete – eine Werkstatt für Feinmechanik und Elektrotechnik in Stuttgart –, bedeutete Hausarbeit noch Schwerstarbeit. Kochstellen wurden mit offenem Feuer beheizt. Geschirr und Wäsche mussten aufwendig von Hand gereinigt werden. Lebensmittel mussten gepökelt oder eingekocht werden. Der Hausputz und der wöchentliche Waschtag kosteten viel Zeit und Kraft.

Robert Bosch erkannte dieses Problem und entwickelte dafür technische Lösungen, wo andere noch nicht einmal Möglichkeiten erahnten. 47 Jahre nach der Gründung des Unternehmens stieg Robert Bosch im Jahr 1933 in die Produktion von Hausgeräten ein, um mit seinen Produkten den Alltag der Menschen zu erleichtern. Und unter dem Eindruck der Weltwirtschaftskrise verordnete er dem inzwischen weltweit im Bereich der Kraftfahrzeug- und Industrietechnik agierenden Bosch-Konzern einen konsequenten Modernisierungs- und Diversifizierungskurs.

Von Anfang an ist es einer der Grundsätze von Robert Bosch, immer bestmögliche Qualität zu produzieren. „Es war mir immer ein unerträglicher Gedanke, es könne jemand bei der Prüfung eines meiner Erzeugnisse nachweisen, dass ich irgendwie Minderwertiges leiste. Deshalb habe ich stets versucht, nur Arbeit hinauszugeben, die jeder sachlichen Prüfung standhielt, also sozusagen vom Besten das Beste war", schreibt Robert Bosch bereits im Jahr 1918. An dieser Haltung orientiert sich das Unternehmen bis heute: Technik und Qualität, Glaubwürdigkeit und Vertrauen sowie Verantwortungsbewusstsein zum Wohle der Menschen.

„Immer habe ich nach dem Grundsatz gehandelt: Lieber Geld verlieren als Vertrauen. Die Unantastbarkeit meiner Versprechungen, der Glaube an den Wert meiner Ware und an mein Wort standen mir stets höher als ein vorübergehender Gewinn", schreibt Robert Bosch im Jahr 1921. Vieles von dem, was der Firmengründer gedacht und vorgelebt hat, übt bis heute eine große Anziehungskraft aus. Robert Bosch ist der Bezugspunkt des Unternehmens, der er bereits früher war und der er auch morgen noch sein wird.

Dem Designteam der Robert Bosch Hausgeräte GmbH gelingt es seit vielen Jahren, die Werte, die bereits ihr Firmengründer verkörperte und die bis heute in den Markenwerten des Unternehmens zum Ausdruck kommen, auf zeitgemäße Art zu interpretieren. Es hat im Hausgeräte-Design wegbereitende Standards gesetzt und einen wesentlichen Beitrag zu mehr Lebensqualität geleistet. Mit Blick auf die gestalterische Leistung des gesamten Teams, das über Jahre hinweg mit qualitativ hochwertigen Produkten und einem kontinuierlich hohen Gestaltungsniveau auf sich aufmerksam gemacht hat, wird der Ehrentitel „Red Dot: Design Team of the Year" im Jahr 2015 an Robert Sachon und das Bosch Home Appliances Design Team verliehen.

"We simply try to give the
appliances the best that there is."
„Wir versuchen einfach, den
Geräten das Beste mitzugeben."

Robert Sachon, Global Design Director Bosch

Red Dot: Design Team of the Year 2015
Interview: Robert Sachon
Global Design Director
Robert Bosch Hausgeräte GmbH

For generations, the Bosch name has stood for groundbreaking technology and outstanding quality. Home appliances from Bosch have been committed to these standards for over 80 years now. Consumers around the world associate the Bosch brand with efficient functionality, reliable quality and design that has won awards at an international level. The brand has won more than 500 awards in the past 10 years alone. The man behind this design success is Robert Sachon. He started his career in 1999 at Siemens-Electrogeräte GmbH, at the time under the leadership of Gerd Wilsdorf. In 2005 he moved to Robert Bosch Hausgeräte GmbH, where he took over from his predecessor Roland Vetter. Together with a team of roughly 40 employees worldwide, he has shaped the design of the Bosch brand in his role as Global Design Director. Burkhard Jacob met with him for an interview in the Red Dot Design Museum Essen.

Mr Sachon, you have been Global Design Director for the Bosch brand for 10 years now. How would you describe your role as head designer?

The term "head designer" is pretty accurate. It sounds a little like a head chef. Similar to the chef de cuisine, who is in charge of the kitchen crew in fine dining establishments, I lead a team of employees who are responsible for the design of the home appliances.

Do you also get stuck in?

I am one of those designers who not only manage other people but are also happy to get their own hands dirty in order to set out the design direction. We have lots of different product categories, but at the end of the day the point is of course to shape the face of the Bosch brand in a similar way to a signature or a common design language. And that's something I like to stay involved in.

Mr Sachon, you exude a calmness that makes me curious as to what your star sign is?

Aries – a healthy mix of diplomacy and stubbornness.

Der Name Bosch steht seit Generationen für wegweisende Technik und herausragende Qualität. Diesem Anspruch sind die Hausgeräte von Bosch seit über 80 Jahren verpflichtet: Mit der Marke Bosch verbinden Konsumenten weltweit effiziente Funktionalität, verlässliche Qualität und ein international ausgezeichnetes Design. Allein in den letzten 10 Jahren wurden mehr als 500 Auszeichnungen gewonnen. Der Mann hinter diesen Design-Erfolgen ist Robert Sachon. Er begann seine Karriere 1999 bei der Siemens-Electrogeräte GmbH, damals unter Gerd Wilsdorf. 2005 wechselte er zur Robert Bosch Hausgeräte GmbH und beerbte seinen Vorgänger Roland Vetter. Gemeinsam mit einem Team von weltweit rund 40 Mitarbeitern hat er als Global Design Director das Design der Marke Bosch geprägt. Burkhard Jacob traf ihn im Red Dot Design Museum Essen zum Interview.

Herr Sachon, seit 10 Jahren sind Sie nun Global Design Director der Marke Bosch. Wie würden Sie Ihre Tätigkeit als Chefdesigner beschreiben?

Der Begriff „Chefdesigner" trifft es schon ganz gut. Es klingt ein wenig wie Chefkoch. Ähnlich dem Chef de Cuisine, der in der gehobenen Gastronomie die Küchenbrigade leitet, führe ich ein Team von Mitarbeitern, die für das Design der Haushaltsgeräte verantwortlich sind.

Greifen Sie auch selbst zum Kochlöffel?

Ich bin einer der Gestalter, die nicht nur managen, sondern auch selbst zum Stift greifen, um die Gestaltungsrichtung vorzugeben. Wir haben viele unterschiedliche Produktkategorien, aber am Ende des Tages geht es natürlich darum, das Gesicht der Marke Bosch im Sinne einer Art Handschrift, einer gemeinsamen Designsprache zu prägen. Und das möchte ich mir gerne erhalten.

Herr Sachon, Sie strahlen eine Ruhe aus, die die Frage provoziert, welches Sternzeichen Sie sein könnten?

Widder – eine gesunde Mischung aus Diplomatie und Dickköpfigkeit.

What characteristics of an Aries are helpful for the role of designer?

There are two characteristics that help me greatly in my work: a tendency to be a perfectionist and a certain amount of tenacity. Both of these things help me to work on different projects of differing durations and scope in order to bring long-term topics to fruition for the Bosch brand.

Do you have a design role model?

That's a difficult question, because I suppose we ultimately always come back to the heroes of design history. I have to admit I have huge respect for Dieter Rams, even though that's probably something that lots of designers say. His design language resonates with me: its clarity, order and meaningfulness. I am of a similar mindset. Without wanting to overstate their importance, his ten principles of good design are still valid today. I am fascinated by this clear design language, which has become very popular again nowadays in particular, making it all the more fitting for Bosch brand values.

What does the Bosch brand stand for?

Bosch has a long tradition spanning more than 125 years, and has always stood for technical perfection and superior quality. In our design, we take these rational values and make them tangible at an emotional level by means of clear design that showcases high-quality materials and their uncompromising finish in a precise manner down to the smallest detail. This is a holistic approach to design which we apply to all of our products

Welche Eigenschaften des Widders sind denn hilfreich für die Tätigkeit als Designer?

Es gibt zwei Eigenschaften, die mir bei der Tätigkeit sehr entgegenkommen: ein gewisser Hang zum Perfektionismus und eine gewisse Beharrlichkeit. Beides hilft mir, zeitgleich an unterschiedlichen Projekten mit unterschiedlicher Laufzeit und Tragweite zu arbeiten, um langfristige Themen für die Marke Bosch durchzusetzen.

Haben Sie ein Vorbild in Fragen der Gestaltung?

Das ist eine schwierige Frage, weil man vermutlich immer bei den Heroen der Designgeschichte landet. Trotzdem – ich habe einen wahnsinnigen Respekt vor Dieter Rams, auch wenn das wahrscheinlich viele Designer sagen. Aber seine Gestaltungssprache liegt mir schon sehr nahe: die Klarheit, die Ordnung, die Sinnhaftigkeit. Da sehe ich eine ähnliche Geisteshaltung. Seine zehn Gebote des Designs haben ja heute noch ihre Gültigkeit, ohne sie religiös überhöhen zu wollen. Mich fasziniert diese klare Gestaltungssprache, die gerade heute wieder hoch im Kurs steht – und die umso mehr zu den Markenwerten von Bosch passt.

Wofür steht die Marke Bosch?

Bosch hat eine lange Tradition von mehr als 125 Jahren und steht seit jeher für die Themen „Technische Perfektion" und „Überlegene Qualität". Diese rationalen Werte machen wir in unserem Design emotional erfahrbar durch eine klare Gestaltung, welche hochwertige Materialien und deren kompromisslose Verarbeitung präzise bis ins kleinste Detail in Szene setzt.

worldwide. In this regard, it is fair to speak of a uniform design language, or DNA, of the Bosch brand.

So you are a brand manager as well as a designer?

Most definitely. Unlike other companies, where design is part of technical development, design at Bosch benefits from the fact that it is a key part of brand management. Our design team plays an important role and has a clear remit, as it gets involved with the development of product concepts at a very early stage – long before any thoughts of marketing for the products or of an advertising campaign.

Do you base your design decisions on the Bosch brand values?

The aim is always to bring the brand values to life. In the area of home appliances, we see a lot of products that pass on certain design features to the next generation. Consequently there is an underlying evolutionary thought process involved. This goes without saying with a brand like Bosch, which has been conveying the same values for over 125 years.

Basing design language on the brand values is one side of the coin. The other, which you have just described, relates to a product's use and benefit for the consumer.

Absolutely. We pursue a user-centred design approach where consumer monitoring plays a major role. We benefit from the fact that we too are all users of home appliances. We therefore can observe ourselves as well as others. And when observing ourselves, it's important to always be aware of our blind spot.

Who is it that ultimately decides whether or not a product goes into production?

As a designer and as Global Design Director, I don't make lonely decisions, relying instead on my team of specialists. Maybe that is one reason why we were awarded the Design Team of the Year title. But the decision of whether a product goes into production is not made by the design team alone. That is a joint decision made by top management. Our task is to convince all of those involved in the process to take a proposed course of action.

What role does market observation play for design?

The competitive environment for home appliances is very tough. There are over 1,000 home appliances brands worldwide. We know some of those brands very well. After all, we meet our competitors regularly at trade fairs. So it is in our mutual interest to set ourselves apart very clearly from our competitors. Obviously we don't just look at the market from the perspective of the competition, but also always with a view to understanding long-term developments. For example, connectivity is one keyword that shows where the journey is headed. As a consequence, the

Das ist ein ganzheitlicher Gestaltungsansatz, den wir auf all unsere Produkte weltweit übertragen. Insofern kann man durchaus von einer einheitlichen Designsprache, einer DNA der Marke Bosch reden.

Sie sind also nicht nur Designer, sondern auch Markenmanager?

Definitiv. Im Unterschied zu anderen Unternehmen, in denen das Design Teil der technischen Entwicklung ist, profitiert das Design bei uns davon, ein wesentlicher Teil des Markenmanagements zu sein. Unser Designteam hat eine wichtige Rolle und eine klare Aufgabenstellung, da es sich bereits sehr früh mit der Entwicklung von Produktkonzepten befasst; und zwar lange bevor über deren Vermarktung oder eine Werbekampagne nachgedacht wird.

Orientieren Sie Ihre gestalterischen Entscheidungen an den Markenwerten von Bosch?

Es geht immer darum, die Markenwerte erfahrbar zu machen. Im Bereich der Hausgeräte finden wir viele Produkte, die bestimmte gestalterische Merkmale an die nächste Generation weitergeben, denen also ein evolutionärer Gedanke zugrunde liegt. Das kann ja auch nicht anders sein bei einer Marke wie Bosch, die seit mehr als 125 Jahren dieselben Werte vermittelt.

Die Designsprache an den Markenwerten zu orientieren, ist eine Seite der Medaille. Sie beschreiben auch noch eine andere Seite: die Seite des Gebrauchs und des Nutzens für den Konsumenten.

Absolut. Wir verfolgen einen nutzerzentrierten Gestaltungsansatz, bei dem die Beobachtung der Konsumenten eine wichtige Rolle spielt. Dabei kommt uns entgegen, dass wir auch alle selbst Nutzer von Hausgeräten sind. Wir haben also die Ebene der Fremd- und der Selbstbeobachtung. Bei der Selbstbeobachtung muss man aber immer auch seinen blinden Fleck im Visier haben.

Wer entscheidet letztlich, ob ein Produkt in Serie geht?

Als Designer und als Global Design Director treffe ich keine einsamen Entscheidungen, sondern vertraue auf mein Team, das aus Spezialisten besteht. Vielleicht ist das auch ein Grund dafür, warum wir mit der Auszeichnung zum Designteam des Jahres bedacht worden sind. Die Entscheidung, ob ein Produkt in Serie geht, trifft das Designteam aber nicht allein. Das ist eine gemeinsame Entscheidung des Topmanagements. Unsere Aufgabe ist es, alle Prozessbeteiligten davon zu überzeugen, einen vorgeschlagenen Weg zu gehen.

Welche Rolle spielt die Marktbeobachtung für das Design?

Im Hausgerätemarkt finden wir ein sehr starkes Wettbewerbsumfeld vor. Es gibt über 1.000 Hausgerätemarken weltweit. Einige davon kennen wir auch sehr gut. Wir treffen unsere Wettbewerber ja regelmäßig auf Messen. Da liegt eine gute Differenzierung im wechselseitigen Interesse.

competition is never the only source of inspiration. We regularly attend trade fairs to scout out new trends. In doing so, we also learn from other industries such as the automotive industry, where the products and innovation cycles are similar in length to those on the home appliances market. In addition, we observe the developments in interior design, architecture and in consumer electronics. We get an insight into a range of vastly different industries, filtering innovations according to whether they constitute short-term or more long-term trends and how they impact on technical and design development on the home appliances market. Ultimately we want to develop products that are attractive not only on the day they are purchased but for many years afterwards.

What topics will most influence the industry for home appliances in the coming years?

There are some developments that have been influencing the home appliances industry for quite a while. For example, visions in relation to the "Internet of Things" have featured at trade fairs for some years now. With the connectivity of the appliances and the digital transmission of data, these visions are now becoming a reality and are taking shape in a very real way.

To what extent does that also change the work within your design team?

Naturally, developments like these also affect our daily work as designers. In the past our team was very heavily influenced by industrial designers, while nowadays we are focusing more on the topic of user interface design. Although the design of home appliances has always involved the design of functions for use as well as operating elements, digitalisation and connectivity mean that we are also pursuing independent design concepts which in turn result in new forms of operation and user guidance.

Is this also a general indication of how the design of home appliances will develop?

Yes, and that is something which ultimately makes a lot of sense. Coming back to the general developments and influencing factors again, it is fair to say that the change in how we use our living space has a significant role to play. Open-plan living concepts have resulted in kitchens themselves becoming a part of the living area. This of course is also reflected in the concepts behind the appliances. The current trend is to design concepts that fit well into the new living environments. Integration as a whole is an important topic.

To what extent are materials a general topic in the design of home appliances?

The choice and quality of materials are very important to Bosch. These topics run through all categories of appliances and can be found in all product groups. The materials used in home appliances often also have to fulfil technical product characteristics, as they sometimes come into contact with food or are exposed to high temperatures.

Man schaut sich den Markt natürlich nicht nur aus der Perspektive des Wettbewerbs an, sondern immer auch mit Blick auf die langfristigen Entwicklungen. Vernetzung ist beispielsweise ein Stichwort, das zeigt, wo die Reise hingeht. Und insofern ist die Konkurrenz niemals die einzige Quelle der Inspiration. Wir sind regelmäßig als Trendscouts auf Messen. Dabei lernen wir auch von anderen Branchen wie beispielsweise der Automobilindustrie, wo wir es mit ähnlich langlebigen Produkten und Innovationszyklen zu tun haben wie im Hausgerätemarkt. Daneben beobachten wir die Entwicklungen in Interior Design, Architektur und dem Bereich der Consumer Electronics. Wir nehmen Einblick in die unterschiedlichsten Branchen und filtern die Innovationen danach, ob es sich um kurzfristige oder eher langfristige Trends handelt und wie sie die technische und gestalterische Entwicklung im Hausgerätemarkt beeinflussen. Wir wollen ja Produkte entwickeln, die nicht nur im Moment des Kaufs, sondern noch viele Jahre danach attraktiv sind.

Welche Themen werden die Branche der Hausgeräte in den kommenden Jahren besonders beeinflussen?

Es gibt einige Entwicklungen, die auch nicht erst seit gestern Einfluss auf die Hausgeräte-Branche nehmen. Unter dem Schlagwort „Internet der Dinge" gibt es ja bereits seit Jahren Visionen, die auf Messen vorgestellt wurden. Mit der Vernetzung der Geräte und der digitalen Übertragung von Daten werden diese Visionen heute Realität und nehmen ganz konkrete Formen an.

Inwieweit verändert das auch die Arbeit innerhalb Ihres Designteams?

Natürlich beeinflussen solche Entwicklungen auch unsere tägliche Arbeit als Gestalter. Wurde unser Team früher sehr stark von Industriedesignern geprägt, so verstärken wir uns heute beim Thema „User Interface Design". Die Gestaltung von Hausgeräten befasst sich zwar seit jeher mit der Gestaltung von Gebrauchsfunktionen und Bedienelementen, durch die Digitalisierung und Vernetzung verfolgen wir aber auch eigenständige Designkonzepte, die wiederum neue Formen der Bedienung und Benutzerführung mit sich bringen.

Zeigt sich auch hier eine generelle Entwicklung im Design von Hausgeräten?

Ja, was auch letztlich konsequent ist. Wenn wir noch einmal auf die generellen Entwicklungen und Einflussfaktoren zurückkommen, dann spielt die Veränderung der Wohnwelten eine wichtige Rolle. Die Küche hat sich durch offene Wohnkonzepte selbst zu einem Teil des Wohnraums entwickelt. Und das spiegelt sich natürlich auch in den Gerätekonzepten wider. Der Trend geht aktuell zu Gestaltungskonzepten, die sich gut in die neuen Wohnwelten einfügen. Integration überhaupt ist ein wichtiges Thema.

Inwieweit sind Materialien ein generelles Thema im Design von Hausgeräten?

Für Bosch haben Materialien und Materialqualität einen hohen Stellenwert. Sie ziehen sich durch alle Gerätekategorien und finden

As Global Design Director at Bosch, Robert Sachon likes to get stuck in. He is a designer and brand manager in one.

Als Global Design Director Bosch greift Robert Sachon auch selbst zum Zeichenstift. Er ist Designer und Markenmanager in einer Person.

As a result, part of our day-to-day work as designers when dealing with these materials also involves familiarising ourselves with the corresponding technologies for processing the materials.

In order to also express the values of the brand through the quality and processing of the materials?

Most definitely. In some cases, we take processing to the very limits of what is technically feasible, even though many consumers may not be aware of that at first. But even if it is not necessarily the first thing they see, they will notice it when using the products.

Design as a non-verbal means of communication?

Yes, consumer perception is also evolving. For example, this is where influences from the field of consumer electronics and mobile communication come to bear. The materials used in smartphones and tablets as well as their finishing quality change how consumers perceive products. After all, they hold the devices in their hands day after day, and that also makes them more discerning of quality in other product segments.

sich bei allen Produktgruppen. Die verwendeten Materialien müssen im Bereich der Hausgeräte vielfach auch technische Produkteigenschaften erfüllen, da sie teilweise mit Lebensmitteln in Kontakt kommen oder hohen Temperaturen ausgesetzt sind. Insofern gehört es im Umgang mit diesen Materialien auch zur täglichen Arbeit des Gestalters, sich mit entsprechenden Verarbeitungstechnologien auseinanderzusetzen.

Um über die Qualität und die Verarbeitung der Materialien auch die Werte der Marke zum Ausdruck zu bringen?

Definitiv. Dabei gehen wir bei der Verarbeitung teilweise bis an die Grenzen der technischen Machbarkeit, auch wenn es vielen Konsumenten im ersten Moment nicht bewusst sein mag. Aber selbst wenn sie es vielleicht nicht vordergründig wahrnehmen, spüren sie es doch im Umgang mit den Produkten.

Design als nonverbales Mittel der Kommunikation?

Ja, auch die Wahrnehmung der Konsumenten entwickelt sich weiter. Hier kommen beispielsweise die Einflüsse aus dem Bereich der Consumer Electronics und der mobilen Kommunikation zum Tragen. Die verwendeten Materialien im Bereich der Smartphones und Tablets sowie deren Verarbeitungsqualität verändern die Wahrnehmung der Konsumenten. Sie haben die Geräte ja täglich in der Hand, und das macht sie auch sensibler für Qualitäten in anderen Produktsegmenten.

So Bosch is also being forced to be innovative with its home appliances by other industries?

You could see it like that. As a team, we have to ask ourselves the question every day of how we can constantly tweak the quality ethos and the technical perfection of the Bosch brand in order to make the brand visible and tangible again and again. Such efforts include gaps and bending radii that have a major effect on development, production and quality management.

As head designer, are you allowed to have one topic that is particularly close to your heart?

For me that topic is cooking, because it involves appliances that make it possible to be creative with the product in a very immediate way.

And what are the questions in relation to cooking that interest you as a designer?

For example how to combine appliances, and the question of how these appliances relate to each other. What codes and what design elements are suitable to be transferred to other products or categories? That can be very helpful when developing a design DNA for the brand.

Do we even need to be able to cook nowadays?

While our mothers knew exactly how to handle their home appliances, the younger generation is perhaps more likely to use automatic programmes. As a result, today's appliances are geared to meet different user requirements. It was not until sensor technology and new digital display and operating technologies were developed that these options for use and operation became possible. The very question of how the new displays are designed and programmed has become a very exciting field for designers. Our team developed dedicated style guides for images and animations for this purpose that had not existed in that form beforehand.

Don't the technical possibilities automatically lead to a complexity of products that is maybe not even desirable?

The new operating and display technologies have simultaneously given rise to more possibilities when using the appliances. Part of our work is also to prevent the appliances from becoming too complex as a result of more operating possibilities. The whole point is that they should be simple and intuitive to use. Even complex technology must remain manageable. In a technology-based company like Bosch, we as designers also have to act as a control instance vis-à-vis the marketing or technical product development departments, because we keep the user in sight and design an important interface informed by user-focused concepts.

Bosch wird also auch durch andere Branchen zur Innovation im Bereich der Hausgeräte gezwungen?

Wenn man das so sehen will. Als Team müssen wir uns täglich mit der Frage auseinandersetzen, wie wir den Qualitätsgedanken und die technische Perfektion der Marke Bosch stetig nachschärfen können, um diese immer wieder sichtbar oder erfahrbar zu machen. Dazu gehören Spaltmaße und Biegeradien, die sich erheblich auf Entwicklung, Produktion und Qualitätsmanagement auswirken.

Darf man als Chefdesigner auch ein Thema haben, dass einem besonders am Herzen liegt?

Für mich ist es das Thema Kochen, weil man es mit Geräten zu tun hat, die es einem unmittelbar erlauben, kreativ mit dem Produkt umzugehen.

Und welche Fragen interessieren Sie als Designer beim Thema Kochen?

Da geht es beispielsweise um die Kombinierbarkeit von Geräten und die Frage, wie diese Geräte miteinander in Beziehung stehen. Welche Codes und welche gestalterischen Elemente eignen sich, um sie auch auf andere Produkte oder Kategorien zu übertragen? Das kann sehr hilfreich sein, um eine Design-DNA für die Marke zu entwickeln.

Muss man denn heute überhaupt noch kochen können?

Während unsere Mütter noch sehr genau wussten, wie sie mit ihren Hausgeräten umzugehen hatten, greift die jüngere Generation vielleicht eher auf Automatik-Programme zurück. Die Geräte werden also inzwischen unterschiedlichen Nutzeranforderungen gerecht. Diese Möglichkeiten im Gebrauch und in der Bedienung wurden erst durch Sensortechnik und neue digitale Anzeige- und Bedientechnologien eröffnet. Allein die Frage, wie die neuen Anzeigen und Displays gestaltet und bespielt werden, ist ein sehr spannendes Betätigungsfeld für Designer geworden. Unser Team entwickelte dafür eigene Style Guides für Bildwelten und Animationen, die es so vorher nicht gab.

Führen die technischen Möglichkeiten nicht automatisch zu einer Komplexität von Produkten, die vielleicht gar nicht wünschenswert ist?

Mit den neuen Bedien- und Anzeigetechniken wachsen zugleich die Möglichkeiten des Gebrauchs von Geräten. Ein Teil unserer Arbeit besteht auch darin zu verhindern, dass durch mehr Bedienmöglichkeiten die Geräte zu komplex werden. Sie sollen ja gerade einfach und intuitiv zu bedienen sein. Selbst komplexe Technik muss beherrschbar bleiben. In einem technisch geprägten Unternehmen wie Bosch haben wir als Designer also auch die Rolle eines Korrektivs gegenüber dem Marketing oder der technischen Produktentwicklung, weil wir den Benutzer im Blick behalten und durch nutzerorientierte Konzepte eine wichtige Schnittstelle gestalten.

As a designer, are you not always destined to have one foot in the present and one in the future?

We simply try to give the appliances the best that there is. And, depending on the product, we have to look far into the future. To this end, we have developed a dedicated process within the company which is known as "Vision Range". It is roughly comparable with the show cars and concept studies used in the automotive industry. We design an ideal future scenario in order to gear our brand, our products and our design to that scenario from a strategic perspective. This guarantees us a competitive lead, as the content can flow directly into future projects. Maybe that is one of the major advantages of being able to work for one company and with one team on a long-term basis. Because it gives us the freedom to look to the future, quite separately from the specific product.

Thank you for speaking with us, Mr Sachon.

Steht man als Designer nicht permanent mit einem Bein in der Gegenwart und mit dem anderen Bein in der Zukunft?

Wir versuchen einfach, den Geräten das Beste mitzugeben. Und je nach Produkt müssen wir weit vorausschauen. Wir haben dafür in unserem Hause einen eigenen Prozess entwickelt, den wir „Vision Range" nennen. Man kann das in etwa mit den Showcars und Konzeptstudien in der Automobilindustrie vergleichen. Wir entwerfen ein zukünftiges Idealbild, um unsere Marke, unsere Produkte und unser Design strategisch danach auszurichten. Das sichert uns einen Vorsprung, da die Inhalte direkt in künftige Projekte einfließen können. Vielleicht ist es einer der großen Vorzüge, langfristig in einem Unternehmen und mit einem Team arbeiten zu können. Denn es gibt uns die Freiheit, losgelöst vom konkreten Produkt, einen Blick in die Zukunft zu werfen.

Vielen Dank für das Gespräch, Herr Sachon.

The Red Dot: Design Team of the Year 2015 around Robert Sachon, Global Design Director Bosch and Helmut Kaiser, Head of Consumer Products Design Bosch.
Das Red Dot: Design Team of the Year 2015 um Robert Sachon, Global Design Director Bosch und Helmut Kaiser, Head of Consumer Products Design Bosch.

Red Dot: Best of the Best
The best designers of their category
Die besten Designer ihrer Kategorie

The designers of the Red Dot: Best of the Best
Only a few products in the Red Dot Design Award
receive the "Red Dot: Best of the Best" accolade.
In each category, the jury can assign this award to
products of outstanding design quality and innovative
achievement. Exploring new paths, these products
are all exemplary in their design and oriented towards
the future.

The following chapter introduces the people who have
received one of these prestigious awards. It features
the best designers and design teams of the year 2015
together with their products, revealing in interviews
and statements what drives these designers and what
design means to them.

Die Designer der Red Dot: Best of the Best
Nur sehr wenige Produkte im Red Dot Design Award
erhalten die Auszeichnung „Red Dot: Best of the
Best". Die Jury kann mit dieser Auszeichnung in jeder
Kategorie Design von außerordentlicher Qualität
und Innovationsleistung besonders hervorheben. In
jeder Hinsicht vorbildlich gestaltet, beschreiten diese
Produkte neue Wege und sind zukunftsweisend.

Das folgende Kapitel stellt die Menschen vor, die diese
besondere Auszeichnung erhalten haben. Es zeigt die
besten Designer und Designteams des Jahres 2015
zusammen mit ihren Produkten. In Interviews und
Statements wird deutlich, was diese Designer bewegt
und was ihnen Design bedeutet.

Mathias Seiler
Girsberger Holding AG

"Every design for a piece of furniture needs a fundamental and legible design concept. Without that concept all new design remains trivial."

„Jeder Möbelentwurf benötigt eine tragende und lesbare Entwurfsidee. Ohne diese Idee bleibt jedes neue Design belanglos."

What do you particularly like about your own award-winning product?
I like the story behind the design concept. G 125 should be understood as a reinterpretation of the traditional wooden swivel chair that has now vanished from the market. It recalls the beginning of Girsberger, when the company built up its reputation as a chair specialist based on its height-adjustable piano stools.

What inspires you?
Thinking ahead. Every new concept is based on the experiences and knowledge gained from previous design concepts. Designing is a never-ending story.

How do you define design quality?
Man does not require much. What we need should, however, be of the best quality. Design can make a contribution by making products better and more desirable.

Was gefällt Ihnen an Ihrem eigenen, ausgezeichneten Produkt besonders gut?
Mir gefällt die Geschichte hinter dem Entwurf. G 125 versteht sich als Neuinterpretation des traditionellen Holzdrehstuhls, der inzwischen vom Markt verschwunden ist. Er erinnert an die Anfänge von Girsberger, als sich das Unternehmen mit höhenverstellbaren Pianohockern einen Ruf als Stuhlspezialist schuf.

Was inspiriert Sie?
Das Weiterdenken. Jeder neue Entwurf basiert auf den Erfahrungen und Erkenntnissen, die man mit früheren Entwürfen gesammelt hat. Entwerfen ist ein immerwährender Prozess.

Wie definieren Sie Designqualität?
Der Mensch braucht nicht viel. Das, was wir benötigen, sollte jedoch von bester Qualität sein. Design kann einen Beitrag dazu leisten, indem es Produkte besser und begehrenswerter macht.

reddot award 2015
best of the best

Manufacturer
Girsberger Holding AG,
Bützberg, Switzerland

G 125
Wooden Swivel Chair
Holzdrehstuhl

See page 80
Siehe Seite 80

Seth Smoot
Humanscale

"Design for strategic differentiation."
„Design als strategische Differenzierung."

What do you particularly like about
your own award-winning product?
We have developed a chair that automatically adapts to the body in a beautiful
and elegant solution.

Is there a certain design approach that
you pursue?
We look for truth in design. Design which
is not obscured by opinions – a unique,
consistent, logical and business-focused
solution.

What inspires you?
Something that I do not understand.

What do you see as being the biggest
challenges in your industry at present?
Design being allowed into a position of
driving the business forward. We believe
design should be strategic. With precise
research and planning it can transform a
business in a meaningful way.

Was gefällt Ihnen an Ihrem eigenen,
ausgezeichneten Produkt besonders
gut?
Wir haben einen Stuhl entwickelt, der sich
in einer schönen, eleganten Produktlösung
automatisch dem Körper anpasst.

Gibt es einen bestimmten Gestaltungsansatz, den Sie verfolgen?
Wir wollen im Design Ehrlichkeit sehen.
Design, das nicht von Meinungen über-
schattet wird. Eine einzige, konsequente,
logische, unternehmensbezogene Lösung.

Was inspiriert Sie?
Etwas, das ich nicht verstehe.

Worin sehen Sie aktuell die größten
Herausforderungen in Ihrer Branche?
Dem Design sollte eine Rolle zugestanden
werden, in der es das Unternehmen vor-
antreiben kann. Wir glauben, dass Design
strategisch sein sollte. Mithilfe von ge-
nauen Recherchen und Planung kann es
ein Unternehmen sinnvoll verwandeln.

reddot award 2015
best of the best

Manufacturer
Humanscale, New York, USA

Trea
Multipurpose Chair
Mehrzweckstuhl

See page 82
Siehe Seite 82

Cornelius Müller-Schellhorn
König + Neurath AG

"Only those who keep trying the impossible achieve the possible."

„Nur wer immer wieder das Unmögliche versucht, wird das Mögliche erreichen."

What do you particularly like about your own award-winning product?
Its delicate overall appearance and added value: dynamic seating through the intelligent use of material properties.

Is there a project that you have always dreamed about realising someday?
An affordable, innovative, modern family home for young families.

Is there a designer that you particularly admire?
Jasper Morrison, because he manages to perfectly reduce the design of a product to the essential.

How do you define design quality?
An innovative idea, that uses materials responsibly and sparingly, that is self-explanatory in its use, whose design is comprehensible and logical as well as passionate and precise in its details.

Was gefällt Ihnen an Ihrem eigenen, ausgezeichneten Produkt besonders gut?
Die filigrane Gesamterscheinung und der Zusatznutzen: dynamisches Sitzen aus der intelligenten Nutzung der Materialeigenschaften heraus.

Welches Projekt würden Sie gerne einmal realisieren?
Ein bezahlbares, innovatives, modernes Einfamilienhaus für junge Familien.

Gibt es einen Designer, den Sie besonders schätzen?
Jasper Morrison, weil er das Design eines Produktes in Perfektion auf das Wesentliche reduziert.

Wie definieren Sie Designqualität?
Innovativ in der Idee, reduziert und verantwortungsvoll im Materialeinsatz, selbsterklärend in der Nutzung, nachvollziehbar und logisch in der Gestaltung, sowie leidenschaftlich und präzise im Detail.

reddot award 2015
best of the best

Manufacturer
König + Neurath AG, Karben, Germany

MOVE.ME
Mobile Conference and Meeting Concept
Mobiles Konferenz- und Meetingkonzept

See page 84
Siehe Seite 84

Flemming Busk, Stephan B. Hertzog
busk+hertzog

"It can be subtle or obvious, but every design we create must have a story."

„Dezent oder offensichtlich, jede unserer Gestaltungen erzählt eine Geschichte."

What do you particularly like about your own award-winning product?
It has a sculptural yet subtle look and users experience a very high degree of seating comfort and a generous sense of space.

Is there a certain design approach that you pursue?
We always start with the analysis of the intended use and architectural space for the design. It is something that is not only based on our own observations, but also on the dialogue we have with the client. This creates a list of priorities that we work hard to fulfil throughout the whole design process.

What inspires you?
We often find our inspiration in sculptural shapes and lines found in nature. But a stunning piece of architecture or a creative plate of food can also ignite an idea.

Was gefällt Ihnen an Ihrem eigenen, ausgezeichneten Produkt besonders gut?
Es sieht plastisch und doch unaufdringlich aus. Der Nutzer erlebt ein hohes Maß an Sitzkomfort und ein großzügiges Platzgefühl.

Gibt es einen bestimmten Gestaltungsansatz, den Sie verfolgen?
Wir beginnen immer mit der Analyse des geplanten Zwecks und architektonischen Raums, der für die Gestaltung zur Verfügung steht. Dabei verlassen wir uns nicht nur auf unsere eigenen Beobachtungen, sondern erarbeiten das im Dialog mit unseren Kunden. So erstellen wir eine Prioritätenliste, die wir während des gesamten Designprozesses bemüht sind zu erfüllen.

Was inspiriert Sie?
Wir finden unsere Inspiration in den plastischen Formen und Linien der Natur. Doch auch eine faszinierende Architektur oder ein kreativ zusammengestelltes Essen kann eine Idee entfachen.

reddot award 2015
best of the best

Manufacturer
+HALLE, Aarhus, Denmark

Stella
Lounge Chair
Lounge-Sessel

See page 98
Siehe Seite 98

Brian Tong
BitsFactory

"Functionality, reliability, aesthetics, efficiency, innovation and efforts – they all shape and contribute to the design quality of the end product."
„Funktionalität, Zuverlässigkeit, Ästhetik, Wirksamkeit, Innovation und Einsatz – all diese Faktoren tragen zu der Designqualität des Endprodukts bei."

What do you particularly like about your own award-winning product?
I like the simplicity and aesthetic purity of the design. Reducing unnecessary parts yet keeping all the essential aspects of the original concept and functionality.

Is there a certain design approach that you pursue?
I create products that surprise and inspire people. I start by looking at existing products from a new angle and think of ways to refine them, make them better for the end user. Good functionality, simplicity and applicability are the main principles on which my designs are based.

Do you have a motto for life?
Passion drives everything. Don't let it burn out.

Was gefällt Ihnen an Ihrem eigenen, ausgezeichneten Produkt besonders gut?
Ich mag die Einfachheit und ästhetische Reinheit der Gestaltung. Die Reduktion unnötiger Teile, während alle wesentlichen Aspekte des ursprünglichen Konzepts und der Funktionalität beibehalten wurden.

Gibt es einen bestimmten Gestaltungsansatz, den Sie verfolgen?
Ich entwerfe Produkte, die Menschen überraschen und inspirieren. Ich beginne damit, dass ich mir bestehende Produkte unter neuen Gesichtspunkten ansehe und überlege, wie ich sie verfeinern kann, wie ich sie für den Endverbraucher besser gestalten kann. Gute Funktionalität, Einfachheit und Anwendbarkeit sind die Hauptprinzipien, auf denen meine Gestaltungskonzepte basieren.

Haben Sie ein Lebensmotto?
Leidenschaft treibt alles an. Lass sie nicht erlöschen!

reddot award 2015
best of the best

Manufacturer
Strong Precision Machinery, Hong Kong

Penxo
Pencil Lead Holder
Bleistiftminenhalter

See page 112
Siehe Seite 112

Taeno Yoon, Joongho Choi –
Joongho Choi Design Studio
Seungyoub Oh – 3M Design

"Design is unique and meaningful, a thoughtful balance between statement
 and purpose."
„Design ist einzigartig und bedeutsam, ein wohl überlegtes Gleichgewicht
 zwischen Aussage und Zweck."

What do you particularly like about your own award-winning product?
The Scotch brand has always been about great products that make smart and creative solutions possible in the world of everyday life. This product shows how design can elevate form and utility to creatively enhance the brand experience for our customers in a way that is not only unique, but also simple, modern and iconic.

How do you define design quality?
Design quality becomes apparent through the opinion expressed by the customers, when it becomes evident that the creative insight, empathy and innovation applied by the designer have successfully translated into an emotional connection and meaningful experience.

Was gefällt Ihnen an Ihrem eigenen, ausgezeichneten Produkt besonders gut?
Bei der Marke Scotch geht es seit jeher darum, großartige Produkte herzustellen, die clevere und kreative Lösungen in der Welt des Alltags ermöglichen. Dieses Produkt zeigt, wie Design Form und Funktionalität aufwerten kann, um das Markenerlebnis unserer Kunden auf eine nicht nur einzigartige, sondern auch einfache, moderne und kultige Weise kreativ zu verbessern.

Wie definieren Sie Designqualität?
Designqualität wird durch das Urteil der Kunden sichtbar, wenn klar wird, dass die kreative Einsicht, die Empathie und die Innovation, die man als Designer angewandt hat, erfolgreich umgesetzt wurden und zu einer emotionalen Verbindung und sinnvollen Erfahrung geführt haben.

reddot award 2015
best of the best

Manufacturer
Motex, Seoul, South Korea

Scotch® Tape Dispenser
Scotch® Klebestreifen-Abroller

See page 122
Siehe Seite 122

Stephan Niehaus
Hilti Corporation

"The symbiosis of design and technology leads to products of lasting added value."

„Die Symbiose von Design und Technik erzeugt Produkte mit dauerhaftem Mehrwert."

What do you particularly like about your own award-winning product?
What is impressive about PR 30-HVS is its design concept through which the four added bumper grips protect the laser measuring technology on the inside from both small and massive blows. This is due to the construction of a technically complex structure consisting of several hard and soft components that deform on impact in a pre-determined way in order to provide the necessary protection for the highly sensitive measuring elements inside. In this way highly sensitive technology is rendered to the highest degree possible thanks to the design concept. The idea arose from a rather casual conversation between a developer and a designer. Both had different perspectives when it came to improving a product like this and both reached the same approach in the course of a creative discussion.

Was gefällt Ihnen an Ihrem eigenen, ausgezeichneten Produkt besonders gut?
Das Beeindruckende am PR 30-HVS ist das Gestaltungskonzept, bei dem die vier additiven „bumper grips" die innenliegende Laser-Messtechnik vor kleinen, aber auch massiven Stößen schützen. Dies geschieht durch den Aufbau einer technisch komplexen Struktur aus mehreren Hart- und Weichkomponenten, die sich beim Aufprall definiert so verformen, wie es der Schutz der hochsensiblen Messelemente im Innern benötigt. Somit wird hochsensible Technologie über das Gestaltungsprinzip maximal baustellengerecht! Die Idee dazu entstand in einem eher beiläufigen Gespräch zwischen einem Entwickler und einem Designer. Beide hatten unterschiedliche Perspektiven bezüglich der Verbesserungen eines solchen Produktes und kamen über eine kreative Diskussion zum gleichen Lösungsansatz.

reddot award 2015
best of the best

Manufacturer
Hilti Corporation, Schaan, Liechtenstein

Hilti PR 30-HVS
Rotating Laser
Rotationslaser

See page 126
Siehe Seite 126

TEAMS Design GmbH

"Products of Empathy."

Is there a certain design approach
that you pursue?
The products should convey the greatest
possible degree of quality and the Bosch
brand values.

What inspires you?
People ... and Italian motorbikes.

Is there a project that you have always
dreamed about realising someday?
An individual transport system for the
21st century.

How do you define design quality?
Democratically: if it helps people, brands
and products advance as well as possible.

What do you see as being the biggest
challenges in your industry at present?
To continue developing the profession
with the help of sophisticated design
and to establish it in the business world
as a key factor and driver for success.

Gibt es einen bestimmten Gestaltungs-
ansatz, den Sie verfolgen?
Die Produkte sollen das größtmögliche
Maß an Qualität und die Firmenwerte der
Marke Bosch transportieren.

Was inspiriert Sie?
Menschen ... und italienische Motorräder.

Welches Projekt würden Sie gerne
einmal realisieren?
Ein individuelles Transportsystem für das
21. Jahrhundert.

Wie definieren Sie Designqualität?
Demokratisch: wenn sie Menschen, Marken
und Produkte bestmöglich weiterbringt.

Worin sehen Sie aktuell die größten
Herausforderungen in Ihrer Branche?
Den Berufsstand durch anspruchsvolles
Design weiterzuentwickeln und in der Busi-
nesswelt als Erfolgstreiber zu etablieren.

reddot award 2015
best of the best

Manufacturer
Robert Bosch Elektrowerkzeuge GmbH,
Leinfelden-Echterdingen, Germany

PLR 50 C
Laser Measure
Laser-Entfernungsmesser

See page 128
Siehe Seite 128

Gustav Landberg
Husqvarna Group

"Collaborating and bouncing ideas off each other always creates the better versions."

„Die Zusammenarbeit und der Gedankenaustausch mit anderen bringen immer bessere Varianten hervor."

Is there a certain design approach that you pursue?
I strive for honesty in expressing the essence of the brand and the product. With that comes an understanding of the design which is then able to speak for itself.

What inspires you?
Creative people in any shape or form. In this context, I am in my element when I find a way together with others.

What does winning the Red Dot: Best of the Best mean to you?
It means more than I first realised. In this business you tend to quickly move on to the next thing so it was great to revisit this project. I am really proud of this product.

Gibt es einen bestimmten Gestaltungs-ansatz, den Sie verfolgen?
Ich bemühe mich um Ehrlichkeit, wenn ich versuche, den Kern einer Marke oder eines Produkts zu erfassen. Das führt zu einem Verständnis des Designs, das dann für sich selbst stehen kann.

Was inspiriert Sie?
Kreative Menschen jeder Art. In diesem Kontext bin ich in meinem Element, wenn ich gemeinsam mit anderen einen Weg finde.

Was bedeutet die Auszeichnung mit dem Red Dot: Best of the Best für Sie?
Sie bedeutet mehr als mir zunächst klar war. In diesem Geschäft ist man meist schnell mit der nächsten Sache beschäftigt. Daher war es toll, nochmals auf dieses Projekt zurückzukommen. Ich bin richtig stolz auf das Produkt.

reddot award 2015
best of the best

Manufacturer
Husqvarna Group, Stockholm, Sweden

Husqvarna Forest Helmet Technical
Forest Helmet
Forsthelm

See page 156
Siehe Seite 156

William Mittelstadt, David Castiglione
3M

"Safety innovation – to give you the confidence and power to protect your world."

„Sicherheit durch Innovation – um das Vertrauen und die Kraft zu geben, sich zu schützen."

What do you particularly like about your own award-winning product?
The Quick Latch mechanism helps users who move in and out of contaminated environments frequently. It also makes it easier to put the mask on and take it off. Users love the extra convenience of this feature.

Is there a certain design approach that you pursue?
We observe users in the workplace to learn the real pain points in their jobs, and then figure out how to help them solve those problems.

What do you see as being the biggest challenges in your industry at present?
To make the products comfortable and easy to use. Then the users will be more inclined to wear them when needed and won't say: "This is only a quick thing, so I don't need to put my mask on."

Was gefällt Ihnen an Ihrem eigenen, ausgezeichneten Produkt besonders gut?
Der Quick-Release-Mechanismus hilft Anwendern, die häufig kontaminierte Bereiche betreten und wieder verlassen. Er erleichtert ihnen auch das schnelle Auf- und Absetzen der Maske. Dieser zusätzliche Vorteil kommt besonders gut an.

Gibt es einen bestimmten Gestaltungs-ansatz, den Sie verfolgen?
Wir beobachten Anwender am Arbeitsplatz, um die wirklichen Schmerzpunkte ihrer Aufgaben zu verstehen. So finden wir heraus, wie wir ihnen helfen können, diese Probleme zu lösen.

Worin sehen Sie aktuell die größten Herausforderungen in Ihrer Branche?
Die Produkte bequem und nutzerfreundlich zu gestalten. Dann sind Anwender eher dazu geneigt, sie zu tragen, wenn es nötig ist, und sagen nicht: „Das geht ganz schnell, da muss ich meine Maske nicht aufsetzen."

reddot award 2015
best of the best

Manufacturer
3M, St. Paul, Minnesota, USA

3M 6500QL Half Mask Series
3M Halbmasken-Serie 6500QL
Respirator
Atemschutzmaske

See page 158
Siehe Seite 158

Thomas Harrit, Nicolai Sørensen
HarritSorensen

"Design matters – it serves a purpose, it stirs emotion, it endures. It can last a lifetime."

„Design ist relevant – es dient einem Zweck, es weckt Emotionen, es bleibt bestehen. Es kann ein Leben lang halten."

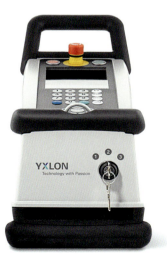

Is there a project that you have always dreamed about realising someday?
The complete and all-encompassing design of hospitals, both in terms of macro solutions for entire departments and ambulances as well as specific products for patients.

Is there a designer that you particularly admire?
Ever since our college days, Scandinavian design in general has been a great inspiration to us. Arne Jacobsen is a shining example of a designer who succeeded in working across the board, both in terms of materials, applications, and industries. He solved design problems beautifully and functionally.

What do you see as being the biggest challenges in your industry at present?
That actual, physical products are disappearing and are being replaced by software.

Welches Projekt würden Sie gerne einmal realisieren?
Die komplette und allumfassende Gestaltung von Krankenhäusern, sowohl in Bezug auf Makrolösungen für gesamte Abteilungen und Krankenwagen als auch spezifische Produkte für Patienten.

Gibt es einen Designer, den Sie besonders schätzen?
Seit unserer Zeit an der Uni hat uns skandinavisches Design allgemein sehr inspiriert. Arne Jacobsen ist ein leuchtendes Beispiel eines Designers, dem es gelungen ist, übergreifend zu arbeiten, sowohl in Bezug auf Materialien und Anwendungen als auch auf Branchen. Er hat Designprobleme funktional und auf sehr schöne Weise gelöst.

Worin sehen Sie aktuell die größten Herausforderungen in Ihrer Branche?
Dass tatsächliche, physische Produkte verschwinden und durch Software ersetzt werden.

reddot award 2015
best of the best

Manufacturer
YXLON Copenhagen, Taastrup, Denmark

SMART EVO
Portable X-Ray System
Tragbares Röntgensystem

See page 168
Siehe Seite 168

Andreas Haug, Harald Lutz,
Bernd Eigenstetter, Tom Schönherr
Phoenix Design

"Good is not good enough."

„Gut ist nicht gut genug."

What do you particularly like about your own award-winning product?
Care-O-bot 4 is a new archetype of a robot. It is a successful symbiosis of design and engineering, function and emotion that will quickly tempt users to interact with it.

Is there a certain design approach that you pursue?
To develop products and entire product ranges that consistently focus on users and are typical of the brand. To be aligned with user needs and aim to achieve self-explanatory interaction with the product. Our goal is always to make the brand identity or the product come alive in a tangible way.

What inspires you?
Observing our surroundings. How do people move within a given environment? What kind of solutions provides nature? Everything is an inexhaustible source of inspiration.

Was gefällt Ihnen an Ihrem eigenen, ausgezeichneten Produkt besonders gut?
Care-O-bot 4 ist ein neuer Archetyp von Roboter: eine gelungene Symbiose von Design, Engineering, Funktion und Emotion, die den Anwender schnell zur Interaktion verführt.

Gibt es einen bestimmten Gestaltungsansatz, den Sie verfolgen?
Konsequent nutzerzentriert und markentypisch, so entwickeln wir Produkte und ganze Sortimente. Ausgerichtet auf die Bedürfnisse des Anwenders, auf die selbsterklärende Interaktion mit dem Produkt. Ziel ist immer, die Identität der Marke oder des Produktes auf berührende Weise erlebbar zu machen.

Was inspiriert Sie?
Das Beobachten unserer Umgebung. Wie bewegen sich Menschen in einem bestimmten Umfeld? Welche Lösungen hat die Natur hervorgebracht? Alles ist eine unerschöpfliche Inspirationsquelle.

reddot award 2015
best of the best

Manufacturer
Fraunhofer IPA, Stuttgart, Germany
Schunk GmbH + Co. KG, Lauffen a.N., Germany

Care-O-bot 4
Service Robot
Serviceroboter

See page 198
Siehe Seite 198

Shinichi Temmo
blayn Co., Ltd.

"Three words that I bear in mind whenever I create a product:
minimalism, proficiency and sophistication."

„Es gibt drei Worte, die ich immer im Sinn habe, wenn ich ein Produkt entwerfe:
Minimalismus, Fertigkeit und Raffinesse."

What do you particularly like about your own award-winning product?
I like the fact that it doesn't really look like a register. Conventional registers take up a lot of space and are quite unsophisticated, but this one has been designed to reinvent and take over the conventional ones.

Do you have a motto for life?
To be independent, to not be controlled by others, to act and think freely.

What do you see as being the biggest challenges in your industry at present?
The hospitality industry is not at the forefront of technology. Therefore the biggest challenge is to boost the use of technologies.

What does winning the Red Dot: Best of the Best mean to you?
As a Japanese manufacturer, I can now see the huge potential for expanding internationally.

Was gefällt Ihnen an Ihrem eigenen, ausgezeichneten Produkt besonders gut?
Mir gefällt die Tatsache, dass es nicht wirklich wie eine Registrierkasse aussieht. Herkömmliche Kassen nehmen viel Platz ein und sind eher schlicht. Unser Produkt wurde gestaltet, diese neu zu interpretieren und zu ersetzen.

Haben Sie ein Lebensmotto?
Unabhängig sein, nicht von anderen kontrolliert werden, frei handeln und denken.

Worin sehen Sie aktuell die größten Herausforderungen in Ihrer Branche?
Das Gastgewerbe hat keine technologische Vorreiterrolle. Die größte Herausforderung liegt deshalb darin, die Anwendung von Technologien zu fördern.

Was bedeutet die Auszeichnung mit dem Red Dot: Best of the Best für Sie?
Als japanischer Hersteller sehe ich nun das enorme Potenzial, international zu expandieren.

reddot award 2015
best of the best

Manufacturer
blayn Co., Ltd., Tokyo, Japan

blayn Register
blayn Registrierkasse

See page 204
Siehe Seite 204

MEDUGORAC Industrial Design
METTLER TOLEDO Design Team

"Clear and simple lines that seamlessly fit into the customer's surroundings and, alongside unpretentious aesthetics, offer a wealth of options in the hidden detail."

„Klare und einfache Linien, die sich nahtlos in die Kundenumgebung einfügen und neben der schlichten Ästhetik im versteckten Detail eine Fülle an Möglichkeiten bieten."

What do you particularly like about your own award-winning product?
Its design transports Mettler-Toledo's pioneering role into the future. Clear, architectural shapes, outstanding user-friendliness, high-quality materials and particularly sophisticated surfaces characterise the timeless design of the scale.

Is there a certain design approach that you pursue?
The interplay of symmetry and asymmetry, clear lines, as well as a focus on the essentials.

What inspires you?
A constant interest in and a critical attitude towards the world.

How do you define design quality?
An honest product has nothing to conceal, nothing to disguise, nothing to hide. It speaks for itself.

Was gefällt Ihnen an Ihrem eigenen, ausgezeichneten Produkt besonders gut?
Das Design transportiert die Pionierrolle von Mettler-Toledo in die Zukunft. Architektonische, klare Formen, hohe Bedienungsfreundlichkeit, hochwertige Materialien und ein besonderer Anspruch an die Oberflächen kennzeichnen die zeitlose Gestaltung der Waage.

Gibt es einen bestimmten Gestaltungsansatz, den Sie verfolgen?
Das Spiel von Symmetrie und Asymmetrie, klare Linien sowie die Fokussierung auf das Wesentliche.

Was inspiriert Sie?
Ein ständiges Interesse und eine kritische Auseinandersetzung mit der Welt.

Wie definieren Sie Designqualität?
Ein ehrliches Produkt muss nichts kaschieren, nichts schminken, nichts verstecken. Es spricht für sich.

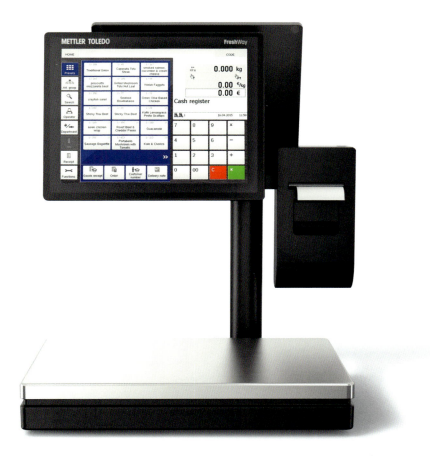

reddot award 2015
best of the best

Manufacturer
METTLER TOLEDO, Mettler-Toledo
(Albstadt) GmbH, Albstadt, Germany

FreshWay
Touchscreen Scale
Touchscreen-Waage

See page 206
Siehe Seite 206

Fabio Baratella, Luca Scalambrin, Marco Pozzati, Alessandro Zen, Giulio Bertasi, Alberto Quadretti, Davide Cattozzo
IRSAP

"IRSAP – Creating your comfort."

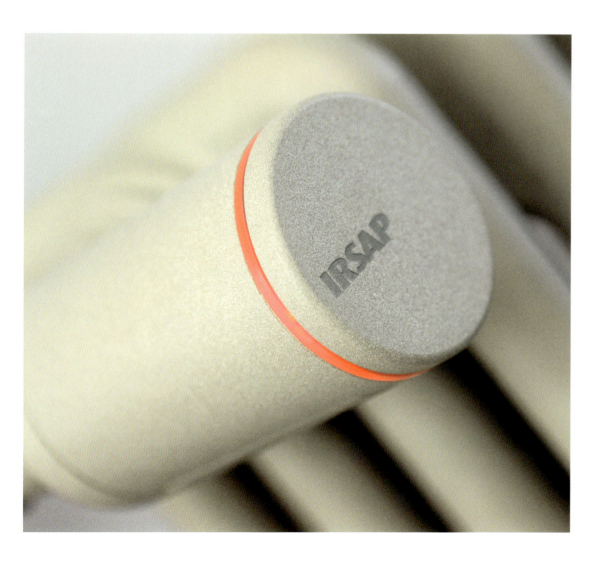

What do you particularly like about your own award-winning product?
That people and their homes are at the centre and that it pays careful attention to the environment. All these things are joined with high technology and connected to a global design language in a unique Italian style.

Is there a certain design approach that you pursue?
Creating a product that improves the life of people who will use it, and that helps to save energy while providing comfort at the same time.

Is there a designer that you particularly admire?
Antonio Citterio, for his simplicity and his extreme attention to innovation and detail.

Was gefällt Ihnen an Ihrem eigenen, ausgezeichneten Produkt besonders gut?
Dass Menschen und ihr Zuhause im Zentrum stehen und es die Umwelt stark berücksichtigt. Diese Aspekte werden mit Spitzentechnologie kombiniert und im einzigartigen italienischen Stil mit der internationalen Designsprache verbunden.

Gibt es einen bestimmten Gestaltungsansatz, den Sie verfolgen?
Ein Produkt zu schaffen, welches das Leben der Menschen, die es verwenden, verbessert und das dazu beiträgt, den Energieverbrauch zu senken, aber gleichzeitig Komfort bietet.

Gibt es einen Designer, den Sie besonders schätzen?
Antonio Citterio, für seine Einfachheit und seine extreme Detail- und Innovationsfreude.

reddot award 2015
best of the best

Manufacturer
IRSAP, Arquà Polesine (Rovigo), Italy

NOW – Smart Radiators System
Heating Control System
Heizungsregelungssystem

See page 216
Siehe Seite 216

Dirk Schumann
Schumanndesign

"Quality through reduction and emotion."

„Qualität durch Reduktion und Emotion."

Is there a certain design approach that you pursue?
Clear and easily understandable forms combined with emotional elements.

Is there a designer that you particularly admire?
Nature. It finds solutions to surmount all obstacles. It meets challenges with creativity and clarity and its solutions are always of high aesthetic quality. It loves experimenting and shuns stagnation.

What do you see as being the biggest challenges in your industry at present?
In making the increasing interconnectedness of the world manageable. In finding ways to overcome the intercultural barriers on an industrial and economical level without influencing the cultural identity and the values of different societies too much.

Gibt es einen bestimmten Gestaltungsansatz, den Sie verfolgen?
Klare und einfach verständliche Formen, verbunden mit emotionalen Elementen.

Gibt es einen Designer, den Sie besonders schätzen?
Die Natur. Sie findet Lösungen, um alle Hindernisse zu überwinden. Sie begegnet Herausforderungen mit Kreativität und Klarheit, und ihre Lösungen sind immer von hoher ästhetischer Qualität. Sie liebt das Experiment und scheut den Stillstand.

Worin sehen Sie aktuell die größten Herausforderungen in Ihrer Branche?
Darin, die zunehmende Vernetzung der Welt handhabbar zu gestalten. Wege zu erkunden, um die interkulturellen Grenzen auf der industriellen und wirtschaftlichen Ebene zu überwinden, ohne die kulturelle Identität und die Wertvorstellungen der unterschiedlichen Gesellschaften zu stark zu beeinflussen.

reddot award 2015
best of the best

Manufacturer
Stiebel Eltron GmbH & Co. KG,
Holzminden, Germany

DHE Connect & DHE Touch
Fully Electronic Controlled
Instantaneous Water Heaters
Vollelektronische Durchlauferhitzer

See page 236
Siehe Seite 236

Ryosuke Ogura
FUJIFILM Corporation

"Once you learn what a 'convention' is, you can become truly 'unconventional'."

„Wenn man einmal gelernt hat, was eine ‚Konvention' ist, kann man wirklich ‚unkonventionell' werden."

What do you particularly like about your own award-winning product?
That in an industry where standard sizes are defined, the product achieves a "design with curved sides" which reduces the physical burden placed on both the patients undergoing an X-ray examination and the technicians conducting it.

Is there a certain design approach that you pursue?
I aim to give a product the shape and function that it must have, while also emphasising the need to make it much better than a conventional one.

What do you see as being the biggest challenges in your industry at present?
In the medical industry, the biggest challenge is bringing about a paradigm shift in true value that is not an extension of the current ideas and products.

Was gefällt Ihnen an Ihrem eigenen, ausgezeichneten Produkt besonders gut?
Dass es in einer Branche, in der Standardgrößen festgelegt sind, ein „Design mit geschwungenen Seiten" erhalten hat. Das reduziert die körperliche Belastung sowohl für Patienten, die einer Röntgenaufnahme unterzogen werden, als auch für Techniker, die die Röntgenaufnahme durchführen.

Gibt es einen bestimmten Gestaltungsansatz, den Sie verfolgen?
Ich versuche, einem Produkt die Form und Funktion zu geben, die es haben muss, und bin gleichzeitig bemüht, es besser als ein herkömmliches Produkt zu gestalten.

Worin sehen Sie aktuell die größten Herausforderungen in Ihrer Branche?
In der Medizin ist die größte Herausforderung, einen wirklichen Paradigmenwandel der Werte herbeizuführen, der nicht lediglich eine Fortsetzung der aktuellen Ideen und Produkte ist.

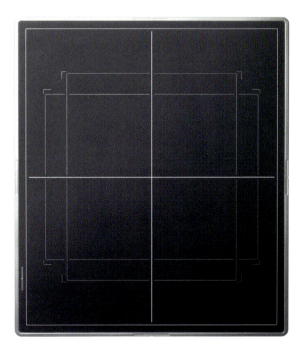

reddot award 2015
best of the best

Manufacturer
FUJIFILM Corporation, Tokyo, Japan

FDR D-EVO II Series (C43, C35, G43, G35)
Cassette-Type Digital X-Ray Imaging Diagnostic Device
Digitales Kassetten-Röntgendiagnosegerät

See page 252
Siehe Seite 252

Coloplast Design Team

"Stepping into the shoes of the consumer is the key to successful product design in any business."

„In die Rolle des Verbrauchers zu schlüpfen, ist in jeder Branche der wesentliche Schlüssel zu erfolgreichem Produktdesign."

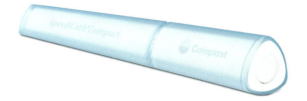

What do you particularly like about your own award-winning product?
The fact that it makes a big difference to the user. The catheter is developed with not just the patient and the medical condition in mind, but considers the "whole human being" using it. It helps the person to lead an active and normal life with a minimum amount of time having to be spent on handling her condition, both physically and mentally.

Is there a certain design approach that you pursue?
We listen a lot and we never assume anything about the women and nurses that use our products.

How do you define design quality?
When every detail has been considered and these details have been converted into a perfect, useful design.

Was gefällt Ihnen an Ihrem eigenen, ausgezeichneten Produkt besonders gut?
Die Tatsache, dass es für die Benutzer einen großen Unterschied macht. Der Katheter wurde nicht nur vor dem Hintergrund der Patienten und ihres Gesundheitszustands entwickelt, sondern zieht den „gesamten Menschen" in Betracht. Es hilft den Anwendern, ein aktives und normales Leben zu führen, bei dem sie ihrem Gesundheitszustand nur minimal Zeit opfern müssen – sowohl physisch als auch psychisch.

Gibt es einen bestimmten Gestaltungsansatz, den Sie verfolgen?
Wir hören viel zu und setzen nie etwas als gegeben voraus, was die Frauen und Krankenschwestern betrifft, die unsere Produkte verwenden.

Wie definieren Sie Designqualität?
Wenn jedes Detail durchdacht ist und diese Details in ein perfektes, sinnvolles Design verwandelt wurden.

reddot award 2015
best of the best

Manufacturer
Coloplast A/S, Humlebæk, Denmark

SpeediCath Compact Eve
Single-Use Female Catheter
Einmal-Katheter für Frauen

See page 272
Siehe Seite 272

Masanori Yonemitsu, Masashi Nomura
IM Business Unit, Yamaha Motor Co., Ltd.

"Technology and design only make sense if they reach out to users."

„Technik und Design ergeben nur einen Sinn, wenn sie die Anwender unterstützen."

Is there a certain design approach that you pursue?
If you design a vehicle, drive it a lot. If you design a tool, use it a lot. Then you will achieve the user's point of view.

What inspires you?
Communication with users. Meeting and talking with them.

Is there a product that you have always dreamed about realising someday?
Any product that makes the user smile, but also people around them. A product that uplifts and encourages users.

What do you see as being the biggest challenges in your industry at present?
Different systems, product specifications, laws and regulations between countries.

Gibt es einen bestimmten Gestaltungsansatz, den Sie verfolgen?
Wenn man ein Fahrzeug gestaltet, sollte man es viel fahren. Wenn man ein Werkzeug gestaltet, sollte man es viel benutzen. So kann man sich in die Betrachtungsweise des Anwenders hineindenken.

Was inspiriert Sie?
Die Kommunikation mit den Anwendern. Sie zu treffen und mit ihnen zu reden.

Welches Produkt würden Sie gerne einmal realisieren?
Ein Produkt, das den Nutzer zum Lächeln bringt, aber auch die Menschen um ihn herum. Ein Produkt, das bereichert und Anwender anregt.

Worin sehen Sie aktuell die größten Herausforderungen in Ihrer Branche?
In den verschiedenen Systemen, Produktspezifikationen, Gesetzen und Vorschriften der unterschiedlichen Länder.

reddot award 2015
best of the best

Manufacturer
Yamaha Motor Co., Ltd., Shizuoka, Japan

JWX-2
Power Assist Unit for Wheelchairs
Elektrische Antriebshilfe für Rollstühle

See page 306
Siehe Seite 306

Asami Yamagishi, Charlotta Franzen
Sony Mobile Communications Inc.

"Do what has never been done before. Create something that puts a smile on people's faces."

„Das noch nie Dagewesene zu schaffen. Etwas zu kreieren, das Menschen zum Lächeln bringt."

What do you particularly like about your own award–winning product?
The format. The proportions are good and it fits the hand amazingly. The phone will not slip out of your hand when you are playing around with it. It is a really handsome companion.

How do you define design quality?
When a product has emotional value, it just feels right, simply works every time, and sometimes it puts a smile on your face. That you find a small detail, even after a long time of using a product.

What do you see as being the biggest challenges in your industry at present?
To always remember to try to exceed the expectations of the consumer. Often that only requires a small thing, but it has to be the right thing.

Was gefällt Ihnen an Ihrem eigenen, ausgezeichneten Produkt besonders gut?
Das Format. Die Proportionen sind gut und die Handhabung phantastisch. Wenn man damit spielt, rutscht das Telefon nicht aus der Hand. Es ist ein wirklich stilvoller Begleiter.

Wie definieren Sie Designqualität?
Wenn ein Produkt einen emotionalen Wert hat, wenn es sich einfach richtig anfühlt, jedes Mal funktioniert und es einen manchmal zum Lächeln bringt. Dass man plötzlich ein kleines Detail entdeckt, selbst wenn man das Produkt schon seit Langem benutzt.

Worin sehen Sie aktuell die größten Herausforderungen in Ihrer Branche?
Immer daran zu denken, dass man versuchen muss, die Erwartungen des Konsumenten zu übertreffen. Häufig reicht etwas ganz kleines, doch muss es das richtige Etwas sein.

reddot award 2015
best of the best

Manufacturer
Sony Mobile Communications Inc.,
Tokyo, Japan

Xperia™ E3
Smartphone

See page 316
Siehe Seite 316

BlackBerry Design Team

"Premium utility."

„Erstklassiger Nutzen."

What do you particularly like about your own award-winning product?
We love how the hardware keyboard and software work in unison in a very unexpected and modern way. The stainless steel I-beam construction, inspired by Mies van der Rohe's work, provides a unique and ultra-strong structure that is intelligent (acting as part of the antenna) and has its own beauty.

What inspires you?
Inspiration is a moment that can be felt when the project accelerates into a certain direction. Sometimes this is a personal moment, but most of the time it is the result of collaboration.

How do you define quality?
Quality for BlackBerry is utility and emotion. A connection must be made between those two, for our customers to feel secure and confident.

Was gefällt Ihnen an Ihrem eigenen, ausgezeichneten Produkt besonders gut?
Wir sind davon begeistert, wie die Hardware-Tastatur und die Software auf sehr unerwartete und moderne Art perfekt zusammenarbeiten. Die I-Träger-Konstruktion aus Edelstahl, die von Mies van der Rohes Arbeit inspiriert wurde, bietet eine einzigartige, äußerst starke Struktur, die sowohl intelligent (sie fungiert als Teil der Antenne) als auch auf ihre eigene Art schön ist.

Was inspiriert Sie?
Inspiration ist ein spürbarer Augenblick, der dann zustande kommt, wenn ein Projekt eine bestimmte Richtung einschlägt und Fahrt aufnimmt. Manchmal ist das ein persönlicher Moment, aber meistens ist er das Ergebnis einer Zusammenarbeit.

Wie definieren Sie Qualität?
Für BlackBerry ist Qualität Nutzen und Emotion. Es muss eine Verbindung zwischen beiden entstehen, damit unsere Kunden sich sicher fühlen und Vertrauen haben.

reddot award 2015
best of the best

Manufacturer
BlackBerry, Waterloo, Canada

BlackBerry Passport
Smartphone

See page 324
Siehe Seite 324

Shawn Bender, Otto Williams, Ken Mak, Tim McCollum
Cisco Systems

"Humans should not have to adapt to technology;
make technology adapt to humans."

„Die Menschen sollten sich nicht der Technik anpassen müssen.
Die Technik sollte sich den Menschen anpassen."

What do you particularly like about your own award-winning product?
We like the invisible lighting panel that magically glows from behind the fabric facade when making a video call. It's a great example of how industrial and mechanical design together can provide a delightful user experience.

Is there a project that you have always dreamed about realising someday?
It would be cool to design a new form of transportation to bring people together physically more easily. Elon Musk's concept for the Hyperloop could be a possibility.

What do you see as being the biggest challenges in your industry at present?
The fragmentation of knowledge and the lack of coherent tools for people to support them in working together is a big problem.

Was gefällt Ihnen an Ihrem eigenen, ausgezeichneten Produkt besonders gut?
Uns gefällt das unsichtbare Leuchtpanel, das hinter einer Textilverkleidung verborgen auf geradezu magische Weise leuchtet während eines Videogesprächs. Es ist ein großartiges Beispiel dafür, wie industrielles und mechanisches Design zusammen zu einer sehr angenehmen Nutzererfahrung führen können.

Welches Projekt würden Sie gerne einmal realisieren?
Es wäre cool, eine neue Transportart zu entwerfen, um so Menschen physisch einfacher zusammenbringen zu können. Das Konzept von Elon Musk für den Hyperloop wäre eine Möglichkeit.

Worin sehen Sie aktuell die größten Herausforderungen in Ihrer Branche?
Die Fragmentierung des Wissens und der Mangel an stimmigen Hilfsmitteln, die Menschen dabei unterstützen zusammenzuarbeiten, ist ein großes Problem.

reddot award 2015
best of the best

Manufacturer
Cisco Systems, San Jose, USA

Cisco TelePresence IX5000
Video Conferencing System
Videokonferenzsystem

See page 344
Siehe Seite 344

Dell Experience Design Group

Is there a certain design approach that you pursue?
We design the Dell computers, monitors and peripherals together as a coherent system versus designing the products individually, one by one. This approach ensures that Dell products are consistent in across-the-line ergonomics, seamless in connectivity and rich in features. We always create designs that are elegantly simple and intuitive.

What inspires you?
There are over seven billion people in the world, and the simple thought of one day touching every one with our thoughtfully designed products greatly inspires all of us.

"Elegant, simple, intuitive designs that perform and are coherent all along the line."

„Elegante, schlichte, intuitive Designs, die funktionieren und auf der ganzen Linie schlüssig sind."

Gibt es einen bestimmten Gestaltungsansatz, den Sie verfolgen?
Wir gestalten Dell-Computer, -Bildschirme und -Peripheriegeräte als ein einheitliches System, anstatt die Produkte einzeln zu gestalten. Dieser Ansatz gewährleistet, dass Dell-Produkte auf der ganzen Linie ergonomisch schlüssig sind und nahtlose Konnektivität sowie eine Fülle von Funktionalitäten bieten. Wir erstellen stets Designs, die elegant, einfach und intuitiv sind.

Was inspiriert Sie?
Es gibt mehr als sieben Milliarden Menschen auf der Welt. Der einfache Gedanke, dass wir eines Tages jeden einzelnen davon mit unseren wohl durchdacht gestalteten Produkten erreichen können, ist für uns alle eine enorme Inspiration.

reddot award 2015
best of the best

Manufacturer
Dell Inc., Round Rock, Texas, USA

Dell UltraSharp 34 Monitor

See page 360
Siehe Seite 360

Kevin Falk
Hewlett-Packard

"We define solutions that enable our customers to be more than they are without us: to be faster, smarter, more productive, saving time and money – simply better."

„Wir definieren Lösungen, die unseren Kunden erlauben, mehr zu sein als sie ohne uns wären: schneller, schlauer, produktiver, Zeit und Geld sparend – einfach besser."

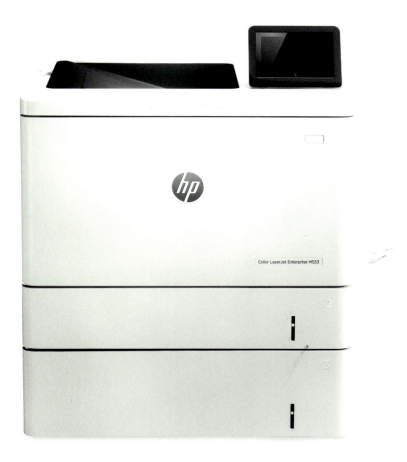

What do you particularly like about your own award-winning product?
It's simple, clean and reductionistic approach to complex functionality.

Is there a certain design approach that you pursue?
We employ and create design tools that define and create meaningful experiences with our partners and for our customers. Our commandments are quality, profitability and creating pleasant customer experiences.

What does winning the Red Dot: Best of the Best mean to you?
It is the first award HP LES has applied for and won in years. It is great to be recognised by global peers for the hard work and sacrifice it takes to become part of the world class.

Was gefällt Ihnen an Ihrem eigenen, ausgezeichneten Produkt besonders gut?
Sein einfacher, klarer und reduktionistischer Lösungsansatz für eine komplexe Funktionalität.

Gibt es einen bestimmten Gestaltungsansatz, den Sie verfolgen?
Wir verwenden und kreieren Gestaltungsmittel, die mit unseren Partnern und für unsere Kunden inhaltsreiche Erfahrungen definieren und schaffen. Unsere Gebote sind dabei Qualität, Rentabilität und das Gestalten von angenehmen Kundenerfahrungen.

Was bedeutet die Auszeichnung mit dem Red Dot: Best of the Best für Sie?
Es ist seit Jahren die erste Auszeichnung, für die HP LES sich beworben hat und die wir gewonnen haben. Es ist großartig, von internationalen Fachleuten Anerkennung zu bekommen für die harte Arbeit und die Opfer, die man bringen muss, um zur Weltklasse zu gehören.

reddot award 2015
best of the best

Manufacturer
Hewlett-Packard, Boise, USA

HP Color LaserJet Enterprise M553
Printer
Drucker

See page 412
Siehe Seite 412

Blackmagic Design Team

"Design quality is an effortless solution to a complex problem."
„Designqualität ist die mühelose Lösung eines komplizierten Problems."

reddot award 2015
best of the best

Manufacturer
Blackmagic Design Pty Ltd,
Melbourne, Australia

Blackmagic Cintel
Film Scanner

See page 418
Siehe Seite 418

Blackmagic Studio Camera
Digital Film Camera
Digitale Filmkamera

Doing
See page 390
Siehe Seite 390

Offices
Büro

G 125
Wooden Swivel Chair
Holzdrehstuhl

Manufacturer
Girsberger Holding AG,
Bützberg, Switzerland

In-house design
Mathias Seiler

Web
www.girsberger.com

reddot award 2015
best of the best

The maxim of simplicity

The office chair has been subject to many interpretations, such as the highly purist designs by the German architect Egon Eiermann. The G 125 wooden swivel chair was developed to mark the 125th anniversary of the Swiss company Girsberger. In its early beginnings, the company was successful in the production of piano stools that combined formal simplicity with sophisticated functionality – both are criteria, which constitute also the key elements of this chair, whereby it also features a highly contemporary appeal. A striking feature of the design is the wooden backrest. It expressively puts emphasis on the perfectly curved form, while the base frame appears well-balanced and elegant. This swivel chair deliberately does without the great array of functions that are often featured in office chairs. As an alternative design to common swivel chairs today, it offers only few possible adjustments. Also inspired by the traditional piano stool, it is however based on a surprisingly intelligent ergonomic concept. The U-shaped seat and backrest support made of quenched and tempered steel function as a flexible torsion rod, allowing for a comfortable rocking movement of seat and backrest. Thus, the G 125 provides sophisticated seating comfort without the need for complex tilting mechanisms and technical expenditure. The highly successful reinterpretation of classic predecessor models fascinates with its precisely tailored simplicity and beauty.

Die Maxime der Simplizität

Der Bürostuhl hat bereits viele Interpretationen erfahren, wie etwa die sehr puristischen Entwürfe des deutschen Designers und Architekten Egon Eiermann. Der Holzdrehstuhl G 125 entstand anlässlich des 125. Jahrestags des Schweizer Unternehmens Girsberger. Dieses war in seinen Anfangszeiten erfolgreich mit Pianohockern, die ihre formale Simplizität mit einer durchdachten Funktionalität verbanden. Kriterien, die auch diesen Stuhl prägen, wobei er zugleich sehr zeitgemäß anmutet. Auffällig ist seine Gestaltung mit einer Holzlehne. Sie bringt die perfekt geschwungene Form ausdrucksvoll zur Geltung, während das Untergestell ausgewogen und elegant wirkt. Bei diesem Holzdrehstuhl wurde bewusst auf die mit Bürostühlen oftmals einhergehende Vielzahl an Funktionen verzichtet. Als Gegenentwurf zu den heute üblichen Drehstühlen bietet er nur wenige Verstellmöglichkeiten. Ebenfalls in Anlehnung an den traditionellen Pianohocker liegt ihm jedoch ein verblüffend intelligentes Ergonomiekonzept zugrunde. Der aus Vergütungsstahl gefertigte U-förmige Sitz- und Lehnenträger fungiert dabei als flexibler Torsionsstab, der eine komfortabel wippende Bewegung von Sitz und Lehne ermöglicht. Der Holzdrehstuhl G 125 bietet deshalb ohne aufwendige Neigemechanik und technischen Aufwand hochentwickelten Sitzkomfort. Durch die sehr gelungene Interpretation klassischer Vorbilder fasziniert er mit seiner auf den Punkt gestalteten Simplizität und Schönheit.

Statement by the jury

Impressive simplicity inspired by European tradition. The G 125 wooden swivel chair shows a highly successful reduction to essential elements. As an alternative design to swivel chairs today, its design combines formal minimalism and a fascinatingly intelligent functionality. The innovative concept of a flexible torsion rod allows for excellent seating comfort. Its high quality is brought to perfection through the choice of fine woods as material.

Begründung der Jury

Bestechende Einfachheit in traditionellem europäischen Stil. Der Holzdrehstuhl G 125 zeigt eine überaus gelungene Reduktion auf das Wesentliche. Als Gegenentwurf zu heutigen Drehstühlen verbindet seine Gestaltung formalen Minimalismus mit einer beeindruckend intelligenten Funktionalität. Das innovative Konzept eines flexiblen Torsionsstabs ermöglicht dabei ausgezeichneten Sitzkomfort. Perfektioniert wird seine hohe Qualität durch die Materialwahl in Form von edlen Hölzern.

Designer portrait
See page 28
Siehe Seite 28

Trea
Multipurpose Chair
Mehrzweckstuhl

Manufacturer
Humanscale, New York, USA

In-house design
Humanscale

Web
www.humanscale.com

reddot award 2015
best of the best

Naturally modelled

A multipurpose chair has to be shaped to offer the most diverse seating options, so that its design complies with certain functional standards right from the start. The Trea multipurpose chair fascinates with its distinctive lines that merge frame and seating shell in a harmonious way. Since the chair is made of only a few parts that assemble in a modular manner, it is also highly durable. Another important aspect is that it boasts outstandingly good seating comfort. The backrest features an innovative aluminium reclining mechanism that offers up to 12 degrees of recline to encourage a healthy posture. Since the backrest follows the body's natural movement, it reduces disc pressure on the back, allowing users to intuitively sit in the most comfortable posture with ease. Fascinatingly, this process is facilitated without needing to adjust manual knobs and levers. The glass-filled nylon backrest is moulded and contoured to cradle the body's natural curves. It thus offers added lumbar support, while the seat edge reduces pressure on the back of the knees. The Trea multipurpose chair is available as a cantilever, four-star base or four-leg base version. Its modular construction allows upgrading for the most diverse seating purposes. It easily blends in with almost any space aesthetics, making it suitable for cafeterias, home offices and dining rooms alike – impressing users and beholders with its distinctive use of form, functionality and ergonomics.

Natürlich modelliert

Ein Mehrzweckstuhl muss äußerst vielfältigen Einsatzzwecken dienen, weshalb sein Design von vornherein bestimmte funktionale Anforderungen verwirklichen muss. Der Mehrzweckstuhl Trea begeistert mit einer prägnanten Linienführung, welche Gestell und Sitzschale harmonisch vereint. Da dieser Stuhl aus nur wenigen Teilen und modular aufgebaut ist, ist er auch sehr langlebig. Ein wichtiger Aspekt ist außerdem sein für einen Mehrzweckstuhl ausgezeichneter Komfort. Die Rückenschale verfügt über eine innovative, aus Aluminium gefertigte Neigemechanik mit einer Neigung von bis zu zwölf Grad, wodurch eine aktive Sitzhaltung gefördert wird. Da sie sich an die natürliche Körperbewegung anpasst, wird der Druck auf die Bandscheibe reduziert und man erreicht intuitiv den besten Sitzkomfort. Faszinierend ist, dass dies gänzlich ohne ein zusätzliches Betätigen von Knöpfen oder Hebeln geschieht. Die mit Glas gefüllte Rückenschale ist zudem nach der natürlichen Körperkontur modelliert. Damit unterstützt sie die Lendenwirbelsäule, wobei die Sitzfläche auch den Druck auf die Kniekehle reduziert. Der Mehrzweckstuhl Trea ist mit Freischwingergestell, vierarmigem Fußkreuz mit Gleitern oder Vierfußgestell erhältlich. Seine modulare Bauweise erlaubt ein Aufrüsten für die unterschiedlichsten Zwecke. Leicht fügt er sich in die allgemeine Raumästhetik ein, wobei er gleichermaßen zu Cafeterien, Home-Offices und Speisezimmern passt – er beeindruckt dort durch seine prägnante Formensprache, Funktionalität und Ergonomie.

Statement by the jury
The futuristic form language of the Trea multipurpose chair is complemented by sophisticated ergonomics. It features a design of beautiful proportions and looks aesthetically pleasing from any angle. Its innovative, state-of-the art seating comfort is fascinating. Functional details such as the sophisticated reclining mechanism and a backrest shell with glass inlay enable an intuitive sitting posture. Users sit on it as comfortably as on a special kind of office chair, although it is a multipurpose chair.

Begründung der Jury
Die futuristische Formensprache des Mehrzweckstuhls Trea geht einher mit einer ausgereiften Ergonomie. Er ist mit sehr schönen Proportionen gestaltet und sieht aus jedem Blickwinkel ästhetisch aus. Beeindruckend ist sein hochentwickelter, innovativer Sitzkomfort. Funktionale Details wie ein raffinierter Neigungsmechanismus und eine mit Glas gefüllte Rückenschale ermöglichen intuitiv die richtige Sitzhaltung. Man sitzt auf ihm so gut wie auf einem speziellen Bürostuhl, obwohl es ein Mehrzweckstuhl ist.

Designer portrait
See page 30
Siehe Seite 30

You can also find this product in
Dieses Produkt finden Sie auch in
Living
Page 80
Seite 80

MOVE.ME
Mobile Conference and Meeting Concept
Mobiles Konferenz- und Meetingkonzept

Manufacturer
König + Neurath AG,
Karben, Germany

In-house design
Cornelius Müller-Schellhorn

Web
www.koenig-neurath.de

reddot award 2015
best of the best

Mobile lightness

Modern work environments often consist of flexible and open work in teams and seminar groups. MOVE.ME is perfectly adapted to this kind of everyday office style, offering a multitude of functional solutions. Impressively incorporating the concept of a mobile conference and meeting scenario, it consists of folding-leg tables, canteen tables, chairs, lecterns for sitting or standing and screens for visual privacy. These can be combined in the most diverse configurations, forming a clear, premium-quality language of form. In line with the flexibility of the meeting and training program, all components can be stacked or folded together to save space. An impressive eye-catcher is embodied in the intelligent and filigree thin steel construction. This makes the chair light as a feather and easy to transport. Furthermore, the moveable mount feature of the seat shell ensures relaxed and dynamic sitting. It is ergonomically well thought-out and supports any alteration in sitting position, something that is particularly important for long periods of sitting. Thanks to its filigree aesthetics and sophisticated modularity, it is suitable for the most diverse tasks, lending itself as an ideal all-rounder for training and conference rooms, project groups, lecture theatres or canteens. In a highly consistent manner, it thus complements a product family that stands for a highly flexible concept.

Mobile Leichtigkeit

In der modernen Arbeitswelt ist ein flexibles und offenes Arbeiten in Teams oder Seminargruppen an der Tagesordnung. Das Konzept von MOVE.ME vollzieht diesen Büroalltag unmittelbar mit und bietet dafür eine Vielzahl überaus funktionaler Lösungen. Als ein beeindruckend mobiles Meeting- und Seminarprogramm integriert es Klapp- und Kantinentische, Stühle, Sitz- und Stehpulte sowie optische Trennelemente. Diese können in den unterschiedlichsten Konstellationen kombiniert werden, wobei sie eine klar und hochwertig anmutende Formensprache vereint. Entsprechend der Flexibilität dieses Meeting- und Seminarprogramms lassen sich alle Komponenten platzsparend zusammenlegen oder stapeln. Ein beeindruckender Blickfänger ist dabei die intelligente, filigrane Konstruktion aus dünnem Stahl. Der Stuhl ist deshalb federleicht und lässt sich ganz einfach transportieren. Gestaltet mit einer beweglich gelagerten Sitzschale, ermöglicht er darüber hinaus ein entspanntes und dynamisches Sitzen. Er ist ergonomisch sehr gut durchdacht und unterstützt jede Änderung der Sitzhaltung, was insbesondere bei langem Sitzen sehr wichtig ist. Mittels seiner filigranen Ästhetik und ausgefeilten Modularität kann er für verschiedenste Aufgaben eingesetzt werden und ist ein idealer Allrounder für Schulungs- und Konferenzräume, Hörsäle oder Kantinen. Auf stimmige Weise wird so eine Produktfamilie komplettiert, die für ein überaus flexibles Konzept steht.

Statement by the jury

The minimalist design of the MOVE.ME mobile conference and meeting concept equipment showcases a fascinating, clear language of form. The design looks filigree and visualises its enormous flexibility at the very first glance. The multifunctional concept offers variable use options for all areas of modern offices, conference and training rooms. And above all, the chair impresses with an intelligent construction comprising an invisible spring effect that guarantees an ergonomically optimal sitting posture at all time.

Begründung der Jury

Das minimalistisch gestaltete mobile Meeting- und Seminarprogramm MOVE.ME zeigt eine faszinierend klare Formensprache. Das Design wirkt filigran und visualisiert auf den ersten Blick die enorme Flexibilität. Das multifunktionale Konzept erlaubt einen variablen Einsatz in allen Bereichen moderner Büro-, Konferenz- und Seminarräume. Der Stuhl beeindruckt vor allem durch die intelligente Konstruktion mit verstecktem Federungseffekt, wodurch eine ergonomisch optimale Sitzhaltung gewährleistet wird.

Designer portrait
See page 32
Siehe Seite 32

FLY
Chair
Stuhl

Manufacturer
Hangzhou Hengfeng Furniture Co., Ltd., Hangzhou, China
Design
Claudio Bellini, Milan, Italy
Web
www.hz-hengfeng.com
www.claudiobellini.com

The design concept of the Fly chair is driven by the consideration that a chair is a mirror of the human body and its needs. The polypropylene shell has been shaped in order to maximise flexibility and comfort, while simultaneously taking strength and durability requirements into account. It is adaptable to a large variety of different leg frames and configurations. A light, dynamic and vibrating visual language supports this colourful, multipurpose chair in enhancing everyday life at home, in the office and during leisure time.

Das Designkonzept des Stuhls Fly basiert auf der Überlegung, dass der Stuhl ein Spiegel des menschlichen Körpers und seiner Bedürfnisse ist. Die aus Polypropylen gefertigte Sitzschale wurde nach dem Prinzip geformt, Flexibilität und Komfort zu maximieren und gleichzeitig Erfordernisse bezüglich Robustheit und Widerstandsfähigkeit zu berücksichtigen. Sie lässt sich an eine Vielzahl von Beinrahmen und Konfigurationen anpassen. Die leichte, dynamische und lebendige Formensprache unterstützt den Mehrzweckstuhl darin, den Alltag zuhause, bei der Arbeit und in der Freizeit zu verbessern.

Statement by the jury
The dynamic chair Fly calls special attention to itself with its large selection of chair legs, which allow a wide range of possible configurations.

Begründung der Jury
Mit seiner großen Auswahl an Stuhlbeinen und damit vielfältigen Konfigurationsmöglichkeiten macht der dynamisch gestaltete Fly besonders auf sich aufmerksam.

Embrace
Chair Series
Stuhlserie

Manufacturer
Kinnarps AB, Kinnarp, Sweden
Design
brodbeck design (Stefan Brodbeck),
Munich, Germany
Web
www.kinnarps.com
www.brodbeckdesign.de

Designed for office and conference rooms, this chair family attracts attention with its soft and reduced forms. The flexible back shell made of bent plywood dispenses with mechanics and lumbar adjustment. Its continuously tapered thickness provides maximum stability with a minimal use of materials. As a cantilever chair, wire-framed, or with a four- or five-star base beam – Embrace combines recyclable and natural materials in an intelligent way.

Statement by the jury
The form of this chair inspires with its organically flowing, straightforward appearance and material aesthetics arising through the combination of natural materials.

Die für Büro- und Konferenzraum entworfene Stuhlfamilie fällt durch eine weiche und reduzierte Gestaltung auf. Die flexible Rückenschale aus Formholz macht eine Mechanik oder Lumbalverstellung verzichtbar, während ihre kontinuierlich sich verjüngende Stärke maximale Stabilität bei minimalem Materialeinsatz ermöglicht. Ob Freischwinger, Drahtgestell, 4- oder 5-strahliges Fußkreuz – Embrace kombiniert recycelbare und natürliche Materialien auf intelligente Art.

Begründung der Jury
Dieser Stuhl begeistert formal durch sein organisch fließendes, schlichtes Erscheinungsbild und materialästhetisch durch die Verbindung natürlicher Werkstoffe.

Tipo–NST, Tipo–MA/MAC
Multifunctional Chair
Multifunktionsstuhl

Manufacturer
AICHI Co., Ltd., Nagoya, Japan
In-house design
Taku Kumazawa
Web
www.axona-aichi.com
Honourable Mention

Tipo-NST is a mesh chair that can be stacked horizontally to save space once the seat is flipped up. Its clear aesthetic design language and the innovative frame construction allow for a variety of styles, for instance as a single chair or as an attractive continuous line of nested chairs. The stackable Tipo-MA/MAC version with a four-leg frame comes with option of glides or casters. The seat and back surfaces are available in polypropylene or mesh fabrics in attractive colours. All resin components of the Tipo series products are made from 100 per cent recycled materials. The mesh is fashioned from fully recycled knitted yarn from reclaimed PET bottles.

Statement by the jury
The use of PET bottles and the fact that it is entirely made from recycled materials make the stylish Tipo-NST mesh chair an item of seating furniture attuned to the times.

Tipo-NST ist ein Netzstuhl, der sich mit hochgeklappter Sitzfläche platzsparend horizontal stapeln lässt. Die klare Formensprache und innovative Rahmenkonstruktion ermöglichen eine große stilistische Vielfalt des Einsatzes z. B. als einzelner Stuhl oder als attraktive Reihe ineinander gestapelter Stühle. Die stapelbare Version Tipo-MA/MAC mit vierfüßigem Rahmen ist optional mit Rollen oder Gleitern erhältlich. Die Sitz- und Rückenfläche ist in PP- oder Netzgewebe in attraktiven Farben wählbar. Sämtliche Harzkomponenten der Tipo-Serie bestehen zu 100 Prozent aus recycelten Materialien, das Netz ist komplett aus wiederverwertetem Strickgarn von PET-Flaschen gefertigt.

Begründung der Jury
Die Verwendung von PET-Flaschen und die Tatsache, dass er vollständig aus recycelten Materialien hergestellt wurde, machen den stilvollen Netzstuhl Tipo-NST zum zeitgemäßen Sitzmöbel.

Zoo
Chair
Stuhl

Manufacturer
Profim sp. z o.o., Turek, Poland
Design
Paul Brooks Design GmbH (Paul Brooks),
Kehl am Rhein, Germany
Web
www.profim.eu
www.paulbrooksdesign.com
Honourable Mention

The Zoo collection is suited for large conference halls and smaller rooms. Due to its wide range of variations, it can be adapted to individual requirements and the chairs connected to form rows. After an event, they can be easily stacked and removed quickly. Various designs are available: with four legs or cantilevered, with a plastic or upholstered seat, with a plastic, mesh or upholstered backrest, with optional armrests and in six different colours.

Statement by the jury
This chair line makes a friendly impression. Thanks to its high level of variability and combinability, a broad spectrum of applications is possible.

Die Kollektion Zoo bewährt sich im großen Konferenzsaal wie in kleineren Räumen. Die Stühle können dank zahlreicher Ausführungen an die Situation angepasst und auch zu Reihen verbunden werden. Nach einem Event lassen sie sich leicht aufeinanderstapeln und wegräumen. Verschiedene Ausführungen sind erhältlich: 4-Fuß oder Kufe, ein Sitz aus Kunststoff oder gepolstert, eine Rückenlehne aus Kunststoff, Netz oder gepolstert, optionale Armlehnen sowie sechs Farbvarianten.

Begründung der Jury
Diese Stuhlfamilie strahlt eine freundliche Anmutung aus und bietet durch den hohen Grad an Variabilität und Kombinierbarkeit viele Einsatzmöglichkeiten.

nooi
Stacking Chair
Reihenstuhl

Manufacturer
Wiesner-Hager Möbel GmbH,
Altheim, Austria
Design
neunzig° design (Barbara Funck),
Wendlingen, Germany
Web
www.wiesner-hager.com
www.neunzig-grad.com
Honourable Mention

nooi is a stacking chair with a clear design, which does not require additional elements to be linked together in rows. The two adjoining legs are stacked on top of each other, linking from left to right. Thus valuable space for a comfortable seat width is gained, which simultaneously gives rise to a visually calm appearance. Thanks to the frame's fast and uncomplicated linking mechanism, nooi is excellently suited for large halls.

Statement by the jury
This stacking chair captivates with its innovative linking mechanism, offering the user extra space and comfort.

nooi ist ein klar gestalteter Reihenstuhl, der ohne ein zusätzliches Verkettungselement auskommt. Die beiden angrenzenden Beine werden einfach übereinander gestapelt und dabei von links nach rechts verkettet. Wertvoller Platz wird so für eine komfortable Sitzbreite gewonnen und zugleich entsteht ein visuell ruhiges Bild. Die schnelle und unkomplizierte Verbindung seiner Gestelle machen nooi hervorragend geeignet für eine Saalbestuhlung.

Begründung der Jury
Dieser schlichte Reihenstuhl besticht durch seinen innovativen Verkettungsmechanismus, der dem Benutzer mehr Platz und Komfort bietet.

Pelikan
Chair
Stuhl

Manufacturer
Profim sp. z o.o., Turek, Poland
Design
Massive Design (Mac Stopa), Warsaw, Poland
Design Ballendat (Martin Ballendat), Braunau, Austria
Web
www.profim.eu
www.massivedesign.pl
www.ballendat.com

This chair has been designed to reflect the typical beak form of a pelican. It proves of value in all places where one has to sit for extended periods, such as in libraries, conference rooms or at reception desks. What serves the bird for the storage of food also provides storage space here for a book, laptop or bag. The chair comes in two versions – with or without armrests, which are available in a pleasant-to-touch material. There are four frames to choose from: four-leg frame, cantilever, round flat base and four-star base.

Dieser Stuhl ist in der charakteristischen Schnabelform des Pelikans gestaltet und bewährt sich überall dort, wo man länger sitzen muss, etwa in der Bibliothek, im Konferenzsaal oder am Empfang. Was bei dem Vogel zur Aufbewahrung von Nahrung dient, bietet sich auch hier als Stauraum für ein Buch, Laptop oder eine Tasche an. Der Stuhl ist in zwei Ausführungen erhältlich – mit und ohne Armlehnen. Diese sind in einem haptisch angenehmen Stoff ausgeführt. Zur Auswahl stehen außerdem vier Gestelle: Füße, Kufen, Flach- und Kreuzgestell.

Statement by the jury
By paying reference to this particular zoomorphic form, the Pelikan chair has a self-contained design that also offers an additional practical benefit.

Begründung der Jury
Durch die Anlehnung an die besondere Tiergestalt gelingt dem Pelikan eine eigenständige Gestaltung, die darüber hinaus zusätzlichen praktischen Nutzen bietet.

Cantata
Seminar Chair
Seminarstuhl

Manufacturer
Koleksiyon, Istanbul, Turkey
In-house design
Faruk Malhan
Web
www.koleksiyon.com.tr
www.koleksiyoninternational.com
Honourable Mention

The reduced yet clear design of the Cantata chair series presents a wide range of functional options, along with a variety of material and colour choices. It is designed for training and seminar rooms and also comes with casters, enabling easy movement. The chair's specially patented writing tablet can be rotated around its body so that the user may decide on which side to position the tablet. Additionally, the seminar chair offers a storage solution with a basket placed under the seat.

Das unauffällige, aber klare Design der Stuhlreihe Cantata bietet eine breite Palette an funktionalen Möglichkeiten sowie verschiedene Materialien und Farben. Für Trainings- und Seminarräume konzipiert, gibt es den Stuhl auch mit Rollen für eine erhöhte Beweglichkeit. Der patentierte Schreibuntersatz lässt sich um den Stuhl herum drehen, sodass der Benutzer wählen kann, auf welcher Seite der Untersatz positioniert werden soll. Zudem hält der Seminarstuhl mit einem unter dem Sitz befindlichen Korb eine Ablagemöglichkeit bereit.

Statement by the jury
The Cantata chair catches the eye with its unobtrusive design and the rotating writing tablet, distinguishing it as a seminar chair with high utilisation value.

Begründung der Jury
Der Stuhl Cantata fällt durch seine schlichte Gestaltung sowie die drehbare Schreibunterlage ins Auge, die ihn als einen Seminarstuhl mit hohem Nutzwert ausweist.

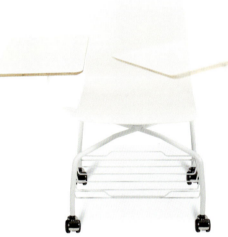

VLEGS
Table with Bench
Tisch mit Bank

Manufacturer
BEGASIT Gauger+Söhne GmbH & Co.KG, Mössingen, Germany
Design
Design Ballendat, Simbach am Inn, Germany
Web
www.begasit.com
www.ballendat.com

The Vlegs table and bench models feature elliptically shaped frame tubes, which are bent at an angle of more than 60 degrees and can thus be stacked closely together. Here, the V-shaped legs are moved outwards from underneath the tabletop using a special mechanism, which allows tables to be stacked without any space between the tabletops. Table and bench form an aesthetic whole and can be easily linked together using conventional row connectors.

Die Vlegs-Modelle Tisch und Bank sind durch elliptisch geformte Gestellrohre gekennzeichnet, die in einem Winkel von über 60 Grad gebogen sind und sich dadurch eng stapeln lassen. Dabei werden die V-Beine über einen neuen Beschlag nach außen neben die Tischplatten geschoben, sodass sich diese ohne Zwischenraum Platte auf Platte stapeln lassen. Tisch und Bank bilden eine ästhetisch harmonische Einheit und können durch die klassische Reihenverbindung einfach miteinander verkettet werden.

giroflex 313
Swivel Chair
Drehstuhl

Manufacturer
Stoll Giroflex AG, Koblenz, Switzerland
Design
Paolo Fancelli, Biasca, Switzerland
Web
www.giroflex.com
www.fancelli.ch

The swivel chair giroflex 313 impresses with its specially developed balance-move system. This means that the user sits down and the chair does everything else, owing to its fully automatic self-adjustment capability. Both the seat and the backrest adjust to the body weight of the user. Only the seat height and, optionally, the locked upright position are manually set. Equally suitable as a work or conference chair, its sophisticated design reflects high mobility by formally alluding to a ball.

Der Drehstuhl giroflex 313 besticht durch ein eigens entwickeltes Balance-Move-System. Das bedeutet, der Benutzer setzt sich und der Stuhl übernimmt dank vollautomatischer Selbsteinstellung den Rest. Sitzfläche und Rückenteil korrespondieren dabei mit dem Körpergewicht. Manuell eingestellt werden lediglich Sitzhöhe und gegebenenfalls die Arretierung in aufrechter Position. Als Arbeits- wie als Konferenzstuhl geeignet, spiegelt sein durchdachtes Design die hohe Beweglichkeit durch die formale Andeutung einer Kugel wider.

Statement by the jury
The special innovation of the giroflex 313 is its automatic capacity for self-adjustment, which turns it into a functionally impressive work chair designed for flexible use.

Begründung der Jury
Die besondere Innovation des giroflex 313 ist seine automatische Selbsteinstellung. Sie macht ihn zu einem funktional überzeugenden und flexibel einsetzbaren Arbeitsstuhl.

Camiro work&meet
Office Swivel Chair
Bürodrehstuhl

Manufacturer
Girsberger Holding AG, Bützberg, Switzerland
Design
Design Ballendat, Simbach am Inn, Germany
Web
www.girsberger.com
www.ballendat.com

The Camiro work&meet was designed specifically for conferences and also as an alternative for desk settings when an easy-to-adjust swivel chair is called for. It features only two adjustment functions: seat height and backrest lock. In terms of form, the seat appears to float on its aluminium supports. Despite its thin seat cushioning, this swivel chair is very comfortable as it features an internal, stretched-mesh suspension and a tilt function that allows movement while seated.

Der Camiro work&meet wurde speziell für Konferenzen sowie als Alternative für den Schreibtisch entworfen, wenn gezielt kein mühsam einzustellender Drehstuhl gewünscht ist. Denn er verfügt nur über die beiden Einstellungsfunktionen Sitzhöhe- und Rückenlehnenarretierung. Der Sitz scheint förmlich auf dem Aluminiumträger zu schweben. Trotz seiner dünnen Polsterung ist dieser Drehstuhl sehr bequem, da ein im Inneren gespanntes Netz für die Unterfederung und eine Neigefunktion für Bewegung beim Sitzen sorgen.

Statement by the jury
With its clear design language and functionality that avoids any unnecessary elements, the Camiro work&meet appears elegant and discreet, which makes it suitable as a chair for both conference and office contexts.

Begründung der Jury
Mit seiner klaren Formgebung und einer alles Überflüssige vermeidenden Funktionalität wirkt der Camiro work&meet elegant zurückhaltend, sodass er sich als Konferenzstuhl wie auch für die Büroarbeit eignet.

Sylphy
Office Swivel Chair
Bürodrehstuhl

Manufacturer
Okamura Corporation, Kanagawa, Japan
In-house design
Design Division (Shoichi Izawa)
Web
www.okamura.jp

The distinctive feature of the Sylphy office swivel chair is the excellent seating comfort for every body type provided by the horizontally adjustable curvature of the backrest in connection with its improved support for the back. The upholstery has a three-dimensionally curved grid shell structure, which ensures flexibility in combination with the elastic back support frame. The seat upholstery is composed of three polyurethane types of different densities, which reduces pressure on the thighs and at the same time supports the hips.

Die Besonderheit des Bürodrehstuhls Sylphy ist, dass seine horizontal einstellbare Krümmung der Rückenlehne in Verbindung mit einer verbesserten Rückenstützung hervorragenden Sitzkomfort für jeden Körperbau bietet. Die Polsterung zeigt einen dreidimensional gekrümmten Gitterschalenaufbau, der zusammen mit dem biegsamen Rückenstützrahmen für Flexibilität sorgt. Die Sitzpolsterung aus drei unterschiedlich dichten Polyurethanen reduziert den Druck auf die Oberschenkel und stützt gleichzeitig die Hüfte.

Statement by the jury
The sophisticated conception of Sylphy, focused on the ergonomic comfort of different body types, is accompanied by a compact appearance that also conveys stability.

Begründung der Jury
Die ausgeklügelte, auf den ergonomischen Komfort verschiedenster Körpertypen ausgerichtete Konzeption des Sylphy geht einher mit einer kompakten Anmutung, die gleichermaßen Stabilität ausstrahlt.

plimode
Office Swivel Chair
Bürodrehstuhl

Manufacturer
Okamura Corporation, Kanagawa, Japan
In-house design
Design Division (Shoichi Izawa)
Web
www.okamura.jp

The plimode office chair features a backrest that is equipped with a highly elastic mesh inside and a top-quality fabric cover. It provides optimised natural seating comfort independent of the body structure. A slide lever enables selection of a locked setting and an easy or firm reclining strength. A synchronised mechanism centred on the base hinge enables seat and back tilt adjustment. The dynamic design is expressed both in the lines and in three fabric covers with 30 colour variations. Thanks to a handle on the backrest, the chair is easy to move and carry.

Der Bürostuhl plimode hat eine Rückenstütze, die mit einem hochelastischen Netz im Inneren und einem hochwertigen Stoffbezug ausgestattet ist und unabhängig vom Körperbau optimalen natürlichen Sitzkomfort bietet. Ein Schiebehebel ermöglicht die Auswahl zwischen verriegelter Einstellung und einer lockeren oder strammen Federung. Ein am Fußgelenk zentrierter Mechanismus synchronisiert die Rückneigung und Sitzverschiebung. Die dynamische Gestaltung des Stuhls kommt nicht nur in seiner Linienführung, sondern auch in den drei Bezugsstoffen mit 30 Farbvariationen zum Ausdruck. Mithilfe eines Griffs an der Rückenlehne lässt er sich leicht bewegen und tragen.

Uneo
Task Chair
Arbeitsstuhl

Manufacturer
Nurus A.Ş., Istanbul, Turkey
Design
Design Ballendat (Martin Ballendat), Simbach am Inn, Germany
Web
www.nurus.com
www.ballendat.com

Uneo is an ergonomic task chair whose look is characterised by its geometric design language and plain functionality, which is due to its simple form accentuated by its integrated controls. Uneo regulates the amount of support according to its user, while allowing finer, personal adjustments. It is available with or without armrests and further radiates lightness as its back is upholstered with mesh fabric. The solidly padded seat ensures elevated seating comfort and provides an interesting aesthetic contrast to the formally reduced backrest.

Uneo ist ein Arbeitsstuhl, dessen Aussehen durch seine geometrische Designsprache und einfache Funktionalität charakterisiert ist, welche dank seiner schlichten Form durch die integrierte Steuerung hervorgehoben werden. Uneo reguliert das Maß der Unterstützung entsprechend seinem Benutzer und erlaubt feinere, persönliche Einstellungen. Er ist mit oder ohne Armlehnen verfügbar und strahlt außerdem Leichtigkeit aus, da er mit Mesh-Material gepolstert ist. Der fest gefüllte Sitz sichert gehobenen Sitzkomfort und bietet einen interessanten ästhetischen Kontrast zu der formal reduzierten Rückenlehne.

Statement by the jury
Thanks to its straight-lined style, the task chair Uneo provides quality and pleasant seating comfort – an all-rounder among office and working chairs.

Begründung der Jury
Mit seiner geradlinigen Formgebung vermittelt der Arbeitsstuhl Uneo Wertigkeit und einen angenehmen Sitzkomfort – ein Allrounder unter den Büro- und Arbeitsstühlen.

Gispen Zinn
Office Chair
Bürostuhl

Manufacturer
Gispen, Culemborg, Netherlands
Design
Justus Kolberg Design (Justus Kolberg),
Hamburg, Germany
Web
www.gispen.com
www.kolbergdesign.com

When developing the Zinn office chair, the aim was to reach a maximum with regard to ergonomics, style and seating comfort. The result is a compact-looking, comfortable chair with an understated design. In accordance with the standards NEN EN 1335 and NPR 1813, Zinn is very versatile in adapting to changes in posture and offering support in every position. The technology of the dynamic-synchronous mechanism promotes a natural sitting experience.

Statement by the jury
The Zinn design succeeds in creating a concept that embodies the prototype of an office chair: a classic in terms of design, while also offering premium-quality, ergonomic comfort.

Der Bürostuhl Zinn wurde mit dem Ziel entwickelt, in puncto Ergonomie, Stil und bequemes Sitzen ein Maximum zu erreichen. Ergebnis ist ein kompakt anmutender und komfortabler Stuhl in zurückhaltendem Design. Erhältlich als Ausführung gemäß EN 1335 und NPR 1813 macht der vielseitige Zinn Veränderungen der Körperhaltung mit und bietet Halt in jeder Position. Der dynamisch-synchrone Mechanismus fördert eine natürliche Sitzhaltung.

Begründung der Jury
Der Gestaltung des Zinn gelingt ein Entwurf, der den Prototyp des Bürostuhls verkörpert: klassisch in der Formgebung, hochwertig und ergonomisch im Komfort.

IN
Office Swivel Chair
Bürodrehstuhl

Manufacturer
Wilkhahn, Wilkening+Hahne GmbH+Co.KG,
Bad Münder, Germany
Design
wiege Entwicklungsgesellschaft mbH,
Bad Münder, Germany
Web
www.wilkhahn.de
www.wiege.com

The office swivel chair IN features the synchronously supporting Trimension mechanism. It allows the right and left parts of the seat to be tilted in any direction by means of independently movable swivel arms. Thus, the natural movement of the pelvis is activated. The 3D-synchronised mechanism is coupled with a highly elastic seat-back system and automatically adapts to these movements so as to support the body in every posture.

Statement by the jury
The high motility enabled by this office chair supports correct body posture and makes it an exemplary ergonomic desk chair.

Mit Trimension verfügt der Bürodrehstuhl IN über einen synchron stützenden Mechanismus, bei dem sich rechte und linke Sitzhälfte mittels unabhängig voneinander beweglicher Schwenkarme in alle Richtungen neigen lassen. Dadurch wird die natürliche Bewegung des Beckens aktiviert. Die mit einem hochelastischen Sitz-Rücken-System gekoppelte 3-D-Synchronmechanik passt sich automatisch an diese Bewegungen an und stützt den Körper in jeder Haltung.

Begründung der Jury
Die hohe Beweglichkeit, die der Bürodrehstuhl ermöglicht, unterstützt eine korrekte Körperhaltung und macht ihn zu einem ergonomisch beispielhaften Schreibtischstuhl.

Stella
Lounge Chair
Lounge-Sessel

Manufacturer
+HALLE, Aarhus, Denmark

Design
busk+hertzog, Lisbon, Portugal

Web
www.plushalle.dk
www.busk-hertzog.com

reddot award 2015
best of the best

Invitingly comfortable

In frantic everyday life or when travelling, the available seating often seems like a long awaited oasis of peace: a popular goal that many approach with a single-minded focus. The Stella lounge chair perfectly embodies just such a haven of peace, although its design challenges viewers in its combination of apparent opposites. The entire structure, which rests on an elegant swivel base made of aluminium, appears narrow and almost delicate – an impression which, however, is qualified by its generous upholstery. The lines of the comfortable chair are clearly defined and give it an air of both lightness and strength. Another fascinating aspect of the chair is its broad back, which also functions as an armrest. It is mounted to the seat at one single, barely noticeable spot, creating the impression that the back is floating above the seat. This design detail offers an enormous advantage for the comfort provided by the chair, as it creates more space on both sides between the seat and the backrest. This allows the sitter to get comfortable and relax in a number of different seating positions. The extremely comfortable Stella lounge chair, conceived for relaxation and communication, almost cries out for someone to sit in it. Many people are sure to wish they could spend more time in it.

Einladender Komfort

Im hektischen Alltag oder auf Reisen erscheinen die angebotenen Sitzgelegenheiten oftmals wie ein langersehnter Ruhepol, den man gerne und zielgerichtet ansteuert. Der Lounge-Sessel Stella verkörpert einen solchen in Perfektion, wobei seine Gestaltung den Betrachter herausfordert, da sie gekonnt scheinbare Gegensätze vereint. Die gesamte, auf einem eleganten Drehfuß aus Aluminium ruhende Konstruktion wirkt schmal und nahezu filigran, wobei dieser Eindruck jedoch durch die großzügige Polsterung wieder relativiert wird. Die Linien des komfortablen Sessels wirken klar definiert und verleihen ihm einen Ausdruck von Leichtigkeit und Schärfe zugleich. Faszinierend für den Betrachter ist zudem die breit gestaltete Rückenlehne, die auch als Armlehne dient. Sie ist am Sitz nur an einer einzigen, kaum auffallenden Stelle montiert, wodurch der Eindruck entsteht, als schwebe die Rückseite des Sessels. Dieses gestalterische Detail besitzt im Hinblick auf den Komfort den enormen Vorteil, dass dadurch mehr Raum an den beiden Seiten zwischen Sitz und Rückenlehne geschaffen wurde. Dies ermöglicht dem Sitzenden, es sich bequem zu machen und unterschiedliche Sitzpositionen einzunehmen. Konzipiert für die Entspannung und Kommunikation, scheint der überaus bequeme Lounge-Sessel Stella geradezu zum Sitzen einzuladen. Gerne will man auch über einen längeren Zeitraum in ihm verweilen.

Statement by the jury

The elegant lines and generous proportions of the Stella lounge chair welcome those in search of relaxation. The chair's design skilfully blends opposites, contrasting the narrow, delicate construction of the frame with the thickness of the upholstery. Stella offers a high degree of ergonomic seating comfort and relaxation. This lounge chair is just as at home in living areas as it is in public spaces.

Begründung der Jury

Mit seiner eleganten Linienführung und großzügigen Proportionen heißt der Lounge-Sessel Stella den Ruhesuchenden willkommen. Gekonnt vereint die Gestaltung des Sessels Gegensätze, wobei etwa die schmale und filigrane Konstruktion des Gestells reizvoll mit der Dicke der Polster korrespondiert. Stella bietet ein hohes Maß an ergonomischem Sitzkomfort und Entspannung. Dieser Lounge-Sessel ist dabei im Wohnbereich ebenso gut einsetzbar wie im öffentlichen Raum.

Designer portrait
See page 34
Siehe Seite 34

Oasis
Lounge Chair
Sessel

Manufacturer
Mobica+ GmbH, Eckenthal, Germany
Design
Design Ballendat, Simbach am Inn, Germany
Web
www.mobicaplus.de
www.ballendat.com

With its flowing shape and continuously curved steel tube covered by a patented, seamlessly knitted fabric, Oasis is a new interpretation of the wingback chair. The double-woven fabric in the area of the lumbar spine and the head provides optimal support and high sitting comfort. Due to its pendulum mechanism, the chair is easy to tilt forwards and backwards. A rotating pad attached to the column can be used as a writing and work surface.

Statement by the jury
The unusual dynamic characterising the design of the Oasis makes it an eye-catching and ergonomically sophisticated piece of sitting furniture.

Mit seiner fließenden Formgebung und dem durchlaufend gebogenen Stahlrohr, das mit einem patentierten nahtlosen Gestrick umspannt ist, stellt Oasis eine Neuinterpretation des Ohrensessels dar. Die doppelt gewebten Bereiche in Lordose- und Kopfhöhe geben guten Halt und sorgen für hohen Sitzkomfort. Mithilfe der Pendelmechanik lässt sich der Sessel bequem nach vorne und hinten neigen. Ein rotierend befestigtes Tablar dient als Schreib- und Arbeitsfläche.

Begründung der Jury
Die ungewöhnliche Dynamik, die die markante Formgebung des Oasis kennzeichnet, macht ihn zum Blickfang und zu einer ergonomisch durchdachten Sitzgelegenheit.

Massaud Work Lounge with Ottoman
Seating
Sitzmöbel

Manufacturer
Coalesse, San Francisco, USA
Design
Studio Massaud (Jean-Marie Massaud), Paris, France
Web
www.coalesse.com
www.massaud.com

Designed for comfort and connecting with technology, the Massaud Work Lounge with Ottoman is an alternative destination to work or relax. The swivelling tablet brings work closer while settling back into a relaxed posture with the ottoman. This piece of furniture features an adjustable headrest with removable pillow insert for ergonomic seating, a swivel base with auto-return and integrated wire management.

Statement by the jury
The design of the Massaud Work Lounge offers high comfort and encourages relaxation. Moreover, the tablet adds a functional touch.

Für Komfort konzipiert und mit technologischen Schnittstellen ausgestattet, stellt der Massaud Arbeitssessel mit Schemel einen alternativen Ort zum Arbeiten oder Entspannen dar. Die schwenkbare Platte bringt die Arbeit näher heran, während sich der Nutzer entspannt in den Sessel zurücklehnt. Das Möbelstück hat eine verstellbare Kopfstütze und ein herausnehmbares Kissen für den ergonomisch passenden Sitz sowie ein drehbares Fußgestell mit automatischer Rückstellung und integriertem Kabelmanagement.

Begründung der Jury
Die Gestaltung des Massaud Arbeitssessels vermittelt hohen Komfort und lädt zum Entspannen ein. Die Ablage stattet ihn zudem mit einer funktionalen Note aus.

LOU
Lounge Chair
Sessel

Manufacturer
BRUNE Sitzmöbel GmbH, Königswinter, Germany
Design
uwe sommerlade design (Uwe Sommerlade), Kassel, Germany
Web
www.brune.de
www.uwe-sommerlade.eu

Lou is a contemporary interpretation of the comfortable single lounge chair for commercial and residential use, mounted on a swivelling cross base. It is characterised by the tension between the generous proportions of the seat and armrests, which flow into a slender, upright backrest, and the tightly upholstered body with its soft seat and neck cushions. The elegantly sculpted surfaces of the upholstered body engender a distinct, circumferential contour.

Statement by the jury
Lou stands out due to its clear aesthetic appearance, which creates original colour highlights with the seat and neck cushions, the latter attached to the top of the backrest.

Lou stellt eine zeitgemäße Interpretation des komfortablen Einzelsessels für den Büro- und Wohnbereich dar – drehbar auf einem Kreuzfuß. Er ist charakterisiert durch die Spannung zwischen den großzügigen Proportionen des Sitz- und Armlehnbereichs, die in einen schlank aufstrebenden Rücken münden, und dem straff gepolsterten Körper mit seinen weichen Kissen für Sitz und Nacken. Die fein modellierten Flächen des Polsterkörpers erzeugen eine ausgeprägte umlaufende Kontur.

Begründung der Jury
Lou fällt durch sein klares ästhetisches Erscheinungsbild ins Auge, das mit Sitzkissen und übergeworfenem Nackenkissen originelle Farbakzente setzt.

Clip
Chair
Sessel

Manufacturer
Materia AB, Tranås, Sweden
Design
Fredrik Mattson Verkstad (Fredrik Mattson),
Stockholm, Sweden
Web
www.materia.se
www.fredrikmattson.se

Clip is an item of furniture that combines a table, chair and armchair in an innovative fashion. The seat and table surface is formed from one continuous element and is characterised by a minimalist, expressive design – encouraging relaxation, reading or working in equal measure. It has a stable swivel base featuring a return mechanism. Connectors are available as optional extras.

Statement by the jury
An individualistic, distinct design with a geometrical silhouette is what characterises this versatile piece of furniture in particular.

Clip ist ein Möbelstück, das Tisch, Stuhl und Armsessel auf innovative Weise miteinander vereint. Die aus einem zusammenhängenden Element geformte Sitz- und Tischfläche ist durch ein minimalistisches, ausdrucksstarkes Design gekennzeichnet und lädt gleichermaßen zum Entspannen wie zum Lesen oder Arbeiten ein. Sie ist auf einem stabilen drehbaren Fuß mit Rückholfederung montiert. Als Zubehör sind Steckverbinder erhältlich.

Begründung der Jury
Eine individuelle, prägnante Gestaltung mit einer geometrischen Silhouette – dadurch zeichnet sich dieses vielseitig nutzbare Möbelstück besonders aus.

Frame
Chair
Sessel

Manufacturer
Materia AB, Tranås, Sweden
Design
Ola Giertz Design Studio (Ola Giertz),
Helsingborg, Sweden
Web
www.materia.se
www.olagiertz.se

The design of the Frame chair was inspired by a picture frame, the rectangular form of which makes it a piece of furniture with a graphic look. Its versatility promotes a creative atmosphere. A person can sit in the middle as on a stool or else on the edge using one side as a backrest. The sides and top of the frame envelop the person sitting there and screen off the surroundings, thus creating the feeling of being in a separate space.

Statement by the jury
The sitting experience and posture of this chair are as pleasant as its design is unusual. This lends the Frame a touch that is both personal and playful.

Die Gestaltung des Sessels Frame ist von einem Bilderrahmen inspiriert, dessen rechteckige Formensprache ihn zu einem grafisch anmutenden Möbelstück macht. Seine Vielseitigkeit fördert eine kreative Atmosphäre. Der Nutzer kann sich in die Mitte wie auf einen Hocker setzen oder auf dem Rand Platz nehmen und eine Seite als Rückenlehne verwenden. Seiten und Dach umschließen ihn dabei und schirmen ihn von der Umgebung ab, sodass eine Art Raumgefühl entsteht.

Begründung der Jury
So ungewöhnlich die Gestaltung dieses Sessels ist, so angenehm sind Sitzhaltung und -gefühl darin. Der Frame erhält dadurch eine persönliche wie spielerische Note.

Ballo
Stool
Hocker

Manufacturer
Humanscale, Piscataway, USA
In-house design
Web
www.humanscale.com
Honourable Mention

This stool is a modern take on traditional ball seats. Thanks to its sturdy central column and air-filled rubber domes, it strengthens balance and musculature. The weight of 6.1 kg and the built-in handles accentuate ease of use. With a seat height of 62.2 cm, the stool is also suitable for informal meetings in the office. Its bright, playful look and vivid colour range allow it to blend in well with any environment.

Der Hocker Ballo ist die moderne Umsetzung des traditionellen Gymnastikballs und stärkt dank seines stabilen Stands und luftgefüllter Gummikuppeln Gleichgewicht und Muskulatur. Ein Gewicht von 6,1 kg und integrierte Haltegriffe stellen die einfache Handhabung in den Vordergrund. Mit einer Sitzhöhe von 62,2 cm eignet er sich auch für informelle Meetings im Büro und lässt sich mit seiner spielerisch-fröhlichen Optik und kräftigen Farbpalette in jede Umgebung integrieren.

pig
Pouffe
Hocker

Manufacturer
Guangzhou Lightspace Furniture Co., Ltd., Guangzhou, China
Design
Favaretto&Partners (Francesco Favaretto), Padua, Italy
Web
www.lightspace.cc
www.favarettoandpartners.com
Honourable Mention

Pig is a zoomorphic pouffe that can be used in a wide range of situations, displaying a simple yet elegant shape coated in either leather or fabric. By lifting the upper half, it becomes a small tea table with an ellipsoidal Plexiglas panel under which little objects and magazines can also be placed.

Pig ist ein Hocker in Tiergestalt. Mit seiner schlichten und gleichzeitig eleganten Form, die entweder mit Stoff oder Leder überzogen erhältlich ist, kann er auf vielfältige Weise genutzt werden. Wird die obere Hälfte abgenommen, verwandelt er sich in einen kleinen Tisch mit einer ellipsenförmigen Platte aus Plexiglas, unter der kleinere Gegenstände und Zeitschriften Platz finden.

Statement by the jury
The playful look of this pouffe not only represents esprit and lightness but also surprises with its diverse applications as table and storage space.

Begründung der Jury
Die verspielte Formgebung dieses Hockers verkörpert nicht nur Witz und Leichtigkeit, sondern überrascht darüber hinaus durch den Mehrfachnutzen als Tisch und Stauraum.

Wyspa
Sofa

Manufacturer
Profim sp. z o.o., Turek, Poland
Design
ITO Design (Mugi Yamamoto, Christopher Schmidt), Cham, Switzerland and Nuremberg, Germany
Web
www.profim.eu
www.ito-design.com

The Wyspa collection offers an original solution for integrated interior and furniture design. Each sofa is embedded in a surrounding wall element. Placing several elements adjacently gives rise to an atmosphere conductive to personal conversation, while a positioning opening out towards the room is suitable for unconventional meetings. The collection includes armchairs and sofas with either two or three seats, along with upholstery in vibrant or pastel colours.

Die Kollektion Wyspa bietet eine innovative Lösung für integrierte Raum- und Möbelgestaltung. Jedes Sofa ist in ein umgebendes Wandelement eingebettet. Mehrere Elemente zusammengestellt schaffen eine gute Atmosphäre für persönliche Gespräche, die offene Positionierung im Raum eignet sich für unkonventionelle Meetings. Die Kollektion besteht aus Sesseln sowie Zweier- und Dreiersofas mit Polsterungen in lebendigen oder matten Farben.

Statement by the jury
Wyspa creates a new look in furniture design and achieves a particularly striking appearance thanks to the integrated slanting wall element.

Begründung der Jury
Wyspa setzt neue Akzente in der Möbelgestaltung und erzielt durch das integrierte schräggestellte Wandelement eine besonders markante Ausstrahlung.

Gate
Seating Furniture
Sitzmöbel

Manufacturer
Offecct AB, Tibro, Sweden
Design
Claesson Koivisto Rune,
Stockholm, Sweden
Web
www.offecct.se
www.claessonkoivistorune.se

Gate is a modular sofa system with rigorous proportions, where both technical and emotional aspects were involved in the development process. The thin seating contrasts with the broad backrest, which features conveniently placed electrical sockets for devices like laptops. This creates an unconventional silhouette. Gate is constructed almost like building elements that can be easily assembled according to need.

Statement by the jury
Gate is more than a modular sofa with a distinct and minimalist design: with sockets along the backrest, it also has a practical benefit.

Gate ist ein modulares Sofasystem mit strengen Proportionen, bei deren Ausarbeitung sowohl technische als auch emotionale Aspekte eine Rolle gespielt haben. Der Sitz ist schmal geschnitten, die hingegen breite Rückenlehne hält an geeigneten Stellen Steckdosen für Geräte wie Laptops bereit. So entsteht eine ungewöhnliche Silhouette. Gate ist ähnlich wie Bauelemente konstruiert, die nach den jeweiligen Bedürfnissen zusammengesetzt werden können.

Begründung der Jury
Gate ist mehr als ein klar und minimalistisch gestaltetes modulares Sofa: Mit Steckdosen auf seinem Rücken bietet es zudem einen praktischen Nutzen.

Airberg
Seating Furniture
Sitzmöbel

Manufacturer
Offecct AB, Tibro, Sweden
Design
Studio Massaud (Jean-Marie Massaud),
Paris, France
Web
www.offecct.se
www.massaud.com
Honourable Mention

Airberg is a furniture concept consisting of an armchair and a sofa, whose forms convey the impression of being filled with air. In reality it is held up by a technically advanced set of elastic ribs and springs. The craftsmanship used in the production of Airberg has created a seemingly deconstructed, highly comfortable piece of furniture that is defined by its contemporary, asymmetrical shape. The furniture is made of sustainable materials.

Statement by the jury
This upholstered furniture is striking thanks to its exceptionally amorphous design vocabulary. At the same time, its bulky proportions facilitate enhanced seating comfort.

Airberg ist ein Möbelkonzept aus Sessel und Sofa, deren Körper wirken, als seien sie mit Luft gefüllt. Tatsächlich werden sie durch eine Reihe von technisch ausgeklügelten elastischen Latten und Federn gehalten. Die zur Herstellung von Airberg angewandte Handwerkskunst ließ ein wie aufgebrochen wirkendes, sehr bequemes Möbelstück entstehen, das durch seine zeitgemäße asymmetrische Form definiert ist. Es wurde aus nachhaltigen Materialien gefertigt.

Begründung der Jury
Diese Polstermöbel fallen durch ihre ungewöhnlich amorphe Formensprache ins Auge. Ihre wuchtigen Proportionen schaffen zudem hohen Sitzkomfort.

TST
Sofa

Manufacturer
Gispen, Culemborg, Netherlands
Design
Michael Young Ltd (Michael Young),
Hong Kong
Web
www.gispen.com
www.michael-young.com

The TST stands out due to its striking appearance with a high, circumferential backrest. It creates a private atmosphere that encourages relaxation or inspires togetherness at home or in the office, while also able to promote communication. Thanks to smooth lines and rounded contours, the sofa provides enhanced seating comfort. A large selection of different upholstery fabrics is available.

Statement by the jury
TST – a sofa that grants visual privacy – enables real relaxing. Its unobtrusive, high-quality design makes you wish for a break or for a diverting conversation.

Das Sofa TST sticht durch sein auffälliges Erscheinungsbild mit der umlaufenden hohen Rückenlehne hervor. Diese lässt eine persönliche Atmosphäre entstehen, die zu Hause oder im Büro zum Relaxen oder anregenden Miteinander einlädt, aber auch die Kommunikation fördern kann. Dank weicher Linien und runder Konturen bietet das Sofa hohen Sitzkomfort. Es steht eine große Auswahl verschiedener Bezugsstoffe zur Verfügung.

Begründung der Jury
TST – ein Sofa mit Sichtschutz – ermöglicht echtes Entspannen. Sein schlichtes hochwertiges Design macht ebenso Lust auf eine Pause wie eine kurzweilige Unterhaltung.

Applica
Docking System

Manufacturer
ASSMANN Büromöbel GmbH & Co. KG, Melle, Germany
In-house design
Gerd Lauszus
Design
B·PLAN (Wolfgang Blume), Nienburg, Germany
Web
www.assmann.de
www.bplan-nienburg.de

Applica is a structural desk element that incorporates power and data lines coming from floor boxes or overhead ducts and distributes them across up to nine connectors for mobile devices. Permanently installed hardware such as monitors or printers can be linked through further plug connectors inside the module. Mini-PCs can also be fitted into the wall element and secured with lockable wall panels. The frame profile comes with a groove that can hold monitor and iPad mounts. The modules are available in different widths, heights and colours and may also be used for partitioning the workspace.

Applica ist ein tragendes Schreibtischelement, das Strom- und Datenleitungen über Bodentanks oder die Deckenzuführung aufnimmt und auf bis zu neun Anschlüsse für mobile Endgeräte verteilt. Fest installierte Hardware wie Monitore oder Drucker können über weitere Steckerleisten im Modulinneren dauerhaft angeschlossen werden. Auch Mini-PCs lassen sich in dem Wandelement unterbringen und über verschließbare Wandfüllungen absichern. Das Rahmenprofil ist mit einer Nut ausgestattet, die Monitor- oder iPad-Halter aufnimmt. Die Module sind in verschiedenen Breiten, Höhen und Farben lieferbar und fungieren auch als Abschirmung des Arbeitsplatzes.

Statement by the jury
Applica convinces with its broad functional scope: uniting the functionality of a load-bearing desk support, intelligent multi-plug connector and fashionable screen, it is an effective piece of office equipment.

Begründung der Jury
Applica überzeugt durch seine funktionale Bandbreite: Zugleich tragendes Tischbein, intelligenter Mehrfachstecker und schicker Raumteiler, zeigt es sich als effektives Büroutensil.

SolarDesk
Solarschreibtisch

Manufacturer
Solstrøm Furniture UG, Singen, Germany
In-house design
Web
www.solstrom-furniture.com
Honourable Mention

SolarDesk has a distinct and puristic design and features a matt glass surface that is impact-resistant and dirt-repellent due to the integrated lotus effect. The tabletop is enclosed in a light, black aluminium frame. This solar desk can charge smartphones and tablets to full capacity, as 60 night-black, square solar cells charge the built-in battery both in daylight and artificial light.

Statement by the jury
SolarDesk is a desk with added value: with a minimalist look, it also acts as an eco-friendly "charger".

Der klar und puristisch gestaltete Solar-Desk hat eine matte Glasoberfläche, die schlagresistent und durch den integrierten Lotuseffekt schmutzabweisend ist. Die Tischplatte wird durch einen leichten schwarzen Aluminiumrahmen eingefasst. Der Solarschreibtisch kann Smartphones oder Tablets vollständig aufladen. 60 nachtschwarze, quadratische Solarzellen befüllen den integrierten Akku sowohl bei Tages- wie auch bei künstlichem Licht.

Begründung der Jury
Der SolarDesk ist ein Schreibtisch mit Mehrwert: Minimalistisch gestaltet, fungiert er zugleich als umweltfreundliches „Ladegerät".

Reverse Conference
Table Series
Tischserie

Manufacturer
Andreu World, Valencia, Spain
Design
Piergiorgio Cazzaniga, Lentate, Italy
Web
www.andreuworld.com
www.piergiorgiocazzaniga.com

Reverse Conference is a series of conference tables, available with one or multiple bases and variable tabletops in 14 sizes. The table displays an elegant look with its 30 mm laminated wood plate in oak veneer or lacquered style, and the contrasting thick bases tapering upward. The base frames are made of 100 per cent recyclable polyethylene and come in matt or glossy versions.

Statement by the jury
The design of this table series draws its appeal from contrasting forms and materials, thus achieving a pronounced aesthetic appearance.

Reverse Conference, ein Programm von Konferenztischen, ist mit einem oder mehreren Füßen und variablen Tischplatten in insgesamt 14 Größen erhältlich. Mit seiner 30 mm dünnen Schichtholzplatte in Eichenfurnier oder lackiert und dazu kontrastierend den mächtigen, nach oben konisch zulaufenden Beinen präsentiert sich der Tisch in elegantem Design. Die Untergestelle aus 100 Prozent recyclebarem Polyethylen werden in matter oder glänzender Ausführung angeboten.

Begründung der Jury
Die Gestaltung dieses Tischprogramms erzielt ein ausgesprochen ästhetisches Erscheinungsbild, das seine Spannung aus den Form- und Materialkontrasten bezieht.

Quickstand
Computer Workstation
Computerarbeitsplatz

Manufacturer
Humanscale, Piscataway, USA
In-house design
Web
www.humanscale.com
Honourable Mention

Quickstand allows a change between sitting and standing while using a computer and thus enables a more dynamic workplace experience. With minimum space requirements, it can be mounted to the rear edge of the working space. The workstation does not require numerous knobs and levers, and it provides a mounting option for attaching up to two monitors. The keyboard is situated on a height-adjustable shelf. The pre-integrated cable management system allows users to simply plug in and go.

Quickstand ermöglicht den Wechsel zwischen Sitzen und Stehen am Computer und damit mehr Dynamik am Arbeitsplatz. Es lässt sich platzsparend an der Hinterkante der Arbeitsfläche anbringen, kommt ohne eine Vielzahl von Knöpfen und Hebeln aus und bietet Befestigungsmöglichkeiten für bis zu zwei Monitore. Die Tastatur befindet sich auf einer höhenverstellbaren Ablage. Der vorinstallierte Kabelstrang ermöglicht einfachen Plug-and-play-Betrieb.

Statement by the jury
Quickstand presents an innovative solution for working flexibly in sitting and standing postures at the desk. The design is functional and reserved.

Begründung der Jury
Quickstand stellt eine innovative Lösung dar, um am Schreibtisch flexibel im Sitzen oder Stehen arbeiten zu können. Die Gestaltung ist dabei funktional und zurückhaltend.

Cable Snake® Orbit®
Cable Guiding System
Kabelführungssystem

Manufacturer
Q-LAB, Prien, Germany
In-house design
Maximilian Rüttiger
Web
www.q-lab.de

The Orbit cable snake is a customisable cable management system. Its backbone is a patented bayonet connecter onto which eight differently shaped chain links can be mounted according to each user's preferences. They are available in colourful bioplastics, metal, glass, wood or other customer-selected materials. When pushed together, the cable chain can be compressed by 25 per cent and then used as rigid cable channel. If the chain links are uncompressed, they snap into a flexible, two-axle position to enable optimal guiding.

Die Kabelschlange Orbit ist ein individualisierbares Kabelführungssystem. Ihr Rückgrat besteht aus einem patentierten Bajonettverbinder, auf den acht verschiedene Kettengliedformen je nach Vorliebe des Nutzers aufgesteckt werden können. Diese sind in buntem Bio-Kunststoff, in Metall, Glas, Holz oder einem anderen Wunschmaterial erhältlich. Die Kabelkette kann durch Zusammenschieben um 25 Prozent gestaucht und als starrer Kabelkanal verwendet werden. Zieht man die Kettenglieder auseinander, rasten sie in einer zweiachsig gelenkigen Position ein, um optimal abzurollen.

Statement by the jury
This cable chain combines high functionality, the individual design of the chain links and their materiality to create a desk aid that is contemporary and useful in equal measure.

Begründung der Jury
Diese Kabelkette verbindet hohe Funktionalität und die individuelle Formgebung der Kettenglieder samt ihrer Materialität zu einem zeitgemäßen wie nützlichen Schreibtischutensil.

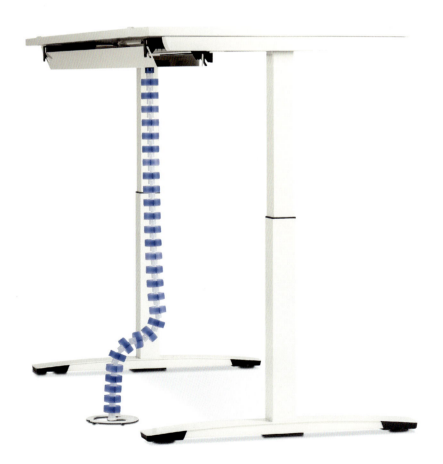

R8.5
Training Furniture System
Möbelsystem für Schulungen

Manufacturer
Aurora (China) Co., Ltd., Shanghai, China
In-house design
Aurora Design Center
Web
www.aurora.com.cn
Honourable Mention

The R8.5 training furniture system presents a simple, contemporary and group-oriented design language and includes a training table, a lectern, a whiteboard and a cart. With its T-legs, the training table provides the user with enough freedom of movement, while the visible braking device and universal castors make it flexible and easy to operate. With the aid of a one-hand folding function, the tabletop can be easily stored.

Statement by the jury
The R8.5 furniture system, designed for training contexts, communicates a distinct and reduced design that attaches great importance to practical suitability.

Die Formensprache des für Schulungen entworfenen Möbelsystems R8.5 ist schlicht, modern und gruppenorientiert. Es beinhaltet einen Übungstisch, ein Lesepult, ein Whiteboard und einen Wagen. Die T-Beine lassen dem Nutzer genügend Bewegungsfreiheit, während der Stuhl dank der sichtbaren Bremsvorrichtung und der Universalrollen flexibel und leicht zu handhaben ist. Mithilfe der Einhandfunktion lässt sich die Tischplatte einfach verstauen.

Begründung der Jury
Das für Schulungen entworfene Möbelsystem R8.5 präsentiert sich in einem klaren reduzierten Design, das hohen Wert auf seine praktische Tauglichkeit legt.

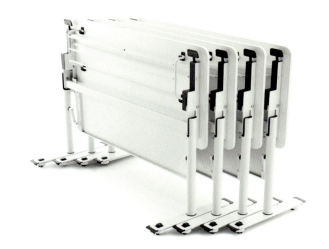

CablePort desk^2
Desktop Connection
Tischanschlussfeld

Manufacturer
Kindermann GmbH, Eibelstadt, Germany
Design
CaderaDesign (Florian Labisch, Daniel Maisch), Würzburg, Germany
Web
www.kindermann.com
www.caderadesign.de

The CablePort desk^2 unites technology, functionality and a mature design. The aluminium body, which is bent at striking angles, gives the casing a distinctive design vocabulary. It adapts to users' needs in the modern office world by offering numerous possible electrical and data port combinations. It can be configured individually with international mains sockets, all standard data interfaces and special multimedia solutions.

Statement by the jury
CablePort desk^2 is the favourable solution for the contemporary workplace: with a high-quality functional design, it constellates all vital connections in one place.

Das CablePort desk^2 vereint Technik, Funktionalität und eine ausgereifte Gestaltung. Der in einem markanten Winkel gebogene Aluminiumkörper verleiht dem Gehäuse eine unverwechselbare Formensprache. Mit zahlreichen Kombinationsmöglichkeiten für Strom- und Datenleitungen passt er sich den Nutzerbedürfnissen der modernen Bürowelt an und ist mit internationalen Steckdosen, den gängigen Datenanschlüssen und Multimedia-Sonderlösungen auch individuell konfigurierbar.

Begründung der Jury
Für einen zeitgemäßen Arbeitsplatz stellt CablePort desk^2 die ideale Lösung dar: In hochwertigem funktionalen Design ermöglicht es die Verknüpfung der wichtigsten Anschlüsse.

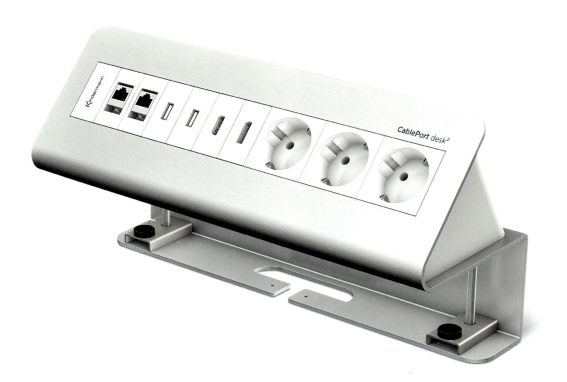

Penxo
Pencil Lead Holder
Bleistiftminenhalter

Manufacturer
Strong Precision Machinery,
Hong Kong

Design
BitsFactory (Brian Tientak Tong),
Dublin, California, USA

Web
www.syitung.com
http://briantongtak.prosite.com
www.penxo.com

reddot award 2015
best of the best

Precise minimalism

Writing with a pencil produces a very particular sensation, which is partly governed by the softness of the pencil lead chosen. This pencil lead holder is, in effect, a special form of pencil with the advantage that it can be "refilled" again and again. The Penxo gives this writing tool a fascinating new twist. It succeeds in combining a form reduced to the essentials with an innovative construction. The strict minimalist approach does away with the usual assembled components. Instead, the casing of the Penxo is carefully made from a single aluminium block. As there are no springs, buttons and other mechanical parts, this pencil lead holder with a diameter of just 2 mm looks seamless and cohesive. It is simple yet stylish and stands out due to its carefully considered functionality. A slot in the centre serves to release the lead and also as a window to show the colour, marking and usage level of the lead. The slit at the tip of the pencil creates torque at the other end so that the resulting counter-balance exerts a clamp-like hold on the pencil lead. This astonishingly simple, but sophisticated functionality makes the Penxo pencil lead holder an impressive tool for everyday use and provides a successful, contemporary reinterpretation of the humble pencil.

Präziser Minimalismus

Das Schreiben mit dem Bleistift hat eine besondere Sinnlichkeit, die auch durch die Weichheit der gewählten Mine bestimmt wird. Der Bleistiftminenhalter ist im Grunde eine spezielle Form des Bleistifts mit dem Vorteil, dass er immer wieder „befüllt" werden kann. Mit dem Penxo erfährt dieses Schreibgerät nun eine faszinierend neue Interpretation. Es gelingt dabei die schwierige Aufgabe, eine formale Reduktion mit einem innovativen Aufbau zu verbinden. Einem streng minimalistischen Ansatz folgend, wurde hier auf die übliche Gestaltung aus zusammengesetzten Einzelteilen verzichtet. Der Penxo entsteht durch das präzise Herausfräsen des Gehäuses aus einem einzigen Aluminiumblock. Da Federn, Knöpfe und andere mechanische Teile fehlen, wirkt der Minenhalter mit einem Durchmesser von nur 2 mm wie aus einem Guss und sehr in sich geschlossen. Er ist ebenso schlicht wie stilvoll und glänzt zudem durch seine perfekt durchdachte Funktionalität. Eine mittige Spalte dient dem Lösen der Mine und zugleich auch als Anzeigefenster für deren Farbe, Kennzeichnungen und Verbrauch. Der vordere Schnittspalt an der Spitze des Stifts erzeugt eine Drehspannung am entgegengesetzten Ende. Das dadurch entstehende Gegengewicht sorgt für einen klammerartigen Effekt, der die Mine hält. Der Bleistiftminenhalter Penxo beeindruckt tagtäglich mit dieser verblüffend einfachen und raffinierten Funktionalität. Die Bedeutung des Bleistifts wird damit sehr erfolgreich zeitgemäß interpretiert.

Statement by the jury

In its simplicity, the Penxo pencil lead holder achieves a fascinatingly clear beauty. Its design is based on the principles of gravity, using it to create a shape that is self-explanatory to the user. The classic pencil has been transported into the computer era in a highly convincing manner. The functionality of Penxo has been carefully considered and offers this particular extra twist that just makes it fun to use.

Begründung der Jury

In seiner Simplizität strahlt der Bleistiftminenhalter Penxo eine faszinierend klare Schönheit aus. Seine Gestaltung nutzt das Prinzip der Schwerkraft und entwickelt daraus eine Formensprache, die sich dem Betrachter bereits auf den ersten Blick erklärt. Der klassische Bleistift wird hier auf sehr überzeugende Weise in das Computerzeitalter transportiert. Die Funktionalität des Penxo ist perfekt durchdacht – sie hat diesen besonderen Clou, der einfach Spaß macht.

Designer portrait
See page 36
Siehe Seite 36

CYBER
Pen
Stift

Manufacturer
Shanghai KACO Industrial Co., Ltd., Shanghai, China
In-house design
Design
Then Creative (Lei Xie), Shanghai, China
Web
www.kacoconcept.com
www.thencreative.com

The design of the Cyber roller pen displays strength and suppleness. Its slim metallic barrel, which is available in various matt colours, ends in a cap that accommodates a USB flash drive in its clip. This makes the pen a useful and efficient tool in conjunction with desks and computers, as one can always keep data and enough memory close at hand.

Die Gestaltung des Stiftes Cyber strahlt gleichermaßen Stärke und Geschmeidigkeit aus. Sein schlanker metallischer Schaft, der in verschiedenen matten Farben wählbar ist, mündet in eine Kappe, in deren Halterung sich ein USB-Stick befindet. Das macht den Stift zu einem nützlichen und effizienten Werkzeug an Schreibtisch und Computer, da man seine Daten und genügend Speicherplatz stets unkompliziert zur Hand hat.

Statement by the jury
This pen impresses with its straightforward aesthetic appearance, but also with the innovative idea to integrate, almost invisibly, a USB flash drive into its cap.

Begründung der Jury
Dieser Stift punktet nicht nur durch seine schlichte ästhetische Anmutung, sondern auch durch die innovative Idee, einen USB-Stick nahezu unauffällig in seine Kappe zu integrieren.

e-motion „pure Black"
Fountain Pen
Füllfederhalter

Manufacturer
A.W. Faber-Castell Vertrieb GmbH, Stein, Germany
Design
Heinrich Stukenkemper Industrial Design Team,
Castrop-Rauxel, Germany
Web
www.faber-castell.com
www.stukenkemper.com

A reduced design with a dynamic silhouette and a matt, deep-black shade turn the e-motion "pure Black" fountain pen into a striking eye-catcher on the desk. Its aluminium barrel with a guilloche pattern has a pleasantly cool feel and rests comfortably in the hand. The PVD-coated, stainless-steel nib ensures high writing comfort. The spring-loaded metal clip on the chromed cap holds the fashionable pen in place, also on the go.

Das reduzierte Design mit dynamischer Silhouette und die matte, tiefschwarze Farbgebung machen den Füllfederhalter e-motion „pure Black" zum markanten Eyecatcher auf dem Schreibtisch. Sein Aluminiumschaft ist guillochiert und weist damit eine angenehm kühle Haptik auf und liegt gut in der Hand. Für hohen Schreibkomfort sorgt die PVD-beschichtete Edelstahlfeder. Der gefederte Metallclip an der verchromten Kappe gibt dem modischen Schreibgerät auch unterwegs sicheren Halt.

Statement by the jury
The distinct shape, the finely patterned aluminium barrel and the rich black hue are the characteristic design features distinguishing this fountain pen.

Begründung der Jury
Die prägnante Formgebung, der fein gemusterte Schaft aus Aluminium und das matte Schwarz sind die charakteristischen Merkmale, die den Füllfederhalter gestalterisch besonders hervorheben.

Cuttlelola Dotspen
Electric Drawing Pen
Elektronischer Zeichenstift

Manufacturer
Dongguan Dots Trading Co., Ltd., Dongguan, China
In-house design
Shuochang Song, Chunhong Cao, Xiaojun Yan
Web
www.cuttlelola.com
www.dotspen.com

Cuttlelola is an innovative electric drawing pen that is mainly used for the professional creation of comics and educational materials. It simulates dotting ten times faster than a traditional pen, which saves a lot of time and allows beginners to draw more easily and precisely within a short period of time. The lithium-ion battery charges the pen during use. An environmentally friendly mechanism replaces expensive screen tones, which are used for comics only once, thus avoiding the excessive consumption of paper.

Cuttlelola ist ein innovativer elektronischer Zeichenstift, der vor allem für professionelle Comic-Kreationen und für die Aufklärungsarbeit benutzt wird. Die Punktierung wird zehnmal schneller als bei einem traditionellen Stift simuliert. Das spart Zeit und ermöglicht Anfängern auf schnellem Weg mehr Leichtigkeit und Präzision beim Zeichnen. Mit dem Lithium-Ionen-Akku wird der Stift während der Benutzung aufgeladen. Ein umweltfreundlicher Mechanismus ersetzt teure Screen Tones, die nur einmal für Comics eingesetzt werden können, und vermeidet somit übermäßigen Papierverbrauch.

Dyson Pen
Stift

Manufacturer
Dyson Ltd, Wiltshire, Great Britain
In-house design
James Dyson
Web
www.dyson.co.uk
www.dyson.com

When designing the Dyson Pen, the aim was to create a stylish and comfortable pen with an unadorned and practical design. Its mechanism is a simple push-down-and-up locking function, which is a straightforward solution that stands out due to its efficient usability. The pen was inspired by the unpretentious geometry of the Dyson Airblade tap and employs the pure, highly polished metal of the pen in the same way.

Statement by the jury
The Dyson Pen impresses with a design that is reduced to the most essential elements and goes hand in hand with an enhanced quality of materials.

Das Ziel bei der Entwicklung des Dyson Pen war ein Stift, der stilvoll und komfortabel ist wie auch ein schlichtes und praktisches Design zeigt. Der Mechanismus seiner Push-down-and-up-Sperrfunktion ist eine einfache Lösung, die sich durch effiziente Nutzbarkeit auszeichnet. Der Stift ist inspiriert von der unverfälschten Geometrie des Dyson-Airblade-Wasserhahns und verwendet reines hochglanzpoliertes Metall in gleicher Weise.

Begründung der Jury
Der Dyson Pen beeindruckt durch seine auf das absolut Wesentliche reduzierte Formgebung, die mit einer hohen Materialqualität Hand in Hand geht.

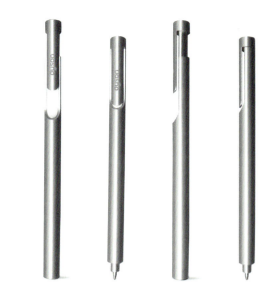

TURNUS
USB Pen
USB-Stift

Manufacturer
Klio-Eterna Schreibgeräte GmbH & Co KG, Wolfach, Germany
Design
briggl industrie design (Hariolf Briggl), Schwäbisch Hall, Germany
Web
www.klio.com
www.briggl-design.de

TURNUS elegantly combines the elements of a ballpoint pen and a USB flash drive in a high-quality plastic cover. The removable cap houses the 2 to 64 GB flash drive with 2.0 or 3.0 high-speed memory, featuring the new chips-on-board technology. A specially developed silk-tech pen tip provides a velvety feel when writing. Numerous possible combinations of colours, materials and surfaces turn this ballpoint pen into an appealing advertising medium.

Statement by the jury
Versatile design possibilities and an advanced USB flash drive in the cap are the attractive features that distinguish the slender TURNUS.

TURNUS vereint die Elemente Kugelschreiber und USB-Speicher auf elegante Art in einem hochwertigen Kunststoffgehäuse. Die abnehmbare Kappe integriert den USB-Stick der neuen COB-Technik von 2 bis 64 GB in 2.0- oder 3.0-Highspeed-Speichertechnologie. Eine eigens entwickelte Silktech-Mine sorgt für ein samtweiches Schreibgefühl. Zahlreiche Kombinationsmöglichkeiten zwischen Farben, Materialien und Oberflächen machen den Kugelschreiber zu einem ansprechenden Werbeträger.

Begründung der Jury
Vielfältige Gestaltungsmöglichkeiten und ein fortschrittlicher USB-Stick in der Kappe sind die attraktiven Merkmale, die den schlanken TURNUS auszeichnen.

Rule/One
Pen + Ruler
Stift + Lineal

Manufacturer
hmm, Taipei, Taiwan
In-house design
Web
www.hmmproject.com
Honourable Mention

Rule/One is a pen and a ruler in one. The unit consists of five aluminium components and is lightweight yet still robust. At a length of 13 cm, it is easy to carry. The cap at the top end of the pen can be replaced with a removable lanyard. There is also a special tip for touchscreen devices – another practical feature of the tool. Neodymium magnets inside the ruler ensure that the pen stays securely in place.

Statement by the jury
The design of Rule/One reflects a simple and straightforward solution to combining the dual functions of pen and ruler. This makes it an expedient desk accessory.

Rule/One ist Stift und Lineal in einem. Die Einheit besteht aus fünf Aluminiumkomponenten, ist daher leicht und dennoch robust. Mit einer Länge von etwas über 13 cm lässt sie sich bequem transportieren. Die Kappe am oberen Ende des Stiftes kann gegen eine abnehmbare Tragekordel ausgetauscht werden. Für Touchscreen-Geräte gibt es eine spezielle Spitze, sodass das Werkzeug einen weiteren praktischen Nutzen erfüllt. Neodymmagnete im Lineal sorgen dafür, dass der Stift an Ort und Stelle bleibt.

Begründung der Jury
Die Gestaltung des Rule/One findet eine einfache und klare Lösung, um die beiden Elemente Stift und Lineal miteinander zu verbinden. Das macht ihn zum zweckdienlichen Schreibtischutensil.

transotype
Cutting Ruler
Schneidelineal

Manufacturer
Holtz Office Support GmbH, Wiesbaden, Germany
In-house design
Tobias Liliencron
Web
www.holtzofficesupport.com

The cutting ruler transotype is made of anodised aluminium and then powder-coated in red, thus making it easy to locate, even on crowded tables. With a centimetre scale, a cutting edge and a non-slip rubber back, the ruler facilitates precise and expedient working. The cutting edge features a straight back, so that the cutter blade can be guided over the paper with precision, which enables exact cutting. It is available in lengths of 30 and 60 cm, with the larger ruler allowing A2 formats to be cut along the longitudinal plane.

Das Schneidelineal transotype wurde aus eloxiertem Aluminium mit roter Pulver-beschichtung gefertigt und lässt sich so selbst auf vollen Tischen schnell wiederfin-den. Durch cm-Skalierung, Schneidekante und rutschfeste Gummierung an der Rück-seite ermöglichst es präzises und schnelles Arbeiten. Dank der Schneidekante mit geradem Rücken lässt sich die Klinge eines Cutters präzise über das Papier führen, was einen exakten Schnitt ermöglicht. Es ist in den Längen 30 und 60 cm erhältlich, das größere Lineal erlaubt es somit, selbst A2-Formate längsseitig zu schneiden.

senseBag by transotype
Pencil Case
Mäppchen

Manufacturer
Holtz Office Support GmbH, Wiesbaden, Germany
In-house design
Tobias Liliencron
Web
www.holtzofficesupport.com
Honourable Mention

The senseBags are designed to transport the necessary instruments for drawing, writing or painting – such as markers, pens, brushes and much more. The pencil cases, including the roll-up variety, are made of natural-coloured, high-grade linen or hard-wearing, black polyester fabric. They are highlighted with red-hued details like pen loops, Velcro and zip fasteners. Both varieties of pencil cases accommodate up to 72 elements.

Mit den senseBags lassen sich die nötigen Utensilien zum Zeichnen, Schreiben oder Malen transportieren. Marker, Stifte, Pinsel und vieles mehr findet darin Platz. Die Rollmäppchen und Etuis sind aus naturfarbenem, hochwertigem Leinen oder schwarzem, strapazierfähigem Polyestergewebe gefertigt und setzen mit ihren roten Details wie Stifteschlaufen, Klettbändern oder Reißverschlüssen Akzente. Je nach Ausführung können in einer Mappe bzw. Etui bis zu 72 Elemente untergebracht werden.

Statement by the jury
These pencil cases unite aesthetics and utilisation value. Thanks to the materials employed, they appear friendly and durable.

Begründung der Jury
Die Stiftehalter und -mäppchen führen Ästhetik und Nutzen zusammen und vermitteln dank ihrer Materialien Sympathie und Langlebigkeit.

Rapid Supreme Omnipress
Stapler
Heftgerät

Manufacturer
Isaberg Rapid AB, Hestra, Sweden
Design
ipdd GmbH & Co. KG, Stuttgart, Germany
Web
http://office.rapid.com
www.ipdd.com

The Supreme stapler featuring Omnipress technology utilises the lever action of existing press-less solutions, thus arriving at a size so compact that it can be operated with one hand and minimal effort. The ergonomic design transfers the stapling pressure regardless of where it is exerted. The device, which was developed in the Scandinavian style with a minimalist and functional look, is available in the base colours of matt black or white.

Statement by the jury
Thanks to its mature technology, Supreme is very easy to operate and also impresses with its ergonomic-dynamic form.

Das Heftgerät Supreme mit Omnipress-Technologie bietet die Hebelwirkung bisheriger Press-Less-Lösungen und gelangt dabei zu einer derart kompakten Größe, dass es mit nur einer Hand sowie geringem Kraftaufwand bedient werden kann. Die ergonomische Formgebung überträgt den Druck beim Heften unabhängig davon, von welcher Stelle er ausgeübt wird. Das nach Art skandinavischen Designs minimalistisch und funktional entworfene Gerät gibt es in Mattschwarz bzw. Weiß als Grundton.

Begründung der Jury
Dank seiner ausgeklügelten Technologie sorgt Supreme für eine spielend leichte Bedienung und überzeugt zudem durch eine ergonomisch-dynamische Gestaltung.

Trim & Stick Scissors
Trim & Stick-Schere
Scissors with Tape
Schere mit Kleberolle

Manufacturer
Optinova GmbH, Hagen, Germany
Design
Ergobionik GmbH (Thomas Klingbeil, Barbara Weck), Bottrop, Germany
Web
www.optinova.de
Honourable Mention

The Trim & Stick Scissors combines scissors and tape in an ergonomically designed 2-in-1 product. The tape holder fits neatly between the thumbhole and pivot pin and does not interfere when cutting. Non-slip TPE inlays in the holes provide a pleasant feel. Smooth transitions between the two units give rise to a harmonious overall impression. The stainless steel blades and the fibreglass-reinforced plastic components are very durable.

Statement by the jury
The design of the Trim & Stick scissors succeeds in combining the two products, scissors and tape, in a harmonious way, so that both their functions are easily achieved.

Die Trim & Stick-Schere kombiniert Schere und Klebeband in einem ergonomisch gestalteten 2-in-1-Produkt. Der Rollenhalter passt genau zwischen Daumenloch und Gelenk und stört nicht beim Schneiden. Rutschfeste TPE-Einlagen in den Ringen sorgen dabei für eine angenehme Haptik. Die gleitenden Übergänge zwischen beiden Utensilien erzeugen einen harmonischen Gesamteindruck. Die Klingen aus rostfreiem Edelstahl und der glasfaserverstärkte Kunststoff sind sehr langlebig.

Begründung der Jury
Der Gestaltung der Trim & Stick-Schere gelingt es, die beiden Produkte Schere und Klebeband harmonisch so miteinander zu verbinden, dass sich ihre Funktionen jeweils problemlos nutzen lassen.

Leitz Style
Notebooks
Notizbücher

Manufacturer
Esselte Leitz GmbH & Co KG,
Stuttgart, Germany
In-house design
Web
www.leitz.com

The Leitz Style notebook series has well-thought-out features, such as a cover design whose surface looks like brushed aluminium. The series is available in five colours and features a useful folding pocket for loose notes, a quick pocket for business cards and the like, two textile ribbons to quickly retrieve important pages and 20 adhesive page markers for labelling significant content. A ruler and a pen loop are likewise included.

Statement by the jury
With a variety of well-considered details and an elegant-looking cover, the Leitz Style notebook provides quality and usefulness in the daily business routine.

Die Notizbuchserie Leitz Style zeigt durchdachte Features, etwa ein Umschlagdesign, dessen Oberfläche wie gebürstetes Aluminium aussieht. Die Serie ist in fünf Farben erhältlich und umfasst eine praktische Falttasche für lose Zettel, ein Quickpocket für Visitenkarten u. Ä., zwei textile Zeichenbändchen, um wichtige Seiten schnell wiederzufinden, sowie 20 Haftmarker zum Kennzeichnen wichtiger Inhalte. Auch Lineal und Stifteschlaufe dürfen nicht fehlen.

Begründung der Jury
Mit einer Vielzahl wohlüberlegter Details und einem edel anmutenden Umschlag vermittelt das Notizbuch Leitz Style Wertigkeit und Nützlichkeit im Businessalltag.

Leuchtturm1917 – Master Slim Dotted
Notebook
Notizbuch

Manufacturer
Leuchtturm Albenverlag GmbH & Co. KG,
Geesthacht, Germany
In-house design
Web
www.leuchtturm1917.de

Inspired by the school of architecture: The Leuchtturm1917 notebook captivates with its puristic design in an oversized A4+ format, ink proof 100g/m² paper and dotted ruling – unobtrusive dots give orientation for drawing and writing. Pagination and a table of content facilitate a structured working. Some pages are detachable and its oversize provides a lot of space for notes as well as for documents. Full and shining cover colours complete the concept.

Statement by the jury
This strikingly large notebook offers ample space for notes, drawings or documents and also impresses with its exclusive paper and appealing colours.

Inspiriert von der Architektenschule: Das Leuchtturm1917 Notizbuch besticht durch sein puristisches Design im Überformat A4+, tintensicheres 100-g/m²-Papier und die Lineatur Dotted – dezente Punkte geben Orientierung beim Zeichnen und Schreiben. Seitenzahlen und ein Inhaltsverzeichnis erleichtern das strukturierte Arbeiten. Einige Seiten lassen sich heraustrennen, und seine Übergröße bietet Platz für Notizen und Dokumente gleichermaßen. Satte, leuchtende Einbandfarben runden das Konzept ab.

Begründung der Jury
Viel Platz für Notizen, Zeichnungen oder Unterlagen bietet dieses auffällig übergroße Notizbuch, das auch mit ausgewähltem Papier und ansprechenden Farben punktet.

ID FRAME
Card Holder
Kartenhalter

Manufacturer
Plus X, Seoul, South Korea
In-house design
Myungsup Shin, Youngin Koh, Hyun Lee,
Dajung Hyeon, Seongwoo Goh
Web
www.plus-ex.com

ID Frame is an identity card holder made of polycarbonate with a distinct and unadorned design, which is also very functional. The length of the neck strap can be adjusted easily on both sides. A reel is seamlessly attached in the frame; the loop is extended as soon as the frame is pulled down, which contributes to making the frame unbreakable. Thanks to a magnet on the body and the strap, the frame springs back into its original position.

Statement by the jury
The ID Frame identity card holder attracts attention with its restrained, elegant design, turning it into a stylish and easy-to-use business companion.

ID Frame ist eine aus Polycarbonat klar und schnörkellos gestaltete Ausweishülle, die zugleich sehr funktional ist. Die Länge des Halsriemens kann an beiden Seiten bequem eingestellt werden. Im Rahmen ist nahtlos eine Rolle angebracht, eine Schlinge darauf verlängert sich, sobald man am Rahmen zieht, was auch die Bruchsicherheit erhöht. Dank einem Magneten an Körper und Riemen rastet der Rahmen einfach wieder in seiner ursprünglichen Position ein.

Begründung der Jury
Die Ausweishülle ID Frame fällt durch ihre formschöne reduzierte Gestaltung ins Auge, die sie als zugleich stilvollen wie einfach zu handhabenden Business-Begleiter ausweist.

Scotch® Tape Dispenser
Scotch® Klebestreifen-Abroller

Manufacturer
Motex, Seoul, South Korea

In-house design
Motex

Design
3M Design, St. Paul,
Minnesota, USA
Joongho Choi Design Studio,
Seoul, Korea

Web
www.motex.co.kr
www.3m.com/design

reddot award 2015
best of the best

Harmony of form and function

Both in the office and at home, adhesive tape and
other desk utensils are always in demand, and
often hard to find. The Scotch Tape Dispenser offers
a seductively new solution to this everyday dilemma.
It turns the functional concept of a combined storage
box and tape dispenser into reality, using harmonious
design vocabulary. A separate, U-shaped compartment
provides storage space for all kinds of desk utensils
such as pencils, notepads and scissors, with the tape
dispenser placed at the other end. Users will initially
be surprised by the shape of the futuristic cogged
wheel that is part of a carefully thought-out, innovative mechanism to cut the adhesive tape. The ingeni-
ous inner structure of this mechanism holds the tape
in place on the spinning wheel, while it is cut at the
lower edge by a childproof blade. The tape can easily
be removed with the finger or be placed back on the
wheel. For effective use, the ergonomic wheel has also
been covered with embossed rubber, pleasant to the
touch while providing good grip as the tape dispenser
rotates. By combining innovative functionality with
creative design, the Scotch Tape Dispenser becomes
an indispensable, elegant desktop accessory.

Harmonie aus Form und Funktion

Sowohl im Büro als auch zu Hause benötigt man
immer wieder Klebeband oder andere Büroutensilien,
die dann jedoch oftmals unauffindbar sind. Für diese
alltägliche Situation bietet die Gestaltung des Scotch
Klebestreifen-Abrollers eine bestechende neue Lösung.
Sie verwirklicht in einer harmonischen Formensprache
das funktionale Konzept einer Einheit aus Aufbewah-
rungsbox und Klebestreifen-Abroller. Ein separates,
U-förmig gestaltetes Fach dient der Aufbewahrung von
allen möglichen Schreibtischutensilien wie Bleistiften,
Zetteln und Scheren, während sich am anderen Ende
der Klebestreifen-Abroller befindet. Der Betrachter ist
dabei zunächst verblüfft von der Form eines futuris-
tisch anmutenden Zahnrades, das jedoch Bestandteil
einer überaus durchdachten, innovativen Mechanik für
das Abtrennen des Klebebands ist. Deren ausgeklügelte
innere Struktur hält das Klebeband auf dem sich drehenden Zahnrad fest, während es im unteren Teil kindersicher von einer Klinge durchtrennt wird. Anschließend kann man es leicht mit einem Finger entfernen
oder wieder auf das Zahnrad legen. Das ergonomische
Rad wurde zudem für eine effektive Nutzung mit
einem geprägten Gummibezug versehen. Es hat eine
haptisch ansprechende Oberfläche und bietet eine sehr
gute Griffigkeit, wenn sich der Klebebandspender dreht.
Indem der Scotch Klebestreifen-Abroller innovative
Funktionalität mit kreativer Gestaltung vereint, wird
er zu einem unverzichtbaren, eleganten Schreibtisch-
Accessoire.

Statement by the jury

This combination of desktop organiser and tape dispenser is extremely modern and practical. The simple
and exciting use of form allows it to fit harmoniously
into the office environment. The both functionally and
ergonomically well-engineered design with its rotating
cogged wheel to dispense the tape is fun to use and
childproof. In addition, thanks to its colouring, the
Scotch Tape Dispenser provides a positive feel.

Begründung der Jury

Diese Kombination aus Tischorganizer und Klebestreifen-Abroller ist äußerst zeitgemäß und praktisch. Mit
ihrer simplen und emotionalisierenden Formensprache
fügt sie sich harmonisch in das Büroumfeld ein. Die
ebenso funktional wie ergonomisch ausgereifte Gestaltung mit einem sich drehenden Zahnrad für die Bereitstellung von Klebeband macht Spaß beim Gebrauch
und ist kindersicher. Auch durch seine Farbgebung
verbreitet der Scotch Klebestreifen-Abroller eine sehr
positive Atmosphäre.

Designer portrait
See page 38
Siehe Seite 38

Components	Anlagen
Electronic tools	Arbeitsplatzsysteme
Equipment	Arbeitsschutz
Hand tools	Betriebstechnik
Home tools	Elektrowerkzeuge
Industrial packaging	Handwerkzeuge
Industrial tools	Heimwerkzeuge
Machines	Industrieverpackungen
Measuring and testing technology	Industriewerkzeuge
Operating technology	Komponenten
Robots	Maschinen
Safety clothing	Mess- und Prüftechnik
Safety technology	Roboter
Timers	Sicherheitskleidung
Work protection	Sicherheitstechnik
Workstation systems	Zeitsysteme

Industry and crafts
Industrie und Werkzeuge

Hilti PR 30-HVS
Rotating Laser
Rotationslaser

Manufacturer
Hilti Corporation,
Schaan, Liechtenstein

In-house design
Hilti Corporation

Web
www.hilti.com

reddot award 2015
best of the best

Compact precision

Historically speaking, surveying was a complicated task, which was accomplished with a measuring chain and groma. Surveying today is made with a rotation laser device: a laser beam creates a reference plane through the rotation of a reflective prism. Perfectly adjusted to the specific requirements of construction engineering, the Hilti PR 30-HVS rotating laser fascinates with a new form language. Its design features well-balanced proportions and compact dimensions. A central aspect is the extraordinary ruggedness of this device. The casing is built extremely solid, while its four hard rubber grips absorb even the strongest of impacts without the device being damaged. Thus, the laser is well protected for use even under harsh jobsite conditions such as in constant rain, dust or if it falls to hard ground. Even in extreme environments, it continues to provide precise readings. The ergonomically designed grips also have a pleasant feel and allow for comfortable transport. Equipped with advanced and sensitive technology, the Hilti PR 30-HVS rotating laser is suitable for any measuring operation such as positioning, levelling and aligning. Measuring results and data are displayed clearly and are easy to read. Its compact, solid and functional design makes this laser a reliable measuring instrument for construction site work.

Kompakte Präzision

Historisch betrachtet war das Vermessungswesen eine komplizierte Angelegenheit, die per Messkette und Winkelkreuz geschah. Heutige Vermessungen erfolgen per Rotationslasergerät, wobei mittels Laserstrahl durch Rotation eines Umlenkprismas eine Bezugsebene erzeugt wird. Perfekt angepasst an die spezifischen Bedingungen im Bauwesen, begeistert der Rotationslaser Hilti PR 30-HVS durch eine neue Formensprache. Gestaltet ist er mit ausgewogenen Proportionen und kompakten Maßen. Ein zentraler Aspekt ist die außerordentliche Robustheit dieses Gerätes. Das Gehäuse ist äußerst solide gefertigt, seine vier Hartgummigriffe fangen auch starke Stöße ohne Schaden ab. Das Gerät ist damit gut gewappnet selbst für schwierige Bedingungen wie Dauerregen, Staub oder das Herunterfallen auf harte Böden. Es liefert auch unter solchen extremen Gegebenheiten weiterhin präzise Messergebnisse. Die ergonomisch gestalteten Handgriffe sind zudem haptisch angenehm und erlauben dem Nutzer ein komfortables Transportieren. Ausgestattet mit einer hochentwickelten, sensiblen Technologie ist der Rotationslaser Hilti PR 30-HVS einsetzbar bei allen Messvorgängen für das Nivellieren, Ausrichten und Planieren. Messergebnisse und Daten werden dabei klar und übersichtlich dargestellt. Durch seine kompakte, solide und funktionale Gestaltung bietet er damit ein sehr verlässliches Messgerät für die Arbeit auf Baustellen.

Statement by the jury

The Hilti PR 30-HVS rotating laser impresses with a well-proportioned and compact design. Its harmonious form language instantly communicates that this device is meant for use on construction sites. With elaborate details such as hard rubber grips and an exceptionally rugged housing, it is very well protected for use even under the harshest conditions. This rotation laser is intuitive and versatile in use.

Begründung der Jury

Der Rotationslaser Hilti PR 30-HVS beeindruckt den Anwender mit einer wohl proportionierten und kompakten Gestaltung. Seine ausgewogene Formensprache kommuniziert auf den ersten Blick die Bestimmung für den Einsatz auf Baustellen. Mit perfekt ausgearbeiteten Details wie Hartgummigriffen und einem ausgesprochen soliden Gehäuse ist er selbst unter den härtesten Bedingungen sehr gut geschützt. Dieser Rotationslaser lässt sich intuitiv bedienen und ist vielseitig einsetzbar.

Designer portrait
See page 40
Siehe Seite 40

PLR 50 C
Laser Measure
Laser-Entfernungsmesser

Manufacturer
Robert Bosch Elektrowerkzeuge
GmbH, Leinfelden-Echterdingen,
Germany

Design
TEAMS Design GmbH,
Esslingen, Germany

Web
www.bosch-pt.com

reddot award 2015
best of the best

Clear navigation

Nowadays, laser measures are used instead of classic folding yardsticks to measure, for example, the distance between two walls. The PLR 50 C has been especially developed for the DIY market and is adjusted to meet the expectations of this target group. Its design includes the benefits of modern consumer electronic products, translating them into a high-quality, functional tool. With clear radii, the well-balanced laser measure rests comfortably in the hand. An innovative feature is the interaction with the device: it is operated via an easy-to-read 2.4" touchscreen, which allows quick and easy access to all information and an intuitive navigation through diverse measurement functions. Another innovation in the DIY sector is the integrated level sensor, which allows for precise measurement results even with furniture positioned against the wall. Based on the distance and the angle of the measurement, the device automatically calculates the actual distance to the wall. The functional housing of this laser measure features a rugged design and, through the use of high-grade materials ensures robustness and a long operating life. Its clear form language and comprehensible operation define the PLR 50 C as a highly user-friendly and easy-to-use tool for do-it-yourselfers.

Klare Navigation

Laser-Entfernungsmesser werden heute statt des klassischen Zollstocks eingesetzt, wenn etwa der Abstand zwischen zwei Wänden gemessen werden soll. Der PLR 50 C wurde speziell für die Zielgruppe der Heimwerker konzipiert und an deren Erwartungen angepasst. Seine Gestaltung greift die Vorteile moderner Produkte aus dem Unterhaltungselektronik-Bereich auf und überführt diese in ein hochwertiges und funktionales Werkzeug. Entworfen mit klar anmutenden Radien, liegt dieser Laser-Entfernungsmesser gut austariert in der Hand des Nutzers. Eine Innovation ist die Art der Interaktion mit dem Gerät: Es kann vom Nutzer per Fingertipp auf einem übersichtlich gestalteten 2,4"-Touchscreen bedient werden. Möglich sind so ein unkomplizierter Zugang zu allen Informationen und eine intuitive Navigation durch die vielfältigen Messfunktionen. Eine weitere Innovation stellt im Heimwerkerbereich zudem ein integrierter Neigungssensor dar, mit dem die Messungen exakt auch dann erfolgen können, wenn beispielsweise Möbel an der Wand stehen. Das Gerät errechnet aus der Entfernung und dem Winkel der Messung dabei automatisch den tatsächlichen Abstand zur Wand. Das funktional gehaltene Gehäuse dieses Laser-Entfernungsmessers ist robust gestaltet und durch den Einsatz hochwertiger Materialien widerstandsfähig und langlebig. Seine klare Formensprache und gut verständliche Art der Bedienung definieren den PLR 50 C als ein sehr nutzerfreundliches und leicht anwendbares Gerät für den Heimwerker.

Statement by the jury

The PLR 50 C laser measure features a design that impresses with extraordinary clarity and conciseness. Every detail has been developed with utmost care to provide users with a measuring tool for precise and absolute reliable operation. The device is controlled easily and intuitively by the touch of a finger on the clearly structured display, which gives quick access to the versatile measuring functions. The PLR 50 C is compact and lies comfortably in the hand.

Begründung der Jury

Der Laser-Entfernungsmesser PLR 50 C zeigt ein Design, das durch seine außerordentliche Klarheit und Prägnanz beeindruckt. Jedes seiner Details wurde sehr sorgfältig entwickelt, wodurch dieser Entfernungsmesser dem Nutzer Präzision und absolute Bedienungssicherheit bietet. Intuitiv und problemlos kann er das Gerät per Fingertipp auf dem übersichtlichen Display bedienen und gelangt leicht an die vielfältigen Messfunktionen. Der PLR 50 C ist dabei kompakt und liegt ausgezeichnet in der Hand.

Designer portrait
See page 42
Siehe Seite 42

GRL 500 H/HV Professional &
LR 50 Professional
Rotation Laser and Laser Receiver
Rotationslaser und Laserempfänger

Manufacturer
Robert Bosch Elektrowerkzeuge GmbH,
Leinfelden-Echterdingen, Germany
Design
TEAMS Design GmbH, Esslingen, Germany
Web
www.bosch-pt.com

This system solution consists of the GRL 500 H/HV Professional rotation laser and the LR 50 Professional intelligent laser receiver, which doubles as the remote control. Together the two components form a cohesive unit. When they are separated, a centrally located, flashing red warning LED signalises the activated theft protection. The built-in lithium-ion batteries ensure an overall space-saving design.

Diese Systemlösung besteht aus dem Rotationslaser GRL 500 H/HV Professional und dem intelligenten Laserempfänger LR 50 Professional, der gleichzeitig die Fernbedienung ist. Zusammen bilden die beiden Komponenten eine geschlossene Einheit. Sind sie voneinander getrennt, signalisiert ein rot leuchtendes und zentral angeordnetes LED-Licht den aktivierten Diebstahlschutz. Im System fest verbaute Lithium-Ionen-Akkus sorgen für eine platzsparende Gesamterscheinung.

Zamo
Laser Measure
Laser-Entfernungsmesser

Manufacturer
Robert Bosch Elektrowerkzeuge GmbH,
Leinfelden-Echterdingen, Germany
Design
TEAMS Design GmbH, Esslingen, Germany
Web
www.bosch-pt.com

The Zamo laser measure features a
centrally located slide control, which not
only gives the device high recognition
value but also makes it intuitive to use.
Moving the slider automatically starts
the measuring process. At the touch of a
button, individual measurements can be
stored in the display. Thanks to its compact housing, the device fits comfortably into a trouser pocket.

Statement by the jury
Zamo's simplified design looks functional
and professional. The interfaces that are
relevant to the user have been clearly
emphasised.

Der Laser-Entfernungsmesser Zamo
verfügt über einen zentral platzierten
Reglungsschieber, den Slider, der nicht
nur für einen hohen Wiedererkennungswert des Geräts sorgt, sondern auch
eine intuitive Bedienung ermöglicht. Denn
mit Aufschieben des Sliders wird der
Messvorgang automatisch gestartet. Per
Knopfdruck können dann einzelne Mess-
werte im Display gespeichert werden.
Dank des kompakten Gehäuses passt das
Gerät bequem in die Hosentasche.

Begründung der Jury
Die vereinfachte Gestaltung von Zamo
mutet sachlich und professionell an. Die
für den Nutzer relevanten Schnittstellen
sind klar betont.

Mk5
Ultrasonic Thickness Gauge
Ultraschall-Dickenmessgerät

Manufacturer
Cygnus Instruments Limited, Dorchester, Great Britain
Design
WDL (Dan Walton), Cambridge, Great Britain
Web
www.cygnus-instruments.com
www.wrightdesign.net
Honourable Mention

The Mk5 ultrasonic thickness gauge measures material thickness using ultrasound. The two-component construction consists of a polycarbonate substrate and a thermoplastic elastomer, making the body very robust, shock-resistant and tightly sealed. The rubberised buttons provide tactile feedback when pressed. The measuring device is available in different versions: with combined end- and front-oriented display, with four or eight buttons and with two connector types.

Statement by the jury
With its premium craftsmanship and choice of materials, the Mk5 ultrasonic thickness gauge makes a very robust impression.

Das Dickenmessgerät Mk5 prüft Materialdicken per Ultraschall. Durch die 2-Komponenten-Bauweise aus Polycarbonatsubstrat und thermoplastischem Elastomer ist das Gehäuse sehr robust, schlagfest und dicht. Die gummierten Tasten geben ein taktiles Feedback. Das Messgerät ist in verschiedenen Ausführungen verfügbar: mit kombiniertem Display, das nach hinten und vorne gerichtet ist, mit vier oder acht Tasten und mit zwei Steckertypen.

Begründung der Jury
Das Dickenmessgerät Mk5 vermittelt aufgrund seiner sorgfältigen Verarbeitung und Materialauswahl einen sehr robusten Eindruck.

TMS-500 TopMap
Topography Measurement System
Oberflächenmesssystem

Manufacturer
Polytec GmbH, Waldbronn, Germany
Design
Scala Design technische Produktentwicklung, Böblingen, Germany
Web
www.polytec.com
www.scala-design.de

The TMS-500 TopMap is a white-light interferometer for the measurement of technical surfaces with high resolution even on large fields of view. The device provides an unobstructed view of the control area. The vertical measurement range of 70 mm makes it possible to also access hard-to-reach areas like deep drill holes. Its compact design and flexibly adaptable software provide quality assurance which is either process-oriented or directly integrated in the manufacturing line.

Statement by the jury
The interferometer impresses with its rounded shapes, which are oriented towards the user and thus artfully direct the attention to the working area.

Das TMS-500 TopMap ist ein Weißlicht-Interferometer zur Messung von Oberflächen mit hoher Auflösung auch bei großen Flächen. Das Messgerät bietet ungehinderte Einsicht in den Bedienbereich. Der vertikale Messbereich von 70 mm erlaubt es, auch schwer zugängliche Stellen, wie Flächen innerhalb von Bohrungen, zu erreichen. Die kompakte Bauform und eine flexibel anpassbare Software sorgen für eine prozessnahe oder direkt in die Fertigungslinie integrierte Qualitätssicherung.

Begründung der Jury
Das Interferometer gefällt durch seine gerundeten Formen, die ganz dem Anwender zugewandt sind und so den Blick geschickt auf den Arbeitsbereich lenken.

Galaxy 1 GNSS System
Satellite Receiver
Satellitenempfänger

Manufacturer
South Navigation Limited, Guangzhou, China
In-house design
Huiming Jiang
Web
www.southgnss.com

The Galaxy 1 GNSS System features the latest technology for global satellite-based navigation and receives signals from all existing satellite constellations. The use of magnesium and aluminium makes this receiver particularly lightweight and robust, protecting it against environmental influences. As one of the first of its kind the device is equipped with a Bluetooth 4.0 module for stable communication with controllers, smartphones and tablet PCs.

Statement by the jury
The compact form of the satellite receiver's housing looks particularly tough. The minimalist design of the control panel conveys professionalism.

Das Galaxy 1 verfügt über die neueste Technik für globale satellitengestützte Navigationssysteme (GNSS) und empfängt Signale aller bestehenden Satellitenkonstellationen. Die Verwendung von Magnesium und Aluminium machen den Empfänger besonders leicht und robust gegen Umwelteinflüsse. Das Gerät ist als eines der ersten seiner Art mit einem Bluetooth-4.0-Modul für eine stabilere Kommunikation mit Controllern, Smartphones und Tablet-PC ausgestattet.

Begründung der Jury
Die kompakte Gehäuseform des Satellitenempfängers wirkt besonders widerstandsfähig. Das reduziert gestaltete Bedienfeld strahlt Professionalität aus.

TKSA 41
Shaft Alignment Tool
Wellenausrichtsystem

Manufacturer
SKF Maintenance Products,
Nieuwegein, Netherlands
Design
Scope Design & Strategy,
Amersfoort, Netherlands
Web
www.skf.com
www.scopedesign.nl
Honourable Mention

The TKSA 41 shaft alignment tool is used to align axles, which helps to increase the lifespan of horizontally rotating industrial machines. The tool features an ergonomic display with touchscreen, Bluetooth connectivity and an integrated QR code scanner for quick access to available machine data. The body made of thermoplastic elastomer is particularly shock-resistant.

Statement by the jury
Distinct design features, such as the integrated blue element, give this shaft alignment tool a robust and independent look.

Das Wellenausrichtsystem TKSA 41 wird zur präzisen Ausrichtung von Wellen per Laser verwendet, um die Laufzeit von horizontal rotierenden Industriemaschinen zu verlängern. Das Gerät verfügt über ein ergonomisches Display mit Touchscreen, eine Bluetooth-Verbindung sowie einen integrierten QR-Code-Scanner für einen schnellen Zugriff auf bereits vorhandene Maschinendaten. Das Gehäuse aus thermoplastischem Elastomer ist besonders stoßfest.

Begründung der Jury
Markante Gestaltungsmerkmale wie die blauen Elemente verleihen dem Wellenausrichtsystem eine robuste Anmutung und ein eigenständiges Profil.

TE 1000-AVR
Breaker
Abbruchhammer

Manufacturer
Hilti Corporation, Schaan, Liechtenstein
In-house design
Web
www.hilti.com

The TE 1000-AVR breaker with its ergonomic handle is used for demolition or correction work. The drive, the impact mechanism and the robust S-chuck are finely synchronised in order to transfer the enormous single-impact energy of 26 joules safely and efficiently. Three separate lubrication chambers keep the dust away from sensitive areas inside the device. The brushless motor is virtually maintenance-free for its entire service life. The cord can be removed from the housing.

Der Abbruchhammer TE 1000-AVR mit seinem ergonomischen Handgriff kommt bei Abriss- oder Korrekturarbeiten zum Einsatz. Das Getriebe, der Schlagmechanismus und die robuste S-Werkzeugaufnahme sind genau aufeinander abgestimmt, um die enorme Einzelschlagenergie von 26 Joule sicher und effizient zu übertragen. Drei voneinander getrennte Schmierkammern halten den Staub von sensiblen Bereichen im Gerät fern. Der bürstenlose Motor ist über seine gesamte Lebensdauer praktisch wartungsfrei. Das Kabel kann vom Gehäuse abgesteckt werden.

Statement by the jury
The streamlined design clearly indicates the direction of the chisel, thus imbuing the TE 1000-AVR with dynamic underlying tension.

Begründung der Jury
Die stromlinienförmige Gestaltung gibt klar die Richtung für den Meißel vor und verleiht dem TE 1000-AVR dadurch eine dynamische Grundspannung.

HKD-TE-CX
Setting Tool
Setzwerkzeug

Manufacturer
Hilti Corporation, Schaan, Liechtenstein
In-house design
Web
www.hilti.com

The HKD-TE-CX setting tool makes it possible to complete two tasks with only one tool: drilling a hole and setting the anchor. Since it is no longer necessary to release the chuck to change tools, the entire installation can be carried out almost twice as quickly as before. The setting tool is simply inserted into the chuck of a rotary hammer. After drilling is completed, the front section of the tool is pulled off and the anchor is set in the wall using the setting section of the tool. The drilled holes always have the correct depth.

Das Setzwerkzeug HKD-TE-CX macht es möglich, mit nur einem Werkzeug zwei Arbeitsschritte zu erledigen: das Bohren des Lochs und das Setzen des Dübels. Da das Ein- und Ausspannen unterschiedlicher Werkzeuge entfällt, wird der gesamte Arbeitsablauf nahezu doppelt so schnell. Das Setzwerkzeug wird einfach in die Werkzeugaufnahme eines Bohrhammers eingesteckt, nach dem Bohren wird der vordere Teil des Werkzeugs ausgesteckt und der Dübel mit dem Setzteil in der Wand angebracht. Die Löcher werden stets mit der korrekten Tiefe gebohrt.

DX 2
Powder-Actuated Fastening Tool
Bolzensetzgerät

Manufacturer
Hilti Corporation, Schaan, Liechtenstein
In-house design
Web
www.hilti.com

The DX 2 is a powder-actuated fastening tool for a wide range of fastening applications on concrete or steel. Thanks to its slim form, it can be used even in difficult-to-reach areas. Moreover, the all-metal housing, in combination with the durable impact pistons, provides consistently high fastening quality even under the most challenging conditions. The practice-oriented ergonomics and optimised recoil forces make the unit particularly safe and convenient to use.

Das Bolzensetzgerät DX 2 ist für verschiedenste Befestigungsanwendungen auf Beton oder Stahl geeignet. Dank seiner schlanken Form lässt es sich jederzeit auch in schwer zugänglichen Bereichen einsetzen. Darüber hinaus ermöglicht das Vollmetallgehäuse in Kombination mit den langlebigen Schlagkolben eine durchgehend hohe Befestigungsqualität auch unter härtesten Bedingungen. Die anwendungsorientierte Ergonomie sowie optimierte Rückstoßwerte sorgen dafür, dass das Gerät besonders komfortabel und sicher im Gebrauch ist.

Statement by the jury
The DX 2 is particularly impressive due to its powerful proportions and elongated form. This makes the powder-actuated fastening tool look highly efficient.

Begründung der Jury
Besonders beeindruckend am DX 2 sind seine kraftvollen Proportionen und die langgezogene Form. Dadurch wirkt das Bolzensetzgerät sehr leistungsstark.

SID 2-A
Impact Driver
Schlagschrauber

Manufacturer
Hilti Corporation, Schaan, Liechtenstein
In-house design
Web
www.hilti.com

The SID 2-A tangential impact driver is characterised by its highly ergonomic design and small dimensions, while featuring a powerful performance provided by the 10.8-volt platform. Integration of LED lights in the base has kept the front section around the chuck very narrow. Thus the full force of the impact driver is available to the user even in difficult to reach areas. Thanks to the rubber surfaces on the housing and battery, the tool can be put down safely. The fibreglass-reinforced plastic enclosure is highly robust.

Der Tangential-Schlagschrauber SID 2-A zeichnet sich durch seine hohe Ergonomie und geringe Abmessungen bei gleichzeitig starker Leistung auf der 10,8-Volt-Plattform aus. Die Integration der LED-Lampen in den Gerätefuß ermöglicht eine sehr schmale Bauweise im Bereich der Bit-Aufnahme. Damit verfügt der Anwender auch an schwer zugänglichen Stellen über die volle Kraft des Schlagschraubers. Durch die gummierten Flächen an Gehäuse und Akku lässt sich das Werkzeug sicher ablegen. Das glasfaserverstärkte Kunststoffgehäuse bietet hohe Robustheit.

Statement by the jury
The contrasting colours of black and red, which are typical of the brand, were cleverly used in the design of the SID 2-A in order to highlight its compact size.

Begründung der Jury
Die markentypischen Kontrastfarben Schwarz und Rot wurden beim SID 2-A geschickt eingesetzt, um die Kompaktheit des Schlagschraubers zu betonen.

Bosch FlexiClick System
Cordless Drill Driver
Akku-Bohrschrauber

Manufacturer
Robert Bosch Elektrowerkzeuge GmbH,
Leinfelden-Echterdingen, Germany
Design
TEAMS Design GmbH, Esslingen, Germany
Web
www.bosch-pt.com
www.teamsdesign.com

The powerful Bosch FlexiClick system consists of a drill driver and four adapters for different tasks: a drill chuck adapter, off-set angle adapter, angle adapter and rotary hammer adapter. The applications range from screwing to drilling in wood, metal and concrete. The adapters can be adjusted without removal from the tool, which saves working steps and reduces set-up time.

Das leistungsstarke Bosch FlexiClick-System umfasst einen Bohrschrauber und vier verschiedene Aufsätze für unterschiedliche Anwendungen: Bohrfutter-, Exzenter-, Winkel- und Bohrhammeraufsatz. Das Anwendungsspektrum reicht vom Schrauben bis zum Bohren in Holz, Metall und Beton. Die Aufsätze müssen zum Justieren nicht vom Gerät genommen werden, was Arbeitsschritte einspart und Rüstzeiten verkürzt.

Statement by the jury
The design concept of the Bosch Flexi-Click system surprises with its particularly versatile adapters and the drill driver's highly compact size.

Begründung der Jury
Das Gestaltungskonzept des Bosch FlexiClick-Systems überrascht durch die Vielseitigkeit der Aufsätze und die hohe Kompaktheit des Bohrschraubers.

SF 2-A / SFD 2-A
Cordless Drill Driver
Akku-Bohrschrauber

Manufacturer
Hilti Corporation, Schaan, Liechtenstein
In-house design
Web
www.hilti.com

The cordless drill drivers from the SF 2-A and SFD 2-A series are based on the 10.8-volt platform. The lithium-ion battery in the compact housing provides lasting drilling performance in both wood and metal. With its optimally balanced centre of gravity and ergonomically shaped handle, each drill driver rests comfortably in the hand. Moreover, there are two LED lights at the bottom of the device, which illuminate the working area. Precision coupling enables precise operation, even for difficult screwing tasks.

Die Akku-Bohrschrauber der Gerätereihen SF 2-A und SFD 2-A basieren auf einer 10,8-Volt-Plattform. Der Lithium-Ionen-Akku in einem kompakten Gehäuse sorgt für eine lange Lebensdauer und hohe Bohrleistung sowohl in Holz als auch in Metall. Mit dem optimal ausbalancierten Schwerpunkt und dem ergonomischen Griff liegen die Bohrschrauber sehr gut in der Hand. Darüber hinaus befinden sich unten am Gerät zwei LED-Lampen, die den Arbeitsbereich in jeder Situation ideal ausleuchten. Eine Präzisionskupplung ermöglicht die exakte Bedienung auch bei schwierigen Schraubarbeiten.

Statement by the jury
Due to their distinctive lines, the cordless drill drivers convey a striking, masculine impression.

Begründung der Jury
Die Akku-Bohrschrauber wirken aufgrund ihrer ausgeprägten Linienführung auf besonders ansprechende Weise markant und maskulin.

QMC 21
Screwdriver
Schraubendreher

Manufacturer
Atlas Copco Industrial Technique AB,
Nacka, Sweden
In-house design
Ola Nyström, Ola Stray, Göran Johansson
Web
www.atlascopco.com

The QMC 21 is an intelligent precision screwdriver for applications with very low torque, such as the assembly of micro screws in electronic components. Thanks to its design and the selection of materials, the tool is lightweight, highly efficient and requires very little maintenance. The screwdriver is precise and easy to program, and it even detects assembly errors such as overtightened screws.

Statement by the jury
The QMC 21 has a delicate and accentuated form. Its seamless transitions from one individual component to the next lend this screwdriver the appearance of being made from one piece.

Der intelligente Präzisionsschraubendreher QMC 21 ermöglicht Anwendungen mit sehr niedrigen Drehmomenten, z. B. die Montage von Kleinstschrauben bei elektronischen Bauteilen. Das Design und die Materialauswahl machen das Werkzeug nicht nur leicht, sondern sorgen auch für seine hohe Produktivität und geringe Wartungsanfälligkeit. Der Schrauber ist einfach und präzise zu programmieren und erkennt selbständig Montagefehler wie überdrehte Schrauben.

Begründung der Jury
Der QMC 21 ist fein und akzentuiert in seiner Form. Durch die nahtlosen Übergänge der Einzelkomponenten wirkt dieser Schraubendreher wie aus einem Stück gefertigt.

QMC 41
Screwdriver
Schraubendreher

Manufacturer
Atlas Copco Industrial Technique AB,
Nacka, Sweden
In-house design
Ola Nyström, Ola Stray, Göran Johansson
Web
www.atlascopco.com

The QMC 41 is an intelligent precision screwdriver for the assembly of micro screws used in electronic devices. The design and choice of materials have resulted in a weight reduction of up to 50 per cent as compared to conventional models. The metal tool identification plate features product information. The torque sensor monitors the screwing process, and the LEDs provide visual feedback about the status of the process.

Statement by the jury
The QMC 41 impresses with its remarkably low weight. The clear use of form communicates that this is a precision tool.

Der QMC 41 ist ein intelligenter Präzisionsschraubendreher für die Montage von Kleinstschrauben, die bei elektronischen Geräten zum Einsatz kommen. Das Design und die Materialwahl ergeben eine Gewichtsersparnis von bis zu 50 Prozent im Vergleich zu herkömmlichen Modellen. Auf dem Metalltypenschild sind Produktinformationen angegeben. Die Drehmomentsensoren überwachen den Schraubvorgang, und die LEDs geben visuelle Rückmeldung über den Prozessstatus.

Begründung der Jury
Der QMC 41 begeistert durch sein leichtes Gewicht. Die klare Formsprache kommuniziert, dass es sich um ein Präzisionswerkzeug handelt.

MT Focus 6000
Controller
Steuereinheit

Manufacturer
Atlas Copco Industrial Technique AB, Nacka, Sweden
In-house design
Magnus Lindström, Ola Nyström
Web
www.atlascopco.com

The MT Focus 6000 controller is connected to a power tool and ensures that micro screws are tightened properly. An SD card slot enables operators to access and configure data, while an intuitive user interface communicates the work results. The compact size and durable materials have been developed for use in a demanding manufacturing environment.

Die Steuereinheit MT Focus 6000 wird an ein Elektrowerkzeug angeschlossen und stellt sicher, dass Mikroschrauben richtig festgezogen werden. Ein SD-Karten-Anschluss ermöglicht dem Benutzer den Zugriff auf die Daten und ihre Konfiguration, während eine intuitive Benutzeroberfläche die Ergebnisse der Arbeit kommuniziert. Die kompakte Größe und langlebige Materialien sind auf die Anwendung in einer anspruchsvollen Fertigungsumgebung abgestimmt.

Statement by the jury
The user interface takes up almost the entire front of this control unit, thus appearing exceptionally large, inviting and user-friendly.

Begründung der Jury
Die Bedienoberfläche füllt fast die gesamte Vorderseite der Steuereinheit aus, dadurch erscheint sie ausnehmend großzügig, einladend und benutzerfreundlich.

SK 2154 PH-06 3C Screwdriver Set with Striking Cap

SK 2154 PH-06 3K-Schrauben-dreher-Satz mit Schlagkappe

Manufacturer
Oplast d.o.o., Slovenske Konjice, Slovenia
Design
GEDORE Tool Center GmbH & Co. KG
(Björn Spahlinger, Damian Schwierz),
Remscheid, Germany
Web
www.oplast.si
www.gedore.com

The 3C screwdrivers have a striking cap at the handle top. Jammed slotted- and cross-head screws can thus be loosened more easily when hammer blows are dealt to the striking cap. The blades are made from hexagonally shaped material and feature an additional hexagon bolster. The handles have an ergonomic design and are made of three components that provide a safe grip. The screwdrivers are available in various different sizes.

Statement by the jury
These screwdrivers have an extremely pleasant feel, and the contrasting colours elegantly complement the form and quality of the handle.

Die 3K-Schraubendreher haben am Ende des Griffs eine Schlagkappe. Durch Hammerschläge auf diese Kappe lassen sich festsitzende Schlitz- und Kreuzschlitz-schrauben besser lösen. Die Klingen sind aus Sechskantmaterial und verfügen über einen zusätzlichen Sechskantansatz. Die Griffe sind ergonomisch gefertigt und bestehen aus drei Komponenten, die für einen sicheren Halt sorgen. Die Schraubendreher sind in verschiedenen Größen erhältlich.

Begründung der Jury
Die Schraubendreher fühlen sich ausgesprochen angenehm an. Die kontrastreiche Farbgebung greift Form und Beschaffenheit des Griffs gekonnt auf.

2162 PH VDE Screwdriver SLIM DRIVE

2162 PH VDE-Schrauben-dreher SLIM DRIVE

Manufacturer
Oplast d.o.o., Slovenske Konjice, Slovenia
Design
GEDORE Tool Center GmbH & Co. KG
(Björn Spahlinger, Damian Schwierz),
Remscheid, Germany
Web
www.oplast.si
www.gedore.com

The VDE screwdriver Slim Drive is characterised by flush insulation at the tapering tip. The blade made from molybdenum vanadium-plus tempered steel thus provides full mechanical and electrical safety at an ultra small diameter. The ergonomically designed, three-component handle enables precise and fatigue-free use. The interlocking combination of handle and blade provides optimal transmission of force.

Statement by the jury
The screwdriver impresses with its distinct, colour-coded surfaces. The well-proportioned handle provides a secure grip.

Der VDE-Schraubendreher Slim Drive zeichnet sich durch eine bündig abschlie-ßende Isolierung an der verjüngten Spitze aus. Dadurch bietet die Klinge aus Molybdän-Vanadium-Plus-Stahl volle mechanische und elektrische Sicherheit bei einem ultrakleinen Durchmesser. Der ergonomisch gestaltete 3-Komponenten-Griff ermöglicht ein präzises und ermüdungsfreies Arbeiten. Die form-schlüssige Verbindung von Griff und Klinge sorgt für eine optimale Kraftübertragung.

Begründung der Jury
Der Schraubendreher besticht durch seine farblich prägnant voneinander abgehobenen Oberflächen. Der wohlproportionierte Griff bietet sicheren Halt.

GARANT MD1
Torque Wrench

Drehmomentschlüssel

Manufacturer
Hoffmann GmbH Qualitätswerkzeuge,
Munich, Germany
Design
Böhler GmbH, Corporate Industrial Design
(Dirk Lenssen, Thomas Breun),
Fürth, Germany
Web
www.hoffmann-group.com
www.boehler-design.de

The Garant MD1 mechanical torque wrench is clearly divided into two functional areas, the handle and the display. The required torque is set manually but displayed digitally, thus guaranteeing high accuracy. For adjusting, the mechanism at the end of the handle is unlocked; an orange circle appears as a result, indicating the unlocked status. The tool is suitable for use in medium- and large-scale serial production.

Statement by the jury
The Garant MD1 has an appealing, particularly long silhouette. At the same time, the relationship between handle and wrench is well balanced.

Der Drehmomentschlüssel Garant MD1 ist klar in die Funktionszonen Griff und Display gegliedert. Die Einstellung des Drehmoments erfolgt manuell, ist aber digital ablesbar und gewährleistet so eine hohe Genauigkeit. Zum Justieren wird der Mechanismus am Griffende entriegelt, wodurch ein orangefarbener Ring als Entriegelungssymbol sichtbar wird. Das Werkzeug ist für die Anwendung bei mittleren bis größeren Fertigungsserien geeignet.

Begründung der Jury
Der Garant MD1 gefällt durch seine besonders langgestreckte Silhouette. Gleichzeitig stehen Griff und Schlüssel in einem wohl austarierten Verhältnis.

GARANT Slipper
Torque Wrench
Drehmomentschlüssel

Manufacturer
Hoffmann GmbH Qualitätswerkzeuge,
Munich, Germany
Design
Böhler GmbH, Corporate Industrial Design
(Dirk Lenssen, Christopher Sens),
Fürth, Germany
Web
www.hoffmann-group.com
www.boehler-design.de
Honourable Mention

The Garant Slipper torque wrench is used for a controlled tightening of screws with a preset torque. It is equipped with an integrated ratchet function and an ejector. The tool is unlocked using a mechanism at the end of the handle, and the locked or unlocked status is displayed by the symbols. A release mechanism ensures that the set torques are reached reliably, yet not exceeded.

Statement by the jury
This torque wrench's compact design is characterised by its user-oriented functionality.

Der Drehmomentschlüssel Garant Slipper dient zum kontrollierten Anziehen von Schrauben, mit voreingestelltem Drehmoment. Er ist mit einer Knarrenfunktion und einem Auswerfer ausgestattet. Per Arretierknopf am Griffende wird das Werkzeug entriegelt, der Zustand der Ver- und Entriegelung ist an Symbolen erkennbar. Ein Auslösemechanismus stellt sicher, dass die eingestellten Drehmomente zuverlässig erreicht, aber nicht überschritten werden.

Begründung der Jury
Der kompakt gestaltete Drehmomentschlüssel zeichnet sich durch eine nah am Nutzer entwickelte Funktionalität aus.

ZEBRA Dustproof Ratchet
Staubgeschützte Umschalt-knarre

Manufacturer
Adolf Würth GmbH & Co. KG,
Künzelsau, Germany
In-house design
Alexander Daus, Andreas Dierolf,
Frank Kollmar, Thomas Rapp
Web
www.wuerth.com

The dustproof reversible ratchet is able to withstand very high torque loads. The tool's most prominent design feature is its double sealing washer. The closed design prevents dirt from entering the mechanism. Moreover, the ratchet was designed with the reversing lever in a protective, recessed position. The optimised choice of material for the handle guarantees increased resistance and durability.

Statement by the jury
The seamless design of the reversible ratchet places a strong focus on excellent craftsmanship and premium materials.

Die staubgeschützte Umschaltknarre hält sehr hohen Drehmomentwerten stand. Das gestalterische Hauptmerkmal des Werkzeugs bildet die zweifache Dichtungsscheibe. Durch die geschlossene Bauweise kann kein Schmutz in die Mechanik eindringen. Die Knarre wurde zudem so konzipiert, dass der Umschalthebel geschützt angebracht ist. Die optimierte Materialauswahl des Handgriffs gewährleistet eine erhöhte Widerstandsfähigkeit und Langlebigkeit.

Begründung der Jury
Bei der nahtlosen Gestaltung der Umschaltknarre wurde in hohem Maße Wert auf eine ausgezeichnete Verarbeitung und qualitätsvolle Materialien gelegt.

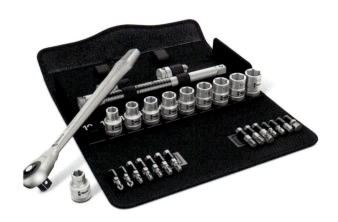

Zyklop Metal Set
Zyklop-Metal-Satz
Ratchets
Knarren

Manufacturer
Hermann Werner GmbH & Co. KG,
Wuppertal, Germany
In-house design
Michael Abel, Werner Tretbar
Web
www.wera.de

The Zyklop Metal Set is manufactured entirely from solid metal. The ratchet is extremely slim and has a long leverage. The secure socket lock mechanism prevents the unintentional loss of sockets. The tools, along with accessories, come in a textile box, which weighs much less than a metal container and makes it easier to transport the set.

Statement by the jury
This ratchet set embodies a successful combination of effective functionality, careful choice of materials and user-friendly convenience.

Der Zyklop-Metal-Satz ist aus geschmiedetem Vollmetall gefertigt. Die Knarre ist extrem schlank und verfügt über einen langen Hebel. Die Nussverriegelungsfunktion verhindert das ungewollte Lösen der Nuss. Das Werkzeug samt Zubehör wird in einer textilen Box geliefert, die sehr viel weniger wiegt als ein Metallkasten. Dadurch ist der Satz leichter zu transportieren.

Begründung der Jury
Der Knarrensatz verkörpert eine gelungene Kombination aus effektiver Funktionalität, sorgfältiger Materialauswahl und benutzerfreundlichem Komfort.

FlipSelector
Bit Box

Manufacturer
Wiha Werkzeuge GmbH, Schonach, Germany
In-house design
Jürgen Stern
Web
www.wiha.com

The FlipSelector is a bit box with twelve slots, which can be equipped according to the field of application. When the red logo button is slid down, the bit-clamping element pivots by 180 degrees and releases the bits for removal. The magnetic bit holder is housed in a separate clamping element. The box consisting of impact-resistant, fibreglass-reinforced polyamide fits into a trouser pocket or can be attached to the corresponding belt clip.

Statement by the jury
The FlipSelector impresses with a highly practical tilt function that facilitates the working process. Colour is aptly employed to clearly emphasise the mechanism.

Der FlipSelector ist eine Bitbox mit zwölf Steckplätzen, die je nach Anwendungsgebiet bestückt werden können. Wenn die rote Logo-Taste nach unten geschoben wird, schwenkt das Bit-Klemmelement um 180 Grad nach außen und gibt die Bits zur Entnahme frei. Der magnetische Bithalter ist in einem gesonderten Klemmelement untergebracht. Die Box aus schlagzähem, glasfaserverstärktem Polyamid passt in die Hosentasche oder an den dazugehörigen Gürtelclip.

Begründung der Jury
Der FlipSelector begeistert mit seiner überaus praktischen Kippfunktion, die das Arbeiten erleichtert. Der Mechanismus ist farblich deutlich hervorgehoben.

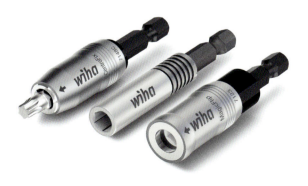

CentroFix & MagicFlip
Bit Holders
Bithalter

Manufacturer
Wiha Werkzeuge GmbH, Schonach, Germany
In-house design
Thomas Schandelmeier
Web
www.wiha.com

The CentroFix and MagicFlip Force bit holders make the tightening of screws easier by reducing the play between bit and chuck to a minimum. The bit is automatically locked when it is inserted. When unlocked, the bit moves forward a few millimetres to enable easy removal. In addition, the MagicFlip Force clip features a strong ring magnet, which also holds long screws securely.

Statement by the jury
Accuracy and functionality are excellently combined in the carefully conceived geometry of these bit holders.

Die Bithalter CentroFix und MagicFlip Force erleichtern das Eindrehen von Schrauben, indem sie das Spiel zwischen Bit und Halter auf ein Minimum reduzieren. Der Bit wird beim Einstecken automatisch verriegelt. Beim Entriegeln schiebt er sich um einige Millimeter nach vorne und lässt sich leicht entnehmen. Der MagicFlip Force verfügt zusätzlich über einen starken Ringmagneten, der auch lange Schrauben sicher festhält.

Begründung der Jury
Akkuratesse und Funktionalität spielen in der wohldurchdachten Geometrie der Bithalter hervorragend zusammen.

L.SP3
Geared Offset Heads
Flachabtriebe

Manufacturer
Johannes Lübbering GmbH,
Herzebrock-Clarholz, Germany
In-house design
Bruno Bergmann, Rudolf Mense
Web
www.luebbering.de
Honourable Mention

The L.SP3 geared offset heads for fastening and unfastening nuts and bolts permit precise and repeatable screw connections, especially in places that are difficult to access. The uniform housing shape of all parts enables use with any type of interface and a simple expansion of the fastening system with additional components. With corresponding angle heads, the geared offset heads can be adapted for assembly tasks.

Statement by the jury
The homogeneous design of the geared offset heads provides well-thought-out flexibility, facilitating a wide spectrum of applications.

Die Flachabtriebe L.SP3 zum Eindrehen und Lösen von Schrauben und Muttern ermöglichen eine präzise und wiederholbare Verschraubung vor allem an schwer zugänglichen Stellen. Aufgrund der einheitlichen Gehäuseform sämtlicher Bestandteile kann das Schraubsystem bei allen Arten von Schnittstellen eingesetzt und einfach um Zusatzkomponenten erweitert werden. Mit entsprechenden Winkelköpfen lassen sich die Flachabtriebe für die Montage anpassen.

Begründung der Jury
Die gleichförmige Bauweise der Flachabtriebe bietet eine gut durchdachte Flexibilität, die ein breit gefächertes Anwendungsspektrum zulässt.

GARANT TQ-Station
Screwdriver and Bit Assortment
Schraubendreher-Bit-Sortiment

Manufacturer
Hoffmann GmbH Qualitätswerkzeuge,
Munich, Germany
Design
Böhler GmbH, Corporate Industrial Design
(Thomas Breun, Dirk Lenssen),
Fürth, Germany
Web
www.hoffmann-group.com
www.boehler-design.de

The Garant TQ-Station is a 27-piece, dynamometric screwdriver and accompanying bit assortment, which presents all of the tools in a well-arranged manner, thus making them easy to remove. The modular storage station consists of a solid sheet-metal body and clearly ordered storage units made of lightweight plastic. The station can be placed safely on a workbench or machine without the risk of it tipping over. Small parts can be stored in the three removable boxes.

Statement by the jury
Each item has its own place in the Garant TQ-Station. Therefore, it always conveys a tidy and ready-to-use impression.

Die Garant TQ-Station ist ein 27-teiliges Drehmoment-Schraubendreher-Bit-Sortiment, das alle Werkzeuge übersichtlich und leicht entnehmbar zur Verfügung stellt. Die modulare Aufnahmestation besteht aus einem stabilen Blechkorpus und klar gegliederten Aufbewahrungseinheiten aus leichtem Kunststoff. Die Station kann kippsicher auf der Werkbank oder Maschine abgestellt werden. Kleinteile lassen sich in den drei herausnehmbaren Boxen aufbewahren.

Begründung der Jury
In der Garant TQ-Station findet jedes Teil einen festen Platz, sodass die Konstruktion stets einen aufgeräumten und einsatzbereiten Eindruck macht.

Chisels HDC
Stemmeisen HDC

Manufacturer
Hultafors Group AB, Bollebygd, Sweden
In-house design
Nikita Golovlev
Web
www.hultafors.com

The Chisels HDC can be positioned particularly flat on a surface, opening up more usage areas. The durable blade is forged as a single piece and runs through the entire shaft; the chisel thus provides very high bending strength and optimal transmission of force. The hard nylon cap reduces impact vibrations transmitted back to the hand through the hammer.

Statement by the jury
The Chisels HDC's ergonomics and functionality demonstrate an excellent sense for fine craftsmanship.

Die Stemmeisen HDC können besonders flach an die zu bearbeitende Oberfläche angesetzt werden, wodurch sie mehr Spielraum in der Anwendung lassen. Die strapazierfähige Klinge ist aus einem Stück gefertigt und durchläuft den gesamten Schaft. Dadurch erreichen die Stemmeisen eine sehr hohe Biegefestigkeit und bieten eine optimale Kraftübertragung. Die harte Nylonkappe reduziert die Stoßvibrationen, die von Hammerschlägen in die Hand übertragen werden.

Begründung der Jury
Auf hohem Niveau zeigt die Ergonomie und Funktionalität der Stemmeisen HDC ein Gespür für handwerkliche Details.

HBX
Hand Saw
Handsäge

Manufacturer
Hultafors Group AB, Bollebygd, Sweden
Design
Veryday (Hans Himbert, Pelle Reinius, Stefan Strandberg), Bromma, Sweden
Web
www.hultafors.com
www.veryday.com

The HBX hand saw features an interchangeable blade system. Thanks to its small tip, areas that are difficult to reach can be accessed comfortably. A newly developed coating ensures that the saw blades are exposed to extremely low friction over the course of their entire service life. In addition, the life span of the saw blades has been increased through a blade guard, which simultaneously lowers the risk of injury.

Die Handsäge HBX hat ein austauschbares Sägeblattsystem. Mit der schmalen Spitze lassen sich auch schwer zugängliche Stellen gut erreichen. Eine neu entwickelte Beschichtung sorgt dafür, dass die Sägeblätter über ihre gesamte Lebensspanne extrem wenig Reibung ausgesetzt sind. Zusätzlich wird die Haltbarkeit der Sägeblätter durch den Klingenschutz erhöht, der gleichzeitig die Verletzungsgefahr verringert.

Statement by the jury
This hand saw impresses with its technical look. Its clear design language meets high ergonomic and safety standards.

Begründung der Jury
Die Handsäge punktet durch ihre technische Anmutung. Ihre klare Formensprache erfüllt hohe Ansprüche an Ergonomie und Sicherheit.

WoodXpert
Wood-Handling Tool Range
Forstgeräte-Serie

Manufacturer
Fiskars Garden Oy Ab, Helsinki, Finland
In-house design
Fiskars R&D Team
Web
www.fiskarsgroup.com

The WoodXpert wood-handling tools are used during the entire process of creating firewood, from felling the tree to chopping the wood into fireplace-ready pieces. The tools' grip areas feature a non-slip surface, which provides a secure hold when processing and moving large logs as well as lifting and piling smaller log pieces. Hook-shaped handle ends ensure additional safety.

Statement by the jury
The seamless design of this wood-handling tool range convinces with its ergonomic qualities and high performance. The tools harmonise well thanks to their homogeneous look.

Die Forstgeräte-Serie WoodXpert hilft beim gesamten Herstellungsprozess von Kaminholz, angefangen beim Fällen des Baumes bis hin zum Zerkleinern des Holzes in kamingerechte Scheite. Die Gerätegriffe haben eine rutschfeste Oberfläche und bieten dadurch sicheren Halt beim Bearbeiten und Transportieren von großen Baumstämmen sowie beim Anheben und Stapeln kleinerer Stammstücke. Hakenförmige Griffenden garantieren zusätzliche Sicherheit.

Begründung der Jury
Das nahtlose Design der Forstgeräte-Serie überzeugt durch Ergonomie und Leistungsstärke. Die Werkzeuge harmonieren durch ihr einheitliches Erscheinungsbild.

X21
Axe
Axt

Manufacturer
Fiskars Garden Oy Ab, Helsinki, Finland
In-house design
Fiskars R&D Team
Web
www.fiskarsgroup.com

The X21 axe has a patented 3D handle, which prevents the development of moisture between the upper hand and the shaft. In addition, the elongated handle protector prevents both hands from slipping. Thanks to this improved control, the swing speed can be increased with the help of the lower hand. Furthermore, the optimised blade weight provides an ideal balance between the axe head and the shaft.

Statement by the jury
This axe impresses with its slimness and straight lines. Handle and shaft are clearly distinguished by the design but nevertheless flow into one another.

Die Axt X21 hat einen patentierten 3D-Griff, der die Reibung und die Entstehung von Feuchtigkeit zwischen der oberen Hand und dem Stiel verhindert. Zudem bewahrt der verlängerte Griffschutz die Hände davor, abzurutschen. Durch die verbesserte Kontrolle kann mithilfe der unteren Hand die Schwunggeschwindigkeit erhöht werden. Darüber hinaus sorgt das optimierte Klingengewicht für eine ideale Gleichgewichtsverteilung zwischen Axtkopf und Stiel.

Begründung der Jury
Die Axt besticht durch ihre Schlankheit und Geradlinigkeit. Griff und Stiel sind gestalterisch klar voneinander abgehoben und fließen dennoch ineinander.

PERFORMA-DT
Stranded Wire Rope
Litzenseil

Manufacturer
Fatzer AG Drahtseilwerk, Romanshorn,
Switzerland
In-house design
Christof Nater
Web
www.fatzer.com
Honourable Mention

The Performa-DT stranded wire rope has
been developed especially for the whis-
per-quiet operation of urban cableways.
To achieve this, the rope features plastic
profiles between the strands. These
profiles fill the surrounding circular area
almost entirely, separate the strands
from each other spatially and stabilise
them in their position. Thanks to the
smooth surface structure the rope runs
practically without vibration and noise-
lessly across rollers and sheaves.

Statement by the jury
The stranded wire rope features an
inspiring innovative construction.
The different-coloured strands create
subtle accents.

Das Litzenseil Performa-DT wurde speziell
für den flüsterleisen Betrieb bei Stadtseil-
bahnen entwickelt. Hierfür wurden Kunst-
stoff-Profile zwischen die Litzen verseilt.
Diese Profile füllen die umgebende
Kreisfläche fast vollständig aus, trennen
die Litzen räumlich voneinander ab und
stabilisieren sie in ihrer Lage. Dank der
glatten Oberflächenstruktur läuft das Seil
praktisch vibrationsfrei und geräuschlos
über Rollen und Scheiben.

Begründung der Jury
Das Litzenseil begeistert durch seine innovative Konstruktionsart. Die verschieden-
farbigen Litzen schaffen subtile Spannungen.

DHAS
Adaptive Gripper
Adaptiver Greifer

Manufacturer
Festo AG & Co. KG, Esslingen, Germany
In-house design
Matthias Wunderling, Tilo Grimm
Web
www.festo.com

Two flexible straps connected to each other by means of crosspieces form the basic structure of the DHAS. This sturdy, yet flexible gripper made of polyurethane readily adapts to the contours of a workpiece without damaging it. It is thus especially well-suited for gripping sensitive objects with irregular shapes such as fruit or vegetables. The basic functional principle is additionally emphasised by the simple design.

Zwei flexible, über Zwischenstege miteinander verbundene Bänder bilden die Grundstruktur des DHAS. Dieser feste, aber gleichzeitig flexible Greifer aus Polyurethan passt sich der Kontur eines Werkstücks problemlos an, ohne es zu beschädigen. Besonders geeignet ist er daher zum Greifen empfindlicher Objekte mit unregelmäßiger Geometrie wie Obst oder Gemüse. Das einfache Wirkungsprinzip wird durch die schlichte Gestaltung zusätzlich unterstrichen.

Statement by the jury
The adaptive gripper translates a bionic principle into a clever design. The open construction provides good visibility.

Begründung der Jury
Der adaptive Greifer überführt ein bionisches Prinzip in ein cleveres Design. Die offene Konstruktion ermöglicht eine sehr gute Einsehbarkeit.

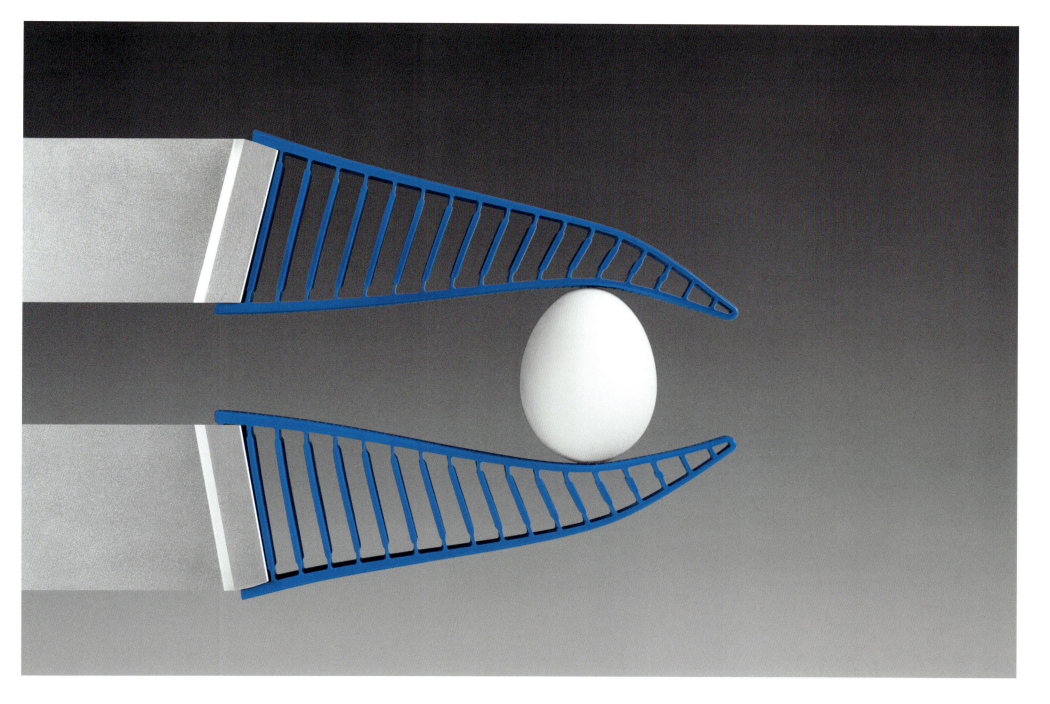

Lucas ProFinish
Paintbrush
Malerpinsel

Manufacturer
S. Lucas Limited, Lucas ProTools, Wrotham, Great Britain
Design
Jedco Product Designers Ltd (Edward Griffiths),
Weybridge, Great Britain
Web
www.lucasprotools.com
www.jedco.co.uk
Honourable Mention

The Lucas ProFinish decorator's paintbrush features a handle made of sustainably sourced, FSC-certified beechwood, stained with a special mix of natural tannin oils. The extruded aluminium ferrule fits flush with the handle, thus allowing the brush to rest comfortably in the hand. Furthermore, the two embossed brush rests on one side allow the paintbrush to be set down without the bristles touching the surface. Thanks to their oval shape, the bristles can absorb more paint than those of conventional brushes.

Der Malerpinsel Lucas ProFinish verfügt über einen Griff aus nachhaltigem, FSC-zertifiziertem Buchenholz. Die Lasur besteht aus einer speziellen Mischung aus natürlichen Tanninölen. Die Zwinge aus stranggepresstem Aluminium schließt bündig mit dem Griff ab, wodurch der Pinsel komfortabel in der Hand liegt. Darüber hinaus kann der Pinsel durch zwei eingeprägte Auflagen auf einer Seite hingelegt werden, ohne dass die Borsten dabei die Oberfläche berühren. Durch die ovale Form der Borsten kann mehr Farbe aufgenommen werden als mit herkömmlichen Pinseln.

Statement by the jury
The environmentally friendly design of this paintbrush demonstrates ecological responsibility in an exemplary way.

Begründung der Jury
Die umweltfreundliche Gestaltung des Malerpinsels zeugt auf vorbildliche Weise von ökologischem Verantwortungsbewusstsein.

bottBox
Storage Box
Aufbewahrungskasten

Manufacturer
Bott GmbH & Co. KG, Gaildorf, Germany
Design
TEDES Designteam GmbH, Roth, Germany
Web
www.bott.de
www.tedes-designteam.de

The bottBox is designed for use as a service tray, an open-fronted storage bin or a shelf box for vehicles or in workshops. The transparent dividers, viewing panes and lids offer a wide range of options for storing materials. The labelling clips available in various colours simplify the systematic organisation of shelves. Two different widths and five depths provide space for all kinds of small components.

Statement by the jury
The high-quality bottBox offers outstanding combination possibilities, thus providing a wide range of design options.

Die bottBox ist als Greifschale, Sichtlagerkasten oder Regalbox in Fahrzeug- oder Betriebseinrichtungen zu verwenden. Die transparenten Trennwände, Frontscheiben und Deckel bieten vielfältige Möglichkeiten, Material zu verstauen. Die Clips zur Beschriftung mit unterschiedlicher Farbkennzeichnung erleichtern das systematische Anordnen im Regal. Zwei verschiedene Breiten und fünf unterschiedliche Tiefen schaffen Platz für Kleinteile jeder Art.

Begründung der Jury
Die hochwertige bottBox lässt sich hervorragend kombinieren und bietet dadurch einen weiten Gestaltungsspielraum.

GARANT EasyFix
Modular Arrangement System
Modulares Ordnungssystem

Manufacturer
Hoffmann GmbH Qualitätswerkzeuge, Munich, Germany
Design
Böhler GmbH, Corporate Industrial Design (Julian Rathmann, Paul Bromme), Fürth, Germany
Web
www.hoffmann-group.com
www.boehler-design.de

Garant EasyFix is an individually configurable modular arrangement system for the storage of tools. The system is premised on basic boards available in two sizes, which can be fastened to a perforated wall. The boards feature wires for different applications as well as multifunctional clips and are the base for attachable special holders. The individual tiles can be arranged into theme-specific islands and horizontal lines.

Statement by the jury
Garant EasyFix offers an inspiringly wide variety of arrangement options, providing a solution for every single requirement using just a few elements.

Garant EasyFix ist ein individuell konfigurierbares Ordnungssystem für die Aufbewahrung von Werkzeugen. Es basiert auf Grundplatten in zwei Größen, die in eine Lochwand eingerastet werden. Die Platten sind mit Drähten für verschiedene Anwendungen sowie mit multifunktionalen Clips ausgestattet und bilden die Basis für aufsteckbare Spezialhalter. Die einzelnen Kacheln lassen sich sowohl zu themenspezifischen Inseln als auch zu horizontalen Bändern anordnen.

Begründung der Jury
Garant EasyFix begeistert mit eine Vielzahl an Ordnungsvariationen, die aus nur wenigen Elementen eine passende Lösung für jedes Bedürfnis ermöglichen.

GARANT Workstation
Modular Workstation
Modularer Systemarbeitsplatz

Manufacturer
Hoffmann GmbH Qualitätswerkzeuge, Munich, Germany
Design
Böhler GmbH, Corporate Industrial Design (Christoph Böhler, Katja Lautenbach), Fürth, Germany
Web
www.hoffmann-group.com
www.boehler-design.de

The Garant Workstation features rotatable, swivelling and height-adjustable elements that can be combined to create a customised workstation. Furthermore, the individual modular workstations can be linked to form larger units. Storage modules and lighting may be mounted to the columns in the front. Inside the workstation, diverse units such as cable ducts and electrical sockets can be installed.

Statement by the jury
This ergonomically shaped workstation accommodates individual configurations and blends harmoniously with any industrial environment.

Die Garant Workstation ist mit drehbaren, schwenkbaren und höhenverstellbaren Elementen ausgestattet, die sich zu einem individuellen Arbeitsplatz zusammenstellen lassen. Zudem lassen sich die einzelnen Systemarbeitsplätze zu größeren Einheiten verketten. Die Fronten der Trägersäulen dienen zum Einhängen von Aufbewahrungsmodulen und Beleuchtungen, während im Inneren verschiedene Einheiten wie Kabeldurchführungen oder Steckdosen angebracht werden können.

Begründung der Jury
Der ergonomisch gestaltete Systemarbeitsplatz eröffnet individuelle Planungsfreiheiten und fügt sich in jedes industrielle Umfeld harmonisch ein.

ZEISS Smartzoom 5
Digital Microscope
Digitalmikroskop

Manufacturer
Carl Zeiss Microscopy GmbH,
Jena, Germany
Design
ID WEBER Industrial Design,
Altensteig, Germany
Web
www.zeiss.com
www.idweber.de

The ZEISS Smartzoom 5 digital microscope is used in failure analysis and quality control of industrial goods. It is suitable for experienced as well as inexperienced users. Besides microscopic sample observation it also performs topographical 3D analyses of surfaces, measurements and the automatic documentation of data. With the help of a pivot arm the samples can be viewed from different angles.

Statement by the jury
Straight lines characterise the look of the digital microscope and effectively emphasise its functions.

Das Digitalmikroskop ZEISS Smartzoom 5 findet Anwendung in der Schadensanalyse und Qualitätskontrolle von Industriegütern. Es ist sowohl für erfahrene als auch ungeübte Benutzer geeignet. Neben der mikroskopischen Betrachtung ermöglicht es auch die topografische 3D-Analyse von Oberflächen, Messungen sowie eine automatisierte Dokumentation von Daten. Mithilfe des Schwenkarms lassen sich die Proben aus unterschiedlichen Blickwinkeln betrachten.

Begründung der Jury
Geradlinige Geometrien bestimmen das Erscheinungsbild des Digitalmikroskops und rücken die Funktionen auf hervorragende Weise in den Fokus.

Leica CS20 Field Controller Series
Field Controller
Feld-Controller

Manufacturer
Leica Geosystems AG,
Heerbrugg, Switzerland
Design
platinumdesign (Matthias Wieser,
Andreas Dimitriadis), Stuttgart, Germany
Web
www.leica-geosystems.com
www.platinumdesign.com
Honourable Mention

The Leica CS20 Field Controller series has an extremely stable and robust two and three-component housing made from high-tech plastics. The clear layout of the ergonomic controls and the emphasis of the primary function keys ensure simple use even with gloved hands. The device's centre of gravity is located in the tapered grip area. The back has been designed with maximum flatness to ensure an effortless grip.

Statement by the jury
The ergonomic form of the field controller is subtly highlighted by the circumferential light-green band. The housing has a high-quality finish.

Die Leica CS20 Field Controller Serie hat ein extrem stabiles und robustes 2- und 3-Komponenten-Gehäuse aus Hightech-Kunststoff. Das klar gegliederte Layout der ergonomischen Bedienelemente und die Hervorhebung funktional wichtiger Tasten erlauben eine einfache, auch handschuhgerechte Bedienung. Der Schwerpunkt des Geräts liegt im taillierten Griffbereich. Die Rückseite wurde möglichst flach konzipiert, um ein kraftsparendes Halten zu ermöglichen.

Begründung der Jury
Die ergonomische Form der Feld-Controller wird durch das umlaufende hellgrüne Band subtil in Szene gesetzt. Das Gehäuse ist hochwertig verarbeitet.

DESK-LABCAR
Electronic Test System
Elektronische Testeinrichtung

Manufacturer
ETAS GmbH, Stuttgart, Germany
Design
Tricon Design AG (Martin Wolf),
Kirchentellinsfurt, Germany
Web
www.pentair.com
www.tricon-design.de
Honourable Mention

Desk-Labcar is a test system for evaluating electronic control systems such as electronic control units in cars. Conventional test systems are typically located only in test laboratories due to their size and technical design. This device, however, combines high performance with a compact size. Thus developers can also use it directly at their workplace, which promotes testing in the early stages of development.

Statement by the jury
The design of this test system is remarkably compact, which is further emphasised visually by the radiused edges and the light side panels.

Die Testeinrichtung Desk-Labcar überprüft elektronische Regelsysteme wie z. B. elektronische Steuergeräte in Fahrzeugen. Herkömmliche Testeinrichtungen sind aufgrund ihrer Größe und Bauweise lediglich im Testlabor verfügbar. Dieses Gerät dagegen verbindet eine hohe Leistung mit einer kompakten Größe. Somit kann der Entwickler es auch direkt an seinem Arbeitsplatz nutzen, wodurch das Testen in frühen Entwicklungsphasen gefördert wird.

Begründung der Jury
Die Bauweise der Testeinrichtung erreicht eine beeindruckende Kompaktheit, die durch abgerundete Kanten und helle Seitenwände visuell verstärkt wird.

P-Scan Stack System
Ultrasonic Inspection Device
Ultraschallprüfgerät

Manufacturer
FORCE Technology, Brøndby, Denmark
Design
Harrit-Sørensen ApS (Nicolai Sørensen,
Thomas Harrit), Holte, Denmark
Web
www.p-scan.com
www.harrit-sorensen.dk

The P-Scan Stack System for the
automated ultrasonic inspection of
materials and components is particularly
lightweight and robust. With the help
of its Click-and-Play stacking system
individual modules can be configured
in any desired order and for any task.
All local cable connections between the
modules have been replaced by electrical
connections. The robust frame made of
anodised aluminium is built to withstand
bumps and drops.

Statement by the jury
The stackable construction of the testing
device provides excellent flexibility. The
modular design is visually emphasised by
the white right-angled lines.

Das P-Scan Stack System für die automatisierte Ultraschallprüfung von Werkstoffen
und Bauteilen ist besonders leicht und
robust. Mithilfe des Click-and-Play-Stapel-
systems lassen sich die einzelnen Module
in beliebiger Reihenfolge und für jegliche
Aufgabenstellung zusammensetzen. Sämtliche lokale Kabelverbindungen zwischen
den Modulen wurden durch elektrische
Anschlüsse ersetzt. Der solide Rahmen
aus eloxiertem Aluminium ist stoßfest und
sturzresistent.

Begründung der Jury
Der stapelbare Aufbau des Prüfgeräts bietet ausgezeichnete Flexibilität. Optisch wird
die Modularität durch die weißen Winkel
prägnant hervorgehoben.

VN-25MX
Pixel Shift Camera
Pixel-Shift-Kamera

Manufacturer
Vieworks, Anyang, South Korea
In-house design
Su Jin Bae, Seung Hyun Yoon,
Chang Woo Kang, Do Hyun Park
Web
www.vieworks.com
Honourable Mention

The VN-25MX pixel shift camera features extremely high image resolution in the nanoscale range. It is used for the inspection of flat screen monitors, image processing in research and science and to scan documents as well as film. The pixel shift technology is based on precise piezoelectronics and has a resolution of up to 236 million pixels. The interface provides quick transfer rates of up to 25 gigabits per second.

Statement by the jury
The pixel shift camera presents itself in a minimalist and reserved design language. It is thus a timeless original.

Die Pixel-Shift-Kamera VN-25MX bietet extrem hohe Auflösungen im Nanobereich. Sie wird z. B. zur Inspektion von Flachbildschirmen, Bildverarbeitung in Forschung und Wissenschaft sowie zum Scannen von Dokumenten und Filmen eingesetzt. Die Pixel-Shift-Technologie basiert auf der präzisen Piezoelektronik und erreicht eine Auflösung von bis zu 236 Millionen Pixel. Die Schnittstelle liefert eine schnelle Datenübertragung von bis zu 25 Gigabits pro Sekunde.

Begründung der Jury
Die Pixel-Shift-Kamera präsentiert sich in einer reduzierten und zurückgenommenen Formsprache. Sie ist damit von zeitloser Originalität.

FLIR C2
Thermal Imaging Camera
Wärmebildkamera

Manufacturer
FLIR Systems AB, Täby, Sweden
Design
Howl Designstudio (Jens Johansson,
Oscar Karlsson), Stockholm, Sweden
Web
www.flir.com
www.howlstudio.se
Honourable Mention

The Flir C2 is a fully equipped thermal imaging camera in a pocket-size format. The 3" LCD display, the buttons and the connectors are all protected by a shell made of thermoplastic elastomer. The built-in LED spotlight enables the camera to also be used in dark places. The large button for saving images can be operated with just one hand. The ergonomically shaped grip areas and the practical eyelet for the camera strap increase its ease of use.

Statement by the jury
This thermal imaging camera features an inspiringly handy and lightweight design, lending it highly versatility.

Die Flir C2 ist eine voll ausgestattete Wärmebildkamera im Taschenformat. Das 3"-LCD-Display, die Tasten und die Anschlüsse sind durch eine Hülle aus thermoplastischem Elastomer geschützt. Ein eingebauter LED-Strahler erlaubt die Benutzung der Kamera auch an dunklen Orten. Die große Taste zum Speichern der Bilder kann mit nur einer Hand bedient werden. Für zusätzlichen Komfort sorgen die ergonomischen Griffe und die praktische Öse für das Kameraband.

Begründung der Jury
Die Wärmebildkamera begeistert durch ihre handliche und leichte Bauform, wodurch sie vielseitig eingesetzt werden kann.

Wi-Fi Endoscope Camera
Wi-Fi Endoskop-Kamera

Manufacturer
Shenzhen Teslong Technology Limited,
Shenzhen, China
In-house design
Yingjie Sun
Web
www.teslong.com
Honourable Mention

The Wi-Fi Endoscope Camera can be connected wirelessly with a smartphone or tablet PC in order to inspect spaces that are difficult to access, such as hollows or pipes. Via an app the images of the Endoscope camera are transferred to the screen of the smartphone or tablet PC, which can be attached to a sturdy mount at the grip of the Endoscope camera. The images can be stored and forwarded as photo or video files.

Statement by the jury
The diverse surfaces of the housing give the Wi-Fi Endoscope Camera excellent haptic qualities and a robust appearance.

Die Wi-Fi Endoskop-Kamera lässt sich kabellos mit einem Smartphone oder Tablet-PC verbinden, um schwer einsehbare Stellen wie Hohlräume oder Rohre zu untersuchen. Über eine App werden die Bilder der Endoskop-Kamera auf den Bildschirm des Smartphones oder Tablet-PCs, die sich an einer stabilen Halterung am Endoskop-Kamera-Griff fixieren lassen, übertragen. Die Bilder können als Foto- oder Videodatei gespeichert oder verschickt werden.

Begründung der Jury
Die unterschiedliche Beschaffenheit des Gehäuses verleiht der Wi-Fi Endoskop-Kamera eine ausgezeichnete Haptik und eine robuste Anmutung.

Cutter for Fibre-Optic Cables
Optisches Schneidgerät für Glasfaserkabel

Manufacturer
The 41st Research Institute of China,
Electronics Technology Group Corporation,
Anhui, China
Design
Shenzhen Artop Design Co., Ltd.,
Shenzhen, China
Web
www.ei41.com
www.artopcn.com

This cutter for fibre-optic cables is equipped with an ultra-hardened blade, which is distinguished by a very long service life. The aluminium-alloy housing ensures reliable quality. The device's geometric shape is modelled on examples from shipping and astronautics. The cut-off sections of the fibre-optic cable are collected in a removable container.

Statement by the jury
The high-quality workmanship of this cutter communicates professionalism and reliability, and its closed form moreover prevents injuries.

Das optische Schneidgerät zum Durchtrennen von Glasfaserkabeln ist mit einer ultragehärteten Klinge ausgestattet, die sich durch eine sehr lange Lebensdauer auszeichnet. Die Aluminiumlegierung des Gehäuses sorgt für eine stabile Qualität. Die geometrische Form des Geräts orientiert sich an Vorbildern aus der Schiff- und Raumfahrt. Die abgetrennten Glasfaserkabel werden in einem herausnehmbaren Sammelbehälter aufgefangen.

Begründung der Jury
Die hochwertige Verarbeitung des Schneidgeräts kommuniziert Professionalität und Verlässlichkeit. Seine in sich geschlossene Form schützt vor Verletzungen.

Visar 650 & Pegasar 500 accu
Stud Welding Equipment
Bolzenschweißgeräte

Manufacturer
HBS Bolzenschweiss-Systeme GmbH & Co. KG,
Dachau, Germany
Design
Indeed Innovation GmbH, Hamburg, Germany
Web
www.hbs-info.de
www.indeed-innovation.com

The Visar 650 and Pegasar 500 accu stud welding equipment from the Quasar model range is distinguished by a particularly low weight. The Visar 650 facilitates stud welding with a 230-volt instead of 400-volt power supply and the Pegasar 500 accu mobile stud welding without an external power supply. Both devices feature internal cooling systems, making them dust-proof from the outside. The bumpers fashioned from shock-resistant polymer ensure impact protection.

Statement by the jury
This stud welding equipment convinces with its expressive design, which is characterised by the distinctive bumpers and the eye-catching handle.

Die Bolzenschweißgeräte Visar 650 und Pegasar 500 accu aus der Gerätebaureihe Quasar zeichnen sich durch ein besonders geringes Gewicht aus. Der Visar 650 ermöglicht das Bolzenschweißen mit einem 230-Volt- statt 400-Volt-Netzanschluss. Mit dem Pegasar 500 accu ist das mobile Bolzenschweißen ohne Stromanschluss möglich. Die Kühlung befindet sich im Inneren der Geräte, sodass sie von außen staubdicht sind. Die Stoßfänger aus schlagfestem Polymer schützen vor Stößen.

Begründung der Jury
Die Bolzenschweißgeräte überzeugen durch ihre ausdrucksstarke Bauform, die von den markanten Stoßfängern und dem auffälligen Griff geprägt ist.

U-NICA reader
Detector

Manufacturer
U-NICA Micronics AG, Global Security
Solutions, Malans, Switzerland
Design
Flink GmbH (Maurin Bisaz, Remo Frei),
Chur, Switzerland
Web
www.u-nica.com
www.flink.ch
Honourable Mention

The U-nica reader is used by customs to distinguish counterfeited products from genuine ones. The device features different technologies that are used for identification and is operated via a smartphone. Verification is by camera system and spectrometer. With the help of its anti-slip pad, the device can be used in an upright position. The housing is made from aluminium and features a soft-touch coating.

Statement by the jury
The tapering shape of the detector lends the device a premium look and a sculptural touch.

Das Detektionsgerät U-nica reader wird vom Zoll verwendet, um gefälschte von Originalprodukten zu unterscheiden. Das Gerät beherrscht unterschiedliche Technologien, die zur Identifikation eingesetzt werden, und wird über ein Smartphone bedient. Die Beurteilung erfolgt durch ein Kamerasystem und einen Spektrometer. Mithilfe des Haftpads kann das Gerät aufrecht stehend verwendet werden. Das Gehäuse besteht aus Aluminium mit partieller Softtouch-Lackierung.

Begründung der Jury
Die sich verjüngende Form des Detektors wirkt sehr edel und verleiht dem Gerät etwas Skulpturales.

Husqvarna Forest Helmet Technical
Forest Helmet
Forsthelm

Manufacturer
Husqvarna Group,
Stockholm, Sweden

In-house design
Gustav Landberg,
Joel Sellstrand,
Rajinder Mehra

Web
www.husqvarnagroup.com

reddot award 2015
best of the best

Fully protected

Forestry work is hard and challenging even in our times. It poses high demands on the equipment, since something unexpected may occur at any time. The Husqvarna Forest Helmet Technical perfectly suits this kind of work. It features an innovative, forward-focusing design in order to provide lumberjacks with maximum comfort and safety. The design is inspired by the shapes of stamped sheet-metal expressing strength, stability and full protection. In addition, the helmet is very light and well balanced. A central aspect of its sophisticated functionality is the ease with which the forest helmet can be adjusted. It can be intuitively adjusted to individual head shape, and hence fits perfectly. The helmet impresses with an entirely conclusive design, which encompasses an innovative visor system, harness, UV indicator, ventilation system and reflectors. The visor system is fully integrated and builds a formal unity with the helmet, while the strong top part forms the connection with the helmet shell, protecting the hinges. A sturdy chin protector, that is also precisely adjustable, furthermore communicates a sense of protection. The design objective to provide a better protection of the face has been implemented in an overall highly successful way. Lumberjacks are provided with an unobstructed view on the task in front of them, allowing them to pursue their work comfortably and fully protected.

Rundum geschützt

Die Arbeit im Wald ist auch in unserer Zeit hart und anspruchsvoll. Sie stellt hohe Anforderungen an die Ausrüstung, da jederzeit mit unvorhergesehenen Situationen zu rechnen ist. Der Husqvarna Forest Helmet Technical wurde dieser Arbeit perfekt angepasst. Seine Gestaltung ist auf innovative Weise nach vorn fokussiert, um so dem Waldarbeiter ein Höchstmaß an Komfort und Sicherheit zu bieten. Inspiriert von den Formen gestanzten Metallblechs, bringt seine Gestaltung die Attribute Stärke, Festigkeit und Rundumschutz zum Ausdruck. Er ist darüber hinaus sehr leicht und gut austariert. Ein wichtiger Aspekt seiner ausgereiften Funktionalität ist die Einfachheit, mit der dieser Forsthelm angepasst werden kann. Intuitiv lässt er sich auf die individuelle Kopfform einstellen und sitzt dann perfekt. Bei diesem Helm beeindruckt die rundum schlüssige Gestaltung mit einem innovativen Visiersystem, Gurt, UV-Indikator, Belüftungssystem und Reflektoren. Das Visiersystem wurde komplett integriert und bildet eine formale Einheit mit dem Helm, das stabile Oberteil stellt die Verbindung zur Helmschale dar und schützt so die Aufhängung. Ein sicheres Gefühl gibt zudem ein robuster Kinnrahmen, der ebenfalls exakt angepasst werden kann. Sehr gut gelöst wurde so insgesamt das Gestaltungsziel eines besseren Schutzes des Gesichts, wobei der Waldarbeiter zugleich einen uneingeschränkten Blick auf den vor ihm liegenden Arbeitsbereich hat. Komfortabel und rundum geschützt kann er seiner Arbeit nachgehen.

Statement by the jury

The Husqvarna Forest Helmet Technical is perfectly tuned for hard forestry work. Its forward-focusing design allows lumberjacks to concentrate on their work. This forest helmet features an excellent fit and can be adjusted comfortably to individual needs. Its harmonious design encompasses well thought-through details such as an innovative visor and an efficient ventilation system in a highly consistent manner.

Begründung der Jury

Der Husqvarna Forest Helmet Technical ist in jeder Hinsicht auf die harte Arbeit im Wald abgestimmt. Durch seine nach vorne ausgerichtete Gestaltung ermöglicht er es dem Arbeiter, konzentriert seiner Arbeit nachzugehen. Dieser Forsthelm besitzt eine hervorragende Passform und er lässt sich komfortabel individuell verstellen. Seine ausgewogene Formgebung schließt gut gelöste Details wie ein innovatives Visier und ein effektives Belüftungssystem schlüssig mit ein.

Designer portrait
See page 44
Siehe Seite 44

3M 6500QL Half Mask Series
3M Halbmasken-Serie 6500QL
Respirator
Atemschutzmaske

Manufacturer
3M, St. Paul, Minnesota, USA

In-house design
William Mittelstadt,
David Castiglione

Web
www.3m.com

reddot award 2015
best of the best

Comfortable safety

Protective masks are worn in many industries to protect the people working there from potentially harmful gases or vapours. With its sophisticated design, the 3M 6500QL Half Mask Series provides excellent wearing comfort. An innovative injection moulding process was used to create the silicone face-piece, ensuring that the mask adapts comfortably to the face and keeps its shape even in high-heat work environments. The design of these half masks unites all necessary elements in a perfect manner, whereby the innovative Quick Latch mechanism is a particularly fascinating functional detail. It allows users to quickly and intuitively take the mask on and off as they move in and out of contaminated areas without the need to remove a hard hat or face shield. This is an enormous advantage, especially considering that these 3M silicone half masks in combination with particle filters can be worn under welding helmets and face shields. The 3M Cool Flow Valve further increases safety. Its specially designed valve cover reduces heat and moisture by directing exhaled breath downward to reduce fogging of the eye guard. An entirely successful re-interpretation thus produced half masks that combine a new level of wearing comfort with maximum safety.

Komfortable Sicherheit

Schutzmasken werden in vielen industriellen Bereichen getragen, um die dort arbeitenden Menschen vor den mitunter auftretenden schädlichen Gasen oder Dämpfen zu schützen. Die 3M Halbmasken-Serie 6500QL bietet durch ihre hochentwickelte Gestaltung ausgezeichneten Tragekomfort. Die Gesichtsabdichtung wird mittels eines innovativen Spritzgussverfahrens aus Silikon gefertigt, weshalb sie sich gut an das Gesicht anschmiegt und selbst in heißer Arbeitsumgebung ihre Formstabilität behält. Auf perfekte Weise vereint die Gestaltung dieser Halbmasken alle nötigen Elemente, wobei als funktionales Detail der innovative Quick-Release-Mechanismus besonders begeistert. Er ermöglicht ein intuitives und schnelles Auf- und Absetzen vor und nach dem Betreten des Gefahrenbereichs, ohne dass der Helm oder der Gesichtsschutz dafür abgesetzt werden muss. Das ist ein enormer Vorteil vor dem Hintergrund, dass diese 3M Silikon-Halbmasken, in Kombination mit Partikelfiltern, auch unter Schweißerhelmen oder Gesichtsschutzvisieren verwendet werden können. Die Ausstattung mit einem 3M Cool Flow Ausatemventil erhöht zusätzlich die Sicherheit. Dank einer speziell konstruierten Ventilabdeckung bewirkt dieses, dass Hitze und Feuchtigkeit beim Ausatmen nach unten abgeleitet werden und dadurch das Beschlagen des Augenschutzes reduziert wird. Eine rundum gelungene Neuinterpretation brachte damit Halbmasken hervor, die einen neuen Tragekomfort mit einem Höchstmaß an Sicherheit verbinden.

Statement by the jury

This silicone half mask series convinces with an extremely user-friendly design concept. The functional Quick Latch mechanism allows for an easy putting on and taking off when entering or leaving hazardous areas. Sophisticated features, including an innovative combination of exhalation valve and valve cover provide additional safety. The flat design based on a special injection moulding process guarantees a very high degree of wearing comfort.

Begründung der Jury

Diese Halbmasken-Serie aus Silikon überzeugt mit einem überaus nutzerfreundlichen Gestaltungskonzept. Der funktionale Quick-Release-Mechanismus erlaubt ein einfaches Auf- und Absetzen beim Verlassen und Betreten des Gefahrenbereichs. Die durchdachte Ausstattung mit einer innovativen Kombination aus Ausatemventil und Ventilabdeckung bietet zusätzliche Sicherheit. Die flache, auf einem speziellen Spritzgussverfahren basierende Gestaltung sorgt für sehr komfortable Trageeigenschaften.

Designer portrait
See page 46
Siehe Seite 46

6X3
Safety Goggles
Schutzbrille

Manufacturer
Univet, Rezzato (Brescia), Italy
In-house design
Fabio Borsani
Web
www.univet-optic.com

The UDC coating of the 6X3 safety goggles is highly resistant to scratching and has an anti-fog effect. The indirect ventilation system reliably protects against drops and splashes. The panorama lenses are easily replaceable and may be comfortably worn over prescription glasses. The safety goggles can be used together with a respiratory mask or a half mask. There is also an additional face shield available, which covers the entire face. On the whole, the goggles are highly adaptable and suitable for a broad range of applications.

Die UDC-Beschichtung der Schutzbrille 6X3 bietet eine hohe Beständigkeit gegen Kratzer und ist beschlagsresistent. Das indirekte Belüftungssystem schützt zuverlässig vor Tropfen und Spritzern. Die Panoramascheiben sind auswechselbar und können problemlos über einer Korrektionsbrille getragen werden. Die Schutzbrille kann zusammen mit einer Atemmaske oder einer Halbmaske genutzt werden. Außerdem steht ein zusätzlicher Gesichtsschutz zur Verfügung, der das gesamte Gesicht bedeckt. Insgesamt ist die Schutzbrille sehr anpassungsfähig und vielseitig einsetzbar.

Statement by the jury
These safety goggles feature convincing ergonomic qualities and universal applicability. Their lightweight construction does not restrict the wearer in any way.

Begründung der Jury
Die Schutzbrille überzeugt durch ergonomische Qualitäten und universelle Einsetzbarkeit. Die leichte Konstruktion behindert den Träger in keinster Weise.

The Halo Light
Safety Light Ring
Sicherheitslichtring

Manufacturer
Illumagear, Seattle, USA
In-house design
Andrew Royal, Max Baker
Design
Pensar (Alex Diener, Kristin Will),
Seattle, USA
Web
www.illumagear.com
www.pensardevelopment.com

The Halo Light is an uninterrupted ring of light that is worn around a hard hat. The ring can be seen from a distance of 400 metres. On the one hand it improves visibility, and on the other it heightens safety since it can be seen from all sides. Due to its robust design, the ring can withstand mud, dust and rain, and it is also energy-efficient. Four different operation modes are available, which allows the light to adjust to any given situation.

Statement by the jury
The design concept of this 360-degree safety light ring is a simple and intelligent solution that enhances work safety.

Die Halo Light ist ein ununterbrochener Lichtring, der um einen Schutzhelm getragen wird. Der Ring leuchtet über eine Entfernung von 400 Metern. Er sorgt zum einen für eine bessere Sicht, zum anderen für mehr Sicherheit, da er von allen Seiten wahrgenommen wird. Aufgrund seiner robusten Gestaltung hält der Ring Schlamm, Staub und Regen aus, des Weiteren ist er energieeffizient. Vier unterschiedliche Betriebsarten lassen sich an die jeweilige Situation anpassen.

Begründung der Jury
Das Gestaltungskonzept des Sicherheitslichtrings stellt eine ebenso simple wie intelligente Lösung dar, um für mehr Arbeitssicherheit zu sorgen.

GZ-2 Series
Work Lights
Arbeitslichter

Manufacturer
Gentos Co., Ltd., Tokyo, Japan
In-house design
Masahiro Saito
Web
www.gentos.jp
Honourable Mention

The rechargeable work lights of the GZ-2 series offer a particularly wide illumination area. In addition, the integrated top light allows for the illumination of gaps and openings. The head can be adjusted within a range of 180 degrees and set to the desired angle. Due to the powerful magnet on the bottom along with the hook, the work lights may be positioned or hung in different places.

Statement by the jury
These work lights make an appealing impression thanks to their slim design, which facilitates the working process in difficult-to-access areas.

Die wiederaufladbaren Arbeitslichter der Serie GZ-2 besitzen eine besonders große Strahlungsfläche. Das integrierte Oberlicht ermöglicht es zudem, auch Zwischenräume und Öffnungen zu beleuchten. Das Kopfteil ist um 180 Grad beweglich und kann auf den gewünschten Winkel eingestellt werden. Durch einen starken Magneten an der Unterseite sowie einen Haken können die Arbeitslichter an verschiedenen Orten positioniert bzw. aufgehängt werden.

Begründung der Jury
Die Arbeitslichter gefallen durch ihre besonders schlanke Bauweise, die das Arbeiten an schwer zugänglichen Stellen ermöglicht.

GZ-1 Series
Work Lights
Arbeitslichter

Manufacturer
Gentos Co., Ltd., Tokyo, Japan
In-house design
Masahiro Saito
Web
www.gentos.jp

The rechargeable work lights of the GZ-1 series can be positioned in a variety of ways in order to keep the hands free. The adjustable magnetic grip at the bottom allows the lights to be placed on a variety of different surfaces. Thanks to the rear clip, it can be attached to a breast pocket or belt; the lamp may also be hung using the hook. The surface emitting LEDs have a wide illumination range, with the level of brightness depending on the respective model.

Statement by the jury
These work lights impress with their versatile usability and friendly design, which is distinguished by gently contoured edges.

Die wiederaufladbaren Arbeitslichter der Serie GZ-1 können auf vielfältige Weise platziert werden, damit die Hände frei bleiben. Mit dem verstellbaren magnetischen Griff an der Unterseite lassen sich die Lichter auf unterschiedlichen Flächen abstellen. Der Rückenclip ermöglicht das Tragen an Brusttasche oder Gürtel, an dem Haken können die Leuchten aufgehängt werden. Die oberflächenemittierenden Leuchtdioden haben eine große Reichweite, die Helligkeit variiert je nach Modell.

Begründung der Jury
Die Arbeitslichter punkten durch ihre vielseitige Einsetzbarkeit und eine freundliche Gestaltung, die von den sanft geformten Kanten bestimmt wird.

DBI-SALA Nano-Lok Edge
Self-Retracting Lifeline
Höhensicherungsgerät

Manufacturer
Capital Safety Group,
EMEA, Hamburg, Germany
In-house design
Web
www.capitalsafety.com
Honourable Mention

The Nano-Lok Edge self-retracting
lifeline has been developed especially
for working at great heights, when the
anchorage point of the lifeline is at
foot level. In case of a fall, the lifeline
cannot be damaged by extremely sharp
edges and is providing the wearer with
a complete assurance that it will not
rip. In addition, the automatic retract-
ion of the unused section of the lifeline
reduces the risk of tripping. The system
has been designed for user weight up
to 141 kg, exceeding the industry norm
of 100 kg.

Statement by the jury
The Nano-Lok Edge is consistently designed to ensure the safety of the wearer
without limiting their range of movement in the process.

Das Höhensicherungsgerät Nano-Lok Edge
wurde speziell für Arbeiten in großer
Höhe konzipiert, bei denen sich der Be-
festigungspunkt des Sicherungsseils auf
Fußhöhe befindet. Im Falle eines Absturzes
hält das Seil auch äußerst scharfen Kanten
stand und gibt dem Anwender die umfassende Sicherheit, dass es nicht reißt. Der
automatische Einzug der nicht genutzten
Gurtbandlänge reduziert darüber hinaus
das Stolperrisiko. Das maximale Anwender-
gewicht des Systems liegt bei 141 kg und
damit über der Norm.

Begründung der Jury
Das Design von Nano-Lok Edge ist konsequent auf die Sicherheit des Anwenders
ausgelegt, ohne ihn dabei in seiner Bewegungsfreiheit einzuschränken.

SECUMAX 350

Safety Knife
Sicherheitsmesser

Manufacturer
MARTOR KG, Solingen, Germany
In-house design
Martin Herlitz
Web
www.martor.com

The Secumax 350 safety knife weighs only 37 grams and is extremely versatile, for instance when used for cutting two-ply cardboard or adhesive tape. The concealed blade protects both the goods and the user. The blade has two cutting edges, enabling the knife to be simply turned over for extended use. In addition, the handle contains a spare blade. The curved outer edges and the soft-grip of the handle make it comfortable to use.

Statement by the jury
The closed housing of the Secumax 350 conveys robustness and stability. The knife's appearance is softened by its slightly rounded form.

Das Sicherheitsmesser Secumax 350 wiegt nur 37 Gramm und ist vielseitig einsetzbar, z. B. zum Schneiden von zweilagigem Karton oder Folie. Die verdeckt liegende Klinge schützt sowohl die Ware als auch den Benutzer. Der Klingenkopf besitzt zwei Schneidkanten, so kann er einfach gedreht und die Klinge länger genutzt werden. Zusätzlich befindet sich eine Ersatzklinge im Griff. Die gewölbten Außenseiten und der Soft-Grip des Handteils sorgen für einen hohen Bedienkomfort.

Begründung der Jury
Das geschlossene Gehäuse des Secumax 350 strahlt Robustheit und Stabilität aus. Aufgelockert wird das Erscheinungsbild des Messers durch die leicht gerundete Form.

SECUNORM 380

Safety Knife
Sicherheitsmesser

Manufacturer
MARTOR KG, Solingen, Germany
In-house design
Florian Segler
Web
www.martor.com

The Secunorm 380 safety knife features an extra-long blade extension of well over 7 cm. The flexible blade automatically retracts into the handle when the non-slip slider is released and the cut is done. A safety catch prevents the blade from opening unintentionally. The knife is suited e.g. for cutting cardboard, sacks, layers of paper and polystyrene.

Statement by the jury
The sophisticated design of the Secunorm 380 enables an unusually long blade extension. The curved handle displays a highly ergonomic design.

Das Sicherheitsmesser Secunorm 380 verfügt über einen extra langen Klingenaustritt von deutlich über 7 cm. Die biegsame Klinge zieht sich automatisch in den Griff zurück, sobald der rutschfeste Schieber losgelassen und der Schnitt beendet wird. Eine Sicherung verhindert, dass die Klinge ungewollt aus dem Griff gelangt. Das Messer eignet sich z. B. zum Schneiden von Kartons, Sackwaren, Papierbahnen und Styropor.

Begründung der Jury
Das durchdachte Design des Secunorm 380 erlaubt einen außergewöhnlich langen Klingenaustritt. Der geschwungene Griff überzeugt durch seine hohe Ergonomie.

Gloves 9000 Series
Safety Gloves
Schutzhandschuhe

Manufacturer
Granberg AS, Bjoa, Norway
In-house design
Virginijus Urbelis
Web
www.granberg.no

These safety gloves provide protection in extreme working conditions. The back of each glove features protective elements to absorb impact while simultaneously ensuring ample freedom of movement. The crease lines on the palms follow anthropometric hand points to prevent fatigue during work. The materials are cut-resistant and breathable. The bright colours are highly visible from afar, even in the dark.

Diese Handschuhe bieten Schutz unter extremen Arbeitsbedingungen. Auf dem Handrücken befinden sich Schlagschutz-elemente, um Stoßeinwirkungen auf-zufangen und gleichzeitig genügend Bewegungsfreiheit zu gewährleisten. Die Biegelinien auf der Handfläche folgen anthropometrischen Handpunkten, um Ermüdungserscheinungen während der Arbeit vorzubeugen. Die Materialien sind schneidfest und atmungsaktiv. Die leuch-tenden Farben sind auch bei Dunkelheit weithin sichtbar.

Statement by the jury
A dynamic tension and distinct sporti-ness characterise the look of these gloves. The materials unite protection and comfort.

Begründung der Jury
Spannungsbetonte Dynamik und aus-geprägte Sportlichkeit bestimmen das Aussehen der Handschuhe. Die Materialien vereinen Schutz und Komfort.

uvex handwerk
custom-made metal
Protective Clothing
Schutzbekleidung

Manufacturer
UVEX Arbeitsschutz GmbH, Fürth, Germany
In-house design
Web
www.uvex-safety.com

The uvex handwerk custom-made metal collection includes jackets and trousers for use in metal construction work. A specially developed coating protects particularly exposed parts against sparks and small drops of molten metal and reduces the garments' wear and tear. The elasticated inserts in shoulder and knee areas increase freedom of movement and allow for a tighter cut. In addition, the fit is supported by an ergonomic design. The ventilation openings provide a comfortable wearing experience, even in warmer temperatures.

Statement by the jury
This protective clothing is inspired by a modern and sporty look down to the smallest detail, thus flattering the wearer and providing enhanced comfort.

Die Bekleidungskollektion uvex handwerk custom-made metal umfasst Jacken und Hosen, die bei Metallbauarbeiten getragen werden. Eine speziell entwickelte Beschichtung schützt besonders exponierte Stellen vor Funken und kleinen Metallspritzern und reduziert den Verschleiß der Bekleidung. Die Stretchzonen im Schulter- und Kniebereich erhöhen die Bewegungsfreiheit und erlauben eine eng anliegende Schnittführung. Zusätzlich wird die Passform durch eine ergonomische Gestaltung unterstützt. Die Ventilationsöffnungen sorgen auch bei warmen Temperaturen für einen hohen Tragekomfort.

Begründung der Jury
Die Schutzbekleidung orientiert sich bis ins Detail an einem modernen und sportlichen Look, der dem Träger schmeichelt und ihn sich wohlfühlen lässt.

DAKOTA D030
Ladies' Safety Shoe
Sicherheitsschuh für Damen

Manufacturer
Maxguard GmbH, Krefeld, Germany
In-house design
Bernhard Goedecke
Web
www.maxguard.de
Honourable Mention

The Dakota D030 safety shoe is characterised by a slim, sporty sole shape. The high flexible, non-metallic penetration resisting midsole is about 40 per cent lighter than a conventional steel sole and leads to a better rolling movement when walking. The MAX-Microtec upper and the lining are both breathable and provide optimised climate control. The sole demonstrates strong anti-slip properties on a variety of surfaces.

Statement by the jury
Thanks to the lightness of the materials employed, this safety shoe provides strongly enhanced wearing comfort.

Der Sicherheitsschuh Dakota D030 zeichnet sich durch eine schlanke, sportive Sohlenform aus. Der hochflexible, nichtmetallische Durchtrittschutz ist um 40 Prozent leichter als herkömmliche Stahlsohlen und verbessert das Laufverhalten und die Abrollsituation. Sowohl das Obermaterial aus MAX-Microtec als auch das Innenfutter sind atmungsaktiv und sorgen für eine optimale Klimaregulierung. Die Laufsohle garantiert eine starke Rutschhemmung auf unterschiedlichen Untergründen.

Begründung der Jury
Der Sicherheitsschuh sorgt durch die Leichtigkeit der verwendeten Materialien für ein sehr komfortables Tragegefühl.

Dunlop Thermo+ Explorer
Safety Boot
Sicherheitsstiefel

Manufacturer
Dunlop Protective Footwear, Hevea BV, Raalte, Netherlands
In-house design
Design
Vibram USA, Concord, USA
Web
www.dunlopboots.com
www.vibram.com

The Dunlop Thermo+ Explorer safety boot is an improved version of previous models. It features a distinctive, geometric look, which is complemented by the newly added Vibram outsole. The bumpers at the front and back, in combination with the strong side cleats, provide protection in extreme conditions. The dark colours of the boot direct attention towards the red lining, the white Dunlop logo and the yellow Vibram logo.

Statement by the jury
The distinctive profile of this safety boot reflects strength and robustness. The materials make a highly durable impression.

Der Sicherheitsstiefel Dunlop Thermo+ Explorer stellt eine Weiterentwicklung früherer Modelle dar. Charakteristisch ist die geometrische Optik, die durch die neu hinzugefügte Vibram-Außensohle ergänzt wird. Die Dämpfer vorne und hinten sorgen zusammen mit den kräftigen Seitenstollen für Schutz unter Extrembedingungen. Die dunklen Farben des Stiefels lenken die Aufmerksamkeit auf das rote Innenfutter, das weiße Dunlop-Logo und das gelbe Vibram-Logo.

Begründung der Jury
Das ausgeprägte Profil des Sicherheitsstiefels spiegelt Stärke und Robustheit wider. Die Materialien machen einen überaus strapazierfähigen Eindruck.

SMART EVO
Portable X-Ray System
Tragbares Röntgensystem

Manufacturer
YXLON Copenhagen,
Taastrup, Denmark

Design
Harrit-Sørensen ApS
(Thomas Harrit, Nicolai Sørensen),
Holte, Denmark

Web
www.yxlon-portables.com
www.harrit-sorensen.dk

reddot award 2015
best of the best

New lightness
In industry, X-rays are used, for example, to identify changes in the thickness of materials or errors in individual assembly details. With this inspection method companies can guarantee the quality of their products. Smart Evo is a portable X-ray system for non-destructive testing, which was designed specifically to meet the different conditions in companies. The result is a technologically sophisticated system that impresses with a conclusive design concept. The heart of the system is the innovative Control Evo unit, which in combination with a redesigned X-ray tube head allows for an intelligent and efficient workflow. This ergonomically designed system couples high performance with ease of operation. All components are perfectly coordinated, and since the device is light and compact, it is also easy to transport. To ensure that the Smart Evo system is durable and fit for long-term use, it is shock-resistant and ruggedly built. The use of high-quality materials also contributes to guaranteeing that it can be used anywhere, even in the harshest and most remote environments. Recomposing the individual elements of such a testing device, Smart Evo is a sophisticatedly designed X-ray system that facilitates the work of all people involved and increases their flexibility.

Neue Leichtigkeit
In der Industrie werden Röntgenstrahlen eingesetzt, um etwa Materialstärke-Veränderungen oder Fehler einzelner Montagedetails zu ermitteln. Mithilfe solcher Prüfungen können Unternehmen die Qualität ihrer Produkte gewährleisten. Smart Evo ist ein tragbares Röntgensystem für zerstörungsfreie Materialprüfungen, welches eng ausgerichtet an den unterschiedlichen Bedingungen in Unternehmen entstanden ist. Das Ergebnis ist ein technologisch hochentwickeltes System, das durch sein schlüssiges Gestaltungskonzept beeindruckt. Im Mittelpunkt steht die innovative Steuereinheit Control Evo, die in Kombination mit einem neu gestalteten Röntgenrohr intelligente und effektive Arbeitsabläufe erlaubt. Dieses ergonomisch entworfene System verbindet dabei eine hohe Leistung mit der Möglichkeit einer einfachen Bedienung. Alle Komponenten sind perfekt aufeinander abgestimmt, und da das Gerät zudem handlich und leicht ist, lässt es sich auch gut transportieren. Um das Smart Evo-System widerstandsfähig für den dauerhaften Einsatz zu machen, ist es robust und stoßsicher gefertigt. Die Verwendung hochwertiger Materialien trägt ebenso dazu bei, dass es überall, unter extremsten Bedingungen und auch an abgelegenen Orten eingesetzt werden kann. Die einzelnen Elemente eines solchen Prüfgerätes neu komponierend, entstand mit Smart Evo ein durchdacht gestaltetes Röntgensystem, welches allen Beteiligten die Arbeit erleichtert und ihre Flexibilität erhöht.

Statement by the jury
The design of Smart Evo impressively combines the innovative technology of the Control Evo unit with a portable system for non-destructive testing. As such, this system sets new standards for the future. It fascinates with a highly user-friendly operation, which allows it to be perfectly adjusted to various workflows even under the most extreme conditions. It is compact and ergonomically well thought-through.

Begründung der Jury
Die Gestaltung von Smart Evo verbindet auf beeindruckende Weise die innovative Technologie der Steuereinheit Control Evo mit einem tragbaren System für zerstörungsfreie Materialprüfungen. Als solches setzt dieses System neue Standards für die Zukunft. Es begeistert mit einer sehr nutzerfreundlichen Handhabung, durch die es unterschiedlichen Abläufen auch unter extremsten Bedingungen ausgezeichnet angepasst werden kann. Es ist dabei handlich und ergonomisch gut durchdacht.

Designer portrait
See page 48
Siehe Seite 48

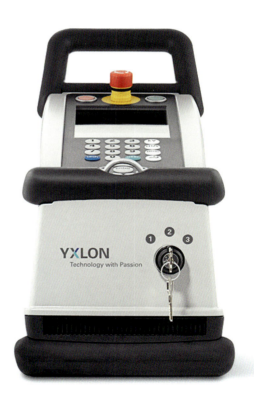

Hygrovision Mini
Dew Point Analyser
Taupunkt-Analysator

Manufacturer
Vympel, Dedovsk (Moscow), Russia
In-house design
Sergey Koshelev
Web
www.npovympel.ru
Honourable Mention

The Hygrovision Mini dew point analyser is a dew point monitor for the manual dew point measurement of water as well as higher hydrocarbons and additives in natural gas. The device consists of a high pressure measuring chamber with integrated condensing mirror, a removable optical system, a control unit featuring an LED display, and a battery which enables independent operation for a period of twelve hours.

Der Hygrovision Mini ist ein Taupunkt-Analysator für die manuelle Taupunktmessung von Wasser sowie höheren Kohlenwasserstoffen und Betriebsstoffen in Erdgasen. Das Gerät besteht aus einer Hochdruckmesskammer mit integriertem Kondensationsspiegel, einem abnehmbaren optischen System, einer Steuerungseinheit inklusive LED-Display und einem Akku, der einen autarken Betrieb von zwölf Stunden ermöglicht.

Statement by the jury
Thanks to its compact size the dew point analyser is ideally suited for mobile use at any location.

Begründung der Jury
Durch seine kompakte Größe ist der Taupunkt-Analysator prädestiniert für den mobilen Einsatz an jedem Ort.

Redesign PN
Pressure Sensor
Drucksensor

Manufacturer
ifm electronic gmbh, Essen, Germany
In-house design
Linda Schmidt
Web
www.ifm.com

The Redesign PN measures the physical pressure in installations. Its high overload protection and robust materials make this sensor also suited for rough environments. The sensor's head and the process connections can be rotated in any desired direction and oriented flexibly. The colour display switches from green to red to visualise dangerous operating ranges. The switching status LEDs, which can be seen from all sides, enable fast and unambiguous readability.

Statement by the jury
Practicability and an elegant design are harmoniously combined in this pressure sensor. The control areas highlighted in orange are clearly emphasised.

Der Redesign PN misst den physikalischen Druck in Anlagen. Seine hohe Überlastfestigkeit und robusten Materialien machen den Sensor auch in einer rauen Umgebung einsetzbar. Der Sensorkopf und die Prozessanschlüsse lassen sich in jede beliebige Richtung drehen und flexibel ausrichten. Das Farbdisplay wechselt von Grün auf Rot, um Gefahrenbereiche zu visualisieren. Auch die von allen Seiten einsehbaren Schaltzustand-LEDs sind schnell und eindeutig lesbar.

Begründung der Jury
Praktikabilität und Ästhetik ergeben beim Drucksensor ein harmonisches Zusammenspiel. Die orange markierten Bedienfelder sind sehr gut hervorgehoben.

HandySCAN 700™
3D Scanner

Manufacturer
Creaform Inc., Lévis, Canada
In-house design
François Lessard, Nicolas Lebrun
Web
www.creaform3d.com

The HandySCAN 700 is a portable 3D scanner, which has been optimised to meet the needs of product developers and engineers. The scanned surfaces appear almost in real-time. The integrated distance meter enables users to detect the appropriate scanning distance without having to divert their eyes from the scanned object. Its multi-grip handle provides great freedom of movement and multi-function buttons facilitate easy interaction with the software.

Statement by the jury
The ergonomically shaped 3D scanner sits comfortably in the hand. The black housing highlights its distinctive lines effectively.

Der HandySCAN 700 ist ein portabler 3D-Scanner, der speziell auf die Bedürfnisse von Produktentwicklern und Ingenieuren abgestimmt wurde. Die eingelesenen Oberflächen werden nahezu in Echtzeit dargestellt. Ein Distanzmesser ermöglicht es Nutzern, den entsprechenden Scanabstand zu erkennen, ohne den Blick vom Objekt abzuwenden. Der Multihandgriff sorgt für eine große Bewegungsfreiheit und die Multifunktionstasten erleichtern die Interaktion mit der Software.

Begründung der Jury
Der ergonomisch geformte 3D-Scanner liegt angenehm in der Hand. Das schwarze Gehäuse lässt die markante Linienführung gekonnt hervortreten.

Freestyle3D
3D Handheld Scanner
3D-Handscanner

Manufacturer
FARO Scanner Production GmbH,
Korntal-Münchingen, Germany
Design
Axel Ruhland Produktgestaltung,
Stuttgart, Germany
Web
www.faro.com
www.axelruhland.de

With the Freestyle3D handheld scanner, all kinds of surfaces and environments can be scanned. Due to its small size and low weight, the scanner may also be used in difficult-to-access areas, for instance in industrial facilities. In addition, the device is characterised by quick data acquisition, real-time visualisation and a high scan volume. This reduces the time required for the scanning process.

Statement by the jury
The Freestyle3D has a surprising slim, triangular shape, which supports its functionality and gives it a distinct appearance.

Mit dem Handscanner Freestyle3D lassen sich alle Arten von Oberflächen und Umgebungen scannen. Aufgrund seiner geringen Größe und seines geringen Gewichts kann der Scanner auch in schwer zugänglichen Bereichen, z. B. in industriellen Anlagen, verwendet werden. Zudem zeichnet sich das Gerät durch eine schnelle Datenerfassung, Echtzeitvisualisierung und ein hohes Scanvolumen aus. Dadurch verkürzt sich die Zeit, die für den Scanvorgang benötigt wird.

Begründung der Jury
Der Freestyle3D überrascht durch seine schlanke dreiwinkelige Form, die seine Funktionalität unterstützt und ihm ein unverwechselbares Äußeres verleiht.

MakerBot Replicator
3D Printer
3D-Drucker

Manufacturer
MakerBot Industries, New York, USA
In-house design
Web
www.makerbot.com

The MakerBot Replicator desktop 3D printer delivers a seamless 3D printing experience for novices and professionals alike. The device has a simple user interface with a rotary knob and a 3.5" display. Through the mobile app, prints can be prepared and initiated. In addition, the built-in camera allows users to remotely monitor prints. The Smart Extruder enables users to swap a worn extruder in minutes and thus minimise printing downtime.

Statement by the jury
This 3D printer features an inspiring, open design. Its robust and technical aesthetics blend perfectly into industrial environments.

Der 3D-Schreibtischdrucker MakerBot Replicator bietet eine nahtlose 3D-Druckerfahrung für Anfänger und Profis. Das Gerät verfügt über eine einfache Benutzeroberfläche mit einem Drehknopf und 3,5"-Display. Mit der mobilen App können Druckaufträge erstellt und initiiert werden. Über die eingebaute Kamera lässt sich der Druckfortschritt auch aus der Ferne überwachen. Der Smart Extruder ist innerhalb von Minuten auswechselbar und minimiert dadurch Ausfallzeiten.

Begründung der Jury
Der 3D-Drucker begeistert durch seine offene Bauweise. Seine robuste und technisch anmutende Ästhetik passt vortrefflich in ein industrielles Umfeld.

Landa S10
Digital Printing Press
Digitale Druckmaschine

Manufacturer
Landa Digital Printing, Rehovot, Israel
Design
I2D Innovation to Design, Tel Aviv, Israel
Web
www.landanano.com
www.i2d.co.il

The Landa S10 digital printing press uses the newly developed Nanography process, which produces ultra-sharp image quality. The inspection table is integrated into the press. The user cockpit consists of a large touchscreen, which visualises work plans and helps monitor the entire printing process. The generous use of black glass communicates the machine's innovative character.

Statement by the jury
The Landa S10 reinterprets the aesthetics of printing presses. The press looks elegantly elongated, which lends it a sculptural appearance.

Die digitale Druckmaschine Landa S10 verwendet ein neu entwickeltes Nanographie-Verfahren, das eine ultrascharfe Bildqualität erreicht. Der Inspektionstisch wurde in die Maschine integriert. Das Benutzer-Cockpit besteht aus großen Touchscreens, die die Arbeitspläne visualisieren und den gesamten Druckprozess überwachen helfen. Die großzügige Verwendung von schwarzem Glas kommuniziert den innovativen Charakter der Maschine.

Begründung der Jury
Die Landa S10 interpretiert die Ästhetik von Druckmaschinen neu. Die Anlage wirkt elegant gestreckt, was ihr eine skulpturale Anmutung verleiht.

ROLAND 700 EVOLUTION
Sheetfed Offset Printing Press
Bogenoffset-Druckmaschine

Manufacturer
Manroland Sheetfed GmbH, Offenbach, Germany
Design
Paul Kruse | Product Design (Paul Kruse), Hildesheim, Germany
Web
www.manrolandsheetfed.com
www.paul-kruse.com

The Roland 700 Evolution sheetfed offset printing press features technological developments, which help to reduce make-ready times and waste. This results in increased productivity and print quality. The PressPilot with touchscreen controls provides high operating efficiency. The light-coloured cover elements distinct the design of the units of this sheetfed offset printing press. The design feature of the large radii is inspired by the bending characteristic of stacked paper when hanging over an edge.

Die Bogenoffset-Druckmaschine Roland 700 Evolution verfügt über fortschrittliche Technologien, die helfen, Rüstzeiten und Makulatur zu reduzieren. Dadurch werden die Produktivität und die Druckqualität gesteigert. Der PressPilot mit Touchscreen-Bedienung sorgt für eine hohe Bedienungseffizienz. Die hellen Abdeckungen prägen das Design der einzelnen Units dieser Bogendruckmaschine. Die Gestaltung der großen Radien ist inspiriert durch die Biegeeigenschaft des Papiers, wenn es im Stapel über eine Kante hängt.

Statement by the jury
The white, rounded front elements of the Roland 700 Evolution achieve a clear, visual structure and convey an impression of immaculate order.

Begründung der Jury
Die weißen abgerundeten Frontelemente der Roland 700 Evolution erzielen eine klare optische Gliederung und vermitteln einen tadellos aufgeräumten Eindruck.

TruBend 5130
Press Brake
Biegemaschine

Manufacturer
TRUMPF GmbH + Co. KG, Ditzingen, Germany
Design
Phoenix Design GmbH + Co. KG, Stuttgart, Germany
Web
www.trumpf.com
www.phoenixdesign.com

TruBend 5130 is a press brake characterised by high precision. An optical laser system measures the bending angle and automatically makes a correction to attain the desired angle. The device's innovative operating element is a working shoe with intelligent sensor technology that triggers the press beam stroke via a foot movement. Without having to carry the conventional foot switch, the operator now has both hands free for the workpiece. The device is operated via a touchscreen.

Die TruBend 5130 ist eine Biegemaschine, die sich durch hohe Genauigkeit auszeichnet. Ein optisches Lasersystem misst den Biegewinkel und korrigiert diesen automatisch auf das gewünschte Winkelmaß. Das innovative Bedienelement der Maschine ist ein mit intelligenten Sensoren bestückter Arbeitsschuh, über den der Hub des Druckbalkens mit dem Fuß ausgelöst wird. Der Benutzer muss daher nicht extra einen Fußtaster mit sich führen und hat beide Hände frei für das Werkstück. Die Steuerung erfolgt über einen Touchscreen.

Statement by the jury
The press brake convinces with its clear geometrical lines and homogeneous front, which convey a clear and clean image.

Begründung der Jury
Die Biegemaschine punktet durch ihre klaren geometrischen Linien und einheitlichen Fronten, die einen sauberen und aufgeräumten Eindruck vermitteln.

TruLaser Tube 5000
Laser Tube Cutting Machine
Laserrohrschneidmaschine

Manufacturer
TRUMPF GmbH + Co. KG,
Ditzingen, Germany
Design
Phoenix Design GmbH + Co. KG,
Stuttgart, Germany
Web
www.trumpf.com
www.phoenixdesign.com

The TruLaser Tube 5000 laser tube cutting machine replaces production processes such as sawing, drilling, milling and deburring. The self-centring chuck and the wide tension rollers enable stable and exact guiding of the tube. The open design with its large sliding door provides optimal access to the working area. The far-reaching swivel arm enables individual positioning of the control panel for operation on both the left and the right.

Statement by the jury
The open architecture of the TruLaser Tube 5000 provides an excellent view into the machine, thus creating a transparent workflow.

Die Laserrohrschneidemaschine TruLaser Tube 5000 ersetzt Fertigungsverfahren wie Sägen, Bohren, Fräsen und Entgraten. Das selbstzentrierende Spannfutter und breite Spannwalzen ermöglichen eine stabile und exakte Führung des Rohrs. Das offene Design mit einer großen Schiebetür sorgt für eine optimale Zugänglichkeit zum Arbeitsbereich. Der weit schwenkbare Arm ermöglicht die individuelle Positionierung des Bedienpanels für Rechts- und Linksbedienung.

Begründung der Jury
Die offene Architektur der TruLaser Tube 5000 erzielt eine hervorragende Einsehbarkeit in die Maschine und schafft dadurch transparente Arbeitsabläufe.

TruLaser 5030 fiber
2D Laser Cutting Machine
2D-Laserschneidmaschine

Manufacturer
TRUMPF GmbH + Co. KG,
Ditzingen, Germany
Design
Phoenix Design GmbH + Co. KG,
Stuttgart, Germany
Web
www.trumpf.com
www.phoenixdesign.com

The TruLaser 5030 fiber is a 2D laser cutting machine for the processing of sheet metal. The Smart Collision Prevention function improves its process reliability by independently creating processing strategies. All keys and handles, as well as the control panel, are integrated into the newly developed function strip, which goes around the machine like a belt. The trim panels are also attached and aligned to this strip, making it much easier to assemble.

Statement by the jury
While the compact form of the TruLaser 5030 fiber communicates its efficiency, the blue colour accents successfully direct attention to key work areas.

Die TruLaser 5030 fiber ist eine 2D-Laserschneidmaschine für die Bearbeitung von Blechen. Die „Smart Collision Prevention"-Funktion erhöht die Prozesssicherheit, indem sie selbständig Abarbeitungsstrategien erstellt. In der neu entwickelten Funktionsleiste, die die Maschine wie eine Taille umgibt, sind alle Tasten, Griffe und das Bedienpult integriert. In diese Leiste werden auch die Bleche eingehängt und darin ausgerichtet, was die Montage vereinfacht.

Begründung der Jury
Während die kompakte Form der TruLaser 5030 fiber ihre Effizienz visualisiert, lenken die blauen Farbakzente den Blick geschickt auf zentrale Arbeitsbereiche.

TruPunch 5000
Punching Machine
Stanzmaschine

Manufacturer
TRUMPF GmbH + Co. KG,
Ditzingen, Germany
Design
Phoenix Design GmbH + Co. KG,
Stuttgart, Germany
Web
www.trumpf.com
www.phoenixdesign.com

The TruPunch 5000 punching machine is equipped with clamps that hold the sheet and can be retracted. This serves to optimise sheet utilisation and reduce material waste. The tool storage provides a wide range of tools in an extremely small space and can be extended by adding modules when needed. Moreover, it has its own operator position and is thus very easy to control.

Statement by the jury
The slim design of the TruPunch 5000 embodies its dynamic performance, while subtly highlighting the machine's economical use of materials at the same time.

Die Stanzmaschine TruPunch 5000 ist mit Pratzen ausgestattet, die das Blech halten und zurückgezogen werden können. Dadurch wird das Blech optimal ausgenutzt, und es entstehen weniger Materialreste. Im Werkzeugspeicher finden viele Werkzeuge auf kleinstem Raum Platz, er kann bei Bedarf modular erweitert werden. Zudem hat er eine eigene Bedienstelle und ist dadurch sehr komfortabel zu steuern.

Begründung der Jury
Die Schlankheit der TruPunch 5000 verkörpert ihre Dynamik und unterstreicht gleichzeitig subtil die Sparsamkeit der Maschine beim Materialverbrauch.

FM200/5AX Linear
Machine Tools
Werkzeugmaschinen

Manufacturer
DOOSAN infracore, Changwon, South Korea
In-house design
Byung Chan Kim, Hee Dong Son
Web
www.doosan.com

The FM200/5AX Linear machine tools are characterised by an ergonomic design which enables easy operation and maintenance. The basic, pentagonal form provides a large interior work space. The low frame construction makes for easy user access to the machine. Two connected metal frames emphasise the distinctive appearance, which is also continued in the design of the accessories.

Die Werkzeugmaschinen FM200/5AX Linear zeichnen sich durch eine ergonomische Gestaltung aus, die eine einfache Bedienung und Wartung ermöglicht. Die fünfeckige Grundform ergibt einen großen Bearbeitungsinnenraum. Die niedrig konzipierte Rahmenkonstruktion erleichtert den Zugang zur Maschine. Zwei verbundene Metallrahmen unterstreichen das charakteristische Erscheinungsbild, das sich auch im Design der Zusatzausstattung wiederfindet.

Statement by the jury
The halves of the FM200/5AX Linear, which are distinct from each other in colour and shape, create fascinating accents and elegantly underline the device's spaciousness.

Begründung der Jury
Die farblich und baulich voneinander abgesetzten Hälften der FM200/5AX Linear setzen spannende Akzente und unterstreichen formschön ihre Geräumigkeit.

KUKA Genius
Friction Welding Machine
Reibschweißmaschine

Manufacturer
KUKA Industries GmbH, Augsburg, Germany
In-house design
Michael Büch er, Otmar Fischer
Design
Grewer Industriedesign (Michael Grewer), Augsburg, Germany
Web
www.kuka-industries.com
www.grewer- ndustriedesign.de

The KUKA Genius friction welding machine features a special hydraulic system and a large speed/load range. A device for swarf removal can be installed either along the side or the back of the machine. Large operation and maintenance doors enable optimal access from above and to the front of the machine. The operation is carried out intuitively via a touch panel that may also be used while wearing safety gloves.

Die Reibschweißmaschine KUKA Genius ist mit einem speziellen Hydrauliksystem und großem Drehzahllastbereich ausgestattet. Eine Vorrichtung für den Abtransport der Späne kann wahlweise seitlich oder auf der Rückseite der Maschine angebracht werden. Große Bedien- und Wartungstüren ermöglichen einen optimalen Zugang von oben und der Maschinenvorderseite. Die Bedienung erfolgt intuitiv über ein Touch-panel, das auch mit Schutzhandschuhen bedienbar ist.

BPT-S Series
BPT-S Serie
Photovoltaic Inverter
Photovoltaik-Wechselrichter

Manufacturer
Bosch Power Tec GmbH, Böblingen, Germany
In-house design
Web
www.bosch-power-tec.com

The photovoltaic inverter of the BPT-S series features a modern control concept including touchless communication interfaces, which are supplemented by an app and an integrated web server with Wi-Fi and Ethernet interfaces. This enables a quick start-up and simplified remote analysis. The user interface with LED concept and integrated display provides important analysis functions.

Statement by the jury
The photovoltaic inverter is characterised by an elegant, minimalist look, which is the result of a seamless design of forms and colours.

Der Photovoltaik-Wechselrichter der BPT-S Serie verfügt über ein modernes Bedienkonzept mit berührungslosen Kommunikationsschnittstellen, die durch eine App und den integrierten Webserver mit Wi-Fi- und Ethernetverbindung ergänzt werden. Dadurch wird eine schnelle Inbetriebnahme und vereinfachte Fernanalyse ermöglicht. Die Bedienoberfläche mit LED-Konzept und integriertem Display liefert wichtige Analysefunktionen.

Begründung der Jury
Der Wechselrichter zeichnet sich durch einen formschönen Minimalismus aus, der sich aus der nahtlosen Gestaltung von Formen und Farben ergibt.

Tool Trolley
Werkzeugwagen

Manufacturer
DMG Mori Seiki AG, Bielefeld, Germany
Design
Dominic Schindler Creations GmbH, Lauterach, Austria
Web
www.dmgmori.com
www.dominicschindler.com
Honourable Mention

The trolley for tools in machine assembly halls is used during the manufacturing process of machine tools. Depending on the machine specification, different tools are required in the work process, which can then be stored in the wagon. It is equipped with an oil tub and swivel castors on the front where it is pulled. Moreover, it has a shelf for folders on the side. To prevent damage it features a circumferential shock protector near the bottom.

Statement by the jury
The no-handle design as well as the contrast of the white surfaces and black details characterise the high design standards of this tool trolley.

Der Werkzeugwagen wird beim Fertigungsprozess von Werkzeugmaschinen eingesetzt. Je nach Maschinenausführung werden unterschiedliche Werkzeuge im Arbeitssystem benötigt, die dann in dem Wagen aufbewahrt werden können. Er ist mit einer Ölwanne versehen und am Zugende mit Schwenkrollen ausgestattet. Zudem ist seitlich ein Ordnerfach integriert. Zur Vermeidung von Beschädigungen ist ein umlaufender Stoßschutz in Bodennähe angebracht.

Begründung der Jury
Die grifflose Konstruktion sowie der Kontrast aus weißen Oberflächen und schwarzen Details prägen den hohen Designanspruch des Werkzeugwagens.

GenCell G5
Generator

Manufacturer
GenCell, Petah Tikva, Israel
Design
I2D Innovation to Design
(Elisha Tal, Mel Bergman), Tel Aviv, Israel
Web
www.gencellenergy.com
www.i2d.co.il
Honourable Mention

The GenCell G5 generator is based on alkaline fuel cell (AFC) technology. It offers an entirely eco-friendly and highly efficient solution for backup power supply – both in extreme natural environments and in urban settings. The design is based on a monolithic, rectangular volume, which is divided into segments. These segments in turn create triangular areas, which frame the different interface functions.

Statement by the jury
This generator is a convincing combination of rigid form and dynamic lines, creating a harmonious whole.

Der Generator GenCell G5 basiert auf alkalischer Brennstoffzellentechnologie. Er bietet eine vollständig umweltfreundliche und hocheffiziente Lösung für die Notstromversorgung – sowohl unter extremen Umweltbedingungen als auch in städtischer Umgebung. Das Design beruht auf einem monolithischen rechteckigen Korpus, der in Segmente unterteilt ist. Diese ergeben dreieckige Flächen, die die verschiedenen Schnittstellenfunktionen umrahmen.

Begründung der Jury
Der Generator stellt ein gelungenes Zusammenspiel aus strenger Form und dynamischen Linien dar, die ein harmonisches Gesamtbild ergeben.

Black OnyX
LED Display

Manufacturer
ROE Visual Co., Ltd., Shenzhen, China
Design
LKK Design Shenzhen Co., Ltd. (Wenbo Li, Gang Wang, Suikai Lin, Yichao Li), Shenzhen, China
Web
www.roevisual.com
www.lkkdesign.com

The Black OnyX LED display weighs less than 7 kg. The magnesium alloy is ultra-light and robust. The weight is distributed evenly onto the frame, therefore the display is easy to assemble and transport and suited for mobile use, e.g. for concerts and exhibitions. The flat construction, which is produced using a highly precise CNC processing method, can be mounted flush to the wall.

Der LED-Display Black OnyX wiegt weniger als 7 kg. Die Magnesiumlegierung ist ultraleicht und stabil. Das Gewicht wird gleichmäßig auf das Rahmenwerk verteilt. Dadurch lässt sich der Display besonders einfach und schnell zusammenbauen und transportieren, z. B. für den mobilen Einsatz bei Konzerten oder Ausstellungen. Die flache Konstruktion, die mit einem hochpräzisen CNC-Verarbeitungsprozess hergestellt wird, lässt sich bündig an die Wand montieren.

CONCEPTLINE C &
STUDIOLINE S/M/L
Milling Machines
Fräsmaschinen

Manufacturer
Kolb Technology GmbH,
Hengersberg, Germany
Design
idukk (Reinhard Kittler, Heinrich Kurz),
Wilhering, Austria
Web
www.kolb-technology.com
www.idukk.at

The milling machines of the Conceptline C and Studioline S/M/L series were specifically developed for model-making in the automotive industry. These machines allow demanding and time-critical milling jobs to be carried out directly in the design studio. The car models are made of soft polystyrene, rigid foam or an industrial plasticine, so-called "clay". The machine is operated via a modern interface.

Statement by the jury
The dynamic, slightly sloping proportions of these milling machines stylishly reveal their speed and precision.

Die Fräsmaschinen der Serien Conceptline C und Studioline S/M/L wurden speziell für den Modellbau in der Automobilindustrie entwickelt. Mit den Maschinen können anspruchsvolle und zeitkritische Fräsaufgaben direkt im Designstudio erledigt werden. Die Automodelle werden aus weichem Styropor, harten Schäumen oder einem Industrieplastilin, dem sogenannten Clay, gefertigt. Die Maschine wird über ein modernes Interface bedient.

Begründung der Jury
In den dynamischen, leicht abgeschrägten Proportionen der Fräsmaschinen offenbart sich auf stilvolle Weise ihre Geschwindigkeit und Präzision.

FCW 150
Boring and Milling Machine
Bohr- und Fräsmaschine

Manufacturer
ŠKODA Machine Tool a.s.,
Pilsen, Czech Republic
Hestego a.s., Vyškov, Czech Republic
In-house design
Design
Martin Tvarůžek Design (Martin Tvarůžek),
Brno, Czech Republic
Web
www.skodamt.com
www.hestego.com
www.tvaruzekdesign.com

The FCW 150 boring and milling machine is primarily intended for customers with a requirement for lighter machining work, such as milling, drilling, boring and threading. Most companies who use the machine focus on working weldments, castings and special fibreglass materials. The control cabin features large glass windows, providing a clear view of the machining process. An outlet in the ceiling of the cabin ensures that the cooling emulsion can escape.

Statement by the jury
The individual areas of the FCW 150 follow a cohesive design language, therefore forming a harmonious unit.

Die FCW 150 Bohr- und Fräsmaschine ist vor allem für Kunden mit einem Bedarf an leichteren Bearbeitungsprozessen wie Fräsen, Bohren und Gewindeschneiden konzipiert. Die Maschine wird insbesondere von Firmen genutzt, die Windturbinen, Maschinenteile oder Schiffsmotorenblätter herstellen. Die Steuerkabine verfügt über große Glasfenster, die freie Sicht auf den Fertigungsprozess ermöglichen. Ein Auslass in der Decke der Kabine gewährleistet, dass die Kühlemulsion entweichen kann.

Begründung der Jury
Die einzelnen Bereiche der FCW 150 sind in einer durchgängigen Designsprache gehalten, und bilden eine harmonische Einheit.

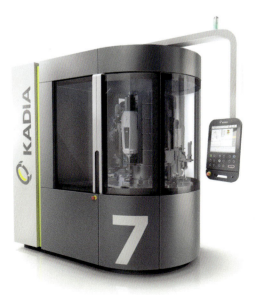

V line
Honing Machine
Honmaschine

Manufacturer
Kadia Produktion GmbH + Co.,
Nürtingen, Germany
Design
Design Tech, Ammerbuch, Germany
Web
www.kadia.de
www.designtech.eu

In the V line honing machine workpieces undergo a highly precise finishing process in order to create dimensional and shape accuracy. The compact structure houses the honing machine and the control boxes in one unit. The large side doors can be fully opened in order to provide optimal access to the working area. The control panel can be swivelled around the entire device.

Statement by the jury
The honing machine impresses with its open architecture and rotatable control panel, which are highly tailored to the user.

In der Honmaschine V line werden Werkstücke einem hochpräzisen Feinbearbeitungsverfahren unterzogen, um sie maß- und formgenau anzupassen. In der kompakten Bauform sind die Maschine und die Schaltschränke in einer Einheit untergebracht. Die großflächigen Seitentüren können komplett geöffnet werden, um einen optimalen Zugang zum Bearbeitungsraum zu gewährleisten. Das Bedienpanel kann um die gesamte Maschine geschwenkt werden.

Begründung der Jury
Die Honmaschine begeistert durch die offene Architektur und das bewegliche Bedienpanel, welche in hohem Maße auf den Anwender zugeschnitten sind.

powRgrip® PGU 9500
Clamping Unit for Toolholders
Spanneinheit für Werkzeug-
aufnahmen

Manufacturer
REGO-FIX AG, Tenniken, Switzerland
In-house design
Web
www.rego-fix.com
Honourable Mention

The powRgrip PGU 9500 automated
clamping and unclamping unit for
toolholders enables fast and safe tool
clamping for machining at the touch of
a button. With its precise concentricity
and innovative vibration damping the
device increases tool life. Furthermore,
the tool clamping does not create heat,
which increases work safety and saves
energy, making the device sustainable
and environmentally friendly.

Statement by the jury
The curved form and eye-catching green
colour accents give this clamping unit an
expressive appearance with high recognition value.

Die automatische Spanneinheit powRgrip
PGU 9500 ermöglicht das schnelle und
sichere Ein- und Ausspannen von Werkzeugen für die mechanische Bearbeitung
per Knopfdruck. Durch einen sehr präzisen
Rundlauf und eine Vibrationsdämpfung
steigert das Gerät die Werkzeugstandzeit.
Zudem erfolgt das Werkzeugspannen,
ohne dass dabei Hitze entsteht, wodurch
die Arbeitssicherheit erhöht und Energie
eingespart wird bei nachhaltiger Rücksicht
auf die Umwelt.

Begründung der Jury
Die geschwungene Form und auffällig
grüne Farbakzente verleihen der Spanneinheit ein expressives Erscheinungsbild mit
hohem Wiedererkennungswert.

Futorque X-1
Tablet Press
Tablettenpresse

Manufacturer
kg-pharma GmbH & Co. KG,
Hamburg, Germany
Design
CaderaDesign (Tom Cadera),
Würzburg, Germany
Web
www.kg-pharma.de
www.caderadesign.de
Honourable Mention

The Futorque X-1 is a rotary tablet press
for R & D and production use. The machine is powered by an advanced torque
motor. All entries are made via the 18.5"
multitouch display connected to the
industrial PC. The simple waterfall design
on the front in white polycarbonate was
combined with sides made from stainless
steel sheets. In addition, the surfaces are
particularly easy and quick to clean.

Statement by the jury
The slim form of the Futorque X-1 not
only looks elegant, but also offers high-
efficiency in a very small footprint.

Die Futorque X-1 ist eine Rundlauf-Tablet-
tenpresse für den wissenschaftlichen und
industriellen Gebrauch. Die Maschine wird
von einem fortschrittlichen Torquemotor
angetrieben. Alle Eingaben werden über
den mit dem Industrie-PC verbundenen
18,5"-Multitouch-Display realisiert. Das
schlichte Waterfall-Frontdesign in weißem
Polykarbonat wurde mit Seiten aus Edel-
stahlblechen kombiniert. Zudem sind die
Oberflächen besonders einfach und schnell
zu reinigen.

Begründung der Jury
Die schlanke Bauform der Futorque X-1
sieht nicht nur elegant aus, sondern bietet
auch eine hohe Effizienz auf sehr kleinem
Raum.

FE75
Tablet Press
Tablettenpresse

Manufacturer
Fette Compacting GmbH,
Schwarzenbek, Germany
Design
Dominic Schindler Creations GmbH,
Lauterach, Austria
Web
www.fette-compacting.com
www.dominicschindler.com

The FE75 tablet press requires very little space. With a footprint of only two square metres and a maximum output of more than 1.6 million tablets per hour it achieves very high productivity. With the help of four pairs of compression rollers even powders that are difficult to compress can be processed. The two-piece frame structure and the air suspension in the feet ensure optimal vibration absorption.

Statement by the jury
With it rounded lines the FE75 appears as if without outer edges. The bright corner panels give the tablet press its visual lightness.

Die Tablettenpresse FE75 benötigt nur wenig Platz. Mit einer Grundfläche von nur zwei Quadratmetern und einem maximalen Ausstoß von mehr als 1,6 Millionen Tabletten pro Stunde erreicht sie eine sehr hohe Produktivität. Mithilfe der vier Druckrollenpaare können auch schwer zu pressende Pulver verarbeitet werden. Die zweiteilige Rahmenstruktur sowie eine Luftfederung in den Standfüßen gewährleisten eine optimale Schwingungsdämpfung der Maschine.

Begründung der Jury
Die FE75 wirkt mit ihren abgerundeten Linien beinahe randlos. Die hellen Kanten verhelfen der Tablettenpresse zu visueller Leichtigkeit.

Roller and Belt Conveyor System
Rollen- und Bandförderer-System

Manufacturer
Avancon SA, Riazzino, Switzerland
In-house design
Design
D'ArteCon sagl (Dieter Specht), Arcegno, Switzerland
Web
www.avancon.com
www.dartecon.net

This roller and belt conveyor system is divided into separate zones, which can be operated independently. The decentralised control system ensures smooth operation when transporting goods and is particularly energy-saving. The system is controlled via a manufacturer-independent fieldbus system. All drives and control elements are integrated into the aluminium profile frame. The system can be assembled and installed without tools.

Das Rollen- und Bandförderer-System ist in verschiedene Zonen unterteilt, die unabhängig voneinander gesteuert werden können. Die dezentrale Steuerung stellt einen reibungslosen Betrieb beim Transport von Gütern sicher und ist besonders Energie sparend. Die Schalttechnik erfolgt über ein herstellerunabhängiges Feldbussystem. Alle Antriebe und Steuerungen sind in den Profilrahmen aus Aluminium verbaut. Die Montage und Installation erfolgt werkzeugfrei.

Demag V-type crane
Demag V-Profilkran
Bridge Crane
Brückenkran

Manufacturer
Terex MHPS GmbH, Düsseldorf, Germany
Terex Material Handling
In-house design
Web
www.demagcranes.com

In contrast to cranes that have enclosed box-section girders, the Demag V-type crane features an eco-friendly V-type design. Tapered diaphragm joints reduce the oscillation characteristics by up to 30 per cent and ensure that pressure and tensile forces are evenly distributed. The open V-shape of the profile-section girder cuts the crane's deadweight by an average of 17 per cent. In addition, less surface area is exposed to the wind and the design allows more light to pass, which makes the crane suitable for both indoor and outdoor operation.

Im Gegensatz zu vergleichbaren geschlossenen Kastenträgerkranen zeichnet sich der Demag V-Profilkran durch eine ressourcenschonende V-Bauweise aus. Verjüngte Membrangelenke reduzieren das Schwingungsverhalten um bis zu 30 Prozent und sorgen für eine gleichmäßige Verteilung der Druck- und Zugkräfte. Durch die offene V-Form der Profilträger wird das Eigengewicht des Krans um durchschnittlich 17 Prozent verringert. Zudem bietet die Konstruktion weniger Windangriffsfläche und mehr Lichtdurchlässigkeit, sodass der Kran sowohl für die Halle als auch für den Außenbereich geeignet ist.

MAN 175D
Marine Engine Series
Schiffsmotorenserie

Manufacturer
MAN Diesel & Turbo SE, Augsburg, Germany
In-house design
THE HighSpeed Team
Design
HYVE Innovation Design GmbH
(Andreas Beer, Grzegorz Dudkiewicz), Munich, Germany
Web
www.175d.man.eu
www.hyve-design.net

The MAN 175D marine engine series was developed for use in workboats and super yachts. The design of this large diesel engine follows strictly the functional application requirements of providing best-in-class power output and efficiency by fulfilling challenging weight and volume restrictions. The overall design is characterised by clarity thereby enabling fast commissioning of engine to vessel and providing easy access and efficient maintenance to all service and operational parts.

Die Motorenserie MAN 175D wurde für den Einsatz auf Arbeitsschiffen und Superjachten entwickelt. Das Design dieses Großdieselmotors folgt dabei den funktionalen Anforderungen der Anwenderapplikationen, um höchste Motorleistung und -effizienz unter Einhaltung von anspruchsvollsten Bauraum- und Gewichtsvorgaben bereitzustellen. Das Design des Motors als Gesamtsystem zeigt eine optische Klarheit, welche eine schnelle Inbetriebnahme und Verbindung vom Motor zum Schiff erlaubt und einen leichten Zugang und eine einfache Wartung aller Service- und Betriebskomponenten ermöglicht.

Statement by the jury
This marine engine series features a fascinating open design, giving it a technoid appearance and making individual functional areas visually experienceable.

Begründung der Jury
Die Schiffsmotorenserie fasziniert durch ihre offene Bauart, die ihr ein technoides Äußeres verleiht und einzelne Funktionsbereiche visuell erfahrbar macht.

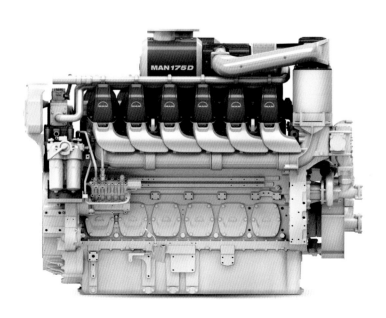

ESBF
Electric Cylinder
Elektrozylinder

Manufacturer
Festo AG & Co. KG, Esslingen, Germany
In-house design
Matthias Wunderling
Web
www.festo.com

The ESBF is a universal electric cylinder for factory automation. It is available in six different sizes and in a special variant for the food industry. The ESBF has a high work speed, a long service life and requires only minimal maintenance. It is designed especially for industrial applications with high feed forces such as materials handling in the wood and plastics processing industries.

Beim ESBF handelt es sich um einen universell einsetzbaren Elektrozylinder für die Fabrikautomation. Er ist verfügbar in sechs Baugrößen sowie in einer Spezialausführung für die Lebensmittelindustrie. Der ESBF arbeitet schnell, ist langlebig und wartungsarm. Er ist speziell für industrielle Anwendungen mit hohen Vorschubkräften konzipiert, z. B. für das Material-Handling in der holz- und kunststoffverarbeitenden Industrie.

Statement by the jury
The electric cylinder's practical design language conveys its functionality in an impressive way. The precise workmanship indicates its high quality standards.

Begründung der Jury
Die sachliche Formensprache des Elektrozylinders vermittelt eindrucksvoll seine Zweckmäßigkeit. Die präzise Verarbeitung zeugt von hoher Qualität.

VSNC
Process Valve
Prozessventil

Manufacturer
Festo AG & Co. KG, Esslingen, Germany
In-house design
Jörg Peschel
Web
www.festo.com

The field of process automation places other demands on components than the field of factory automation. Process valves are used in order to control these processes accurately. Applications for the VSNC are found primarily in water treatment as well as the production of chemicals and petrochemicals, where it is used to control the mixing or fermenting of gases, liquid media and bulk materials.

In der Prozessautomation gelten andere Anforderungen an die Komponenten als in der Fabrikautomation. Um diese Prozesse präzise steuern zu können, werden Prozessventile eingesetzt. Zum Einsatz kommt das VSNC hauptsächlich bei der Wasseraufbereitung sowie in der chemischen und petrochemischen Produktion. Dort übernimmt es das kontrollierte Mischen oder Fermentieren von Gasen, Flüssigkeiten oder Schüttgütern.

Statement by the jury
The process valve convinces with its clear structure and high-quality materials, which protect it from untimely wear and tear.

Begründung der Jury
Das Prozessventil überzeugt mit einer klar strukturierten Gliederung und einer hochwertigen Materialauswahl, die vor vorzeitiger Abnutzung schützt.

DDLI
Pneumatic Linear Drive
Pneumatischer Linearantrieb

Manufacturer
Festo AG & Co. KG, Esslingen, Germany
In-house design
Jörg Peschel
Web
www.festo.com

A wide variety of guides are utilised in manufacturing systems. The DDLI pneumatic linear drive adapts optimally to these guides. It is used in nearly every industry sector as a component for automation tasks. Depending on the mounting position, loads of up to 180 kg are positioned with precise accuracy. The special feature of the DDLI is its integrated displacement decoder, which performs absolute measurements in a contactless way.

In Fertigungsanlagen kommen unterschiedlichste Führungen zum Einsatz. Der pneumatische Linearantrieb DDLI passt sich diesen Führungen optimal an. Er findet Einsatz in nahezu allen Branchen als Komponente für Automatisierungsaufgaben. Je nach Einbaulage werden Massen bis zu 180 kg punktgenau positioniert. Das Besondere des DDLI ist das integrierte Wegmesssystem, welches berührungslos und absolut messend arbeitet.

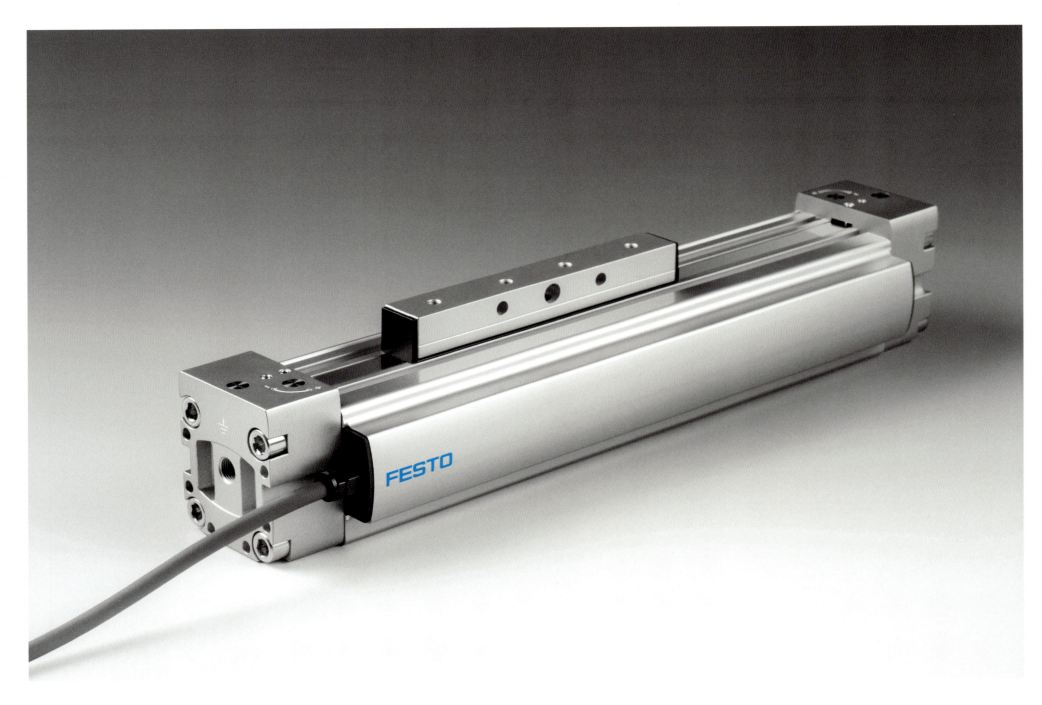

Tec2Screen®
Multimedia Learning Environment
Multimediale Lernumgebung

Manufacturer
Festo Didactic GmbH & Co. KG, Denkendorf, Germany
In-house design
Festo Product Design
Web
www.festo.ccm

Tec2Screen creates a logical connection between the virtual and the actual application, thus making intuitive learning possible. With the help of the multimedia learning companion, the learner is able to comprehend the effects of control commands on the system, as well as how various components interact with each other. The mobile learning companion consists of a tablet, the basic unit, the connections and the online courses.

Tec2Screen schafft die logische Verbindung von virtueller und realer Anwendung und ermöglicht so ein intuitives Lernen. Durch diesen multimedialen Lernbegleiter kann der Lernende nachvollziehen, was Kontroll- und Steuerungsbefehle in einer Anlage bewirken und wie verschiedene Komponenten miteinander interagieren. Der mobile Lernbegleiter besteht aus einem Tablet, der Basiseinheit, den Connects und den Online-Kursen.

Statement by the jury
What makes the Tec2Screen special is how the various components are visually set off from each other while at the same time supplementing each other excellently.

Begründung der Jury
Besonders am Tec2Screen ist, wie die verschiedenen Komponenten visuell voneinander abgesetzt sind und sich dabei gleichzeitig hervorragend ergänzen.

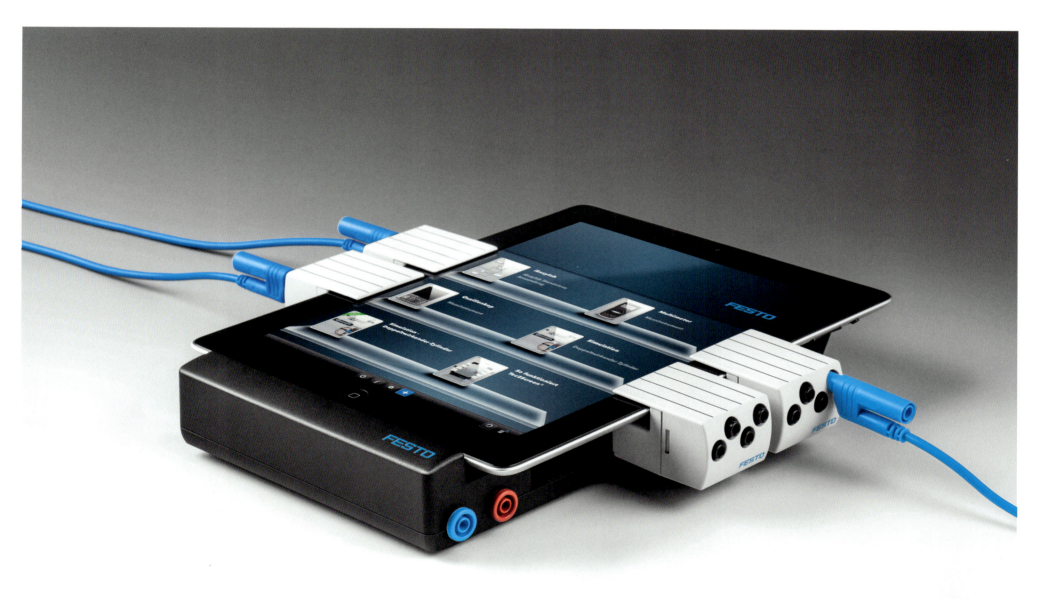

GARANT High-Pressure Centric Clamp
GARANT Hochdruck-Zentrischspanner

Manufacturer
Hoffmann GmbH Qualitätswerkzeuge, Munich, Germany
Design
Böhler GmbH, Corporate Industrial Design
(Julian Rathmann, Katja Lautenbach), Fürth, Germany
Web
www.hoffmann-group.com
www.boehler-design.de

The Garant High-Pressure Centric Clamp is a flexible, modular clamping system. The overall system is extremely stiff and has a high repetitive accuracy. Due to a large range of modular chuck jaws and the possibility to position it universally at its base, the clamping system is very flexible and multifunctional. The clamping force is adjusted via a planetary transmission with slipping clutch. Therefore, only the desired clamping force affects the workpiece.

Beim Garant Hochdruck-Zentrischspanner handelt es sich um ein flexibles modulares Spannsystem. Das Gesamtsystem besitzt eine extreme Steifigkeit sowie Wiederholgenauigkeit. Ein modulares Spannbackensortiment und die universelle Positionierung auf der Grundschiene machen das Spannsystem variabel und multifunktional. Die Krafteinstellung erfolgt über ein Planetengetriebe mit Rutschkupplung, wodurch nur die gewünschte Spannkraft auf das Werkstück aufgebracht wird.

EVOS Ci

Multifunctional Valve for Industrial Gases
Multifunktionsventil für Industriegase

Manufacturer
Linde AG, Pullach, Germany
Design
Indeed Innovation GmbH (Heiko Tullney, Stefan Freitag), Hamburg, Germany
Web
www.linde-gas.com
www.indeed-innovation.com

The Evos Ci is a multifunctional valve for industrial gases. The lever mechanism with automatic locking system allows for simple and safe opening and closing within a fraction of a second. The respective operating status is even clearly discernible from afar, while the pressure gauge informs the user at a glance about the remaining pressure in the gas cylinder. Moreover, the durable metal guard provides maximum security and facilitates handling with a variety of grip options.

Das Evos Ci ist ein Multifunktionsventil für Industriegase. Die Hebelmechanik mit automatischem Verriegelungssystem ermöglicht ein einfaches und sicheres Öffnen und Schließen im Bruchteil einer Sekunde. Der jeweilige Betriebszustand ist schon von weitem eindeutig erkennbar. Auch das Manometer zeigt dem Nutzer auf einen Blick den verbleibenden Druck im Gaszylinder. Zudem bietet der stabile Metallschutzbügel maximale Sicherheit und erleichtert die Handhabung durch vielfältige Griffmöglichkeiten.

Statement by the jury
The design of this multifunctional valve reflects a successful combination of self-explanatory operability, ergonomics and functionality.

Begründung der Jury
Die Gestaltung des Multifunktionsventils stellt eine gelungene Kombination aus selbsterklärender Bedienbarkeit, Ergonomie und Funktionalität dar.

X-TRUCK
Electric Pallet Truck
Elektrohubwagen

Manufacturer
Ningbo Ruyi Joint Stock Co., Ltd.,
Ninghai, China
Design
Ningbo ICO Creative Design Co., Ltd.
(Yu Liao, Yingbo Sun), Ningbo, China
Web
www.xilin.com
www.ico-d.com

The X-Truck electric pallet truck is char-acterised by an extremely small turning circle. The handlebar can be positioned upright in order to save additional space. The pallet truck can thus be easily manoeuvred in very tight spaces and is suited for use in containers or narrow corridors. Moreover, a guide wheel at the tip of the fork enables easy loading and unloading of pallets.

Statement by the jury
The X-Truck combines safe use with a friendly appearance, which is the result of the rounded shapes and the accents in deep yellow.

Der elektrische Hubwagen X-Truck zeichnet sich durch einen extrem kleinen Wende-kreis aus. Die Lenkstange ist senkrecht aufstellbar, um zusätzlichen Platz einzu-sparen. Der Hubwagen lässt sich somit auf sehr engem Raum manövrieren und ist beispielsweise für den Einsatz in Contai-nern oder schmalen Gängen geeignet. Des Weiteren erlaubt ein Führungsrad an der Gabelspitze das reibungslose Be- und Entladen von Paletten.

Begründung der Jury
Der X-Truck verbindet ein einfaches Hand-ling mit einem freundlichen Charakter, der sich aus den gerundeten Formen und den Akzenten in sattem Gelb ergibt.

Golden Eagle
Telecom Power Supply
Netzgerät für
Telekommunikation

Manufacturer
Delta, Taoyuan, Taiwan
In-house design
Web
www.deltaww.com

The Golden Eagle is a 600-watt power supply that requires very little space. Thanks to its innovative technology and natural cooling design, the unit achieves an energy conversion efficiency of 96 per cent. This makes it possible for telecom providers to lower their oper-ational expenditures considerably. Due to its wide input voltage range and high IP65 protection rating, the device is suit-able for safe outdoor use.

Statement by the jury
This power supply unit captivates with its self-contained design. Its exception-ally small and flat form allows it to fit almost anywhere.

Der Golden Eagle ist ein 600-Watt-Netzgerät, das wenig Platz benötigt und dank innovativer Technologie und einem natürlichen Kühlungsdesign eine Ener-gieumwandlungseffizienz von 96 Prozent erreicht. Dadurch können Netzbetreiber der Telekommunikationsindustrie ihre Betriebskosten deutlich senken. Aufgrund des weiten Eingangsspannungsbereichs und der hohen Schutzart IP65 ist das Gerät für den sicheren Betrieb im Außen-bereich geeignet.

Begründung der Jury
Das Netzgerät besticht durch seine in sich geschlossene Bauweise. Mit seiner au-ßergewöhnlich kleinen und flachen Form findet es nahezu überall Platz.

TAPCON®
Voltage Regulator
Spannungsregler

Manufacturer
Maschinenfabrik Reinhausen GmbH,
Regensburg, Germany
Design
Design Tech, Ammerbuch, Germany
Web
www.reinhausen.com
www.designtech.eu

The Tapcon voltage regulator is used for voltage regulation of transformers to ensure a stable power grid. The user interface on the interaction field, which is designed in a contrasting colour, forms the centrepiece of the design. The track-ing wheel provides easy communication between user and system. The clear user interface system at the back of the device makes for easy installation.

Statement by the jury
The design of the voltage regulator has been deliberately kept simple and under-standable, thus meeting high quality standards of intuitive use.

Der Spannungsregler Tapcon dient zur Spannungsregelung von Transformatoren, um ein stabiles Stromnetz zu gewähr-leisten. Die Bedienoberfläche auf dem farblich abgesetzten Interaktionsfeld bildet den gestalterischen Mittelpunkt. Für eine vereinfachte Kommunikation zwischen Anwender und System sorgt das Drehrad. Das übersichtlich gestaltete Schnittstel-lensystem auf der Rückseite des Gerätes erleichtert die Installation.

Begründung der Jury
Die Gestaltung des Spannungsreglers ist bewusst einfach und nachvollziehbar gehalten, wodurch sie hohe Ansprüche an eine intuitive Bedienung erfüllt.

PY X-MA-10
Flashing Light Sounder
Blitzschallgeber

Manufacturer
Pfannenberg Europe GmbH,
Hamburg, Germany
In-house design
Philipp Müller
Web
www.pfannenberg.com

The PY X-MA-10 flashing light sounder is designed for simultaneous or separate visual and acoustic alerting in the areas of machinery construction and building safety. The Xenon technology provides a large signalling area. The design tapers to the top, thus allowing the flashing light sounder to nestle closely against any surface contour. The integrated sound outlets are inspired by the eyes of Japanese cartoon figures.

Statement by the jury
The flashing light sounder impresses with its unusual pyramid form, which is elegant as well as functional.

Der Blitzschallgeber PY X-MA-10 dient zur gleichzeitigen oder separaten optischen und akustischen Alarmierung, z. B. in den Bereichen Maschinenbau und Gebäudesicherheit. Die Xenon-Technologie sorgt für einen großen Signalisierungsbereich. Durch das sich nach außen bzw. oben verjüngende Design schmiegt sich der Alarmgeber an jede Oberflächenkontur an. Die integrierten Schallaustrittsöffnungen sind den Augen von japanischen Comic-Figuren nachempfunden.

Begründung der Jury
Der Blitzschallgeber besticht durch seine ungewöhnliche Pyramidenform, die sowohl formschön als auch praktisch ist.

Nicoll acoustic system
Waste Water Evacuation System
Abwasserrohrleitung

Manufacturer
Nicoll, Cholet, France
Design
Graphic Identité & Co (Serge Hilbey),
Nantes, France
Web
www.nicoll.fr
www.graphic-identite.com
Honourable Mention

The Nicoll acoustic system is a plumbing waste water evacuation system that prevents noise pollution caused by fluids circulating through pipes. Softly curved radii, special acoustic membranes and sound absorbing accessories make it possible to highly reduce vibrations without having to use additional insulation. The pipe is made of recyclable material. Thanks to its modular design it can be adapted on site.

Statement by the jury
The waste water evacuation system's well-engineered insulation concept not only contributes to a high quality of living, it also has a particularly environmentally conscious design.

Das Nicoll acoustic system ist eine Abwasserrohrleitung, die die Lärmbelästigung durch zirkulierende Flüssigkeiten in den Rohren eindämmt. Sanft gebogene Radien, spezielle Akustikmembranen und schallabsorbierendes Zubehör sorgen dafür, dass Vibrationen stark gemindert werden, ohne dass dafür zusätzliche Dämmstoffe aufgewendet werden müssten. Die Leitung besteht aus recycelbaren Materialien. Sie kann durch ihre modulare Bauweise vor Ort angepasst werden.

Begründung der Jury
Das ausgereifte Dämmkonzept der Abwasserrohrleitung trägt nicht nur zu einer hohen Wohnqualität bei, sondern ist auch besonders umweltbewusst gestaltet.

eControl+
Control Unit for Sewer Cleaning Machine
Steuereinheit für die Bedienung von Kanalreinigungsmaschinen

Manufacturer
Rioned, Tilburg, Netherlands
Design
GBO, Helmond, Netherlands
Web
www.rioned.com
www.gbo.eu

The eControl+ control unit has been developed especially for the use of professional sewer cleaning machines. With its modular interface and practical rotary button, operating machines with diverse functionalities is easy. The anti-glare LCD display ensures good readability also in bright daylight and in combination with the LED corona provides additional status information about the machine.

Statement by the jury
The control unit has an exceptionally clear structure. The LED corona gives it a visually convincing frame.

Die Steuereinheit eControl+ wurde eigens für die Bedienung professioneller Kanalreinigungsmaschinen entwickelt. Mit der modularen Benutzeroberfläche und dem praktischen Drehknopf kann sie problemlos für die Bedienung ganz verschiedener Maschinentypen eingesetzt werden. Das blendfreie LCD-Display lässt sich auch im hellen Tageslicht gut ablesen und liefert in Kombination mit dem LED-Kranz zusätzliche Statusinformationen über die Maschine.

Begründung der Jury
Die Steuereinheit ist ausgesprochen klar und übersichtlich strukturiert. Der LED-Kranz verleiht ihr einen optisch gelungenen Rahmen.

Series 880/881
Serien 880/881
Institutional Castor Series
Apparaterollen-Serien

Manufacturer
STEINCO Paul vom Stein GmbH,
Wermelskirchen, Germany
In-house design
Thorsten Rödel, Tobias Weichbrodt,
Guido Arndt
Design
F. G. Schulz Industrial Design,
Remscheid, Germany
Web
www.steinco.de

The modular design of series 880/881 castors allows to supply wheel and brake separately to provide fast and safe assembly of castor to unit. Brake pedal and wheel can be attached tool-free. The castor series feature an interlocking gear brake instead of a conventional friction brake. It is reliable even on dirty or oily floors and protects the tyre. Dents on tyre surface after longer periods of brake actuation are prevented entirely.

Statement by the jury
In this institutional castors, technical expertise has been translated into a sophisticated design concept, which ensures high safety during use.

Die Apparaterollen der Serien 880/881 sind modular aufgebaut. Rad- und Feststell-baugruppe lassen sich von Hand montieren bzw. demontieren. Zur sicheren und schnellen Befestigung der Rolle am Gerät wird das Rad erst nachträglich eingefügt. Die Feststellvorrichtung wirkt formschlüssig auf Rad und Schwenklager und ist in ihrer Wirkung unabhängig von äußeren Einflüssen. Druckstellen der Reifenoberfläche sind auch nach längerem Betätigen des Feststellers ausgeschlossen.

Begründung der Jury
Bei den Apparaterollen wurde technisches Know-how in ein ausgereiftes Designkonzept überführt, das hohe Sicherheit in der Benutzung gewährleistet.

ETH
Tubular Handle
Rohrgriff

Manufacturer
ELESA SpA, Monza, Italy
In-house design
Web
www.elesa.com

The ETH tubular handle combines ergonomics and design. The smooth shape guarantees an excellent level of safety. The special assembly system prevents the rotation of the tube during handling. The tube is fashioned from anodised aluminium and is available in natural or grey-black colouring. The matt grey-black handle shanks are made of fibre-glass-reinforced technopolymer. The side caps in technopolymer with glossy finish are available in six colours.

Statement by the jury
The tubular handle meets high ergonomic standards. Different colour combinations provide a wide range of design options.

Der Rohrgriff ETH verbindet Ergonomie mit Design. Angenehme Formen gewährleisten hohe Sicherheit. Ein Montagesystem verhindert ein Rotieren des Griffrohres während der Handhabung. Das Rohr besteht aus eloxiertem Aluminium in Naturfarbe oder Grau-Schwarz. Die Endstücke sind aus glasfaserverstärktem Thermoplast, grauschwarz und matt. Die Seitenkappen aus Thermoplast mit glänzender Oberfläche sind in verschiedenen Farbvarianten erhältlich.

Begründung der Jury
Der Rohrgriff erfüllt hohe Ansprüche an die Ergonomie. Verschiedene Farbkombinationen eröffnen einen weiten Gestaltungsspielraum.

ENYCASE®
Cable Junction Box
Kabelabzweigkasten

Manufacturer
Gustav Hensel GmbH & Co. KG,
Lennestadt, Germany
In-house design
Web
www.hensel-electric.de

The Enycase cable junction box offers a wide range of mounting options and multistage membranes for cable glands. It features a variety of new functions, such as variable clamping positions, two clamping points per pole, safe and simple closing options, as well as the option to feed in cables through the rear of the enclosure. The robust housing made of plastic is available in eight different sizes in the colours grey and black.

Statement by the jury
Its flattened edges give this cable junction box an aesthetically pleasing profile, which also provides a clear view of the connections.

Der Kabelabzweigkasten Enycase bietet vielfältige Befestigungsmöglichkeiten und mehrstufige Membranen für Kabelverschraubungen. Er ist mit zahlreichen Funktionen ausgestattet, z. B. mit variablen Klemmenpositionen, zwei Klemmstellen je Pol, sicheren und einfachen Verschlussoptionen sowie der Möglichkeit, Kabel durch den Gehäuseboden einzuführen. Das robuste Gehäuse aus Kunststoff steht in acht Größen in den Farben Grau und Schwarz zur Auswahl.

Begründung der Jury
Durch die abgeflachten Kanten erhält der Kabelabzweigkasten ein ästhetisch ansprechendes Profil, das zudem eine gute Sicht auf die Anschlüsse gewährt.

Mailing AIRPCS
Inflatable Mailing Package
Aufblasbares Briefpaket

Manufacturer
Airbag Packing Co., Ltd., New Taipei, Taiwan
In-house design
Mike Liao, Oscar (Yaw-Shin) Liao
Web
www.airpcs.com

The inflatable Mailing Airpcs package consists of air cushions that work independently of each other. If one air tube is damaged, it will not affect the overall protection since the other tubes remain intact. The parcel is inflated by mouth or pump and can simply be torn open to release the air, so that its original, flat form is restored. The material used is biodegradable.

Statement by the jury
The clever design of the Mailing Airpcs combines a practical idea with environmentally conscious implementation.

Das aufblasbare Briefpaket Mailing Airpcs besteht aus Luftpolstern, die unabhängig voneinander arbeiten. Wird eine Luftkammer beschädigt, bleibt der Inhalt des Pakets durch die anderen weiterhin geschützt. Die Kammern werden mit dem Mund oder einer Pumpe aufgeblasen. Zum Öffnen lässt sich das Paket einfach aufreißen und die Luft entweicht, sodass der ursprüngliche, flache Zustand wieder hergestellt ist. Das Material ist biologisch abbaubar.

Begründung der Jury
Die pfiffige Gestaltung des Mailing Airpcs verbindet eine praktische Idee mit einer umweltbewussten Umsetzung.

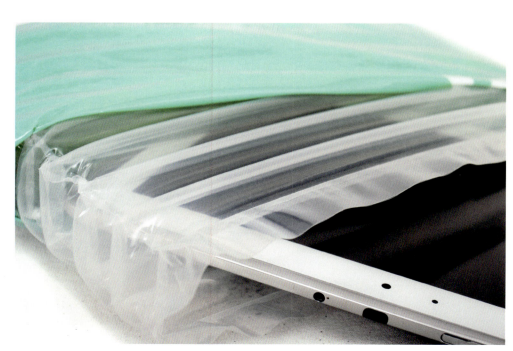

Care-O-bot 4
Service Robot
Serviceroboter

Manufacturer
Fraunhofer IPA,
Stuttgart, Germany
Schunk GmbH + Co. KG,
Lauffen a.N., Germany

Design
Phoenix Design GmbH + Co. KG,
Stuttgart, Germany

Web
www.ipa.fraunhofer.de
www.schunk.com
www.phoenixdesign.com

reddot award 2015
best of the best

Service with care
Robots in the service segment fulfil different tasks than those in other industries. As assistants to human beings, service robots have to reproduce human actions in order for us to accept them. Care-O-bot 4 embodies an exciting new archetype of such a service robot, since it has a particular human appeal. It moves smoothly and elegantly through rooms. Featuring elaborate gestures and facial expressions, it evokes the emotions of everybody involved. In addition, this robot can navigate autonomously, automatically detect obstacles and pick up requested objects. A highly fascinating feature for users is the interaction with the Care-O-bot 4, which happens either via speech recognition or with a clearly designed, almost playfully appealing interface on the robot's head. A modular design concept allows for a customised set-up for different tasks. This makes the robot versatile in use, opening up new areas of application. In an impressive symbiosis of research, engineering and design, this service robot combines an intelligent interface with a design quality that is also appealing to the touch – interaction with it becomes a cognitive-sensual experience, evoking positive emotions.

Service mit Gefühl
Roboter für den Servicebereich erfüllen andere Aufgaben als solche, die in der Industrie eingesetzt werden. Als Dienstleister des Menschen müssen Serviceroboter humane Handlungsabläufe abbilden, denn nur so können wir sie akzeptieren. Der Care-O-bot 4 verkörpert einen spannenden neuen Archetyp eines derartigen Serviceroboters, da er auf besondere Weise menschlich anmutet. Fließend und elegant bewegt er sich durch den Raum. Weil er über Eigenschaften wie eine ausgefeilte Mimik und Gestik verfügt, weckt er dabei die Emotionen aller Beteiligten. Dieser Roboter kann zudem autonom navigieren, Hindernisse selbst erkennen und auch gesuchte Objekte selbständig ergreifen. Für den Nutzer ausgesprochen fesselnd ist die Art der Interaktion mit dem Care-O-bot 4. Sie geschieht mittels Spracherkennung oder über ein klar gestaltetes Interface auf seinem Kopf, wobei dies nahezu spielerisch erscheint. Das Gestaltungskonzept ist modular ausgerichtet und ermöglicht einen individuellen Aufbau für unterschiedliche Aufgaben. Dieser Roboter ist deshalb vielseitig einsetzbar und es erschließen sich neue Bereiche der Anwendung. In einer beeindruckenden Symbiose aus Forschung, Engineering und Design entstand so ein Serviceroboter, dessen intelligentes Interface sich mit einer auch haptisch ansprechenden Designqualität verbindet – die Interaktion mit ihm wird zu einem kognitiv-sinnlichen Erlebnis, das positive Gefühle weckt.

Statement by the jury
The Care-O-bot 4 is the expression of a new understanding of a robot's ability to interact with its environment. Equipped with an intelligent interface, it possesses human-like qualities that exceed imitation and lend it originality and character. Its form language as well as its gestures and facial expressions rouse emotion. Designed with gently rounded edges and soft proportions, this robot fascinates with subservient kindness.

Begründung der Jury
Der Care-O-bot 4 ist Ausdruck eines neuen Verständnisses der Interaktionsfähigkeit eines Roboters mit der Umwelt. Ausgestattet mit einem intelligenten Interface, besitzt er menschenähnliche Qualitäten, die über das Imitieren hinausgehen und ihm Eigenständigkeit verleihen. Seine Formensprache wie auch seine Gestik und Mimik emotionalisieren alle Beteiligten. Gestaltet mit sanft gerundeten Ecken und weichen Proportionen, fasziniert dieser Roboter mit seiner dienstbaren Freundlichkeit.

Designer portrait
See page 50
Siehe Seite 50

KR 6 Agilus
Small Robot
Kleinroboter

Manufacturer
KUKA AG, Augsburg, Germany
In-house design
Annette Steinacker
Design
Selic Industriedesign (Mario Selic), Augsburg, Germany
Web
www.kuka.com
www.selic.de
Honourable Mention

The KR 6 Agilus small robot is suited for safe use in direct cooperation with humans and can be operated without any protection fences, for example in the medical industry. The robot is designed for particularly high operating speeds. Since its entire energy supply including pneumatic tubes is integrated in the robot, it has very slim contours. The consistently radiused edges help avoid injuries.

Der Kleinroboter KR 6 Agilus ist für den sicheren Einsatz in direkter Zusammenarbeit mit Menschen geeignet und kann ohne Schutzzaun betrieben werden, z. B. in der Medizintechnik. Der Roboter ist auf besonders hohe Arbeitsgeschwindigkeiten ausgelegt. Da sich die gesamte Energieversorgung inklusive Druckluftleitungen im Inneren des Roboters befindet, sind seine Konturen sehr schlank. Die durchgängig abgerundeten Kanten vermeiden Verletzungsgefahren.

Statement by the jury
Its balanced proportions give this small robot an athletic look, visualising its manoeuvrability and speed.

Begründung der Jury
Die ausgewogenen Proportionen verleihen dem Kleinroboter eine athletische Anmutung, die seine Wendigkeit und Schnelligkeit visualisiert.

ecoZ
Robot System
Roboter-System

Manufacturer
Handling Tech, Automations-Systeme GmbH,
Steinenbronn, Germany
Design
Design Tech, Ammerbuch, Germany
Web
www.handlingtech.de
www.designtech.eu

The ecoZ robot system takes care of automated loading and unloading of machine tools. Thanks to its small footprint and compact design it can be flexibly docked with the machine tools in a space-saving way. The six axle robot's oblique installation increases its range of operation. The drawers, from which it takes the workpieces, can be continuously reloaded from the other side during operation.

Statement by the jury
The design of the robot system displays an extraordinary feel for ergonomics, mobility and flexibility.

Das Roboter-System ecoZ sorgt für die automatisierte Be- und Entladung von Werkzeugmaschinen. Durch die kleine Stellfläche und die kompakte Bauweise lässt es sich sehr flexibel und raumsparend an die Werkzeugmaschinen andocken. Der geneigt montierte 6-Achs-Roboter bedient einen größeren Aktionskreis; die Schubladen, aus denen er die Werkstücke entnimmt, können im laufenden Betrieb von der anderen Seite kontinuierlich bestückt werden.

Begründung der Jury
Bei der Bauweise des Roboter-Systems wurde ein außergewöhnliches Gespür für Ergonomie, Mobilität und Flexibilität unter Beweis gestellt.

Six-rotor UAV
Unmanned Aerial Vehicle
Drohne

Manufacturer
Beijing Zhongfei Aiwei Aviation
Technology Co., Ltd., Beijing, China
Design
LKK Design Beijing Co., Ltd., Beijing, China
Web
www.lkkdesign.com
Honourable Mention

The unmanned aerial vehicle has its centre of gravity at the bottom of the platform featuring the six rotors, thus giving it more stable flight characteristics in wind. In addition, the centre of the rotor has a 5-degree inward tilt to improve the steering speed. The gear box has been integrated in the camera head, so that the landing gear rotates with the camera. This allows taking aerial shots without the landing gear being visible.

Statement by the jury
The unmanned aerial vehicle translates innovative technical ideas into a convincing design. The aerodynamic form is highlighted by the black and white contrast.

Die Drohne hat ihren Schwerpunkt auf der Unterseite der Plattform mit den sechs Rotoren, wodurch sie stabiler im Wind fliegt. Außerdem ist das Rotorzentrum um fünf Grad nach Innen geneigt, um die Lenkgeschwindigkeit zu verbessern. Das Getriebe ist im Kamerakopf integriert, sodass das Fahrwerk mitrotiert, wenn die Kamera rotiert. Dadurch lassen sich Flugaufnahmen machen, ohne dass das Fahrwerk mit im Bild zu sehen ist.

Begründung der Jury
Die Drohne überführt innovative technische Ideen in ein schlüssiges Design. Die aerodynamische Form wird durch die Schwarz-Weiß-Kontraste betont.

Crossbar Robot 4.0
Crossbar Roboter 4.0
Industrial Robot for Press to Press Transfer
Industrieroboter für die Verkettung von Pressen

Manufacturer
Schuler Pressen GmbH, Göppingen, Germany
Design
designship (Piero Horn, Thomas Starczewski), Ulm, Germany
Web
www.schulergroup.com
www.designship.de

The Crossbar Robot 4.0 is used for automated mechanical, servo-mechanical and hydraulic presses. The white plastic covering and the substructure form a functional unit and prevent any interfering contours which could lead to a collision in the limited space between the press punches. The easy-to-clean surfaces protect the electrical components. LED light strips communicate the functions of the vacuum pumps.

Der Crossbar Roboter 4.0 wird zur Automation mechanischer, servo-mechanischer und hydraulischer Pressenlinien eingesetzt. Die weiße Kunststoffverkleidung bildet mit der Unterkonstruktion eine Einheit und vermeidet zudem Störkonturen, die zu einer Kollision in dem begrenzten Raum zwischen den Pressestempeln führen könnten. Die gut zu reinigenden Oberflächen schützen die elektrischen Bauteile. LED-Leuchtstreifen kommunizieren die Funktion der Vakuumpumpen.

Statement by the jury
The smooth surfaces of the industrial robot appear immaculate and fluid. The bright white communicates clear workflows.

Begründung der Jury
Die glatten Oberflächen des Industrieroboters erscheinen makellos und fließend. Das strahlende Weiß kommuniziert sauber ausgeführte Arbeitsabläufe.

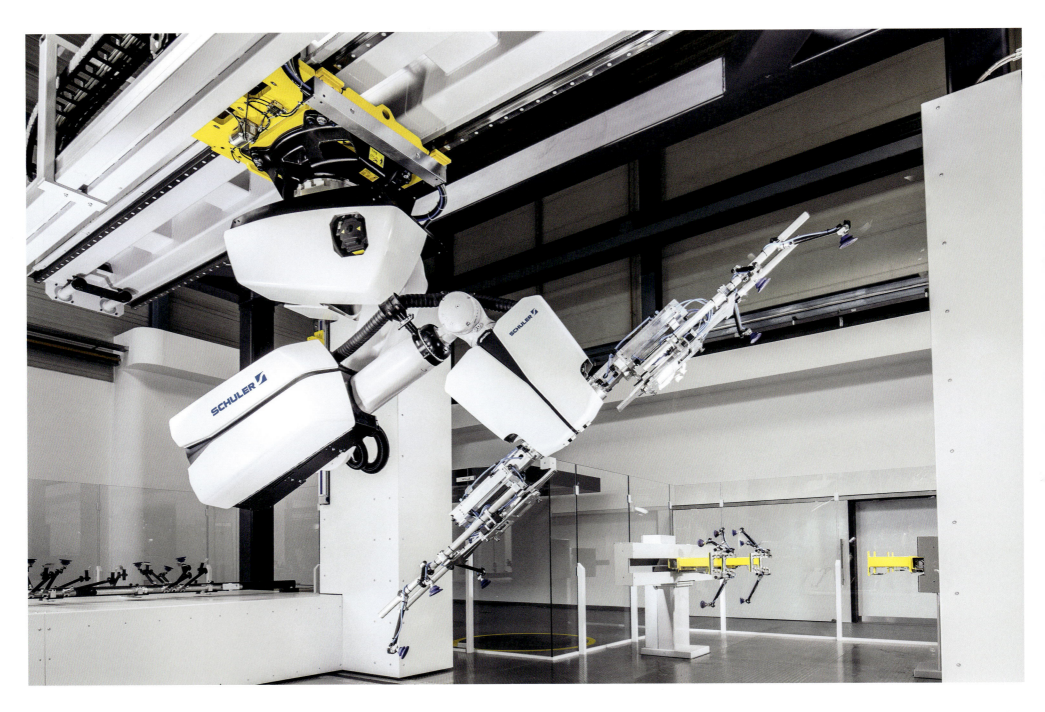

blayn Register
blayn Registrierkasse

Manufacturer
blayn Co., Ltd., Tokyo, Japan

In-house design
Shinichi Temmo

Web
www.blayn.co.jp

reddot award 2015
best of the best

Contemporary interpretation

Managing customer data and purchase orders is a highly complex process that has to be performed quickly and effectively. The blayn register is above all targeted at small restaurants and, as such, represents a contemporary tool that offers an optimisation of operations to all people working with it. Following an innovative approach, this register merges the three elements tablet, price display and receipt printer into one functional unit. The patented concept of an all-in-one register through the use of a tablet is highly convincing and perfectly adapted to the often daily hectic work environment of small businesses. The unit is compact and features a design of clear lines so that it blends in easily with any surroundings. The easy-to-understand user interface is intuitively structured. This makes customer management and operations an easy task, as all information is clearly arranged and presented. Since this register was conceived to be durable even under continuous use, it is manufactured using high-quality materials, while all design details have been carefully implemented. Based on a consistent concept, the register thus delivers a contemporary solution especially for small restaurants – showcasing an elegant design vocabulary and well thought-out functionality, it turns all work processes into pleasure.

Zeitgemäß interpretiert

Die Verwaltung von Kundendaten oder das Erfassen von Bestellungen ist eine komplexe Angelegenheit, da sie schnell und effektiv vonstattengehen muss. Die blayn Registrierkasse richtet sich vor allem an die Zielgruppe kleiner Restaurants und gibt als solche allen Beteiligten ein zeitgemäßes Arbeitsinstrument an die Hand. Auf innovative Weise vereint diese Registrierkasse die Elemente Tablet, Preisanzeige und Kassenbondrucker in einer funktionalen Einheit. Dieses patentierte Prinzip einer tabletbasierten All-in-one-Registrierkasse ist überaus schlüssig und dem oftmals hektischen Alltag in kleinen Betrieben perfekt angepasst. Das Gerät ist kompakt und mit einer klaren Linienführung gestaltet, sodass es sich leicht in jede Umgebung einfügen kann. Die Nutzeroberfläche ist einfach und selbsterklärend strukturiert. Die Bedienung ist deshalb unkompliziert und alle Informationen werden gut nachvollziehbar und übersichtlich dargeboten. Da diese Registrierkasse für einen langlebigen Dauereinsatz ausgelegt wurde, ist sie aus hochwertigen Materialien gefertigt und die Details sind sorgfältig ausgearbeitet. Auf der Basis eines stimmigen Konzepts bietet sie damit eine sehr zeitgemäße Lösung insbesondere für kleine Restaurants – mit ihrer eleganten Formensprache und durchdachten Funktionalität optimiert und ästhetisiert sie dort die Arbeitsabläufe.

Statement by the jury

The design of the blayn register boasts an innovative and highly convincing approach of blending the elements tablet, price display and receipt printer into one unit. It integrates into the point of sale with unobtrusive elegance and appeal, while its high-quality manufacturing lends it robustness and durability. Since all information is represented in a clearly structured manner, it is easy and intuitive to use. This elegant register lends itself to versatile use in almost any environment.

Begründung der Jury

Auf innovative und sehr schlüssige Weise lässt die Gestaltung der blayn Registrierkasse die Elemente Tablet, Preisanzeige und Kassenbondrucker zu einer Einheit verschmelzen. Am Point of Sale zeigt sie eine zurückhaltende Anmutung und Eleganz, während eine qualitativ hochwertige Fertigung ihr Langlebigkeit verleiht. Da die Informationen klar strukturiert dargestellt werden, ist sie einfach und intuitiv zu bedienen. Diese elegante Registrierkasse ist überaus vielseitig einsetzbar.

Designer portrait
See page 52
Siehe Seite 52

FreshWay
Touchscreen Scale
Touchscreen-Waage

Manufacturer
METTLER TOLEDO,
Mettler-Toledo (Albstadt) GmbH,
Albstadt, Germany

Design
MEDUGORAC Industrial Design
(Kostas Medugorac),
Stuttgart, Germany

Web
www.mt.com
www.medugorac.com

reddot award 2015
best of the best

Functional variety

Sales situations in retailing are subject to constant change. Floor layouts change frequently and new collections need extra space. The FreshWay is a retail scale that translates these changes into a perfectly thought-out design concept. Its touchscreen, the customer display as well as the printer unit are all fixed to the scale's chassis as separate functional modules. Based on this innovative modular approach, its construction combines a highly flexible design with a high degree of investment protection. They can be adjusted according to the specific needs of the retail service counter by allowing it to be altered into an array of different configurations. As the sales situation changes over time, the unit can be changed accordingly. Thanks to the focus on high functionality, the touchscreen scale showcases a purist design of clear lines and a slim-line chassis that is particularly fascinating. Taking into account issues of hygiene, the device features an easy-to-clean plain housing surface as well as a weighing plate that can be effortlessly removed, and does away with all unnecessary edges or ridges. The slim silhouette of the device, combined with its flexibly adjustable customer display, avoids visual barriers at the service counter and supports an open atmosphere during sales talks. A highly convincing modular interpretation of a register thus leads to a product with a functionality that embodies the lightness of good design.

Variable Funktionalität

Die Verkaufssituation im Einzelhandel ist einem stetigen Wandel unterworfen. Räume verändern sich oder neue Kollektionen brauchen mehr Platz. FreshWay ist eine Waage, die diese Veränderungen in ein perfekt durchdachtes Gestaltungskonzept überführt. Ihr Touch-Bedienfeld, der Kundenbildschirm wie auch die Druckereinheit sind als eigenständige modulare Funktionsmodule am Chassis der Waage angebracht. Auf der Grundlage dieses innovativen Modulkonzepts ermöglicht ihre Bauform eine situationsnahe Flexibilität und damit ein hohes Maß an Investitionsschutz. Sie kann sich dem jeweiligen Bedarf des Einzelhändlers jederzeit perfekt anpassen und lässt sich in einer Vielzahl von Konfigurationen zusammenstellen. Ändert sich die Verkaufssituation, kann auch die Bauform entsprechend angeglichen werden. Dank des gestalterischen Fokus auf hohe Funktionalität ist diese Touchscreen-Waage puristisch und mit klaren Linien entworfen, wobei insbesondere ihr schlankes Chassis begeistert. In Fragen der Hygiene gut durchdacht sind ihre glatten, leicht zu reinigenden Oberflächen wie auch die abnehmbare Wägeplatte und der Verzicht auf unnötige Ritzen und Kanten. Ihre schlanke Silhouette, kombiniert mit dem flexibel verstellbaren Kundenbildschirm, vermeidet Sichtbarrieren an der Bedientheke und unterstützt eine offene Atmosphäre im Verkaufsgespräch. Eine sehr gelungene modulare Interpretation der Registrierkasse führt so zu einem Produkt, dessen Funktionalität die Leichtigkeit guten Designs verkörpert.

Statement by the jury

The design of the FreshWay touchscreen scale is based on a well-conceived and convincing overall concept. Its touchscreen, the customer display and the printer unit are all fixed to the scale's chassis as separate functional modules, thus offering retailers a high degree of flexibility and ample leeway for individual configurations. The scale impresses not only with its clear and purist design vocabulary, but also with its well-structured, easy-to-operate user interface.

Begründung der Jury

Das Design der Touchscreen-Waage FreshWay basiert auf einem wohl durchdachten, stimmigen Gesamtkonzept. Ihr Touch-Bedienfeld, der Kundenbildschirm und die Druckereinheit sind als eigenständige modulare Funktionsmodule am Chassis der Waage angebracht, wodurch sie Einzelhändlern ein Höchstmaß an Flexibilität für eine individuelle Konfiguration bietet. Diese Waage beeindruckt mit ihrer klaren und puristischen Formensprache ebenso wie mit ihrem gut strukturierten, leicht bedienbaren User-Interface.

Designer portrait
See page 54
Siehe Seite 54

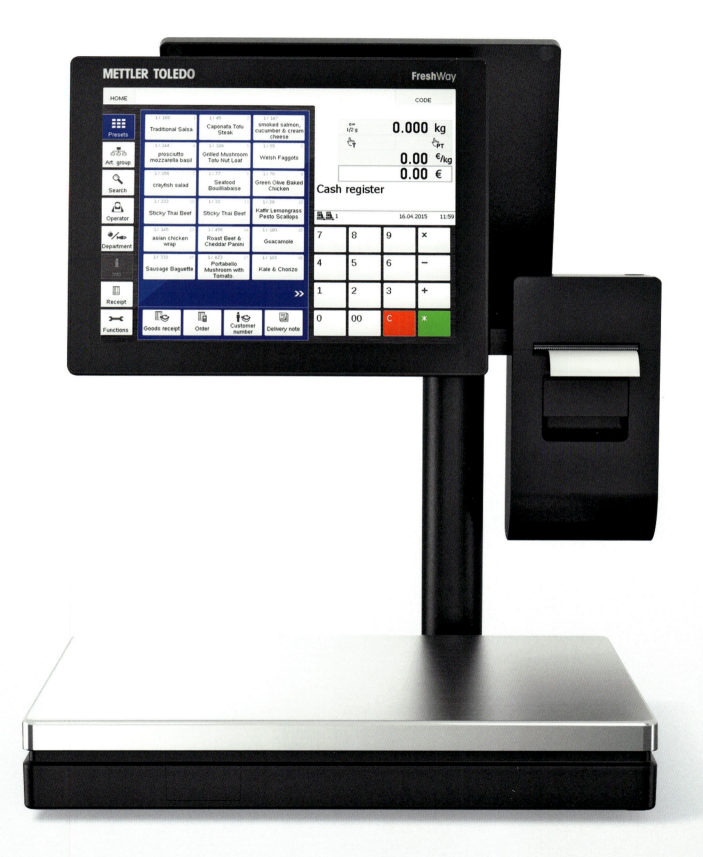

LifeXpress
Content Kiosk

Manufacturer
Medion AG, Essen, Germany
In-house design
Web
www.medion.com

The LifeXpress content kiosk is a customer terminal where vouchers or digital products such as games, software, e-books and prepaid cards can be purchased. The customer chooses the desired product via the interactive touchscreen and a corresponding receipt is printed out, which is then taken to the checkout. After payment has been concluded, the customer receives an active code to be redeemed online.

Statement by the jury
This content kiosk captivates with its particularly flat and elongated silhouette, which requires little space and blends in with any environment.

Der Content-Kiosk LifeXpress ist ein Kundenterminal, an dem Gutscheine oder digitale Produkte wie Spiele, Software, E-Books oder Prepaid-Karten erworben werden können. Der Käufer wählt sein gewünschtes Produkt über den interaktiven Touchscreen aus und bekommt einen entsprechenden Beleg ausgedruckt. Damit geht er zur Kasse und erhält nach der Bezahlung einen aktiven Code, den er online einlösen kann.

Begründung der Jury
Der Content-Kiosk besticht durch seine ausgesprochen flache und gestreckte Silhouette, die wenig Raum in Anspruch nimmt und sich in jede Umgebung einfügt.

MT-4008W
Mobile POS Terminal

Manufacturer
Posiflex Technology Inc.,
New Taipei, Taiwan
In-house design
Web
www.posiflex.com

The MT-4008W is a mobile POS terminal developed for retailers and hospitality clienteles, equipped with the functionality of a stationary POS in one hybrid system. The device consists of an 8" touchscreen tablet, which can be held using a detachable pistol grip or a hand strap, or else placed in a charging docking station. The pistol grip features a built-in barcode scanner as well as an integrated, removable battery, thus ensuring uninterrupted use.

Statement by the jury
This POS terminal meets exacting ergonomic and mobility requirements. Its classic design radiates professionalism and value.

Das MT-4008W ist ein mobiles POS-Terminal für den Einzelhandel und das Gastgewerbe. Es wurde mit den Funktionen eines stationären POS in einem Hybridsystem ausgestattet und besteht aus einem 8"-Touchscreen-Tablet. Es kann mit einem abnehmbaren Pistolengriff oder einer Handschlaufe getragen oder in einer Docking- und Ladestation installiert werden. Der Pistolengriff enthält einen Barcode-Scanner sowie einen herausnehmbaren Akku, der einen ununterbrochenen Betrieb garantiert.

Begründung der Jury
Das POS-Terminal erfüllt hohe Ansprüche an Ergonomie und Mobilität. Sein klassisches Design strahlt Professionalität und Wertigkeit aus.

Digital Signage X

Digitales Beschilderungssystem

Manufacturer
TPV Technology Group
Top Victory Electronics Co., Ltd., New Taipei, Taiwan
In-house design
Yao-Hsing Tsai, Jia-Sheng Wong, Yi-Ting Chung,
Li-Shu Fu, Wen-Chih Wang
Web
www.tpv-tech.com

This digital signage system for small, specialised shops and showrooms creates an interactive customer experience. The 27" display provides a reading situation similar to that of books or magazines and can switch between landscape and portrait formats. All cables are discreetly concealed in the frame. Furthermore, the system features a central pillar which can simultaneously support multiple displays ranging from two- and four-sided to circular arrangements.

Dieses Beschilderungssystem für kleinere Fachgeschäfte oder Ausstellungsflächen erzeugt ein interaktives Kundenerlebnis. Das Betrachten der Inhalte auf dem 27"-Display ähnelt dem Lesen von Büchern oder Zeitschriften. Das Display kann zwischen Hoch- und Querformat wechseln, die Kabel sind diskret im Rahmen verborgen. Darüber hinaus verfügt das System über eine zentrale Säule, an der mehrere Displays gleichzeitig montiert werden können – zweiseitig, vierseitig oder auch ringförmig angeordnet.

Statement by the jury
Digital Signage X captivates with its flat design and generous display, which allows viewers to become fully immersed in the media content.

Begründung der Jury
Digital Signage X besticht durch seine flache Bauform und das großzügig gestaltete Display, das es ermöglicht, in die Medieninhalte regelrecht einzutauchen.

MC18
Enterprise Mobile Computer
Mobiler Einkaufsassistent

Manufacturer
Zebra Technologies, New York, USA
In-house design
Edward Hackett, Mark Fountain
Design
Thinkable Studio, Oxfordshire, Great Britain
Web
www.zebra.com
www.thinkable.co.uk

The MC18 enterprise mobile computer
is a next generation personal shop-
per solution used within supermarkets,
designed to present information on
products and to simplify scanning and
checkout procedures. A large display,
simple touchscreen interface and one
control key provide an intuitive, smart-
phone-like experience. Through a clever
use of modularity and flexibility, the
charger accommodates a wide range of
room situations whilst also offering easy
serviceability.

Statement by the jury
With its minimalist appearance, the
MC18 corresponds to the current aesthetic zeitgeist. Its rounded contours
provide pleasant haptic qualities.

Der mobile Einkaufsassistent MC18 verkörpert die nächste Generation der
Personal-Shopping-Lösungen für den
Supermarkt. Er wurde entwickelt, um
Produktinformationen anzuzeigen sowie
den Scan- und Bezahlvorgang zu vereinfachen. Das große Display, die einfache
Touch-Benutzeroberfläche und eine
einzige Bedientaste bieten eine intuitive
Benutzererfahrung ähnlich wie bei einem
Smartphone. Durch seine Modularität
und Flexibiltät fügt sich das Ladegerät in
verschiedenste Raumsituationen ein und
besitzt eine hohe Servicefreundlichkeit.

Begründung der Jury
Der MC18 entspricht mit seinem minimalistischen Erscheinungsbild dem ästhetischen
Zeitgeist. Seine gerundeten Formen sorgen
für eine angenehme Haptik.

RA21
Industrial Computer
Industriecomputer

Manufacturer
EMBUX, New Taipei, Taiwan
In-house design
Lin-Yu Kao
Web
www.embux.com

RA21 is an industrial computer developed for use in difficult working conditions. Its microprocessor has a low power consumption, while the OLED display provides a quick view of key data. With the help of its screw-on connections a wide variety of sensors and wireless modules can be added, further increasing the range of possible applications.

Statement by the jury
The RA21 features a fascinating geometric form and metallic accents, which successfully underline its industrial character.

RA21 ist ein Industriecomputer, der für den Betrieb unter schwierige Arbeitsbedingungen entwickelt wurde. Sein Mikroprozessor erreicht einen geringen Energieverbrauch, während der OLED-Display eine schnelle Übersicht der wichtigsten Daten liefert. Mithilfe der anschraubbaren Verbindungen können eine Vielzahl von Sensoren und Funkmodulen hinzugefügt werden, die die Anwendungsmöglichkeiten zusätzlich erweitern.

Begründung der Jury
Der RA21 fasziniert durch seine geometrische Form und metallische Akzente, die seinen industriellen Charakter gekonnt unterstreichen.

PA31
Industrial Computer
Industriecomputer

Manufacturer
EMBUX, New Taipei, Taiwan
In-house design
Lin-Yu Kao
Web
www.embux.com
Honourable Mention

The PA31 industrial computer can be used like a tablet PC. It can be used on a flat surface or mounted to the wall. The device has been developed especially for the use with the EiS software platform and serves as a control centre for a wide variety of technical processes. The computer is equipped with a 10" screen and a much smaller OLED display on the side for displaying data.

Statement by the jury
This industrial computer impresses with its highly flat form. The minimalist design highlights its high functionality.

Der Industriecomputer PA31 lässt sich wie ein Tablet-PC verwenden. Er kann auf einem flachen Untergrund genutzt oder an die Wand montiert werden. Das Gerät wurde speziell für die Verwendung mit der Software-Plattform EiS konzipiert und dient als Steuerungszentrale für vielfältige technische Prozesse. Der Computer ist mit einem 10"-Bildschirm sowie mit einem sehr viel kleineren OLED-Bildschirm an der Seite für die Darstellung von Daten ausgestattet.

Begründung der Jury
Der Industriecomputer besticht durch eine sehr flache Bauweise. Die reduzierte Gestaltung betont seine hohe Funktionalität.

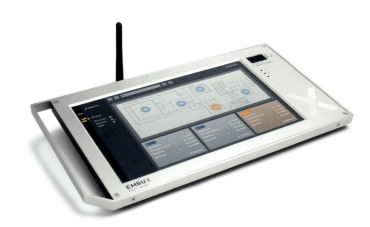

MMT8017 & MMT8024
Industrial PC

Manufacturer
ads-tec GmbH, Nürtingen, Germany
In-house design
Jochen Füllemann, Matthias Bohner
Web
www.ads-tec.de
Honourable Mention

MMT8017 and MMT8024 are industrial PCs for use in the pharmaceutical, food and beverage industries. The housing features no screws on the outside. Its smooth contours make it easy to clean. All seams have been minimised and fully sealed. The patented technology for the housing, a stainless steel aluminium composite, ensures the highest chemical resistance on the outside and perfect heat dissipation for cooling the PC components.

Statement by the jury
The high-quality design of the industrial PC is consistently geared to the hygiene standards of a professional environment.

MMT8017 und MMT8024 sind Industrie-PCs für die Pharma- und Lebensmittelindustrie. Das Gehäuse ist außen schraubenfrei gehalten. Glatte Konturen ermöglichen die einfache Reinigung. Die Fugen sind minimiert und komplett gedichtet. Eine patentierte Gehäuse-Technologie aus einem Edelstahl-Aluminium-Komposit garantiert außen höchste chemische Beständigkeit und eine perfekte Wärmeverteilung für die Kühlung der PC-Komponenten.

Begründung der Jury
Die hochwertige Gestaltung der Industrie-PCs ist konsequent auf die hygienischen Ansprüche in einem professionellen Umfeld ausgelegt.

IT 8200 / IT 8200 FP
Time Recording Terminal
Zeiterfassungsterminal

Manufacturer
ISGUS GmbH, Villingen-Schwenningen, Germany
Design
everdesign (Stephan Hauser), Freiburg, Germany
Web
www.isgus.de
www.everdesign.de

The IT 8200 / IT 8200 FP terminal series provides time recording and access control functions. The series supports all contact-free reading procedures and biometric identification via fingerprinting. The device communicates via Internet or Intranet, thus allowing centralised data analysis. The glass front with integrated keyboard and backlit colour graphic display is designed to withstand intensive use and dirt.

Statement by the jury
With its flat and seamless design the terminal series successfully meets modern technical as well as aesthetic needs.

Die Terminalserie IT 8200 / IT 8200 FP dient der intelligenten Zeiterfassung und Zutrittskontrolle. Die Serie unterstützt alle berührungslosen Leseverfahren sowie die biometrische Identifikation über den Fingerabdruck. Das Gerät kommuniziert über Internet oder Intranet; die Datenauswertung erfolgt zentral. Die Glasfront mit flächenintegrierter Tastatur und hinterleuchtetem Farbgrafik-Display hält intensivem Gebrauch und Verschmutzungen stand.

Begründung der Jury
Die flache und bündig gestaltete Terminalserie überzeugt, weil sie voll und ganz auf moderne technische sowie ästhetische Bedürfnisse ausgerichtet ist.

VOXIO® Touch
RFID Access Control Reader
RFID-Zutrittskontrollleser

Manufacturer
phg Peter Hengstler GmbH + Co. KG, Deißlingen, Germany
Design
HEIKU Design (Georg Heitzmann), Lauterbach, Germany
Web
www.phg.de
www.heikudesign.com

VOXIO Touch is an RFID access control reader with a glass surface and a capacitive PIN-code keyboard. As a status indicator, the light bar is divided into three fields. The symbolism in the form of a square indicates the position of the RFID reader field. The backlight allows the sure application in a dark surrounding. In addition, the low height provides the ideal opportunity for seamless integration into existing architecture.

Statement by the jury
The VOXIO Touch impresses with its remarkable simplicity. The cubical form looks stylish and unobtrusive.

VOXIO Touch ist ein RFID-Zutrittskontrollleser mit einer Glasoberfläche und kapazitiver PIN-Code-Tastatur. Der Leuchtbalken als Statusanzeige ist in drei Felder geteilt. Die Symbolik in Form eines Quadrats zeigt die Position des RFID-Lesefeldes an. Die Hintergrundbeleuchtung ermöglicht den sicheren Einsatz in dunkler Umgebung. Darüber hinaus bietet die geringe Aufbauhöhe die ideale Möglichkeit zur nahtlosen Integration in bestehende Architektur.

Begründung der Jury
Der VOXIO Touch besticht durch bemerkenswerte Einfachheit. Die kubische Form wirkt stilvoll und unaufdringlich.

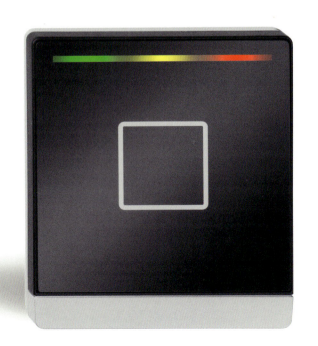

FlyPOS
Mobile POS Terminal

Manufacturer
FLYPOS, San Sebastián de los Reyes (Madrid), Spain
In-house design
Web
www.flypos.es

The FlyPOS mobile point-of-sale terminal is one of the smallest and most powerful devices in its class. It meets all required security standards and accepts all major credit and debit cards. The transactions are carried out either via a magnet stripe reader, a Wi-Fi or Bluetooth connection, or the integrated card reader. For contact-free payment, customers place their cards in front of the device and the data is captured.

Statement by the jury
FlyPOS impresses with its comprehensive, technological features, but also with its particularly small size, which facilitates use anywhere.

Das mobile Point-of-Sale-Terminal FlyPOS ist eines der kleinsten und leistungsstärksten Geräte seiner Klasse. Es erfüllt alle notwendigen Sicherheitsstandards und akzeptiert die gängigen Kredit- und Debitkarten. Die Transaktionen erfolgen entweder über den Magnetstreifenleser, eine Wi-Fi- bzw. Bluetooth-Verbindung oder über den integrierten Kontaktlosleser. Bei der kontaktlosen Zahlung hält der Kunde die Karte vor das Gerät, und die Daten werden erfasst.

Begründung der Jury
Beeindruckend am FlyPOS ist sowohl seine umfassende technische Ausstattung als auch seine besonders kleine Größe. Dadurch ist das Terminal überall einsetzbar.

BIS VM Read/Write Head
BIS VM Schreib-/Lesekopf

Manufacturer
Balluff GmbH, Neuhausen auf den Fildern, Germany
In-house design
Winfried Kunzweiler
Web
www.balluff.com

The read/write head from the BIS VM series combines RFID technology with control sensors for assembly lines. Thanks to its design approach, the device does not require additional mounts and can even be directly affixed to metal surfaces. Its metal housing allows the unit to also be used in harsh manufacturing environments. The LEDs are visible from any angle, enabling monitoring directly at the system.

Statement by the jury
This read/write head impresses with its minimalist design language. The generously large, black display has a modern and valuable look.

Der Schreib- und Lesekopf der Baureihe BIS VM kombiniert RFID-Technik mit Sensorik zur Steuerung von Montagelinien. Das Gerät ist so konstruiert, dass zusätzliche Halterungen komplett entfallen und gleichzeitig eine Montage direkt auf Metall möglich ist. Das Metallgehäuse sorgt dafür, dass das Gerät auch in rauer Fertigungsumgebung eingesetzt werden kann. Die rundum sichtbaren LEDs gestatten die Kontrolle direkt an der Anlage.

Begründung der Jury
Der Schreib- und Lesekopf besticht durch seine minimalistische Formensprache. Das großzügig gestaltete schwarze Display sieht modern und edel aus.

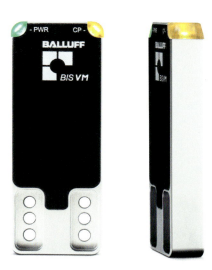

BUTTON GOURMET
Self-Service Open Restaurant
Selbstbedienungsrestaurant

Manufacturer
Grupo Azkoyen, Coffetek, UK
Design
Mormedi, Madrid, Spain
Web
www.buttongourmet.com
www.mormedi.com

The Button Gourmet is a self-service open restaurant that has been designed for the supply of high-quality food and drinks. Meals and drinks are selected via an integrated touchscreen. The modular structure enables a high degree of customisation with regard to the products offered, use of advertising surfaces and overall machine size.

Statement by the jury
The generously radiused edges and the contrast of the natural wood and dark glass surfaces give this vending machine a luxurious and modern look.

Der Button Gourmet ist ein Selbstbedienungsrestaurant für gehobene Ansprüche, der für die Bereitstellung qualitativ hochwertiger Lebensmittel und Getränke konzipiert wurde. Die Menüauswahl erfolgt über einen integrierten Touchscreen. Die Modulstruktur gewährleistet eine Vielzahl von Anpassungsmöglichkeiten, was Produktangebot, Nutzung von Werbeflächen oder die Maße des Automaten angeht.

Begründung der Jury
Die großzügig gerundeten Kanten sowie der Kontrast aus natürlichem Holz und dunklen Glasflächen verleihen dem Automat ein luxuriöses und modernes Äußeres.

Air conditioning systems
Air purifiers
Heating and
energy systems
Heating pumps
Humidifiers
Radiators
Room controllers
Solar technology

Heizkörper
Heizungs- und
Energiesysteme
Klimaanlagen
Luftbefeuchter
Luftreiniger
Raumregler
Solartechnik
Wärmepumpen

Heating and air conditioning
Heiz- und Klimatechnik

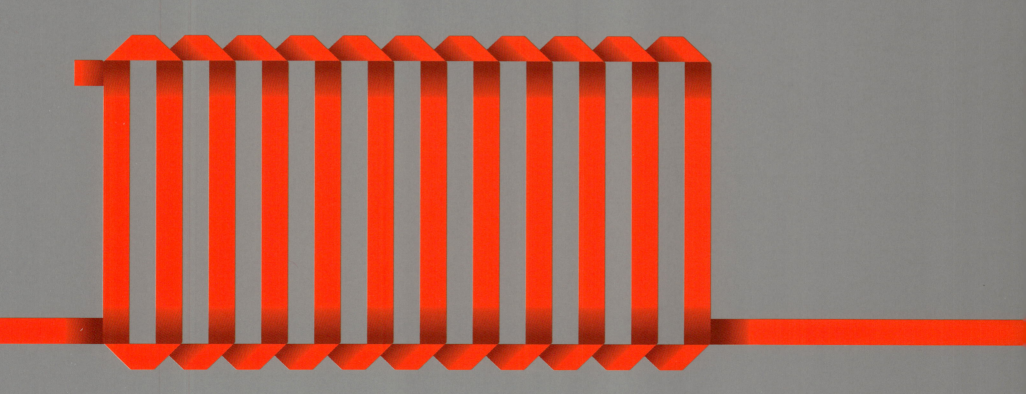

NOW – Smart Radiators System
Heating Control System
Heizungsregelungssystem

Manufacturer
IRSAP, Arquà Polesine (Rovigo),
Italy

In-house design
IRSAP

Web
www.irsap.it

reddot award 2015
best of the best

Right heat in the right place

As everybody perceives ambient temperature in a different way, it needs to be adjusted in every room to fit individual needs. NOW is an innovative heating control system which achieves this in a very well thought-through way. Unlike thermostats which just turn a radiator on or off, this system makes it possible to individually adjust every radiator. Its special algorithms allow it to save up to 40 per cent of energy. The way the various elements have been designed to function together is impressive. NOW integrates a central touchscreen control panel for the setting of the wireless electronic thermostat heads installed on every radiator. It also regulates a generator control which turns the boiler, or the heat pump, on and off. This system is very comfortable to use by means of a colour temperature display. As every level of temperature is given a different colour, the user can quickly see when the temperature changes and only needs to adjust the temperature up or down until the appropriate colour appears. In combination with tasteful design, the NOW heating control system offers simple, individualised temperature control. Through its appealing form language and practical functionality, it will brighten up everyday life.

Besser geregelt

Da das Empfinden der Raumtemperatur individuell unterschiedlich ist, gleicht man sie in den Räumen den jeweiligen Bedürfnissen an. NOW ist ein innovatives Heizungsregelungssystem, welches dies auf eine sehr gut durchdachte Weise gestattet. Anders als bei Temperaturreglern, bei denen nur das Ein- oder Ausschalten der Heizkörper und das generelle Herunter- oder Hinaufregeln erfolgen kann, ermöglicht dieses System die spezifische Regelung jedes einzelnen Heizkörpers. Da es mit speziellen Algorithmen arbeitet, können mit ihm bis zu 40 Prozent Energie eingespart werden. Beeindruckend sind seine gestalterisch perfekt aufeinander abgestimmten Elemente. Es integriert ein zentrales Touchscreen Control Panel für die Regulierung der drahtlosen, an jedem Heizkörper installierten elektronischen Thermostatköpfe sowie ein Steuersystem, das Heizkessel oder Wärmepumpe ein- und ausschaltet. Der hohe Komfort dieses Systems basiert auch auf der Bedienung über ein Farbtemperatur-Display. Nach dem Prinzip einer dem numerischen Wert der Temperatur zugeordneten Farbe sieht der Nutzer sofort, wenn sich die Temperatur ändert. Um sie zu erhöhen oder zu senken, muss er nur in die Richtung der entsprechenden Farbe herunter- oder hinaufregeln. Einhergehend mit einer dezenten Gestaltung, bietet das Heizungsregelungssystem NOW so die Möglichkeit einer einfachen, individuellen Temperatursteuerung. Durch seine sinnlich ansprechende Formensprache und sehr zweckmäßige Funktionalität bereitet es zudem viel Freude im Alltag.

Statement by the jury

This heating control system combines a moving form language with an impressive new way of controlling temperature. Well-designed down to the smallest detail, everything is beautifully coordinated. This system can adjust the temperature of every individual radiator and thereby not only make everyone comfortable, but also function in an environmentally friendly way. An intuitive colour temperature display also gives the user feedback that is easily understood.

Begründung der Jury

Eine emotionalisierende Formensprache verbindet sich bei diesem Heizungsregelungssystem mit einem beeindruckend neuen Konzept der Temperatursteuerung. Jedes seiner Details wurde gestalterisch sehr gut ausgearbeitet und perfekt abgestimmt. Selektiv kann dieses System die Temperatur jedes einzelnen Heizkörpers regeln und für ein individuelles und umweltgerechtes Wohlbefinden sorgen. Ein intuitiv bedienbares Farbtemperatur-Display gibt dem Nutzer dabei ein gut verständliches Feedback.

Designer portrait
See page 56
Siehe Seite 56

Memory
Radiator
Heizkörper

Manufacturer
Varis Top d.o.o., Lendava, Slovenia
In-house design
Silvijan Čivre
Web
www.varis-lendava.si

Memory can be used as an electric radiator, a hot water radiator or as a combination of the two in a single heating unit. It features a striking, flat steel construction. Situated between the front and rear faces is the heating element, whose tubes are embedded in clay and straw. All materials are fully recyclable. Memory can be equipped with a digital thermostat head allowing control via smartphone.

Memory lässt sich als Elektro-, Warmwasser- oder kombinierter Elektro-Warmwasser-Heizkörper nutzen. Er besteht aus einer markanten, flach gearbeiteten Stahlkonstruktion. Zwischen der Vorder- und der Rückfront befindet sich das Heizelement, dessen Rohre in Lehm und Stroh eingebettet sind. Alle Materialien sind vollständig recycelbar. Memory kann mit einem digitalen Thermostatkopf, der die Nutzung via Smartphone ermöglicht, ausgestattet werden.

Tavola, Tavoletta
Radiator
Heizkörper

Manufacturer
Antrax IT S.r.l., Resana (Treviso), Italy
Design
Andrea Crosetta, Resana (Treviso), Italy
Web
www.antrax.it

Tavola and Tavoletta are two radiators displaying minimalist, geometric design. They can be mounted both vertically and horizontally. Thanks to a notch or the application of a special support, they may also serve as a towel or bathrobe hook. Tavola is also optionally available with an integrated mirror. Tavoletta is smaller; this radiator element is also available with a low-consumption electric plate.

Tavola und Tavoletta sind zwei Heizkörper mit minimalistischem, geometrischem Design. Sie können sowohl vertikal als auch horizontal montiert werden. Beide lassen sich dank eines Schlitzes oder einer zusätzlich angebrachten Halterung zum Aufhängen von Handtüchern oder Bademänteln nutzen. Tavola ist darüber hinaus optional mit integriertem Spiegel erhältlich. Tavoletta ist kleiner; dieses Heizelement gibt es auch in einer Ausführung mit elektrischer Energiespar-Heizplatte.

Statement by the jury
With their clear, purist design, the Tavola and Tavoletta radiator elements meet high demands with regard to aesthetic quality.

Begründung der Jury
Die Heizkörper Tavola und Tavoletta werden mit ihrer klaren, puristischen Gestaltung auch gehobenen Ansprüchen an ästhetische Qualität gerecht.

Vitodens 300-W
Wall-Mounted Gas-Fired Condensing Boiler
Gas-Brennwert-Wandgerät

Manufacturer
Viessmann Werke GmbH & Co. KG,
Allendorf/Eder, Germany
Design
Phoenix Design GmbH + Co. KG,
Stuttgart, Germany
Web
www.viessmann.com
www.phoenixdesign.com

High efficiency and a broad range of applications, made possible by integrated sensor technology, characterise the Vitodens 300-W wall-mounted, gas-fired condensing boiler. Its reduced design extends to the easy-to-use interactive colour touchscreen. Moreover, the device can also be controlled comfortably via an app. Distinguishing design features include the clear line management, powder-coated surfaces in the corporate colour Vito White and the orange-silver logotype.

Statement by the jury
The sophisticated design of this wall-mounted, gas-fired condensing boiler convincingly communicates its high functionality.

Eine hohe Effizienz und Einsatzbreite, ermöglicht durch integrierte Sensorik, kennzeichnen das Gas-Brennwert-Wandgerät Vitodens 300-W. Seine reduzierte Gestaltung setzt sich im interaktiven, einfach zu bedienenden Farb-Touchscreen fort. Zusätzlich lässt sich das Gerät komfortabel über eine App steuern. Charakteristische Gestaltungsmerkmale sind die klare Linienführung, die pulverbeschichteten Flächen in der Corporate-Farbe Vitoweiß und die orange-silberfarbene Wortmarke.

Begründung der Jury
Die anspruchsvolle Gestaltung dieses Gas-Brennwert-Wandgeräts kommuniziert überzeugend die hohe Funktionalität des Geräts.

Tungsten Smart-Heat
Portable Gas Heater
Transportabler Gas-Heizstrahler

Manufacturer
Bromic Heating, Silverwater, Australia
In-house design
Web
www.bromicheating.com

The Tungsten Smart-Heat portable gas heater is up to 300 per cent more efficient than traditional devices. Its patented ceramic technology focuses the heat, thus increasing energy conversion. The device is low-cost in operation, has improved wind resistance and spreads the warmth more gently and evenly. The wave-shaped design conveys a harmonious image.

Statement by the jury
The Tungsten Smart-Heat gas heater impresses with its successful combination of modern technology, enhanced functionality and straightforward design.

Der transportable Gas-Heizstrahler Tungsten (Wolfram) Smart-Heat arbeitet bis zu 300 Prozent wirtschaftlicher als traditionelle Geräte. Seine patentierte Keramiktechnologie bündelt die Wärme und erhöht damit die Energieumwandlung. Das Gerät verursacht nur geringe Betriebskosten und zeichnet sich durch verbesserte Windbeständigkeit und behutsame Wärmeverbreitung aus. Das Design mit der geschwungenen Linienführung vermittelt ein harmonisches Bild.

Begründung der Jury
Der Gas-Heizstrahler Tungsten Smart-Heat überzeugt mit einer gelungenen Kombination aus moderner Technik, hoher Funktionalität und einer schnörkellosen Gestaltung.

2in1 – Heat & Light
Infrared Heating Element/Lighting Fixture
Infrarotheizung/Beleuchtungskörper

Manufacturer
Redwell Manufaktur GmbH, Oberwart, Austria
In-house design
Michael Buschhoff
Web
www.redwell.com

2in1 – Heat & Light is a combination of heating and lighting systems with an energy- and space-saving as well as eco-friendly design. The specially structured infrared heating technology is located in the central component of the round cover elements, available in different sizes with a power consumption ranging from 1,200 to 3,400 watts. Both the infrared heating system and the LED lens technology can be powered by green electricity, thus achieving carbon neutrality.

2in1 – Heat & Light ist eine Kombination von Heiz- und Lichtsystemen, die energie- und platzsparend sowie umweltschonend gestaltet sind. Die speziell konzipierte Infrarotheiztechnik liegt im mittleren Bauteil der runden Deckenelemente, die in unterschiedlichen Größen mit Leistungen von 1.200 bis 3.400 Watt erhältlich sind. Das Infrarotheizsystem, wie auch die LED-Linsentechnik, können mit Ökostrom betrieben werden und sind somit CO_2-neutral.

Statement by the jury
This innovative combination device convinces with its resource-saving technology. Due to its straightforward design, it can be used in both business and private environments.

Begründung der Jury
Dieses innovative Kombigerät punktet durch seine ressourcenschonende Technologie. Dank seiner schnörkellosen Gestaltung lässt es sich im geschäftlichen wie im privaten Bereich nutzen.

Kirigamine Zen
MSZ-EF Series
Room Air Conditioner
Raumklimagerät

Manufacturer
Mitsubishi Electric Corporation, Tokyo, Japan
In-house design
Takayuki Nishiguchi
Web
www.mitsubishielectric.co.jp

The name of the Kirigamine Zen room air conditioner says it all: available in white, black or silver, it conveys Japanese Zen aesthetics with a design that also blends harmoniously into modern European interiors. Its minimalist body shape shows a combination of straight lines, flat surfaces and a high-quality finish. The aerodynamic design enables a significant energy-saving performance. It also contributes to a clean overall appearance since the front panel can stay closed, which is not always possible with thin profile models.

Statement by the jury
The Kirigamine Zen presents a convincing symbiosis of energy-saving technology and minimalist, elegant design.

Bei diesem Raumklimagerät ist der Name Kirigamine Zen Programm: Wahlweise in Weiß, Schwarz oder Silber repräsentiert es die Ästhetik des japanischen Zen. Damit passt es gut zu modernen europäischen Wohneinrichtungen. Seine minimalistische Gehäuseform weist eine Kombination von geraden Linien, ebenen Oberflächen und einem hochwertigen Finish auf. Die aerodynamische Gestaltung ermöglicht eine hohe Energiesparleistung und sorgt für ein aufgeräumtes Erscheinungsbild, da die Frontabdeckung, was bei flachen Modellen nicht selbstverständlich ist, geschlossen bleiben kann.

Begründung der Jury
Bei Kirigamine Zen gehen eine energiesparsame Technik und ein minimalistisches wie elegantes Design eine überzeugende Symbiose ein.

Electrolux PureO2

Air Conditioner

Klimagerät

Manufacturer
Electrolux AB. Stockholm, Sweden
In-house design
Web
www.electrolux.com

This air conditioner featuring the PureO2 system is designed to equally integrate well into both European and Asian households. The unit features a soft silhouette, with the slightly curved front edge expressing power and performance. Thanks to elements crafted from stainless steel, the unit has a high-quality appeal. The air conditioner is operated using a remote control that is stored in a magnetic wall holder.

Dieses Klimagerät mit PureO2 System ist so gestaltet, dass es sich gleichermaßen gut in europäische wie in asiatische Haushalte integriert. Das Gerät hat eine weiche Silhouette, die leicht gebogene Vorderkante drückt Kraft und Leistung aus. Durch die Edelstahl-Elemente wirkt es sehr hochwertig. Gesteuert wird das Klimagerät mit einer Fernbedienung, die in einer magnetischen Wandhalterung aufbewahrt wird.

Statement by the jury
With its balanced, harmonious design, this air conditioner blends stylishly into a wide variety of interiors.

Begründung der Jury
Durch seine ausgewogene, harmonische Formensprache fügt sich dieses Klimagerät stilvoll in unterschiedlichste Interieurs ein.

Midea Q2000 Series
Air Conditioner
Klimagerät

Manufacturer
Midea Air-Conditioning Equipment Co., Ltd., Foshan, China
In-house design
Wei Yu
Web
www.midea.com

This air conditioner from the Midea Q2000 series has a high EER value (energy efficiency ratio). With a depth of only 140 mm, it is extremely slim and thus requires only little space. The metal body and brushed-aluminium panel express high aspirations with regard to quality. The generously designed LED display and air outlet can be covered with the front panel when the device is turned off – when closed, the air conditioner renders a purist overall impression.

Dieses Klimagerät der Serie Midea Q2000 weist einen hohen EER-Wert (Energy Efficiency Ratio) auf. Mit seinen 140 mm Tiefe ist es ausgesprochen schmal gebaut und beansprucht daher nur wenig Platz. Der Metallkorpus sowie das aus gebürstetem Aluminium gefertigte Paneel drücken einen hohen Qualitätsanspruch aus. Das großzügig gestaltete LED-Display sowie der Luftaustritt können, wenn das Gerät nicht in Betrieb ist, mit der Frontblende verdeckt werden – im geschlossenen Zustand vermittelt das Klimagerät dann einen puristischen Gesamteindruck.

Statement by the jury
A minimalist design language and the use of high-grade materials lend this air conditioner from the Midea Q2000 series an elegant appearance.

Begründung der Jury
Eine minimalistische Formensprache und der Einsatz hochwertiger Materialien verleihen diesem Klimagerät der Serie Midea Q2000 eine elegante Anmutung.

LG Ceiling-Embedded Air Conditioner
LG Deckenmontiertes Klimagerät

Manufacturer
LG Electronics Inc., Seoul, South Korea
In-house design
Yong Kim, Myungshik Kim, Kyeongchul Cho,
Jinwon Kang, Inhyeuk Choi
Web
www.lg.com

This ceiling-embedded air conditioner includes service covers on both sides so that the device does not have to be removed from the ceiling for inspection and repair. A three-colour LED display indicates the operational status through changes in colour. To satisfy both the residential and commercial markets, the outlet vanes are designed to extend the air current as far as possible, while the grille strongly facilitates air intake for good heating and cooling results.

Dieses in die Zimmerdecke integrierte Klimagerät bietet auf beiden Seiten Wartungsabdeckungen, so muss es für Inspektionen und Reparaturen nicht von der Decke abgenommen werden. Ein dreifarbiges LED-Display zeigt den Betriebsstatus durch Farbwechsel an. Das Gerät ist sowohl für Wohn- als auch für Geschäftsräume konzipiert. Die Auslassstellen sind so gestaltet, dass sie den Luftstrom möglichst weit verteilen, während die Luftaufnahme über das Gitter für sehr gute Heiz- und Kühlungsergebnisse sorgt.

Statement by the jury
This air-conditioner convinces with its well-conceived functionality. Its deliberately reduced design conveys solidity.

Begründung der Jury
Dieses Klimagerät überzeugt durch seine wohldurchdachte Funktionalität. Seine bewusst reduzierte Gestaltung vermittelt Solidität.

CHiQ Airloop
Air Purifier
Luftreiniger

Manufacturer
Sichuan Changhong Electric Co., Ltd.,
Mianyang, China
In-house design
Homwee Technology (Sichuan) Co., Ltd.,
Chengdu, China
Web
www.changhong.com

The innovative CHiQ Airloop air purifier is suspended from the ceiling, thus not taking up any valuable space. Its striking loop construction additionally includes a security camera. With its elegant and unobtrusive design, the CHiQ Airloop blends well into different environments. The device can be operated using a wall-mounted control panel or directly via a smartphone.

Statement by the jury
The CHiQ Airloop convinces with its sophisticated, space-saving design and its classic, elegant appearance.

Der innovative Luftreiniger CHiQ Airloop hängt von der Decke herunter und beansprucht so keinen wertvollen Stellplatz. Die markante Konstruktion in Form einer Schlaufe enthält zusätzlich eine Sicherheitskamera. Mit seiner eleganten, unaufdringlichen Gestaltung integriert sich CHiQ Airloop gut in seine jeweilige Umgebung. Das Gerät lässt sich über eine an der Wand befestigte Bedienungseinheit oder direkt über das Smartphone steuern.

Begründung der Jury
CHiQ Airloop überzeugt durch seine raffinierte, platzsparende Gestaltung und seine klassisch-elegante Anmutung.

P500
Air Purifier
Luftreiniger

Manufacturer
BONECO AG, Widnau, Switzerland
In-house design
Manfred Fitsch
Web
www.boneco.com

The key feature of the P500 air purifier is simple and intuitive handling. The front panel can be removed from the device so that the filters are easily accessible and replaceable from the front. The purist design underscores the simplicity of the purifier. Clear lines and the interplay of white and black with matt and gloss finishes give the device a characteristic appearance.

Statement by the jury
The P500 air purifier cleverly combines a user-friendly product solution with shapely design.

Das Hauptmerkmal des Luftreinigers P500 ist eine einfache und intuitive Handhabung. Die Frontblende kann vom Gerät getrennt werden, wodurch die Filter von vorne zugänglich und leicht einzusetzen sind. Die puristische Gestaltung unterstreicht die Einfachheit des Luftreinigers. Klare Linien, das Wechselspiel von Weiß und Schwarz sowie von matten und glänzenden Oberflächen verleihen dem Gerät ein charakteristisches Erscheinungsbild.

Begründung der Jury
Der Luftreiniger P500 kombiniert geschickt eine nutzerfreundliche Produktlösung mit einer formschönen Anmutung.

Powercube AC4620/ ACP610/AC4600
Air Purifier
Luftreiniger

Manufacturer
Royal Philips, Eindhoven, Netherlands
In-house design
Web
www.philips.com

The Powercube is an air purifier developed for the Chinese market. An aerodynamic airway enables high performance at lower noise levels and reduced power consumption. The TwinPower system, featuring 2.4 kg NanoProtect Plus filters, generates a clean air capacity that is 60 per cent higher than comparable models in the range. A rotary knob provides ease of use, while the "Auto" button in the middle of the knob enables one-touch control. The status of the indoor air quality is indicated by four different colours.

Statement by the jury
The Powercube pleases with its interplay of high performance and user-friendly design.

Der Powercube ist ein für den chinesischen Markt entwickelter Luftreiniger. Ein aerodynamischer Luftweg ermöglicht eine hohe Leistung mit geringer Geräuschentwicklung und niedrigem Verbrauch. Das TwinPower-System mit den 2,4 kg NanoProtect Plus Filtern sorgt für eine 60 Prozent höhere Reinluftkapazität. Für Bedienkomfort sorgt ein Drehknopf, der „Auto"-Taster in dessen Mitte für eine One-Touch-Steuerung. Der Status der Raumluftqualität wird in vier Farben angezeigt.

Begründung der Jury
Der Powercube gefällt durch das Zusammenspiel von großer Leistungsfähigkeit und benutzerfreundlicher Gestaltung.

Air Max® Ambiance
Moisture Absorber
Luftentfeuchter

Manufacturer
Bolton Adhesives, Bison International B.V., Rotterdam, Netherlands
Design
FLEX/the INNOVATIONLAB B.V., Delft, Netherlands
Web
www.boltonadhesives.com
www.flex.nl

The Air Max Ambiance Moisture Absorber is reminiscent of a lamp. With its discreet, grey colouring, the device adapts to different living environments. Its innovative operating principle based on effective air circulation achieves an enhanced level of moisture absorption and thus a comfortable room climate. The absorber is convenient to empty without spilling, easily refillable and features a safe anti-leak system, making it simple and comfortable to operate the device.

Statement by the jury
With its harmonious line management, the Air Max Ambiance Moisture Absorber proves to be a useful home accessory with a high-quality appearance.

Der Luftentfeuchter Air Max Ambiance erinnert an eine Lampe. In einem dezenten Grau gehalten passt er sich verschiedenen Wohnumgebungen an. Sein innovatives Wirkungsprinzip beruht auf einer effektiven Luftzirkulation, die für eine erhöhte Feuchtigkeitsabsorption und damit für ein angenehmes Raumklima sorgt. Das saubere Ausgießen der entstehenden Flüssigkeit, einfaches Neubefüllen sowie ein sicheres Anti-Leckage-System stehen für eine angenehm einfache Handhabung des Geräts.

Begründung der Jury
Durch seine harmonische Linienführung wird der Luftentfeuchter Air Max Ambiance zu einem nützlichen Wohnaccessoire mit hochwertiger Anmutung.

LG WHISEN (LD-139DDL)
Dehumidifier
Luftentfeuchter

Manufacturer
LG Electronics Inc., Seoul, South Korea
In-house design
Sehwan Bae, Chinsoo Hyun, Jinsu Kim,
Jaeyong Park, Miju Kim, Yunseo Jang
Web
www.lg.com

The design of this dehumidifier, which accommodates 13–15 litres, was inspired by the portability of a travel bag. The fresh two-colour design particularly appeals to younger customers. In addition, the device is easy to operate with its intuitive user interface and ergonomically well-conceived handle. The water level can be easily ascertained through the enlarged window.

Statement by the jury
This dehumidifier attracts attention with its unique design and use of fresh colours.

Inspiriert wurde die Gestaltung dieses Luftentfeuchters mit einem Fassungsvermögen von 13–15 Litern von der Tragemöglichkeit einer Reisetasche. Durch seine frisch wirkende Zweifarbigkeit spricht er besonders jüngere Kunden an. Darüber hinaus verfügt das Gerät über hohen Bedienkomfort, eine intuitiv zu handhabende Benutzeroberfläche und einen ergonomisch gut durchdachten Griff. Den Wasserstand kann der Nutzer ganz einfach durch das vergrößerte Fenster sehen.

Begründung der Jury
Mit seiner außergewöhnlichen Formensprache und der Verwendung frischer Farben zieht dieser Luftentfeuchter die Aufmerksamkeit auf sich.

LG WHISEN (LD-179DRC)
Dehumidifier
Luftentfeuchter

Manufacturer
LG Electronics Inc., Seoul, South Korea
In-house design
Sehwan Bae, Chinsoo Hyun, Jinsu Kim,
Jaeyong Park, Miju Kim, Yunseo Jang
Web
www.lg.com

The design of this premium dehumidifier with a capacity of 17 litres is focused on the needs of users. A 3.2'' LCD display, lighting and a large water-level indicator inform the user of the respective operating status. Comfortable features are the one-touch, slide-in-type water tank and the open handles along the top. The side decoration patterns accentuate the superior quality of the device.

Statement by the jury
A high degree of user-friendliness and expressive design go hand in hand with this dehumidifier.

Die Gestaltung dieses Premium-Luftentfeuchters mit 17 Litern Fassungsvermögen stellt die Bedürfnisse des Nutzers in den Mittelpunkt. Ein 3,2''-LCD-Display sowie eine Beleuchtung und eine große Wasserstandsanzeige informieren über den jeweiligen Betriebszustand. Komfortabel sind der leicht einschiebbare One-Touch-Wasserbehälter sowie die offenen Griffe am oberen Rand. Die seitlichen Deko-Muster betonen die Hochwertigkeit des Geräts.

Begründung der Jury
Ein hohes Maß an Nutzerfreundlichkeit sowie ein ausdrucksstarkes Aussehen gehen bei diesem Luftentfeuchter Hand in Hand.

LG WHISEN (LD–159DQV)
Dehumidifier
Luftentfeuchter

Manufacturer
LG Electronics Inc., Seoul, South Korea
In-house design
Sehwan Bae, Chinsoo Hyun, Jinsu Kim, Jaeyong Park,
Miju Kim, Yunseo Jang
Web
www.lg.com

This dehumidifier with its capacity of 13–15 litres is designed for private use. Its compact design harmonises with many interiors. The device is rectangular on top, changing to an oval shape towards the bottom. The water tank is located on the side of the unit and easily removed. Another feature is the ergonomically shaped handle, allowing the relatively heavy device to be pulled or lifted without bending over.

Mit seinen 13–15 Litern Fassungsvermögen ist dieser Luftentfeuchter für den Privatgebrauch gemacht, sein kompaktes Design harmoniert mit vielen Inneneinrichtungen. Oben ist das Gerät rechteckig, nach unten hin geht die Form in ein Oval über. Seitlich befindet sich der Wasserbehälter, der sich leicht entnehmen lässt. Ein weiteres Merkmal ist der ergonomisch geformte Griff, der es ermöglicht, das relativ schwere Gerät zu ziehen und anzuheben, ohne sich zu bücken.

Statement by the jury
This dehumidifier blends well into various interiors and meets the demands of an ergonomically sensible design.

Begründung der Jury
Dieser Luftentfeuchter integriert sich gut in verschiedene Interieurs und wird dabei den Ansprüchen an eine ergonomisch sinnvolle Gestaltung gerecht.

AP-1015A
Air Purifier
Luftreiniger

Manufacturer
Coway Co., Ltd., Seoul, South Korea
In-house design
Coway Design Lab
Web
www.coway.com

The AP-1015A air purifier is equipped with a four-step filtration system for removing fine dust, bacteria, harmful gases and unpleasant odours. It is inclined slightly forward by six degrees so that the purified air flows upward at an angle, circulating the air evenly. A ring-shaped LED display indicates the current air quality with four-step changes in colour. Using NFC technology and a smartphone app, the user can control the status of the device.

Der Luftreiniger AP-1015A ist mit einem vierstufigen Filtersystem ausgestattet, das Feinstaub, Bakterien, schädliche Gase und unangenehme Gerüche entfernt. Er ist um sechs Grad nach vorn geneigt, sodass die aufbereitete Luft schräg nach oben strömt und gleichmäßig zirkulieren kann. Eine ringförmige LED-Anzeige gibt mit vier Farbstufen die Luftqualität an, die auch mittels NFC-Technologie über eine Smartphone-App kontrolliert werden kann.

Statement by the jury
This air purifier is compelling with its clever, angled construction. With its app functionality, the device meets the demands of contemporary home technology.

Begründung der Jury
Dieser Luftreiniger besticht durch seine pfiffige, abgewinkelte Konstruktion. Durch die App wird das Gerät den Ansprüchen an zeitgemäße Haustechnik gerecht.

CF-6600
Air Washer
Luftwäscher

Manufacturer
Comefresh Electronic Industry Co., Ltd.,
Xiamen, China
Design
JEI Design Works (Kyuhong Han, Chul Jin),
Seoul, South Korea
Web
www.comefresh.com
www.designjei.com

The compact CF-6600 air washer is suitable for simultaneous air filtration and humidification. It features an intuitive LED display, a waterproof control button and storage space for the cable. The black-and-white device, manufactured to 90 per cent from recyclable ABS plastics, renders a smooth impression with its rounded edges. The ventilation opening situated at the top of the device underlines the harmonious overall impression.

Statement by the jury
The CF-6600 is characterised by user-friendly technology and a coherent appearance.

Der kompakte Luftwäscher CF-6600 bietet sich zur gleichzeitigen Luftfilterung und -befeuchtung an. Er verfügt über ein intuitiv zu benutzendes LED-Display, einen wasserdichten Bedienknopf und Stauraum für das Kabel. Das schwarz-weiße, zu 90 Prozent aus recycelbarem ABS-Kunststoff gefertigte Gerät strahlt mit seinen abgerundeten Kanten eine sanfte Anmutung aus. Die Belüftungsöffnung befindet sich an der Geräteoberfläche, was den harmonischen Gesamteindruck unterstreicht.

Begründung der Jury
Der CF-6600 zeichnet sich durch eine benutzerfreundliche Technik und ein stimmiges Erscheinungsbild aus.

Air Purifier
Luftreiniger

Manufacturer
Shanghai ENL Newtechnologis Co., Ltd.,
Shanghai, China
Design
Yongin Songdam College (Kun Pyo Hong),
Yongin, South Korea
Web
www.enlnewtech.com
www.ysc.ac.kr

Contrary to conventional models, this air purifier has a cylindrical shape, allowing air to be taken in and released not just from one or two sides, but from all around in a 360-degree radius. To attain the highest possible air quality, the device works with a high-performance HEPA, deodorisation and antibacterial filter. All display functions can be controlled with a smartphone app.

Statement by the jury
The striking characteristic of this air purifier is its well-conceived design language, which visualises the aspiration to high standards of functionality.

Dieser Luftreiniger hat im Gegensatz zu herkömmlichen Modellen eine zylindrische Form. Dies ermöglicht ihm, Luft nicht nur von einer oder zwei, sondern von allen Seiten im 360-Grad-Radius anzusaugen und wieder auszustoßen. Um eine möglichst hohe Luftqualität zu erzielen, arbeitet das Gerät mit einem leistungsstarken HEPA-, Desodorierungs- und antibakteriellen Filter. Alle Anzeigen auf dem Display lassen sich mit einer App über das Smartphone kontrollieren.

Begründung der Jury
Das Bemerkenswerte an diesem Luftreiniger ist seine gut durchdachte Formensprache, die den Anspruch an hohe Funktionalität visualisiert.

APMS-1014D
Air Purifier and Humidifier
Luftreiniger und -befeuchter

Manufacturer
Coway Co., Ltd., Seoul, South Korea
In-house design
Coway Design Lab
Design
V2 Studios, London, Great Britain
Web
www.coway.com
www.v2studios.com

The APMS-1014D is an innovative air purifier and humidifier that sprays fine water particles with the air released by the device. Electrolytic sterilisation ensures a high degree of hygiene by removing 99.99 per cent of all bacteria and viruses from the water. The slim design of the device is characterised by soft curves. The function keys on the curved top slightly slant towards the user and are laid out ergonomically.

Der APMS-1014D ist ein innovativer Luftreiniger und -befeuchter, der feine Wassertröpfchen in die nach außen abgegebene Luft versprüht. Für einen hohen Hygienestandard sorgt die elektrolytische Sterilisation, durch die 99,99 Prozent aller Bakterien und Viren aus dem Wasser entfernt werden. Das schlanke Erscheinungsbild des Geräts ist durch weiche Kurven geprägt. Die Funktionstasten an der geschwungenen Oberseite sind leicht geneigt und ergonomisch positioniert.

Statement by the jury
A combination of innovative technology and organic design language distinguishes the APMS-1014D.

Begründung der Jury
Die Kombination von innovativer Technologie und einer organisch anmutenden Formensprache kennzeichnet den APMS-1014D.

APM-1514G
Air Purifier and Humidifier
Luftreiniger und -befeuchter

Manufacturer
Coway Co., Ltd., Seoul, South Korea
In-house design
Coway Design Lab
Web
www.coway.com

The APM-1514G is a combination of air purifier and humidifier specially developed for the US market. The device takes in air from two sides, the front and the back, and lets it out through the top. Despite its small dimensions, the unit achieves high humidification output thanks to its 3.8 litre water reservoir. The partially exposed reservoir at the front allows the user to easily check the amount of water remaining.

Bei dem speziell für den US-Markt entwickelten APM-1514G handelt es sich um einen kombinierten Luftreiniger und -befeuchter. Er saugt von zwei Seiten, von vorne und hinten, Luft an und stößt sie oben wieder aus. Trotz seiner geringen Ausmaße zeichnet sich das Gerät dank seines 3,8 Liter fassenden Tanks durch eine hohe Befeuchtungsleistung aus. Der teilweise freiliegende Wassertank an der Vorderseite ermöglicht es dem Benutzer, problemlos den Wasserstand zu kontrollieren.

Statement by the jury
This high-performance air purifier and humidifier is characterised by a compact and efficient construction style.

Begründung der Jury
Eine kompakte und effiziente Bauweise zeichnet diesen leistungsstarken Luftreiniger und -befeuchter aus.

Fonterra Smart Control
Area Heating Control
Flächenheizungsregelung

Manufacturer
Viega GmbH & Co. KG, Attendorn, Germany
Design
ARTEFAKT industriekultur, Darmstadt, Germany
Web
www.viega.com
www.artefakt.de

Fonterra Smart Control offers the advantage of a consistently self-optimising network. The product family consists of a room thermostat, a base unit and a heating circuit distributor with return flow temperature sensors at each distributor outlet. The permanent, dynamic, hydraulic comparison of the heating circuits guarantees a high degree of functionality. In addition, the system reacts quickly to temperature changes in an energy-saving manner. The desired room temperature – up to six temperature levels per day and room – can be defined individually.

Fonterra Smart Control bietet den Vorteil eines sich ständig selbst optimierenden Netzwerks. Die Produktfamilie besteht aus Raumthermostat, Basiseinheit und Heizkreisverteiler mit Rücklauftemperaturfühlern an jedem Verteilerabgang. Hohe Funktionalität gewährleistet der permanente dynamische und hydraulische Abgleich der Heizkreise untereinander, zudem reagiert das System schnell und energiesparend auf Temperaturveränderungen. Die gewünschte Raumtemperatur – bis zu sechs Niveaus pro Tag und Zimmer – kann individuell definiert werden.

Statement by the jury
Intelligent technology, simple and intuitive operation, along with an elegant design make the Fonterra Smart Control a state-of-the-art area heating control.

Begründung der Jury
Eine intelligente Technik, die einfache, intuitive Bedienung sowie elegante Gestaltung machen Fonterra Smart Control zu einer zeitgemäßen Flächenheizungsregelung.

Selina by Stadler Form
Selina von Stadler Form
Hygrometer/Thermometer

Manufacturer
Stadler Form Aktiengesellschaft, Zug, Switzerland
In-house design
Matti Walker
Web
www.stadlerform.com

The hygrometer Selina by Stadler Form, with its precise sensor, measures both relative humidity and temperature with high precision. The modern technology is fit in a slim enclosure with a thickness of just 4 mm, defined by its harmonious line management. The design of the display is deliberately kept simple and reduced, which underscores the elegant overall impression.

Das Hygrometer Selina von Stadler Form misst dank seines präzisen Sensors sowohl die relative Luftfeuchtigkeit als auch die Temperatur mit großer Genauigkeit. Untergebracht ist die moderne Technologie in einem schlanken Gehäuse, das nur 4 mm misst und durch seine harmonische Linienführung definiert ist. Die Anzeige auf dem Display ist bewusst einfach und reduziert gestaltet, was den eleganten Gesamteindruck unterstreicht.

Statement by the jury
With its purist design, this particularly slim hygrometer blends discreetly into different living environments.

Begründung der Jury
Durch seine puristische Gestaltung integriert sich dieses besonders schlanke Hygrometer unaufdringlich in verschiedene Wohnumgebungen.

DHE Connect & DHE Touch
Fully Electronic Controlled
Instantaneous Water Heaters
Vollelektronische Durchlauferhitzer

Manufacturer
Stiebel Eltron GmbH & Co. KG,
Holzminden, Germany

Design
Schumanndesign
(Dirk Schumann),
Münster, Germany

Web
www.stiebel-eltron.de
www.schumanndesign.de

reddot award 2015
best of the best

User-friendly and interconnected

Many homes are equipped with an instantaneous water heater that supplies hot water as and when required. DHE Connect & DHE Touch redefine the possible applications of this piece of everyday equipment in an extraordinarily comfortable and state-of-the-art manner. The central feature is a logical and innovative user interface which makes it immediately clear how everything works. The unity afforded by the touch display and touch wheel make the use of the equipment intuitive. DHE Connect also has noteworthy multimedia functions. It can be easily connected by Wi-Fi to the domestic Internet network, thus allowing Internet radio to be listened to on the built-in speaker and weather conditions to be seen on the colour display. Furthermore, the supply of hot water may be initiated using an app on your smartphone. This instantaneous water heater supplies hot water at exactly the right temperature using the powerful and highly efficient 4i technology. A further, well thought-through feature is that the control element can be detached for use away from the equipment. The oval form and striking lines give DHE Connect & DHE Touch flowing elegance which integrates very well into its surroundings and thereby harmoniously unites technology and design.

Komfortabel verbunden

Als wichtiger Bestandteil vieler Haushalte stellt der Durchlauferhitzer Warmwasser dann bereit, wenn es gerade benötigt wird. Die Einsatzmöglichkeiten dieses alltäglichen Gerätes neu definierend, bieten DHE Connect & DHE Touch eine überaus komfortable und zeitgemäße Art des Gebrauchs. Im Mittelpunkt steht ein klar gestaltetes, innovatives User-Interface, welches sich dem Nutzer in seiner Logik sofort erschließt. Als überzeugende Einheit aus einem Touchdisplay und einem Touchwheel erlaubt es eine intuitive Bedienung. Der DHE Connect besitzt zudem bemerkenswerte multimediale Qualitäten. Er kann über Wi-Fi unkompliziert mit dem heimischen Netzwerk verbunden werden und ermöglicht z. B. das Abspielen von Internetradio mittels integriertem Lautsprecher ebenso wie die Präsentation der Wetterlage auf einem farbigen Display. Auch die Warmwasserbereitung kann bequem mit einer für das Mobiltelefon verfügbaren App gesteuert werden. Die Durchlauferhitzer sind für die Warmwasserversorgung ausgestattet mit der leistungsfähigen 4i Technologie, welche äußerst effizient die exakte Wassertemperatur bereitstellt. Sehr gut durchdacht ist dabei außerdem, dass sich das Bedienteil für eine externe Nutzung herausnehmen lässt. Mit ihrer ovalen Grundform und markanten Linienführung besitzen DHE Connect & DHE Touch die Anmutung schwebender Eleganz. Sie fügen sich perfekt in die Architektur ein und vereinen dabei sehr stimmig Technologie und Design.

Statement by the jury

Aside from the user-friendliness of a fully electronic controlled instantaneous water heater, DHE Connect offers a new multimedia experience of Internet radio and the use of a weather app via Wi-Fi. The innovative user interface renders use intuitive and, furthermore, the detachable control element is highly practical. Powerful lines and an overall clear design make visible the technical competence of this piece of equipment.

Begründung der Jury

Neben der Nutzerfreundlichkeit eines vollelektronischen Durchlauferhitzers bietet der DHE Connect zusätzlich neue multimediale Erfahrungen wie Internetradio und die Nutzung einer Wetter-App über Wi-Fi. Das innovative User-Interface ermöglicht eine intuitive Bedienung, wobei auch das herausnehmbare Bedienteil ausgesprochen praktikabel ist. Eine kraftvolle Linienführung und insgesamt klare Gestaltung visualisieren die technologische Kompetenz dieser Geräte.

Designer portrait
See page 58
Siehe Seite 58

I8
Instantaneous Water Heater
Durchlauferhitzer

Manufacturer
Guangdong Macro Gas Appliance Co., Ltd.,
Foshan, China
Design
East-Innovation Shunde GD Co., Ltd.
(Sheng Fang, Zhijian Liang), Foshan, China
Web
www.chinamacro.cn
www.east-innovation.com

The I8 electric instantaneous water heater is particularly user-oriented: to simplify operation, it dispenses with unnecessary functions. The user recognises at one glance the amount of residual hot water thanks to a corresponding, centrally placed display. Wireless operation of the unit via Wi-Fi is deliberately kept simple and thus guarantees easy interaction between user, technology and product.

Statement by the jury
With its user-friendliness and clear, well-proportioned design, the I8 electric instantaneous water heater renders a very appealing overall impression.

Der Elektro-Durchlauferhitzer I8 gibt sich betont kundenorientiert: Um die Bedienung zu vereinfachen, verzichtet dieses Gerät auf unnötige Funktionen. Mit einem Blick erkennt der Nutzer, wie viel warmes Wasser vorhanden ist, denn eine entsprechende Anzeige ist an zentraler Stelle platziert. Die kabellose Bedienung des Geräts über Wi-Fi ist bewusst einfach gehalten und gewährleistet dadurch eine gute Interaktion zwischen Nutzer, Technik und Produkt.

Begründung der Jury
Mit seiner Nutzerfreundlichkeit und seiner klaren, wohlproportionierten Gestaltung hinterlässt der Elektro-Durchlauferhitzer I8 einen sehr ansprechenden Gesamteindruck.

DX 45 E
Instantaneous Water Heater
Durchlauferhitzer

Manufacturer
Stiebel Eltron Asia Ltd., Ayutthaya, Thailand
Design
Schumanndesign (Dirk Schumann),
Münster, Germany
Web
www.stiebeleltronasia.com
www.schumanndesign.de

The main focus of this instantaneous water heater is centred on its control elements. The features are intuitively understandable and ensure easy handling in the immediate shower area. Flowing forms lend elegance to the appliance, and they help to prevent the user from colliding with the device when installed in wet areas. An Aerojet system provides efficiency, while a special tank reduces water and energy consumption.

Statement by the jury
Its elegant design turns the DX 45 E instantaneous water heater into an object that enhances any bathroom environment.

Die Steuerungselemente stehen im Mittelpunkt dieses Durchlauferhitzers. Ihre Funktionen sind für eine problemlose Handhabung im unmittelbaren Duschbereich intuitiv erfassbar. Fließende Formen verleihen dem Gerät Eleganz, und sie verhindern, dass sich der Nutzer an dem Gerät stößt, wenn es im Nassbereich installiert ist. Ein Aerojet-System sorgt für Effizienz, ein spezieller Tank reduziert Energie- und Wasserverbrauch.

Begründung der Jury
Seine elegante Gestaltung macht den Durchlauferhitzer DX 45 E zu einem Objekt, das eine Bereicherung für jedes Badezimmer ist.

Therm 6000i S
Instantaneous Gas Water Heater
Gas-Durchlauferhitzer

Manufacturer
Bosch Termotecnologia, S.A., Aveiro, Portugal
Design
TEAMS Design GmbH, Esslingen, Germany
Web
www.bosch.pt
www.teamsdesign.com

The Therm 6000i S instantaneous gas water heater offers high warm-water comfort combined with favourable energy efficiency. The device features an intuitive touch user interface with a TFT colour display, and its special safety glass front renders the technology working inside visible on the outside. The Therm 6000i S also includes integrated Bluetooth connectivity so that the user can change settings or retrieve information with mobile devices.

Statement by the jury
Its appealing design turns the Therm 6000i S into an eye-catcher at home, and its sophisticated technology is another convincing feature.

Der Gas-Durchlauferhitzer Therm 6000i S bietet hohen Warmwasserkomfort bei gleichzeitig guter Energieeffizienz. Das Gerät wird über ein TFT-Farbdisplay mit Touch-Funktion bedient, seine Front aus Sicherheitsspezialglas macht die im Inneren arbeitende Technik außen sichtbar. Weiterhin verfügt der Therm 6000i S über eine integrierte Bluetooth-Schnittstelle, sodass der Nutzer auch auf mobilen Endgeräten Einstellungen vornehmen oder Informationen abrufen kann.

Begründung der Jury
Seine aufmerksamkeitsstarke Gestaltung macht den Therm 6000i S zu einem Blickfang im Haus, zudem überzeugt seine ausgereifte Technik.

Casarte CV
Gas Water Heater
Gas-Warmwasserbereiter

Manufacturer
Haier Group, Qingdao, China
Design
Haier Innovation Design Center
(Dai Nanhai, Song Lei, Wang Haoxing, Hu Xiaodong,
Sun Yan, Liu Haibo), Qingdao, China
Web
www.haier.com

The body of the Casarte CV gas water heater combines dark stainless steel with a glass display that is also dark and imitates its metal counterpart. In this way, the heater appears to be made from one single material. This impression is reinforced by the fact that there are no screws visible anywhere. The user interface of the display is easily handled, and the device is equipped with safe, eco-friendly technology. A particularly user-friendly feature is the prevention of fluctuations in the water temperature.

Das Gehäuse des Gas-Warmwasserbereiters Casarte CV kombiniert dunklen Edelstahl mit einem Display aus ebenfalls dunklem Glas, welches das Metall imitiert. Auf diese Weise wirkt das Gerät wie aus einem Stück gefertigt. Dieser Eindruck wird dadurch verstärkt, dass keinerlei Schrauben zu erkennen sind. Die Benutzeroberfläche des Displays lässt sich einfach bedienen. Das Gerät ist mit umweltfreundlicher, sicherer Technik ausgerüstet. Besonders benutzerfreundlich ist, dass Schwankungen der Wassertemperatur verhindert werden.

Statement by the jury
With its seamless design and the use of sophisticated materials, the Casarte CV gas water heater has an elegant appearance and integrates well into diverse interiors.

Begründung der Jury
Durch seine nahtlos anmutende Gestaltung und die Verwendung edler Materialien wirkt der Gas-Warmwasserbereiter Casarte CV elegant und integriert sich gut in verschiedene Interieurs.

RSJ-N15
Heat Pump
Wärmepumpe

Manufacturer
Midea Heating & Ventilating Equipment
Co., Ltd., Foshan, China
In-house design
Weitao Chen, Mingyuan Lu
Web
www.midea.com

The RSJ-N15 is an all-in-one heat pump for water heaters. The robust device shows a slim and elegant design language. The energy-saving heat pump is equipped with a control interface for iPhone, iPad or PC and can thus be operated remotely. Based on everyday usage habits, the pump is capable of suggesting the most energy-efficient operation mode for any given time.

Statement by the jury
The solid and high-grade design of this heat pump expresses reliability. Its control interface for communication devices pays respect to a technology-friendly society.

RSJ-N15 ist eine All-in-one-Wärmepumpe für Warmwasserbereiter. Das robuste Gerät zeigt eine schlanke und elegante Formensprache. Die energiesparende Wärmepumpe ist mit einer Schnittstelle für iPhone, iPad oder PC ausgestattet und ist so über eine Fernsteuerung zu bedienen. Basierend auf der alltäglichen Nutzung ist das Gerät in der Lage, den jeweils energieeffizientesten Betriebsmodus vorzuschlagen.

Begründung der Jury
Die solide wie hochwertige Gestaltung dieser Wärmepumpe drückt Zuverlässigkeit aus. Ihre Schnittstelle für Kommunikationsgeräte zollt einer technikfreundlichen Gesellschaft Respekt.

Vitosorp 200-F
Gas Adsorption Heat Pump
Gas-Adsorptionswärmepumpe

Manufacturer
Viessmann Werke GmbH & Co. KG,
Allendorf/Eder, Germany
Design
Phoenix Design GmbH + Co. KG,
Stuttgart, Germany
Web
www.viessmann.com
www.phoenixdesign.com

The Vitosorp 200-F gas adsorption heat pump enables economical and eco-friendly heating. Its zeolite module is hermetically sealed and thus completely maintenance-free. A low level of operating noise and non-harmful pairing of the materials zeolite and water allow the pump to be operated even directly in the living area. The clear design language and high-quality metal surfaces with the orange-silver logotype render it an unmistakable member of the manufacturer's product family.

Statement by the jury
The convincing combination of environmentally friendly technology and timeless aesthetics characterises the Vitosorp 200-F gas adsorption heat pump.

Die Vitosorp 200-F Gas-Adsorptionswärmepumpe ermöglicht sparsames und umweltfreundliches Heizen. Ihr Zeolithmodul ist hermetisch abgeschlossen und damit komplett wartungsfrei. Niedrige Betriebsgeräusche und die unschädliche Stoffpaarung, bestehend aus Zeolith und Wasser, erlauben den Betrieb direkt im Wohnbereich. Die klare Linienführung sowie die hochwertigen metallischen Flächen mit der orange-silberfarbenen Wortmarke ordnen das Gerät unverkennbar der Produktfamilie des Herstellers zu.

Begründung der Jury
Die bestechende Kombination aus umweltfreundlicher Technologie und zeitloser Ästhetik zeichnet die Vitosorp 200-F Gas-Adsorptionswärmepumpe aus.

neeoQube
Energy Storage System
Energiespeicher

Manufacturer
AKASOL GmbH, Darmstadt, Germany
In-house design
Design
Dennis Redmonds, Munich, Germany
Web
www.akasol.com

The neeoQube is a solar energy storage system with which households can cover up to 70 per cent of their energy needs. With its environmentally friendly and durable lithium-ion battery technology, the device has a storage capacity of 5.5 kWh. The system can be fully charged in just one hour and releases energy incrementally as needed. The passive cooling system enables silent operation. With its purist design, the unit can be mounted to the wall in a space-saving way.

neeoQube ist ein Solarspeicher, mit dem Haushalte bis zu 70 Prozent ihres Energiebedarfs decken können. Mit seinem umweltschonenden und langlebigen Lithium-Ionen-Speicher besitzt das Gerät eine Speicherkapazität von 5,5 kWh. Der Speicher ist innerhalb von einer Stunde aufgeladen und gibt die Energie ganz nach Bedarf ab. Das passive Kühlsystem ermöglicht einen geräuschlosen Betrieb. Das puristisch gestaltete Gerät kann platzsparend an einer Wand aufgehängt werden.

Statement by the jury
Efficient, environmentally friendly technology and discreet, modern design go hand in hand in the case of the neeo-Qube energy storage system.

Begründung der Jury
Bei dem Energiespeicher neeoQube gehen effiziente, umweltfreundliche Technologie und dezente, moderne Gestaltung Hand in Hand.

auroFLOW
Solar Station
Solarstation

Manufacturer
Vaillant GmbH, Remscheid, Germany
In-house design
Web
www.vaillant.com

The functional construction of the auro-FLOW solar station facilitates – through the interplay of high-quality, innovative components and a power-saving, high-efficiency pump – the efficient operation of a solar station with reduced energy consumption. The colouring of the round instruments in the speedometer design describes their function intuitively; against the high-gloss surface they stand out as a highlight. The casing is made of eco-friendly materials.

Die funktionale Bauweise der Solarstation auroFLOW ermöglicht im Zusammenspiel mit hochwertigen, innovativen Bauelementen sowie einer stromsparenden Hocheffizienzpumpe den effizienten Betrieb einer Solaranlage bei gesenktem Energieverbrauch. Die Farbgebung der Rundinstrumente im Tachodesign beschreibt intuitiv deren Funktion, auf der Hochglanzoberfläche treten sie als Highlight hervor. Das Gehäuse ist aus umweltverträglichen Materialien gefertigt.

Statement by the jury
The auroFLOW solar station meets high demands for sophisticated technology and attracts attention with its eye-catching design.

Begründung der Jury
Die Solarstation auroFLOW befriedigt auch hohe Ansprüche an eine ausgefeilte Technik. Mit ihrer aufmerksamkeitsstarken Gestaltung zieht sie die Blicke auf sich.

ecoTEC exclusive
Wall-Mounted Gas Boiler
Wand-Gasheizgerät

Manufacturer
Vaillant GmbH, Remscheid, Germany
In-house design
Web
www.vaillant.com

The ecoTEC exclusive wall-mounted condensing gas boiler of the Green iQ product range has very low energy consumption due to intelligent technology and sensors. The eco-friendly device can be powered by renewable energy. It features an integrated, Internet-based interface enabling cross-linking and convenient remote operation via an app. The classic, discreet design of this white device sets eye-catching accents with the green decorative strip and the Green iQ label.

Statement by the jury
The design of this wall-mounted condensing gas boiler is focused on eco-friendliness. In addition, the device is highly user-friendly.

Das Wand-Gasheizgerät ecoTEC exclusive der Produktreihe Green iQ bietet durch den Einsatz intelligenter Technik und Sensoren einen sehr geringen Energieverbrauch. Das umweltfreundliche Gerät kann mit erneuerbaren Energien betrieben werden. Es verfügt über eine integrierte Internet-schnittstelle, was eine Vernetzung sowie die bequeme Fernsteuerung über eine App ermöglicht. Die klassisch-dezente Gestaltung des weißen Geräts setzt mit der grünen Schmuckleiste und dem Label Green iQ aufmerksamkeitsstarke Akzente.

Begründung der Jury
Der Fokus der Gestaltung liegt bei diesem Wand-Gasheizgerät auf ökologischer Verträglichkeit. Zudem ist es in hohem Maße nutzerfreundlich.

S10 MINI
Power Station
Hauskraftwerk

Manufacturer
E3/DC GmbH, Osnabrück, Germany
In-house design
Prof. Marian Dziubiel
Web
www.e3dc.com

The S10 MINI power station is an all-in-one device combining the self-production and storage of electrical energy for houses by employing lithium-ion storage capability. The unit measures 1 x 1 metres and is particularly flat and compact. Its energy management concept includes an integrated power station which, in terms of energy storage and performance, can be used for photovoltaic installations and CHP systems. Due to its compact, square form, the station blends well into any given space – either mounted to the wall or equipped with a base stand.

Statement by the jury
With sophisticated functionality as well as compact and timeless design, the S10 MINI power station renders a convincing impression.

Das Hauskraftwerk S10 MINI ist ein All-in-one-Gerät, welches durch einen Lithium-Ionen-Speicher die Eigenproduktion und Speicherung elektrischer Energie für Häuser vereint. Das Gerät misst 1 x 1 Meter und ist besonders flach und kompakt gebaut. Durch sein Energiemanagement stellt es ein integriertes Kraftwerk dar, das in Bezug auf Speicher und Leistung für Photovoltaik- und KWK-Anlagen verwendbar ist. Dank seiner kompakten quadratischen Form integriert sich das Gerät gut in die gegebenen Räumlichkeiten, es wird entweder an einer Wand befestigt oder mit einem Standfuß aufgestellt.

Begründung der Jury
Mit ausgereifter Funktionalität, kompakter Bauweise und zeitloser Anmutung überzeugt das Hauskraftwerk S10 MINI.

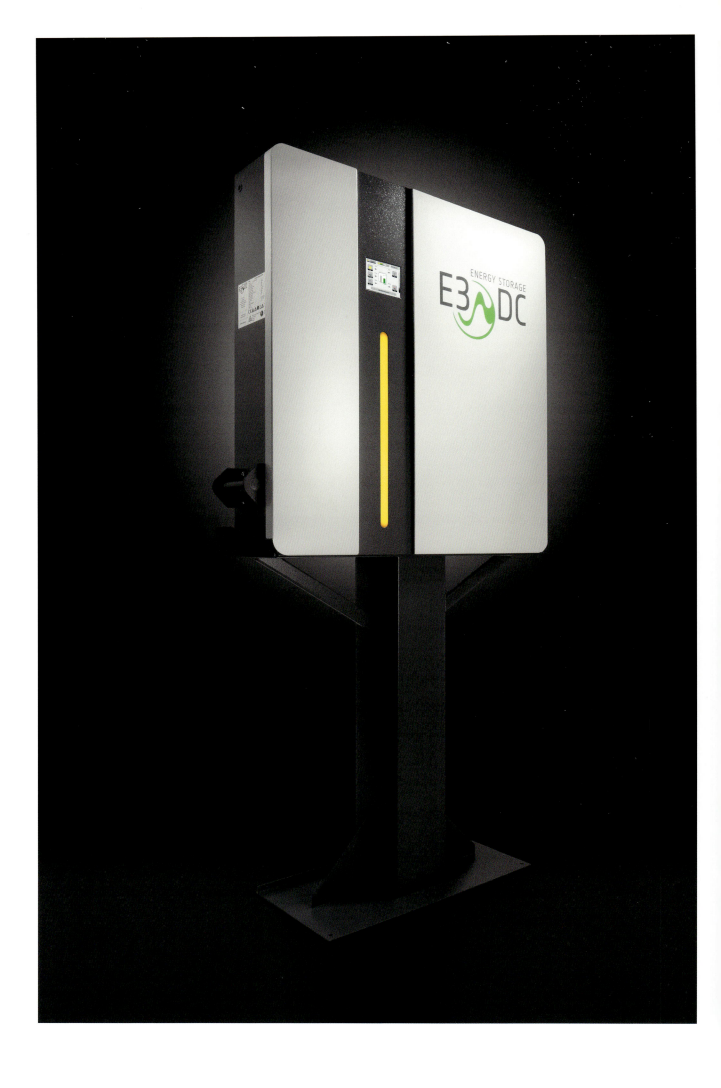

SBBE 301 WP
Hot Water Storage Tank
Warmwasserspeicher

Manufacturer
Stiebel Eltron GmbH & Co. KG,
Holzminden, Germany
Design
burmeister industrial design (Kay Burmeister),
Hannover, Germany
Web
www.stiebel-eltron.de
www.burmeister-id.de

The strength of the SBBE 301 WP lies
in domestic hot water production. In
combination with heat pumps, its 300
litre storage tank is used in single-family
homes. The reinforced, directly applied
insulation provides good energy efficiency. With its rectangular base form,
the striking design faceplate and matching colouring, the device unmistakably
belongs to the manufacturer's Renewable Energies product range.

Statement by the jury
This hot water storage tank combines the
application of efficient technology with
a classically plain design language.

Die Stärke des SBBE 301 WP liegt in der
Trinkwarmwasserbereitung im privaten
Bereich. In Verbindung mit Heizungswär-
mepumpen kommt sein 300-Liter-Speicher
in Einfamilienhäusern zum Einsatz. Die
verstärkte Wärmedämmung sorgt für eine
gute Energieeffizienz. Mit seiner rechteckigen Grundform, der prägnanten Design-
blende und der abgestimmten Farbgebung
gehört das Gerät unverkennbar zum
Produktbereich „Erneuerbare Energien"
des Herstellers.

Begründung der Jury
Dieser Warmwasserspeicher verbindet
den Einsatz effizienter Technik mit einer
klassisch-schlichten Formensprache.

BALLOFIX®
Ball Valve
Kugelhahn

Manufacturer
BROEN A/S, Assens, Denmark
In-house design
Design
Attention, Copenhagen, Denmark
Web
www.broen.com
www.attention-group.com

The new design of the Ballofix ball valve aims at improving energy efficiency, reducing installation time and cutting costs. Traditional materials and production methods were deliberately dispensed with to make the valve more compact, lightweight and material-efficient. It appears like a natural part of the piping system and is equipped with a spring-supported sealing system. Its handle communicates basic functions, for instance indicating when the valve is open or closed.

Die Neugestaltung des Ballofix-Kugelhahns zielt darauf ab, die Energieeffizienz zu erhöhen, die Einbauzeiten zu verringern sowie die Kosten zu reduzieren. Traditionelle Materialien und Herstellungsmethoden wurden bewusst ersetzt, um ihn kompakter, leichter und materialeffizienter zu gestalten. Das Ventil, das wie ein natürlicher Teil einer Rohrinstallation wirkt, ist mit einem federunterstützten Dichtungssystem ausgestattet. Sein Griff kommuniziert grundlegende Funktionen, etwa wann es geöffnet oder verschlossen ist.

Statement by the jury
The design of this ball valve is geared to high functionality and, at the same time, presents a sophisticated appearance.

Begründung der Jury
Die Gestaltung dieses Kugelhahns ist auf hohe Funktionalität und Effizienz bei gleichzeitig formschöner Anmutung ausgerichtet.

SLB130
Valve
Ventil

Manufacturer
ESBE AB, Reftele, Sweden
In-house design
Filip Celander, Martin Kårhammer
Design
Per Liljeqvist Design AB, Anderstorp, Sweden
Web
www.esbe.eu

The Superflow valve is based on an innovative product platform that allows for easy adjustments to customers' wishes in a broad range of applications. Despite its compact size, it possesses remarkable flow characteristics. The flow value is twice as high when compared to conventional valves, yet without compromising controllability. The integrated actuator features intelligent functions.

Das Superflow-Ventil basiert auf einer innovativen Produktplattform, die unkomplizierte Anpassungen an Kundenbedürfnisse in einer großen Applikationsbandbreite ermöglicht. Trotz der kompakten Größe besitzt es beachtliche Fließeigenschaften. Der Volumenstrom ist doppelt so hoch wie bei konventionellen Ventilen, jedoch ohne Kompromisse in der Regulierbarkeit. Der integrierte Stellmotor verfügt über intelligente Funktionen.

Statement by the jury
The powerful design of the SLB130 valve expresses reliability and security, while communicating high-quality standards.

Begründung der Jury
Die kraftvolle Formgebung des Ventils SLB130 drückt Zuverlässigkeit sowie Sicherheit aus und kommuniziert einen hohen Qualitätsanspruch.

GRA111
Circulation Unit
Pumpengruppe

Manufacturer
ESBE AB, Reftele, Sweden
In-house design
Torbjörn Lönkvist
Design
Per Liljeqvist Design AB, Anderstorp, Sweden
Web
www.esbe.eu

GRA111 embodies a series of circulation units for heating and cooling applications. Unique, innovative technology serves to control the risk of over-dimensioning or low valve authority. The unit is completely assembled, tightness-tested and heat-insulated. Its installation is simply achieved using plug-and-play connectivity. With its high-efficiency circulation pump and tailor-made installation, the GRA111 works in an economical and eco-friendly way.

Hinter GRA111 verbirgt sich eine Pumpengruppe für Heiz- und Kühlanwendungen. Durch neuartige Technologie lässt sich hier das Risiko für Überdimensionierung oder geringe Ventilautorität kontrollieren. Die Einheit ist komplett montiert, auf Dichtheit geprüft und wärmeisoliert. Ihre Installation erfolgt einfach durch Plug-and-play. Aufgrund ihrer Hocheffizienzpumpe und der maßgeschneiderten Isolierung arbeitet GRA111 wirtschaftlich und umweltschonend.

Statement by the jury
Well-conceived functionality, easy handling and a classic, elegant look are the design features of this group of circulation units.

Begründung der Jury
Gut durchdachte Funktionalität, einfache Handhabung und ein klassisch-elegantes Aussehen sind die Gestaltungsmerkmale dieser Pumpengruppe.

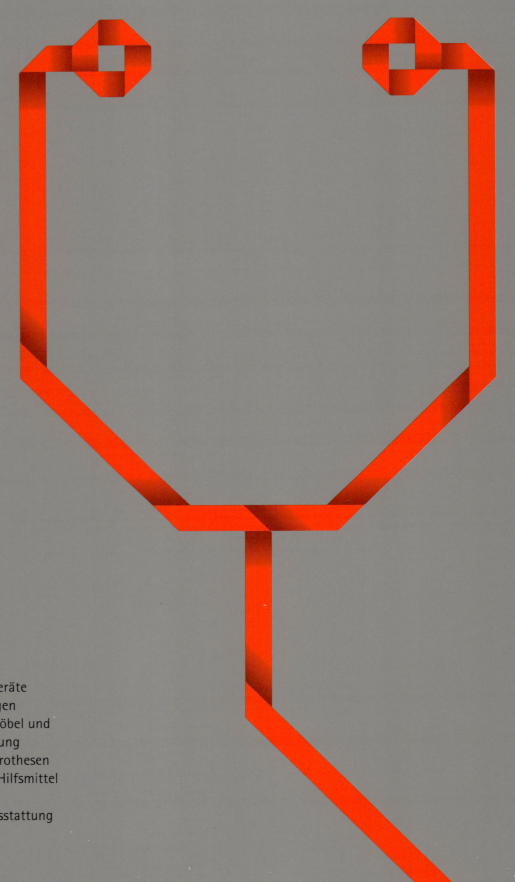

General practice and
hospital fittings
Laboratory technology
and furniture
Medical devices
and equipment
Mobility, care and
communication aids
Orthopaedic aids
Orthoses and prostheses
Rehabilitation

Labortechnik
und -mobiliar
Medizinische Geräte
und Ausrüstungen
Medizinische Möbel und
Sanitärausstattung
Orthesen und Prothesen
Orthopädische Hilfsmittel
Praxis- und
Krankenhausausstattung
Rehabilitation

Life science and medicine
Life Science und Medizin

FDR D-EVO II Series (C43, C35, G43, G35)
Cassette-Type Digital X-Ray Imaging Diagnostic Device
Digitales Kassetten-Röntgendiagnosegerät

Manufacturer
FUJIFILM Corporation,
Tokyo, Japan

In-house design
Ryosuke Ogura

Web
www.fujifilm.com

reddot award 2015
best of the best

A symbiosis of form and function

There are many uses for X-ray machines in the medical world. Every clinical specialist area has its own requirements. With this in mind, the FDR D-EVO II cassette-type digital X-ray imaging diagnostic device is configured with an innovative shape and function. The elegantly designed equipment is extremely light and easy to transport. As it can be used both wirelessly and with cable connection, it suits a wide range of situations. The rear corners are narrow and curved, so that users find the device ergonomically easy and pleasant to use. The sophisticated functionality of the device is enhanced by its high compatibility with other equipment and its performance. As it has been designed in standard IEC size, it can be installed in existing facilities at minimal cost. Its original image receiver and noise-reduction circuit offer high-quality images at a low radiation dose, thus keeping patients' exposure to a minimum. Its waterproof casing and anti-bacterial coating fulfil the highest hygiene standards, and the image size and quality can be adapted to perfectly meet the different demands of the radiology and other clinical departments. This X-ray diagnostic device works thanks to the successful symbiosis of form and function, which serves and enhances everyday medical work in an outstanding way.

Symbiose von Form und Funktion

Im medizinischen Bereich werden Röntgengeräte sehr vielseitig eingesetzt. Jede klinische Fachrichtung hat dabei ihre eigenen Ansprüche. Das digitale Kassetten-Röntgendiagnosegerät FDR D-EVO II bietet vor diesem Hintergrund eine innovative Form und Funktionalität. Dieses mit eleganter Anmutung gestaltete Gerät ist außerordentlich leicht und lässt sich komfortabel transportieren. Da es sowohl kabellos wie auch verkabelt verwendet werden kann, erlaubt es weitreichende Möglichkeiten des Einsatzes. Die hinteren Ecken sind schmal und gebogen geformt, weshalb es von allen Beteiligten auch in ergonomischer Hinsicht als sehr angenehm empfunden wird. Die ausgefeilte Funktionalität des Gerätes geht zudem einher mit einer hohen Kompatibilität und Leistungsfähigkeit. Da es in IEC-Standardgröße gestaltet ist, kann es auch gut in vorhandenen Anlagen zu minimalen Kosten installiert werden. Sein Originalbildempfänger und die Rauschunterdrückungsschaltung bieten hochwertige Bilder bei niedriger Strahlungsdosis, was auch die Belastung der Patienten verringert. Mit seinem wasserdichten Gehäuse sowie einer antibakteriellen Beschichtung erfüllt es zudem höchste Hygienestandards. Bildgröße und Bildqualität lassen sich den unterschiedlichen Anforderungen der Radiologie und anderer klinischer Einrichtungen perfekt anpassen. Dieses Röntgendiagnosegerät begeistert dabei immer wieder durch seine überaus gelungene Symbiose von Form und Funktion, die den Alltag in der Medizin hervorragend begleitet und bereichert.

Statement by the jury

The design of the FDR D-EVO II cassette-type digital X-ray imaging diagnostic device represents a successful combination of technology and user-friendliness. The novel ways in which this item can be used redefine such a product. Its light weight and the special shape of its edges are particularly impressive and also leave nothing to be desired from an ergonomic point of view for neither patient nor user.

Begründung der Jury

Die Gestaltung des digitalen Kassetten-Röntgendiagnosegeräts FDR D-EVO II verwirklicht eine äußerst gelungene Kombination aus Technologie und Benutzerfreundlichkeit. Die Art und Weise, wie dieses Instrument eingesetzt werden kann, definiert ein solches Produkt neu. Beeindruckend sind dabei das sehr leichte Gehäuse und die besondere Form der Kanten. Es lässt daher auch in ergonomischer Hinsicht bei Anwendern und Patienten keine Wünsche offen.

Designer portrait
See page 60
Siehe Seite 60

uMR 570
Magnetic Resonance Imaging System
Magnetresonanztomograph

Manufacturer
Shanghai United Imaging Healthcare Co., Ltd.,
Shanghai, China
In-house design
Corporate Design Innovation Center
Web
www.united-imaging.com

The uMR 570 magnetic resonance imaging system offers high ease of use. The touchscreen and the keypad are integrated in the portal, so that the operator is close to the patient and can better focus on them. Ventilation, ambient light and music can be controlled via the touchscreen in accordance with the patient's wishes. The aluminium control panel, which is crafted from one piece and has radiused edges, is slightly angled and therefore easy to use. An ultra-wide bore with ambient light supported by advanced fiber technology helps release claustrophobia anxiety. Moreover, the device does not require any helium during normal use, which makes it environmentally friendly.

Der Magnetresonanztomograph uMR 570 bietet einen hohen Bedienkomfort. Touchscreen und Tastenfeld sind in das Portal integriert, sodass der Anwender nah am Patienten ist und sich besser auf ihn konzentrieren kann. Über den Touchscreen lassen sich die Belüftung, das Umgebungslicht sowie die Musikeinstellungen nach den Wünschen des Patienten einrichten. Das aus einem Stück gefertigte Aluminiumbedienfeld mit gerundeten Kanten ist leicht angeschrägt und dadurch bequem zu betätigen. Ein breites Loch mit durch Fasertechnologie erzeugtem Umgebungslicht hilft, den Platzangststress des Patienten zu lindern. Zudem verbraucht das Gerät bei normalem Betrieb kein Helium, was die Umwelt schont.

Statement by the jury
The uMR 570 captivates with its innovative touchscreen functionality and ergonomic qualities, which correspond well to a modern medical environment.

Begründung der Jury
Der uMR 570 begeistert durch seine innovativen Touchscreen-Funktionen und ergonomischen Qualitäten, die sehr gut in ein modernes Medizinumfeld passen.

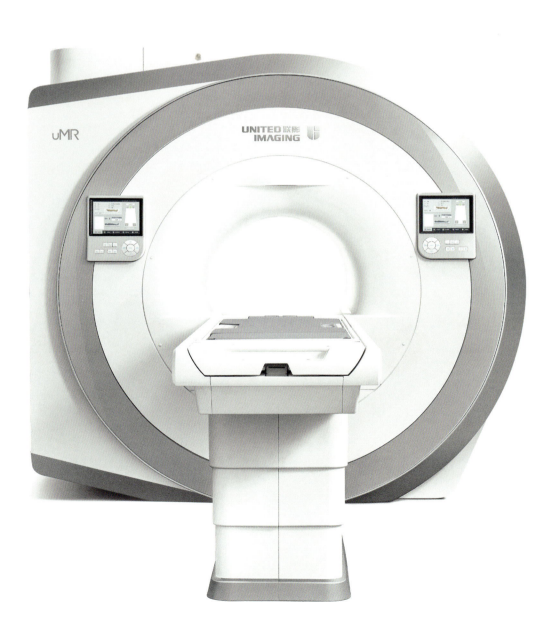

uMI 510
PET/CT System

Manufacturer
Shanghai United Imaging Healthcare Co., Ltd.,
Shanghai, China
In-house design
Corporate Design Innovation Center
Web
www.united-imaging.com

The uMI 510 combines positron emission tomography (PET) with computed tomography (CT). Compared to conventional systems, both the radioactive tracers and the X-ray dose are reduced by more than 50 per cent. Its ultra-fast scan speed minimises the time patients have to spend in the tunnel, which equates to less stress, especially for those suffering from claustrophobia. The warm ambient light enhances relaxation further. The patient table contains sensors that ensure collision-free operation.

Das uMI 510 kombiniert die Positronen-Emissions-Tomographie (PET) mit der Computertomographie (CT). Im Vergleich zu herkömmlichen Systemen werden radioaktive Tracer und Röntgenstrahlen um mehr als 50 Prozent reduziert. Dank der ultraschnellen Scangeschwindigkeiten verkürzt sich der Zeitraum, in dem der Patient in der Röntgenröhre liegen muss, was besonders für Menschen mit Platzangst weniger Stress bedeutet. Das warme Umgebungslicht wirkt zusätzlich beruhigend. Im Patiententisch befinden sich Sensoren, die einen kollisionsfreien Betrieb sicherstellen.

Statement by the jury
This PET/CT system conveys an impression of openness and lightness. The recessed base gives the table a floating appearance.

Begründung der Jury
Das PET/CT-Gerät vermittelt den Eindruck von Offenheit und Leichtigkeit. Der eingerückte Sockel verleiht dem Tisch eine schwebende Anmutung.

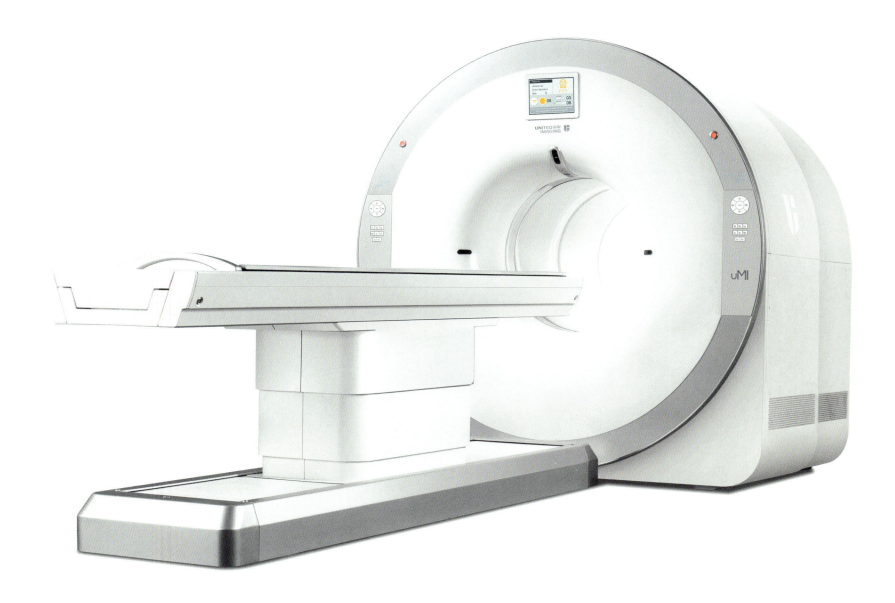

uCT 510
Computed Tomography Scanner
Computertomograph

Manufacturer
Shanghai United Imaging Healthcare Co., Ltd.,
Shanghai, China
In-house design
Corporate Design Innovation Center
Web
www.united-imaging.com

The uCT 510 is a space-saving, low-energy computed tomography scanner that delivers high image quality with a low X-ray dose. The system features advanced technology that greatly reduces the preparation time prior to each scan. Moreover, the design ensures a quiet scanning process while the tube rotates at high speed. The gantry features a nano-coating, which contributes to improved hygiene. The buttons are covered with a ceramic-like finish that prevents wear and tear.

Der Computertomograph uCT 510 ist platzsparend, verbraucht wenig Energie und liefert eine hohe Bildqualität bei geringer Strahlendosis. Das System verfügt über eine hochentwickelte Technologie, welche die Vorbereitungszeit auf die Untersuchung stark verkürzt. Darüber hinaus ermöglicht sie einen leisen Scanvorgang, während die Röhre sich mit hoher Geschwindigkeit dreht. Die Öffnung ist mit einer Nanobeschichtung versiegelt, was zu einer besseren Hygiene beiträgt. Die Tasten sind mit einer keramikähnlichen Veredelung überzogen, die Verschleiß vorbeugt.

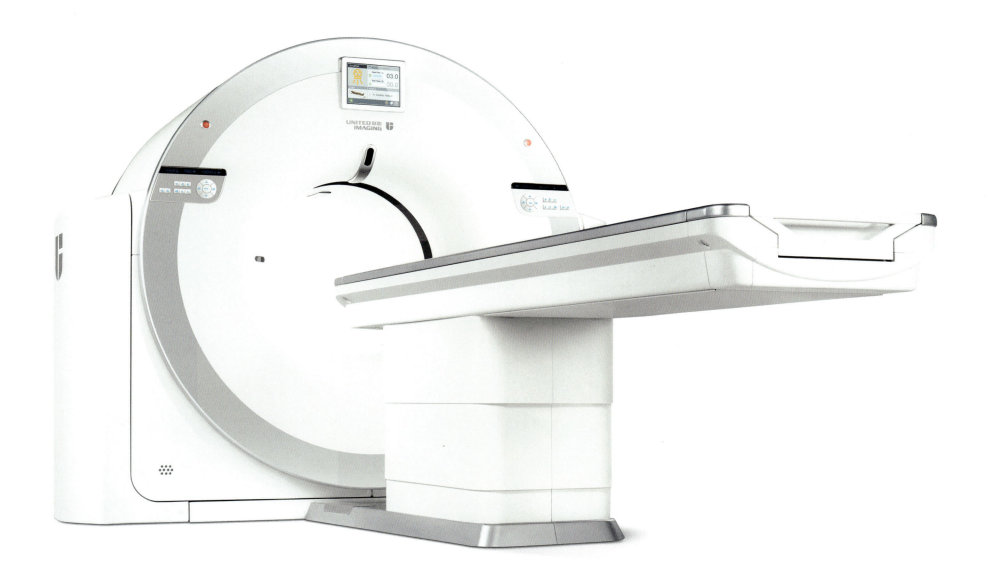

PRIMUS 200
Optical Coherence Tomography Scanner
Optischer Kohärenztomograph

Manufacturer
Carl Zeiss Suzhou Co. Ltd.,
Suzhou, China
Design
Shanghai MOMA Industrial Product Design
(Rong Rong Chen),
Shanghai, China
Web
www.meditec.zeiss.com
www.designmoma.com

The PRIMUS 200 is an imaging device for optical coherence tomography, which is used by ophthalmologists for the diagnostic examination of the patient's retina. The manual joystick control makes it particularly easy to use. The alignment of the patient and the triggering of the imaging process are carried out with one hand via the joystick and the integrated button. Thanks to the optical head's slim housing, the ophthalmologist has a clear view of the patient at all times. The slim swivel arm is a seamless connection between the optical head and the cross table.

Der PRIMUS 200 ist ein optischer Kohärenztomograph, der in der Augenheilkunde zur diagnostischen Untersuchung der Netzhaut eingesetzt wird. Durch die manuelle Joystick-Steuerung ist das Bildgebungssystem besonders einfach zu bedienen. Das Ausrichten und Auslösen der Bildaufnahme erfolgt mit nur einer Hand über den Joystick und den integrierten Knopf. Dank der schmalen Gehäuseform des Optikkopfes hat der Arzt den Patienten stets gut im Blick. Der schlanke Schwenkarm stellt eine nahtlose Verbindung zwischen Optikkopf und Kreuztisch her.

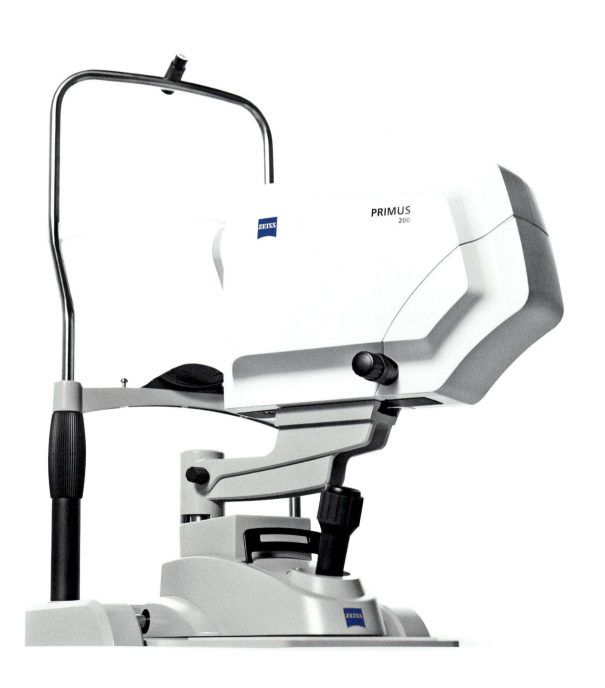

ORTHOPHOS SL
Dental X-Ray Unit
Dentales Röntgengerät

Manufacturer
Sirona Dental Systems GmbH,
Bensheim, Germany
Design
einmaleins (Jan Möller, Judith Tenzer),
Burgrieden, Germany
Web
www.sirona.com
www.einmaleins.net

The Orthophos SL 2D/3D X-ray unit supports the user in the dental office through excellent image quality combined with patented patient positioning aids and intuitive software. On the 2D side, the advanced DCS sensor and Sharp Layer technology provide panoramic images with unparalleled sharpness. For 3D imaging, the user can take advantage of volume sizes up to 11 x 10 cm in order to capture images suitable for any treatment situation.

Statement by the jury
The Orthophos SL shines in glossy white, which in combination with its smooth contours expresses a sense of purity.

Das 2D-/3D-Röntgengerät Orthophos SL unterstützt den Anwender in der Zahnarztpraxis durch eine exzellente Bildqualität kombiniert mit patentierten Positionierungshilfen und einer intuitiven Software. Auf der 2D-Seite liefern der fortschrittliche DCS-Sensor und die Sharp-Layer-Technologie Panoramaaufnahmen mit einer sehr hohen Zeichenschärfe. Im 3D-Bereich besteht die Möglichkeit, die Volumengröße bis 11 x 10 cm auszuwählen und somit jeder Behandlungssituation anzupassen.

Begründung der Jury
Das Orthophos SL erstrahlt in hochglänzendem Weiß, das zusammen mit den glatten Konturen Reinheit ausdrückt.

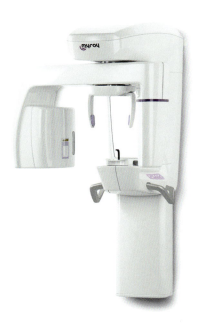

MyRay Hyperion X5
Dental
Panoramic X-Ray System
Dentales
Panoramaröntgengerät

Manufacturer
Cefla s.c., Imola, Italy
Design
Giulio Mattiuzzo Architetto, Rimini, Italy
Web
www.cefla.com
www.myray.it

The MyRay Hyperion X5 is a space-saving dental panoramic X-ray system, which gives the patient unrestricted freedom of movement and is also suitable for installation in emergency vehicles. Its mirror makes the experience of image acquisition less claustrophobic for the patient. Moreover, it allows the practitioner to see the patient's face better. The metal handles and the head positioning system ensure that the patient remains stable.

Statement by the jury
Even though the MyRay Hyperion X5 has a very compact and flat design, its proportions appear remarkably luxurious and generous.

Das MyRay Hyperion X5 ist ein platzsparendes dentales Panoramaröntgengerät, das den Patienten nicht einengt und sich auch für den Einbau in Rettungsfahrzeuge eignet. Der Spiegel macht die Bildaufnahme für den Patienten zu einem weniger beklemmenden Erlebnis. Zudem kann der Behandler das Gesicht des Patienten besser im Blick behalten. Die Metallgriffe und das Positioniersystem für den Kopf sorgen dafür, dass der Patient stabil bleibt.

Begründung der Jury
Obwohl das MyRay Hyperion X5 so kompakt und flach gestaltet ist, wirkt es durch seine Proportionen außergewöhnlich luxuriös und großzügig.

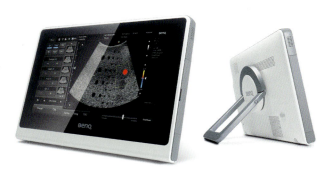

Viva pro
Ultrasound Tablet
Ultraschall-Tablet

Manufacturer
Qisda Corporation, Taipei, Taiwan
Design
BenQ Lifestyle Design Center, Taipei, Taiwan
Web
www.qisda.com
www.benq.com

Viva pro is an ultrasound tablet with a 10" touchscreen base station. The device weighs only 1.9 kg and is particularly energy efficient. The waterproof and impact-resistant housing is suited for mobile use outdoors or in ambulances. The handle at the base can be pulled out and employed as either a stand or a handle. For complex examinations, the tablet may be attached to a trolley that accommodates probes, plastic bottles and cables.

Statement by the jury
Viva pro captivates with its touchscreen, which does not require additional control elements and yields a self-contained design.

Viva pro ist ein Ultraschall-Tablet mit einer 10"-Touchscreen-Basisstation. Das Gerät wiegt nur 1,9 kg und ist besonders energieeffizient. Das wasserdichte und stoßfeste Gehäuse ist für den mobilen Einsatz im Freien oder im Krankenwagen geeignet, der Griff an der Basis kann herausgezogen und als Ständer oder Handgriff verwendet werden. Für komplexe Untersuchungen wird das Tablet an einem Rollwagen fixiert, der Sonden, Plastikflaschen und Kabel aufnehmen kann.

Begründung der Jury
Das Design von Viva pro besticht durch den Touchscreen, der ohne weitere Bedienelemente auskommt und eine in sich geschlossene Bauform ermöglicht.

Voluson E10
Ultrasound System
Ultraschallsystem

Manufacturer
GE Healthcare, Zipf, Austria
In-house design
Aurelie Boudier
Web
www.ge.com

The Voluson E10 ultrasound system was developed for modern gynaecology, and for prenatal diagnostics in particular. The device features very high image quality along with advanced technologies, which help the clinician to detect and diagnose complex medical issues at an early stage. The ultrasound images are displayed on a 23" widescreen monitor, on which small details are visible. The 12.1" operating panel with multitouch functionality can be electronically height-adjusted in order to ergonomically adapt to the position of the examiner.

Das Ultraschallsystem Voluson E10 wurde für die moderne Gynäkologie entwickelt, insbesondere für die pränatale Diagnostik. Das Gerät verfügt über eine sehr hohe Bildqualität und fortschrittliche Technologien, die dem Kliniker dabei helfen, komplexe medizinische Fragestellungen frühzeitig zu beantworten. Die Ultraschallbilder werden auf einem 23"-Breitbildmonitor dargestellt, auf dem auch kleine Details erkennbar sind. Das 12,1"-Bedienpanel mit Multitouch-Funktion ist elektromotorisch höhenverstellbar, damit es ergonomisch an die Position des Untersuchers angepasst werden kann.

Statement by the jury
The generous dimensions of the Voluson E10 appear very inviting, with rounded forms additionally softening its appearance.

Begründung der Jury
Die großzügigen Dimensionen des Voluson E10 wirken sehr einladend. Seine gerundeten Formen lockern das Erscheinungsbild zusätzlich auf.

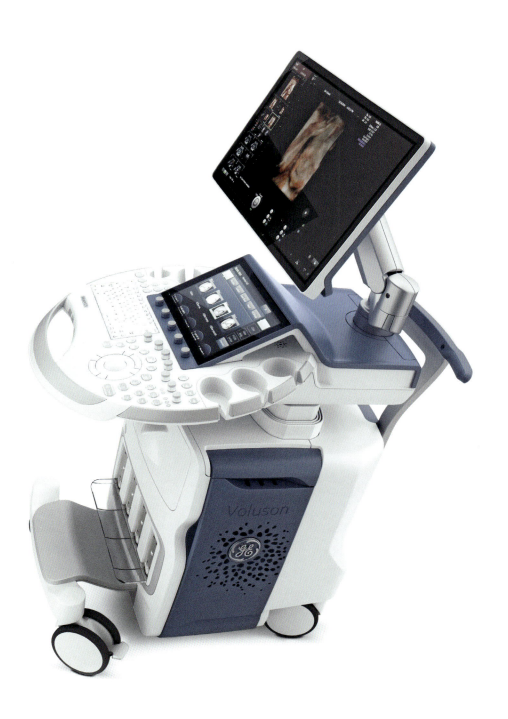

elisa 800 VIT
Intensive Care Ventilator
Intensivrespirator

Manufacturer
SALVIA medical GmbH & Co. KG, Kronberg, Germany
Design
WILDDESIGN GmbH & Co. KG (Dennis Kulage, Oliver Koszel),
Gelsenkirchen, Germany
Web
www.salvia-medical.de
www.wilddesign.de

The elisa 800 VIT is a ventilator for use in intensive care units. All life-support functions are integrated into this compact device and centrally controlled via the large user interface. The ventilator is one of the first systems worldwide to enable fully integrated, non-invasive respiratory monitoring. Ventilation is thus more transparent and intuitively controllable from the touchscreen. Moreover, the interface and the equipment cart form an aesthetic whole due to their homogeneous design.

Der Intensivrespirator elisa 800 VIT ist ein Beatmungsgerät für die klinische Intensivstation. Alle lebenserhaltenden Funktionen sind kompakt integriert und werden zentral über die großzügige Benutzeroberfläche geregelt. Der Respirator ermöglicht als eines der ersten Systeme weltweit ein voll integriertes nichtinvasives Lungenmonitoring. Die Beatmung wird dadurch transparenter und unmittelbar intuitiv über den Touchscreen steuerbar. Das Interface und der Gerätewagen ergeben durch ihr homogenes Design eine ästhetische Einheit.

Statement by the jury
Thanks to the monochrome white and minimalist design language, this intensive care ventilator has a puristic look that is modern and sophisticated.

Begründung der Jury
Durch das monochrome Weiß und die reduzierte Formensprache erzielt der Intensivrespirator eine puristische Optik, die zeitgemäß und wertig ist.

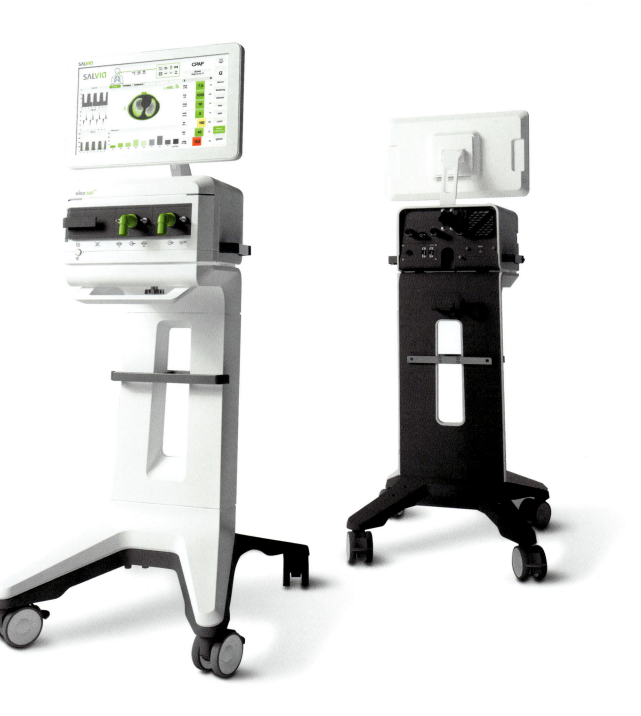

SERVO-n®
Medical Ventilator for Neonatology
Beatmungsgerät für die Neonatologie

Manufacturer
Maquet Critical Care AB, Solna, Sweden
In-house design
Design
Veryday, Bromma, Sweden
Web
www.maquet.com
www.veryday.com

The SERVO-n ventilator was developed especially for the intensive care of newborn babies who cannot breathe on their own. The system takes into account the protection of the lungs, brain and other developing organs and supports the practitioner in making medical decisions. A major focus in the design process was to ensure that the device does not have an intimidating effect on the baby's relatives.

Statement by the jury
The decorative design elements of the SERVO-n take the situation of both child and parents into account in an exceptional way.

Das Beatmungsgerät SERVO-n ist speziell auf die Intensivpflege von Neugeborenen abgestimmt, die nicht selbständig atmen können. Das System berücksichtigt den Schutz von Lunge, Hirn und anderer Organe, die sich noch in der Entwicklung befinden, und unterstützt den Anwender bei seinen medizinischen Entscheidungen. Bei der Gestaltung wurde großer Wert darauf gelegt, dass das Gerät auf die Angehörigen der Säuglinge nicht einschüchternd wirkt.

Begründung der Jury
Bei den dekorativen Gestaltungselementen des SERVO-n wurde auf herausragende Weise Rücksicht auf die Situation von Kind und Eltern genommen.

SERVO-U®
Medical Ventilator
Beatmungsgerät

Manufacturer
Maquet Critical Care AB, Solna, Sweden
In-house design
Karin Blomquist, Anette Sunna, Annelise Müller, Helena Stone, Arne Lindy, Johan Reuterholt, David Jergefalk, Tobias Bodin, Lars Danielsen
Design
Veryday
(Madlene Lindström, Siamak Tahmoresnia, Rahul Sen, Daniel Höglund, Aydin Mert, Anna Carell, Thomas Nilsson, Fredrik Ericsson, Clayton Cook, Lena Edman, Magnus Gyllenswärd, Olof Bendt), Bromma, Sweden
Web
www.maquet.com
www.veryday.com

The SERVO-U medical ventilator was developed to ensure safety of use for individualised patient treatment. Context-based guidance on the touchscreen and suggestions with conveniently positioned shortcut keys help the medical practitioner to make well-founded decisions. The device can be mounted in many different ways and, due to its ergonomic design, may also be positioned on either side of the patient's bed.

Statement by the jury
SERVO-U has an unconventionally slim design for a medical ventilator. Its minimalist form radiates a strong measure of restraint and discretion.

Das mechanische Beatmungsgerät SERVO-U wurde für eine hohe Anwendungssicherheit bei einer individuell angepassten Behandlung des Patienten entwickelt. Kontextbezogene Anleitungen auf dem Touchscreen und Empfehlungen mit praktisch positionierten Tastenkürzeln helfen dem Anwender dabei, fundierte Entscheidungen zu treffen. Das Gerät lässt sich auf vielfältige Weise befestigen und kann aufgrund seines ergonomischen Designs links oder rechts vom Krankenbett positioniert werden.

Begründung der Jury
Servo-u ist für ein Beatmungsgerät ungewöhnlich schmal gebaut. Durch seine reduzierte Form strahlt es ein hohes Maß an Zurückhaltung und Diskretion aus.

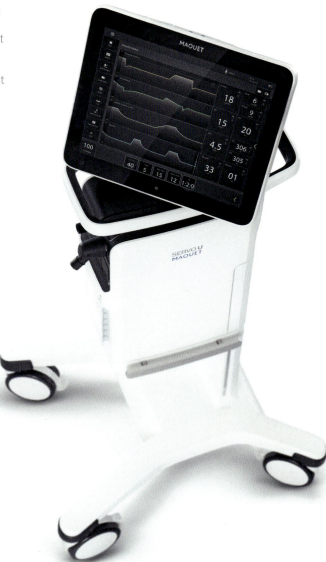

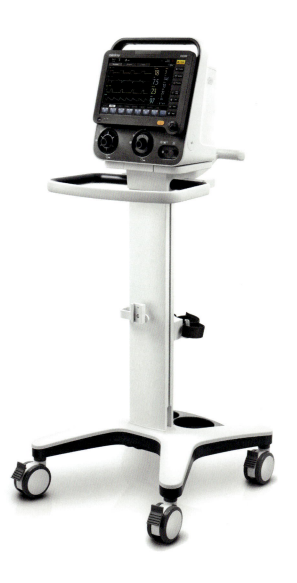

SV 300
Ventilator
Beatmungsgerät

Manufacturer
Shenzhen Mindray
Bio-Medical Electronics Co., Ltd.,
Shenzhen, China
In-house design
Lijuan He, Haibo Chai, Wenhui Zhou,
Xiang Zhou
Web
www.mindray.com

The SV 300 ventilator is suited for mobile and stationary use in any emergency situation. The particularly lightweight and small device can be picked up by the handle and transported easily. The angle of its touchscreen can be adjusted to provide an optimal perspective. The multiview alarm light allows medical staff to react quickly. The inspiration and expiration valves are very simple to attach and remove, thus making them easy to disinfect.

Statement by the jury
The design of this ventilator convinces with its clear and composed structure. The housing makes a very sturdy impression.

Das Beatmungsgerät SV 300 ist für alle Akutsituationen im mobilen und stationären Bereich geeignet. Das besonders leichte und kleine Gerät kann an dem Griff aufgehoben und transportiert werden. Der Touchscreen ist verstellbar für eine optimale Sicht. Das Alarmlicht mit Multisicht erlaubt dem medizinischen Personal schnell zu reagieren. Die In- und Exspirationsventile lassen sich sehr leicht an- und abmontieren und sind dadurch einfach zu desinfizieren.

Begründung der Jury
Die Gestaltung des Beatmungsgeräts überzeugt durch eine klare und aufgeräumte Struktur. Das Gehäuse macht einen sehr stabilen Eindruck.

STARSystem
Medical Equipment
Medizinisches Zubehör

Manufacturer
Adept Medical Ltd,
Auckland, New Zealand
In-house design
Murray Fenton, Mark Webster,
Matt Lazenby, Mike Oxborough
Web
www.adeptmedical.com
Honourable Mention

The STARSystem is a support device that helps create radial access for a cardiac catheter via the wrist. The STARBoard raises the right or left arm to a comfortable working height and positions the wrist in an optimally extended position. The STARSupport is designed to facilitate better access to the left arm. The STARTable is an adjustable work surface for medical staff providing space for catheters and tubes.

Statement by the jury
The fully flexible arms and joints of the STARSystem are extensively adjustable and thus meet high ergonomic standards.

Das STARSystem ist eine Stützvorrichtung, die dabei hilft, einen Herzkatheter über das Handgelenk zu legen. Das STARBoard hebt den rechten oder linken Arm in eine bequeme Arbeitshöhe und positioniert das Handgelenk in eine optimal ausgestreckte Haltung. Der STARSupport erleichtert zusätzlich den Zugang zum linken Arm. Der STARTable bietet dem medizinischen Personal eine verstellbare Arbeitsfläche für Katheter und Schläuche.

Begründung der Jury
Die voll beweglichen Arme und Gelenke des STARSystems lassen sich sehr gut anpassen und erfüllen dadurch hohe Ansprüche an die Ergonomie.

THUNDERBEAT
Open Extended Jaw
Surgical
Tissue Management System
Chirurgisches
Gewebemanagementsystem

Manufacturer
Olympus Medical Systems Corporation,
Tokyo, Japan
Design
Olympus Imaging Corporation Design Center
(Chikayoshi Meguro, Koji Sakai),
Tokyo, Japan
Web
www.olympus-global.com
www.olympus-europa.com

The Thunderbeat Open Extended Jaw surgical tissue management system has been designed for open surgical procedures in gynaecology and urology in particular. It cuts quickly, seals blood vessels with a diameter of up to 7 mm and enables very precise tissue dissection without any unwanted bleeding. The ergonomics of the handle have also been adjusted to reduce hand fatigue and enable surgeons to carry out precise operations.

Statement by the jury
The form with its excellent grip and the well-balanced size of the Thunderbeat Open Extended Jaw allow for almost perfect control over the instrument.

Das Gewebemanagementsystem Thunderbeat Open Extended Jaw eignet sich insbesondere für offene chirurgische Eingriffe in der Gynäkologie und Urologie. Es schneidet schnell, versiegelt Blutgefäße mit einem Durchmesser von bis zu 7 mm und ermöglicht eine sehr präzise Zerteilung des Gewebes ohne unerwünschte Blutungen. Außerdem wurde die Ergonomie des Handgriffs angepasst, was einer Ermüdung der Hand vorbeugt und dem Chirurgen ein präzises Operieren ermöglicht.

Begründung der Jury
Durch die griffige Form und die sorgfältig austarierte Größe des Thunderbeat Open Extended Jaw hat der Anwender nahezu uneingeschränkte Kontrolle über das Instrument.

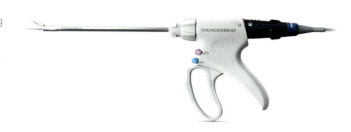

THUNDERBEAT
Open Fine Jaw
Surgical
Tissue Management System
Chirurgisches
Gewebemanagementsystem

Manufacturer
Olympus Medical Systems Corporation,
Tokyo, Japan
Design
Olympus Imaging Corporation Design Center
(Reisuke Osada), Tokyo, Japan
Web
www.olympus-global.com
www.olympus-europa.com

The Thunderbeat Open Fine Jaw surgical tissue management system combines a high-frequency current with ultrasonic technology and can cut as well as seal tissue and vessels in a single step. The scissor grip ensures precise operation. The reduced weight and optimised point of balance within the palm of the hand prevents fatigue. The instrument's curved, fine tip enables precise tissue dissection.

Statement by the jury
The individual areas of this tissue management system are clearly delineated, and sophisticated ergonomics significantly reduce the load on the hand.

Das Gewebemanagementsystem Thunderbeat Open Fine Jaw verbindet Hochfrequenzstrom mit Ultraschalltechnologie und kann in einem Schritt Gewebe und Gefäße sowohl schneiden als auch versiegeln. Der Scherengriff gewährleistet eine präzise Handhabung. Das geringe Gewicht und der optimierte Balancepunkt innerhalb der Handfläche beugen Ermüdungserscheinungen vor. Die gebogene, feine Spitze des Instruments erlaubt ein präzises Zerteilen des Gewebes.

Begründung der Jury
Die einzelnen Bereiche des Gewebemanagementsystems sind differenziert herausgearbeitet. Die ausgeklügelte Ergonomie entlastet die Hand signifikant.

Lumenis Pulse 120H
Surgical Laser
Chirurgischer Laser

Manufacturer
Lumenis Ltd., Yokneam, Israel
Design
Taga, Tel Aviv, Israel
Web
www.lumenis.com
www.tagapro.com

The Lumenis Pulse 120H is a surgical laser for all applications in the field of urology. The device is characterised by a slim, low profile, which requires little space and is thus ideal for use in a fully occupied operating theatre. The large multitouch display may be operated using gloves and is clearly visible even from a distance. Both the screen and the fibre-optic arm can be folded up when not in use.

Statement by the jury
The self-contained design of the Lumenis Pulse 120H appears very robust. The central arrangement of the controls is geared to the user.

Der Lumenis Pulse 120H ist ein chirurgischer Laser für alle Anwendungen in der Urologie. Das Gerät hat ein schlankes, niedriges Profil, das wenig Platz benötigt und somit ideal für den Einsatz in einem vollbesetzten OP-Saal ist. Das große Multitouch-Display kann mit Handschuhen bedient werden und ist auch von weitem erkennbar. Sowohl der Bildschirm als auch der Faseroptik-Arm können eingeklappt werden, wenn sie nicht in Gebrauch sind.

Begründung der Jury
Mit seiner geschlossenen Bauweise wirkt der Lumenis Pulse 120H sehr robust. Die zentrale Anordnung der Bedienelemente ist auf den Anwender zugeschnitten.

Kick
Surgical Navigation System
Chirurgisches Navigationssystem

Manufacturer
Brainlab AG, Feldkirchen, Germany
In-house design
Elmar Schlereth, Michael Voss
Web
www.brainlab.com

The Kick mobile navigation system is equipped with an infrared camera and an electromagnetic field generator. Surgeons can choose which tracking technology they prefer in each individual case. The camera can be adjusted to a height of 2.25 metres, and the field generator offers three predefined positions for optimal coverage of the surgical area. The monitor unit consists of a 21.5" touchscreen with a sterile cover.

Statement by the jury
Bright colour accents and rounded shapes give the navigation system a friendly appearance. Its high mobility strongly enhances ease of use.

Das mobile Navigationssystem Kick ist mit einer Infrarotkamera und einem elektro-magnetischen Feldgenerator ausgestattet. Der Chirurg kann wählen, welche Tracking-technologie er für den Einzelfall bevorzugt. Die Kamera lässt sich bis zu einer Höhe von 2,25 Metern ausrichten, während der Feld-generator drei vordefinierte Positionen für eine optimale Abdeckung des OP-Bereichs bietet. Die Monitor-Einheit besteht aus ei-nem steril abdeckbaren 21,5"-Touchscreen.

Begründung der Jury
Helle Farbkontraste und gerundete Formen verleihen dem Navigationssystem ein freundliches Aussehen. Die hohe Mobilität erleichtert die Handhabung enorm.

Stryker PROFESS™ System
Surgical Navigation System
Navigationssystem für die
Chirurgie

Manufacturer
Stryker Leibinger GmbH & Co. KG,
Freiburg, Germany
In-house design
Sascha Kubis, Martin Stangenberg,
Matthias Wapler
Design
Erdmann Design, Brugg, Switzerland
Brennwald Design, Kiel, Germany
everdesign, Freiburg, Germany
Formpark Industrie Design,
Freiburg, Germany
Web
www.stryker.com/navigation
www.erdmann.ch
www.brennwald-design.de
www.everdesign.de
www.formpark.de
Honourable Mention

The compact Stryker Profess surgical navigation system assists surgeons with intranasal and sinus surgery by tracking and displaying the location of navigated instruments. The intuitive software, tools and registration kit allow for ac-curate, safe and cost effective surgery. The touchscreen enabled software leads through the workflow by instructional videos. The lightweight single use tools offer the ability to control the software from the sterile field.

Statement by the jury
With its use of individually adjustable tracking stickers, Stryker Profess pursues an innovative approach in medical technology.

Das kompakte Navigationssystem Stryker Profess unterstützt Chirurgen bei Nasen- und Nasennebenhöhlenoperationen und zeigt die Position von navigierten Ins-trumenten an. Die intuitive Bedienung des Systems ermöglicht präzise, sichere und kosteneffektive Eingriffe. Die per Touch-screen bedienbare Software führt mittels Instruktionsvideos durch den Arbeitsablauf. Der Chirurg kann die Software mithilfe der federleichten Einweginstrumente auch selbständig steuern.

Begründung der Jury
Durch die Verwendung von individuell anpassbaren Tracking-Aufklebern verfolgt Stryker Profess einen innovativen Ansatz in der Medizintechnik.

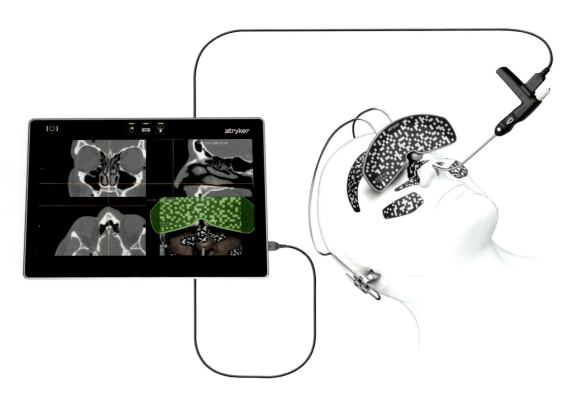

da Vinci Xi Surgical System
da-Vinci-Xi-Chirurgiesystem
Surgical Robot
Operationsrobotor

Manufacturer
Intuitive Surgical, Sunnyvale, USA
In-house design
Intuitive Surgical Design and Engineering (Mike Hanuschik)
Design
Bould Design, Mountain View, USA
Scott Waters, Sunnyvale, USA
Web
www.intuitivesurgical.com
www.bould.com
www.scott-waters.com

The da Vinci Xi Surgical System expands the surgeon's options during minimally invasive procedures. The wearer is presented with a three-dimensional HD view of the surgical field, with magnification ten times greater than the human eye. The hand controls allow the surgeon to rotate and tilt tiny instruments with high precision in more degrees of freedom than otherwise possible with the human wrist. The newly developed telescopic arm design increases the range of the system and frees up space for the surgeon to work.

Das da-Vinci-Xi-Chirurgiesystem erweitert die Möglichkeiten des Operateurs während eines minimal-invasiven Eingriffs. Dem Benutzer wird eine dreidimensionale HD-Ansicht des Operationsfeldes präsentiert, die das Sichtfeld des menschlichen Auges um ein Zehnfaches vergrößert. Die manuelle Steuerung erlaubt es, kleinste Instrumente mit hoher Präzision in mehr Freiheitsgraden zu drehen und zu beugen, als es dem menschlichen Handgelenk möglich ist. Die neu entwickelte Teleskoparm-Architektur vergrößert die Reichweite des Systems und verschafft dem Chirurgen mehr Platz.

Statement by the jury
The da Vinci Xi Surgical System captivates with its extremely slim and flexible design, which sets new standards in the field of medical robotics.

Begründung der Jury
Das da-Vinci-Xi-Chirurgiesystem beeindruckt durch seine überaus schlanke und gelenkige Konstruktion, die eigene Maßstäbe in der Medizinrobotik setzt.

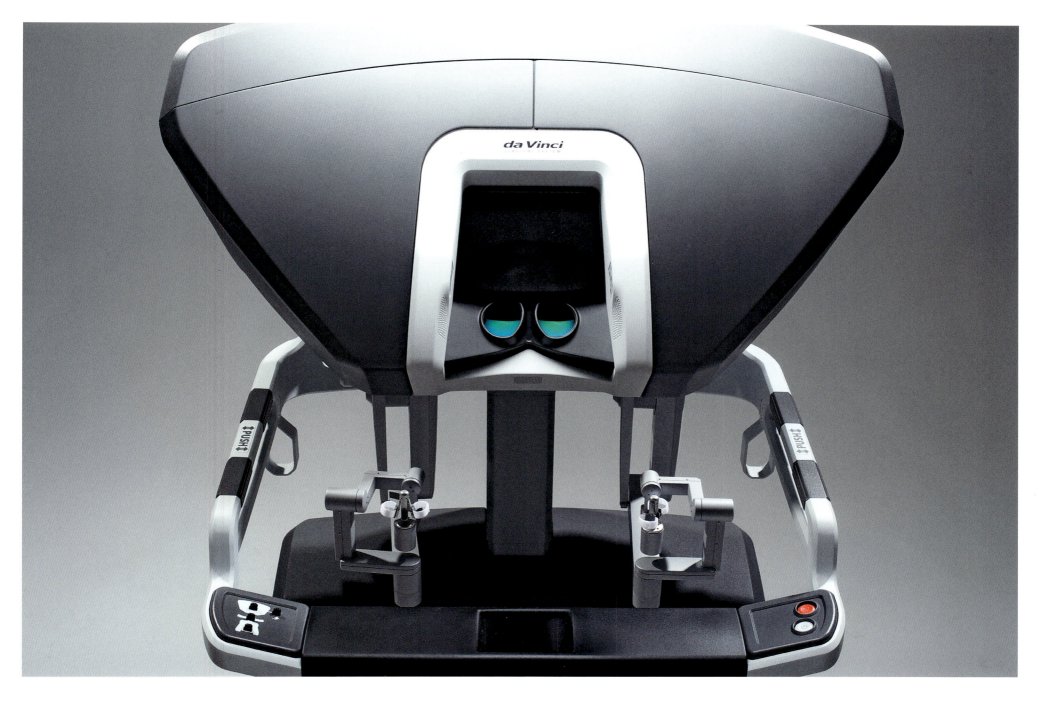

Vivideo
ENT Video Endoscope
HNO-Video-Endoskop

Manufacturer
Pentax Medical, Friedberg, Germany
In-house design
Christoph Rilli
Web
www.pentaxmedical.com

The Vivideo endoscope with an integrated LED light provides high-resolution video images for diagnostic examinations of the entire ear, nose and throat area. The combination of ultra-flexible material and the small diameter of the insertion tube increase patient comfort. The hummingbird-like shape of the handle is coated with silicone and helps the physician adopt a natural and comfortable hand position during the examination. Furthermore, the endoscope is easy to clean and very durable thanks to its high-quality materials.

Das Endoskop Vivideo mit integriertem LED-Licht liefert hochauflösende Videobilder bei diagnostischen Untersuchungen im gesamten Hals-Nasen-Ohren-Bereich. Die Kombination von ultraflexiblem Material und einem kleinen Durchmesser des Einführschlauchs bietet einen verbesserten Patientenkomfort. Der Kolibri-ähnliche Handgriff ist mit Silikon ummantelt und unterstützt den Arzt dabei, eine natürliche und schonende Handhaltung während der Untersuchung einzunehmen. Das Endoskop ist zudem einfach zu reinigen und durch seine qualitativ hochwertigen Materialien sehr langlebig.

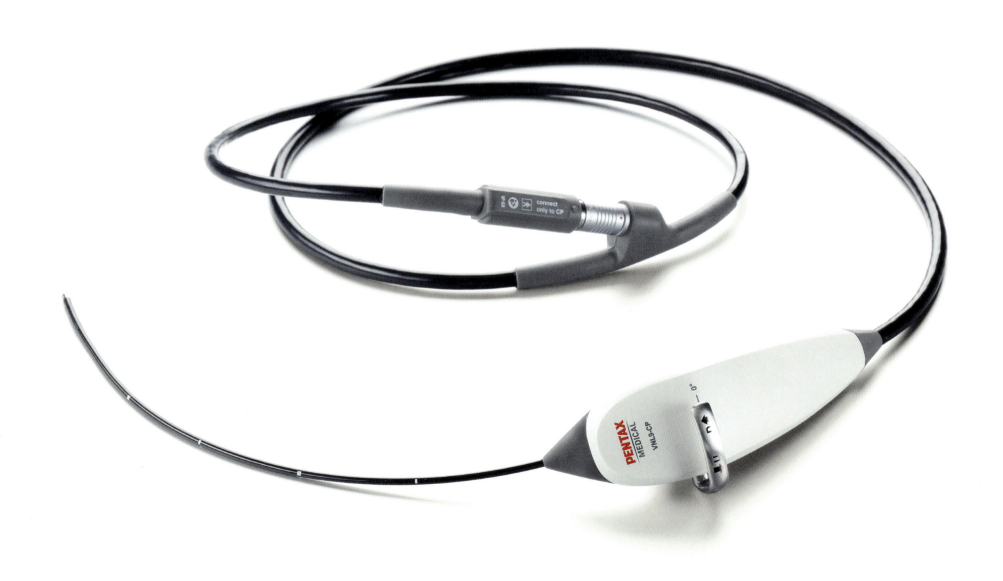

HMS-3000MT
Medical Head Mounted Display System
Medizinischer Datenhelm

Manufacturer
Sony Corporation, Tokyo, Japan
In-house design
Yoshiaki Kumagai
Web
www.sony.net

The HMS-3000MT head mounted display system projects 3D colour images from a surgical camera system directly in front of the user's eye. The device thus gives surgeons more freedom of movement and posture during endoscopic or laparoscopic interventions, because they no longer have to work in a specific position to view images on an external monitor. At the same time, they can continue to easily look down at the operating table. The weight of the head-mounted display is distributed evenly, and the head padding makes it comfortable to wear.

Der Datenhelm HMS-3000MT projiziert 3D-Farbbilder von einem chirurgischen Kamerasystem direkt vor das Auge des Anwenders. Dadurch bietet das Gerät dem Operateur während eines endoskopischen bzw. laparoskopischen Eingriffs mehr Bewegungs- und Haltungsfreiheit, weil er nicht wie bisher in einer bestimmten Position arbeiten muss, um die Bilder auf einem externen Monitor zu betrachten. Zugleich kann er weiterhin problemlos nach unten auf den Operationstisch schauen. Das Gewicht des Helms ist gleichförmig verteilt, und die Kopfpolster sorgen für angenehmen Tragekomfort.

Statement by the jury
This head mounted display system has an airy and light design. Its dynamic lines embody strong ergonomic qualities along with a modern aesthetic.

Begründung der Jury
Der Datenhelm ist sehr luftig und leicht gestaltet. Seine dynamische Linienführung steht nicht nur für eine hohe Ergonomie, sondern auch für eine moderne Ästhetik.

LABTAP Control
Laboratory Tap
Laborarmatur

Manufacturer
Lab Tap, Eindhoven, Netherlands
Design
LabelJOEP (Joep Verheijen),
Eindhoven, Netherlands
Web
www.labtap.nl
www.labeljoep.nl
Honourable Mention

The laboratory taps of the Labtap Control series consist of four components: the cylindrical base, the actual fitting, the operating handles attached to it, and the curved spout. The brass bodies are coated with a pearlescent nickel-chrome finish. The position of the handle indicates whether the valve is open or closed at any given time. The spout and the operating handle have the same colour coding, so that they may be clearly assigned to a liquid or gaseous medium.

Die Laborarmaturen der Serie Labtap Control bestehen aus vier Komponenten: der zylindrischen Basis, der eigentlichen Armatur, den daran befindlichen Betätigungsgriffen und dem gebogenen Auslauf. Die Grundkörper aus Messing sind mit einer Perlglanz-Nickelchrom-Oberfläche beschichtet. Über die Stellung des Betätigungsgriffs kann zu jeder Zeit erkannt werden, ob das Ventil geöffnet oder geschlossen ist. Schlauchtülle und Betätigungsgriff sind mit derselben Farbkodierung gekennzeichnet, damit sie eindeutig einem flüssigen oder gasförmigen Medium zugeordnet werden können.

Statement by the jury
With the shiny matt surface coating, these laboratory taps appear both sturdy and valuable.

Begründung der Jury
Durch die matt glänzende Oberflächenbeschichtung wirken die Laborarmaturen sowohl stabil als auch hochwertig.

AUTOMAGE (gooseneck type)
Automatic Faucet
Automatischer Wasserhahn

Manufacturer
LIXIL Corporation, Tokyo, Japan
In-house design
Miyuki Hashimoto
Web
http://global.lixil.co.jp

The Automage automatic faucet is operated without physical contact, by placing the hands underneath the spout. It was developed for use in washrooms and lavatories, where hygiene and barrier-free access are particularly important, such as in hospitals or welfare facilities. The gooseneck pipe is well suited not only to handwashing, but also to filling water into vases and other large containers. The smooth and elegant shape without any projections blends in with the surroundings and is easy to clean.

Der automatische Wasserhahn Automage wird berührungslos betätigt, indem die Hand unter den Auslass gehalten wird. Die Armatur wurde speziell für den Gebrauch in Waschräumen und Toiletten entwickelt, wo Hygiene und Barrierefreiheit besonders wichtig sind, z. B. im Krankenhaus oder in Fürsorgeeinrichtungen. Die Schwanenhalsform ist nicht nur zum Händewaschen, sondern auch zum Befüllen von Blumenvasen und anderen größeren Behältern gut geeignet. Die glatte und elegante Form ohne Vorsprünge passt sich der Umgebung gut an und ist leicht zu reinigen.

Statement by the jury
With its balanced proportions and functional aesthetics, this faucet embodies restraint and presence simultaneously.

Begründung der Jury
Mit seinen ausgewogenen Proportionen und der schlichten Ästhetik verkörpert der Wasserhahn Zurückhaltung und Präsenz zugleich.

TWIP
Pregnancy Test
Schwangerschaftstest

Manufacturer
Unicoms AG, Zürich, Switzerland
Design
Ilian Milinov, Sofia, Bulgaria
Web
www.unicoms.com
www.ilian-milinov.com

TWIP pregnancy test has been developed specifically for women with irregular periods. It can detect a pregnancy as early as 17 days after the probable conception date without having to calculate the date for the next expected period. The test swivels out and back into the handle effortlessly for easy and hygienic use. TWIP is made of biodegradable material.

Der Schwangerschaftstest TWIP wurde speziell für Frauen mit einer unregelmäßigen Periode entwickelt. Er weist eine Schwangerschaft bereits 17 Tage nach dem Zeitpunkt der Empfängnis nach, ohne dass der Termin der nächsten Periode errechnet werden muss. Der Tester wird aus dem Griff heraus- und wieder hineingeschwenkt, wodurch die Anwendung besonders bequem und hygienisch ist. TWIP ist aus biologisch abbaubarem Material hergestellt.

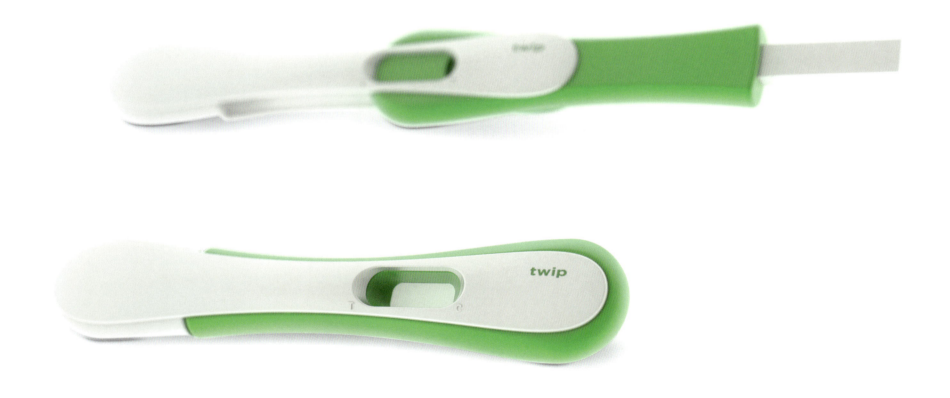

Moocall
Calving Sensor
Sensor für die Kälbergeburt

Manufacturer
Moocall Sensors, Dublin, Ireland
In-house design
Niall Austin
Design
Dolmen
(Christopher Murphy, Lyndsey Bryce),
Dublin, Ireland
Web
www.moocallsensors.com
www.dolmen.ie

Moocall is an innovative device that alerts the farmer when a cow is close to calving. The robust sensor is attached to the cow's tail, and the sensors detect movements indicating that calving is about to begin. The farmer then automatically receives a text message. The device is attached using a ratchet system, which is protected by a special rubber material on the inside and provides a secure and comfortable grip.

Moocall ist ein innovatives Gerät, das den Landwirt alarmiert, wenn eine trächtige Kuh kurz vorm Kalben steht. Der robuste Sensor wird am Schwanz der Kuh angebracht und registriert Bewegungen, die auf das Einsetzen der Geburt hindeuten. Der Landwirt erhält daraufhin automatisch eine SMS. Das Gerät wird mit einer Sperrvorrichtung angebracht, die durch ein spezielles Innenmaterial aus Gummi geschützt ist und einen sicheren, komfortablen Sitz ermöglicht.

SpeediCath Compact Eve
Single-Use Female Catheter
Einmal-Katheter für Frauen

Manufacturer
Coloplast A/S,
Humlebæk, Denmark

In-house design
Coloplast A/S

Web
www.coloplast.com

reddot award 2015
best of the best

A sensitive interpretation

Bladder problems are an everyday concern for many people. The use of intermittent self-catheterisation (ISC) allows sufferers to lead an independent life without limitations. The bladder thus regularly empties itself by means of a single-use catheter. SpeediCath Compact Eve is such a catheter specifically designed for women, which meets their needs with great sensitivity. Its design takes its inspiration from cosmetic products. It therefore looks less like a medical device and more like a mascara bottle or a concealer that one would have in one's handbag; an impression which is reinforced by the turquoise tube used to transport it. The catheter is triangular, offers good ergonomic grip and stable placement. It is also easy and intuitive to use, simple to open and a secure cover ensures that it stays sterile. Once the device is opened, the integrated urine bag becomes visible. After use, the catheter is easily sealed, thus guaranteeing hygienic and discrete disposal. The bold design ensures that this single-use catheter takes the drama out of a sensitive medical problem and clearly demonstrates that design can contribute to a positive attitude towards life.

Sensibel interpretiert

Blasenfunktionsstörungen gehören für viele Menschen zum Alltag. Dank der Praxis des leicht erlernbaren intermittierenden Selbstkatheterismus (ISK) können viele von ihnen jedoch ohne Einschränkungen selbstbestimmt leben. Die Blase wird bei dieser Technik mithilfe eines Einmal-Katheters regelmäßig selbst entleert. Der SpeediCath Compact Eve ist ein solcher Katheter für Frauen, der mit viel Sensibilität für deren Bedürfnisse gestaltet wurde. Sein Design ist maßgeblich inspiriert von Produkten aus der Kosmetikbranche. Er wirkt deshalb nicht wie ein medizinisches Hilfsmittel, sondern eher wie ein Mascara oder Abdeckstift, den man in der Handtasche trägt. Eine Anmutung, die durch eine Gestaltung mit einem außen liegenden türkisfarbenen Röhrchen noch verstärkt wird. Der Katheter ist dreieckig geformt, er liegt gut in der Hand und kann nicht wegrollen. Seine Handhabung ist zudem unkompliziert: Die Frau kann ihn intuitiv verwenden, er ist leicht zu öffnen und ein sicherer Verschluss gewährleistet, dass das Produkt stets steril ist. Beim Öffnen erscheint der Urinbeutelanschluss, nach der Verwendung lässt sich der Katheter wieder verschließen und somit hygienisch und diskret entsorgen. Eine mutige Gestaltung führte bei diesem Einmal-Katheter dazu, dass ein medizinisches Problem weniger stigmatisierend ist, und zeigt deutlich, dass Design zu einer positiven Lebenseinstellung beitragen kann.

Statement by the jury

The SpeediCath Compact Eve single-use catheter is a perfect solution to the burdensome problem of female bladder dysfunction. With its feminine appearance, it looks just like an attractive, up-market cosmetic accessory. Women can use it without shame and can unobtrusively carry it in their handbag. This catheter is thereby very functional, as well as safe and easy to use.

Begründung der Jury

Der Einmal-Katheter SpeediCath Compact Eve stellt eine perfekt gestaltete Lösung für die belastende Problematik weiblicher Blasenfunktionsschwäche dar. Mit seiner femininen Anmutung wirkt er wie ein attraktives und hochwertiges Kosmetikprodukt. Frauen können ihn ohne Schamgefühle verwenden und unkompliziert in der Handtasche mit sich tragen. Dieser Katheter ist dabei überaus funktional und lässt sich ebenso sicher wie komfortabel handhaben.

Designer portrait
See page 62
Siehe Seite 62

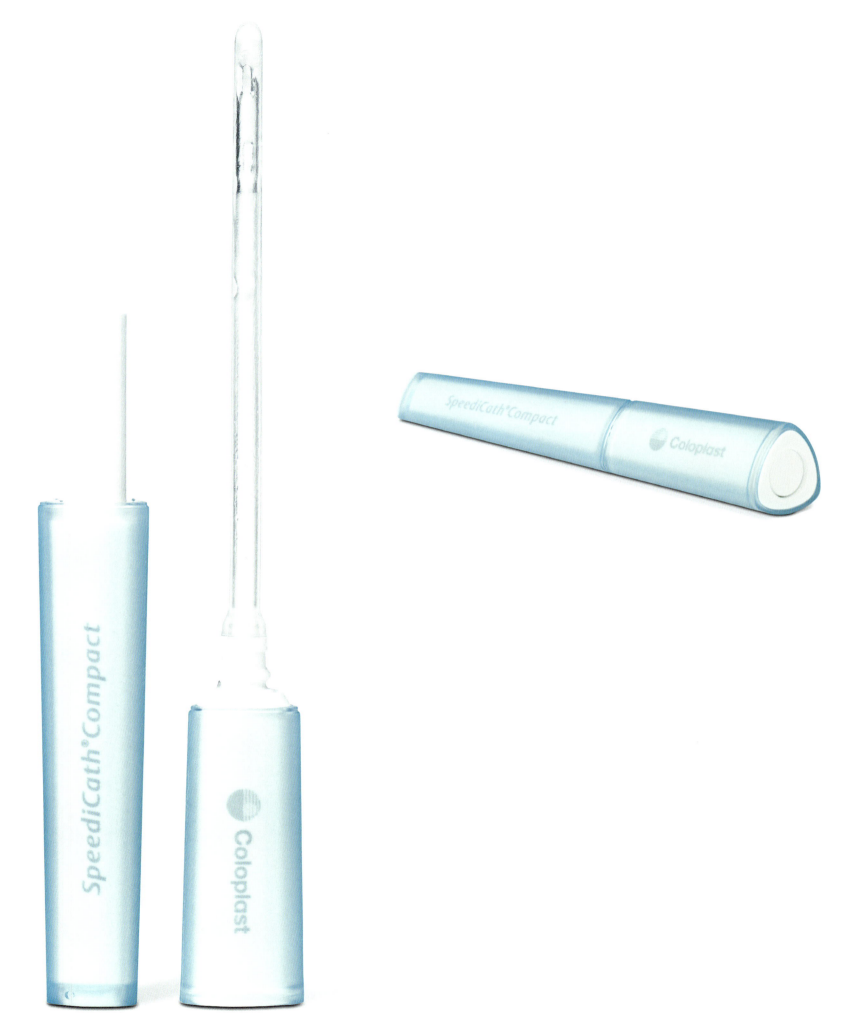

QS-M2
Needle-Free Injector
Nadelfreier Injektor

Manufacturer
Beijing QS Medical Technology Co., Ltd.,
Beijing, China
Design
Suning Chen, China
Web
www.qsjet.com
www.chensuning.com

The QS-M2 needle-free injector is designed for insulin therapy in diabetic patients. The device uses high pressure to inject insulin into the skin without creating an injection hole. This reduces pain and risk of infection to a minimum and avoids a hardening of soft tissue. In addition, the bioavailability of insulin is improved. The system allows for multiple insulin injections after filling the injectors just once. The device must first be unlocked by pressing a button in order for the injection to take place, ensuring that the process is carried out safely.

Der nadelfreie Injektor QS-M2 wird für die Insulintherapie bei Diabetes eingesetzt. Das Gerät erzeugt einen Hochdruck, der das Insulin unter die Haut spritzt, ohne dass dabei ein Einstichloch entsteht. Somit werden Schmerzen und die Infektionsgefahr auf ein Minimum reduziert, und Verhärtungen werden vermieden. Zudem verkürzt sich die Zeitspanne, bis das Insulin wirkt. Mit dem System kann nach einmaligem Aufladen mehrmals Insulin gespritzt werden. Das Gerät muss erst per Knopfdruck entsperrt werden, bevor die Injektion durchgeführt werden kann. Dies gewährleistet eine sichere Anwendung.

Statement by the jury
The extraordinary softness of the QS-M2 is conveyed by its round shapes. The colour-coded control elements improve the device's safety of use.

Begründung der Jury
Die außergewöhnliche Sanftheit des QS-M2 wird durch seine Rundungen kommuniziert. Die farbliche Kennzeichnung der Bedienelemente verbessert die Sicherheit.

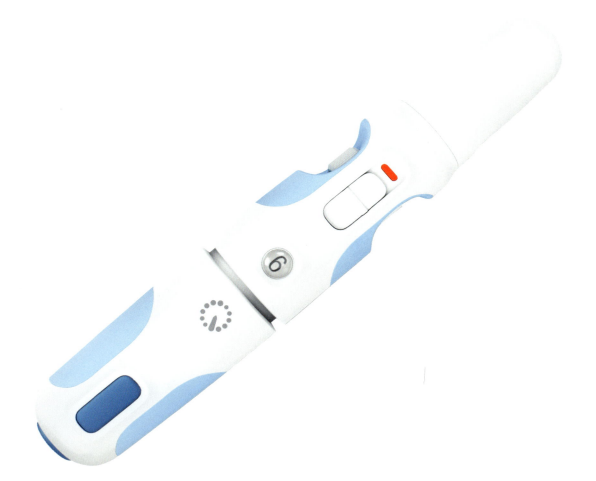

Merck Serono
New Fertility Pen 1.0/1.1
Pen Injectors
Pen-Injektoren

Manufacturer
Merck Serono S.A., Coinsins, Switzerland
Design
Haselmeier GmbH (Heiko Müller),
Stuttgart, Germany
Web
www.merckserono.com
www.haselmeier.com

The Merck Serono New Fertility Pen 1.0/1.1 pen injectors are used during the hormone treatment of patients trying to have children. With the help of a dosing scale, the amount of medication can be selected. The dose window displays only a single dose at any moment in time, making it immediately discernible as to whether the complete dose has been delivered, and thereby reducing the risk of dosing errors. The pens can be distinguished thanks to their colours and labels.

Statement by the jury
The sophisticated design of these pen injectors is particularly user-friendly, since the dosage is clearly visualised.

Die Pen-Injektoren Merck Serono New Fertility Pen 1.0/1.1 werden bei der Hormonbehandlung von Patienten mit unerfülltem Kinderwunsch eingesetzt. Über eine Dosierungsskala kann die Medikamentenmenge eingestellt werden. Im Dosierfenster wird lediglich eine einzelne Dosis angezeigt, sodass sofort zu erkennen ist, ob die komplette Dosis verabreicht wurde. Dadurch wird das Risiko von Dosierungsfehlern reduziert. Die Pens lassen sich mittels Farben und Etikettierung voneinander unterscheiden.

Begründung der Jury
Die durchdachte Gestaltung der Pen-Injektoren ist besonders anwenderfreundlich, da die Dosierung deutlich sichtbar dargestellt wird.

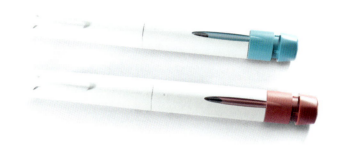

Merck Serono
New Fertility Pen 2.0
Pen Injector
Pen-Injektor

Manufacturer
Merck Serono S.A., Coinsins, Switzerland
Design
Haselmeier GmbH (Heiko Müller),
Stuttgart, Germany
Web
www.merckserono.com
www.haselmeier.com

The Merck Serono New Fertility Pen 2.0 pen injector is used in hormone therapy for women undergoing fertility treatment, supporting their family planning. The pen is designed in such a way that each dose may be individually set and easily identified through a dose window. If the dose has not been completely administered, then the missing amount is immediately displayed, thereby reducing the risk of dosing errors. This provides the user of the pen with reassurance and confidence.

Statement by the jury
The pen injector's clear and functional design language inspires confidence and conveys reliability.

Der Pen-Injektor Merck Serono New Fertility Pen 2.0 wird bei der Hormontherapie für Frauen eingesetzt und unterstützt sie bei der Familienplanung. Der Pen wurde so entwickelt, dass jede Dosis einzeln eingestellt und einfach im Dosierfenster identifiziert werden kann. Falls nicht die komplette Dosis verabreicht worden ist, wird die fehlende Menge sofort angezeigt, wodurch das Risiko von Dosierungsfehlern sinkt. Dies bietet dem Anwender des Pens Zuversicht und Vertrauen.

Begründung der Jury
Die klare und funktionale Formensprache des Pen-Injektors wirkt in hohem Maße vertrauenerweckend und zuverlässig.

MuteDot
Lancing Device
Stechhilfe

Manufacturer
Medifun Corporation, Taichung, Taiwan
In-house design
Matt Huang
Web
www.medifun.com.tw

The MuteDot lancing device aids in measuring blood glucose levels in diabetics. By reducing the vibrations during pricking, the size of the puncture wound is minimised, which in turn causes less pain. A safety mechanism prevents the spring from being cocked while it is open. In turn, when the spring is cocked, the lancing device remains sealed. The used lancet can be ejected by pushing the plunger, using only one hand.

Statement by the jury
This seamlessly designed lancing device impresses with its simplicity, which enables particularly discreet use.

Die Stechhilfe MuteDot wird für die Blutzuckermessung bei Diabetes verwendet. Durch eine Reduzierung der Vibrationen beim Stechen wird die Größe der Einstichwunde minimiert, was weniger Schmerz verursacht. Ein Sicherheitsmechanismus verhindert das Aufziehen der Feder bei geöffneter Verschlusskappe. Bei gespannter Feder kann die Stechhilfe wiederum nicht geöffnet werden. Die benutzte Lanzette kann durch Drücken des Kolbens mit nur einer Hand ausgeworfen werden.

Begründung der Jury
Die nahtlos verarbeitete Stechhilfe punktet durch ihre Schlichtheit, die einen besonders diskreten Gebrauch ermöglicht.

Vetter-Ject®
Closure System for Syringes
Verschlusssystem für Spritzen

Manufacturer
Vetter Pharma-Fertigung GmbH & Co. KG,
Ravensburg, Germany
In-house design
Petra Hund, Tilman Roedle
Web
www.vetter-pharma.com

Vetter-Ject is a closure system for pre-filled syringes. It consists of a piston that is crafted from three different plastic components and the steel injection needle glued inside. The piston with the needle is attached to a syringe and serves, for one, as a safety catch, but also as an indicator that the syringe is unused. The closure system's grey plastic component is permeable so that the superheated steam reaches the needle during the sterilisation process. At the same time, the material functions as a non-slip layer when opening the closure system.

Vetter-Ject ist ein Verschlusssystem für vorgefüllte Fertigspritzen. Es besteht aus einem Kolben, der aus drei verschiedenen Kunststoffkomponenten gefertigt ist, und der darin eingeklebten Kanüle aus Stahl. Der Kolben mit Kanüle wird auf eine Spritze aufgesetzt und dient einerseits zur Sicherung, andererseits als Signal dafür, dass die Spritze unbenutzt ist. Die grauen Kunststoffkomponenten des Verschlusses sind durchlässig, damit beim Sterilisationsprozess der Heißdampf bis zur Nadel durchdringt. Gleichzeitig bieten sie den Fingern Halt beim Öffnen des Verschlusses.

Statement by the jury
With its integrated grip surface and finger groove, Vetter-Ject displays a decidedly ergonomic design.

Begründung der Jury
Vetter-Ject weist durch die eingearbeitete Grip-Fläche und die eingekerbte Fingermulde eine ausgesprochen ergonomische Gestaltung auf.

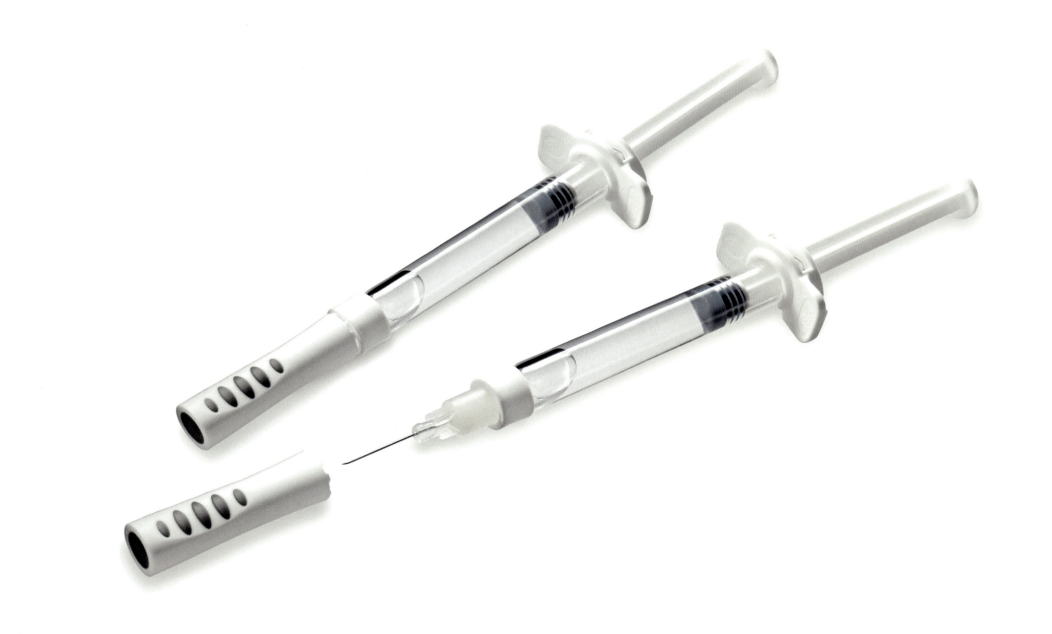

FlowSens
Contrast Agent Injection System
Kontrastmittel-Injektionssystem

Manufacturer
Medex, Saint-Priest, France
In-house design
Anne-Marie Milcent, Nathalie Maneuf
Design
Mézière IDC (Cyril Mézière), Villeurbanne, France
Web
www.medexbvguerbet.com
www.meziere.fr

FlowSens is an innovative system solution for the injection of contrast agents used in CT scan examinations. The solution combines hydraulic syringe-free injection technology with intuitive and safe use during every step of the injection process. Due to its ergonomics and simple handling, the injector meets the high requirements of radiological practice. The system is compatible with pre-filled contrast agent bags as well as conventional contrast-agent and saline bottles.

Bei FlowSens handelt es sich um eine innovative Systemlösung für die Injektion von Kontrastmitteln im Rahmen einer CT-Untersuchung. Dabei wird eine hydraulische spritzenfreie Injektionstechnik mit einer intuitiven und sicheren Anwendung bei jedem Injektionsschritt kombiniert. Aufgrund seiner Ergonomie und einfachen Handhabung entspricht der Injektor den hohen Anforderungen in der radiologischen Praxis. Das System ist sowohl mit vorgefüllten Kontrastmittelbeuteln als auch mit den üblichen Kontrastmittel- und Kochsalzflaschen kompatibel.

Statement by the jury
This injection system impresses with the conical form of its body, lending it a sculptural appeal and visual lightness.

Begründung der Jury
Das Injektionssystem imponiert durch die konische Form des Korpus, die ihm eine skulpturale Anmutung und visuelle Leichtigkeit verleiht.

Unifine Pentips Plus
Pen Needle
Pen-Nadel

Manufacturer
Owen Mumford Ltd,
Woodstock, Great Britain
In-house design
Web
www.owenmumford.com

The Unifine Pentips Plus pen needle features an integrated container in which a used needle can be disposed of after injection. This provides a more convenient option for the immediate disposal of needles after injection and reduces the incidence of needle reuse. The extra-large surfaces provide sufficient grip, even for users with restricted dexterity, when attaching or removing needles from the injector pen.

Die Pen-Nadel Unifine Pentips Plus besitzt einen integrierten Behälter, in dem die benutzte Nadel nach der Injektion entsorgt werden kann. Dies fördert eine sofortige Beseitigung der Nadeln nach der Injektion und verringert die Wahrscheinlichkeit, dass Nadeln wiederverwendet werden. Die extra großen Oberflächen bieten auch Anwendern mit eingeschränkter Greiffähigkeit genügend Halt beim Auf- und Absetzen der Nadel vom Injektor-Pen.

Statement by the jury
The innovative design of this pen needle is decidedly user-friendly. The transparent disposal chamber provides a clear view of its contents.

Begründung der Jury
Die innovative Gestaltung der Pen-Nadel ist ausgesprochen anwenderfreundlich. Der transparent gestaltete Entsorgungsbehälter ist sehr gut einsehbar.

GL 50 evo
Blood Glucose Monitor
Blutzuckermessgerät

Manufacturer
Beurer GmbH, Ulm, Germany
Design
pulse design (Matthias Kolb),
Munich, Germany
Web
www.beurer.com
www.pulse-design.de

The GL 50 evo is a blood glucose monitor, lancing device and plug-in USB flash drive rolled into one. The plain text display ensures that measurements are shown clearly. Average values for 7, 14, 30 or 90 days can be specified. With the help of three different symbols, the user can define whether measuring has taken place before or after eating. The measurements can be transferred to a smartphone or computer via Bluetooth or USB connectivity.

Das GL 50 evo ist Blutzuckermessgerät, Stechhilfe und Plug-in-USB in einem. Auf dem Klartext-Display werden die Werte übersichtlich dargestellt. Es können Durchschnittswerte für 7, 14, 30 oder 90 Tage angegeben werden. Durch drei unterschiedliche Symbole kann definiert werden, ob die Messung vor oder nach dem Essen vorgenommen wurde. Die gemessenen Werte können per Bluetooth oder USB auf ein Smartphone oder einen Computer übertragen werden.

Statement by the jury
The GL 50 evo combines three different functions in a surprisingly compact design, giving it a calm and professional aesthetic.

Begründung der Jury
Beim GL 50 evo wurden drei verschiedene Funktionen in ein überraschend kompaktes Design überführt, dadurch besitzt es eine ruhige und seriöse Ausstrahlung.

BC 80
Blood Pressure Monitor
Blutdruckmessgerät

Manufacturer
Beurer GmbH, Ulm, Germany
Design
pulse design (Matthias Kolb),
Munich, Germany
Web
www.beurer.com
www.pulse-design.de

The BC 80 fully automatic blood pressure monitor is characterised by an easy-to-read XL display and an extremely flat design. The three operating buttons ensure comfortable and simple use for determining blood pressure and pulse. The colourful scale on the device allows for quick classification of the results, while the position indicator ensures correct positioning of the blood monitor at heart level.

Das vollautomatische Blutdruckmessgerät BC 80 zeichnet sich durch ein leicht ablesbares XL-Display und eine extra flache Bauform aus. Die drei Bedienknöpfe bieten für alle Funktionen eine komfortable und einfache Handhabung, um Blutdruck und Puls zu ermitteln. Die farbige Skala auf dem Gerät sorgt für eine schnelle Einordnung der Werte. Die Positionierungsanzeige garantiert die richtige Platzierung des Blutdruckmessgeräts auf Herzhöhe.

Statement by the jury
This blood pressure monitor has a clear design that is easy to use, along with a slim, space-saving construction.

Begründung der Jury
Das Blutdruckmessgerät ist sehr übersichtlich und leicht zu bedienen. Seine schlanke Konstruktion spart viel Platz.

Vion.BP700X
Blood Pressure ECG Monitor
Blutdruck- und EKG-Monitor

Manufacturer
MD Biomedical, Inc., Taipei, Taiwan
In-house design
Web
www.mdbiomedical.com
Honourable Mention

The Vion.BP700X blood pressure ECG monitor takes blood pressure and electrocardiogram measurements and is suitable for long-term monitoring at home. It can be synchronised with mobile devices; the real-time data are transmitted via advanced Bluetooth Smart technology. Attributed to the intuitive user interface and the alarm function, the device is easy and friendly to use. The large storage capacity allows the creation of two different user accounts and storage of up to 128 data records at the same time.

Der Überwachungsmonitor Vion.BP700X führt umfassende Blutdruck- und EKG-Messungen durch und ist für die häusliche Langzeitüberwachung geeignet. Er kann mit mobilen Endgeräten synchronisiert werden, die Echtzeitdaten werden dabei über eine fortschrittliche Bluetooth-Smart-Technologie übermittelt. Durch die intuitiv zu bedienenden Benutzerschnittstellen und die Alarmfunktion ist das Gerät einfach und benutzerfreundlich zu handhaben. Die hohe Speicherkapazität ermöglicht es, zwei unterschiedliche Benutzerkonten und bis zu 128 Datensätze gleichzeitig anzulegen.

Statement by the jury
The blood pressure ECG monitor captivates with its extremely flat design, which makes it appear almost rimless.

Begründung der Jury
Der Blutdruck- und EKG-Monitor gefällt durch seine überaus flache Bauform, wodurch er beinahe randlos erscheint.

SIEMENS

10:00 am 10.05.14

Xprecia Stride

Xprecia Stride™
Coagulation Analyser
Blutgerinnungsmessgerät

Manufacturer
Siemens Healthcare Diagnostics,
Norwood, USA
In-house design
Prabhu Ramachandran, Philip Reavis, Arnol S. Rios
Design
Design + Industry, Melbourne, Australia
Web
www.siemens.com/xprecia

The Xprecia Stride coagulation analyser enables quick diagnosis at the point of care thanks to its practical size. The colour touchscreen safely guides the user through the testing procedure using step-by-step instructions. The test-strip eject button enables the contact-free disposal of used test strips, protecting both patient and staff against biohazard exposure. The protective caps are available in a range of colours for personalisation of the device. The integrated barcode scanner simplifies data entry.

Das Blutgerinnungsmessgerät Xprecia Stride erlaubt aufgrund seiner handlichen Größe eine schnelle Vor-Ort-Diagnostik. Der Farb-Touchscreen leitet den Anwender anhand von Schritt-für-Schritt-Anweisungen sicher durch den Testvorgang. Die Auswurftaste ermöglicht eine berührungslose Entsorgung der Teststreifen, sodass Patient und Personal vor einer biologischen Gefährdung geschützt sind. Die Schutzkappen sind in verschiedenen Farben verfügbar und erlauben eine individuelle Anpassung des Geräts. Der integrierte Barcodescanner vereinfacht die Dateneingabe.

Statement by the jury
The Xprecia Stride captivates with its gentle contours along the top, which indicate where the test strip is inserted and lend the device its distinct appearance.

Begründung der Jury
Das Xprecia Stride besticht durch den sanft gerundeten Ausläufer, der anzeigt, wo der Teststreifen eingeführt wird, und dem Gerät ein prägnantes Aussehen verleiht.

ResMed S+
Sleep Monitor
Schlafmonitor

Manufacturer
ResMed, Dublin, Ireland
Design
Design Partners
(Lukas Fuchs, Mathew Bates),
Bray, Ireland
Web
https://sleep.mysplus.com
www.designpartners.com

The ResMed S+ sleep monitor tracks and analyses sleep patterns using contactless sensor technology. The system records sleep behaviour and data from the bedroom environment to provide personalised feedback, which aims to help improve the quality of sleep and life. The monitor works in conjunction with a free app, which can be downloaded to a smartphone or tablet.

Statement by the jury
The sculptural, floating form of ResMed S+ is an aesthetically pleasing addition to any modern bedroom interior.

Der Schlafmonitor ResMed S+ überwacht und analysiert mithilfe einer berührungslosen Sensortechnologie den Schlaf. Das System erfasst sowohl das eigene Schlafverhalten als auch die Raumumgebung und generiert daraus ein personalisiertes Feedback, das dabei unterstützen soll, die Schlaf- und Lebensqualität zu verbessern. Der Monitor arbeitet zusammen mit einer kostenlosen App, die auf das Smartphone oder Tablet heruntergeladen wird.

Begründung der Jury
Die skulptural anmutende, schwebende Form von ResMed S+ stellt eine ästhetische Bereicherung für jedes moderne Schlafzimmer-Interieur dar.

ResMed AirSense™ 10 & AirCurve™ 10
Sleep Therapy Device
Schlaftherapiegerät

Manufacturer
ResMed, Sydney, Australia
In-house design
Web
www.resmed.com
Honourable Mention

The AirSense 10 and AirCurve 10 sleep therapy devices are particularly small and lightweight. They feature an integrated heated humidifier and intelligent algorithms, which together deliver comfortable therapy for patients suffering from sleep apnoea. The wireless technology enables improved patient care and business efficiencies for health providers. If required, the caregiver can remotely manage the device settings and therapy data via the telemonitoring function.

Statement by the jury
The humidifier of both the AirSense 10 and the AirCurve 10 are discreetly highlighted and form a harmonious whole together with each respective device.

Die Schlaftherapiegeräte AirSense 10 und AirCurve 10 sind besonders klein, leicht und mit einem beheizbaren Atemluftbefeuchter ausgestattet, der gemeinsam mit intelligenten Algorithmen zu einer komfortablen Therapie bei Schlafapnoe beiträgt. Die Funktechnologie erlaubt eine optimierte Vernetzung zwischen Gesundheitsdienstleistern und Patienten und damit eine effizientere Patientenbetreuung. Bei Bedarf kann der Arzt über die Telemonitoring-Funktion die Geräte aus der Ferne einstellen.

Begründung der Jury
Der Atemluftbefeuchter ist sowohl beim AirSense 10 als auch beim AirCurve 10 sehr dezent betont und bildet doch jeweils eine harmonische Einheit mit dem Gerät.

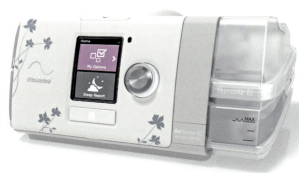
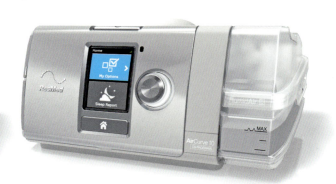

Efficia Patient Monitor Series
Patientenmonitore

Manufacturer
Royal Philips, Eindhoven, Netherlands
In-house design
Web
www.philips.com

The Efficia range of patient monitors is suited for both fixed bedside and portable patient monitoring. Due to the compact size and integrated handles, the monitors are convenient to carry. The intuitive user interface simplifies operation of the devices, also by untrained staff. The visual alarm indicators provide clear feedback, which enhances patient safety.

Statement by the jury
These patient monitors feature a very clear user interface. Thanks to the almost seamless design, they convey a look of high quality and are easy to clean.

Die Patientenmonitore der Serie Efficia sind sowohl für die Überwachung des Patienten am Krankenbett als auch für den mobilen Einsatz geeignet. Aufgrund ihrer kompakten Größe und des integrierten Handgriffs können die Monitore bequem getragen werden. Die intuitive Benutzeroberfläche ermöglicht die Bedienung der Geräte auch durch ungeschultes Personal. Die visuellen Alarmanzeigen geben eindeutige Rückmeldungen, was die Patientensicherheit erhöht.

Begründung der Jury
Die Patientenmonitore bieten eine sehr übersichtliche Benutzeroberfläche. Durch die nahezu fugenlose Gestaltung wirken sie hochwertig und sind einfach zu reinigen.

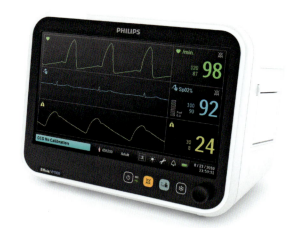

HP ElitePad 1000 Healthcare Jacket
Tablet-Gehäuse

Manufacturer
Hewlett-Packard, Houston, USA
In-house design
Web
www.hp.com

The HP ElitePad 1000 Healthcare Jacket was developed for use in the health sector. It features an antibacterial finish, which protects the HP ElitePad and facilitates easy cleaning. The tablet jacket is IP54 rated and can be used in any clinical environment. Furthermore, it features intelligent functions like a 3D barcode scanner, which enables the accurate identification of administrative and patient data as well as medications.

Statement by the jury
This tablet jacket combines a high-quality finish with additional convincing technical functions. The white colour scheme indicates its use in a clinical setting.

Das Gehäuse HP ElitePad 1000 Healthcare Jacket wurde für das Gesundheitswesen entwickelt. Es wurde antimikrobiell bearbeitet, um das HP ElitePad zu schützen und eine einfache Reinigung zu ermöglichen. Das Tablet-Gehäuse besitzt die Schutzart IP54 und kann in jeder klinischen Umgebung eingesetzt werden. Es verfügt zudem über intelligente Funktionen wie einen 3D-Barcode-Scanner, mit dem sich Verwaltungs- und Patientendaten sowie Medikamente einwandfrei identifizieren lassen.

Begründung der Jury
Das Tablet-Gehäuse verbindet eine hochwertige Verarbeitung mit überzeugenden technischen Zusatzleistungen. Das Weiß deutet auf die klinische Verwendung hin.

HP ElitePad 1000 Rugged Jacket
Tablet-Gehäuse

Manufacturer
Hewlett-Packard, Houston, USA
In-house design
Web
www.hp.com

The HP ElitePad 1000 Rugged Jacket is compatible with the HP ElitePad tablet and is characterised by particularly high robustness and shock-resistance. Its IP65 rating means that the jacket is protected against dust and low-pressure jets of water, allowing it to be used in demanding environments such as the pharmaceutical industry. Thanks to its shoulder strap, the tablet is easy to transport.

Statement by the jury
This jacket is able to withstand high stress, and its robustness is visually highlighted by the rich matt effect.

Das HP ElitePad 1000 Rugged Jacket ist mit dem Arbeits-Tablet HP ElitePad kompatibel und zeichnet sich durch besonders hohe Robustheit und Stoßfestigkeit aus. Es entspricht der Schutzklasse IP65, ist also gegen Staub und Strahlwasser geschützt, sodass es auch unter erschwerten Bedingungen wie beispielsweise in der Pharmaindustrie eingesetzt werden kann. Mithilfe des Schultergurtes kann das Tablet bequem transportiert werden.

Begründung der Jury
Das Gehäuse hält hohen Belastungen stand. Seine Beständigkeit wird durch den satten Matt-Effekt visuell unterstrichen.

amor H1
Heart Rate and Activity Monitor
Herzfrequenz- und Aktivitätsmonitor

Manufacturer
Leadtek Research Inc., New Taipei, Taiwan
In-house design
Prof. Chan Chi Hua
Web
www.leadtek.com

The amor H1 heart rate and activity monitor is used in conjunction with a fitness app. It continuously and precisely measures biometric data for the monitoring of the autonomic nervous system and displays various parameters, such as number of steps, sleep cycle analysis and calories burned. Therefore, it gives wearers a quick overview of their health status. Furthermore, the device provides feedback on personal fitness and exercise routines.

Statement by the jury
The amor H1 features an appealing, practical size and a large display which takes up the entire front, thus representing the data with high clarity.

Der Herzfrequenz- und Aktivitätsmonitor amor H1 wird zusammen mit einer Fitness-App verwendet. Er liefert mit hoher Genauigkeit und Kontinuität biometrische Daten zur Überwachung des vegetativen Nervensystems und zeigt unterschiedliche Parameter wie Schrittzahl, Schlafzyklus-Analyse und Kalorienverbrauch an. So erhält der Anwender einen schnellen Überblick über seinen Gesundheitszustand. Zudem gibt das Gerät Rückmeldung zur persönlichen Fitness und zur Trainingsroutine.

Begründung der Jury
Der amor H1 gefällt durch seine handliche Form und das großzügige Display, das die gesamte Front ausfüllt und die Daten klar und deutlich darstellt.

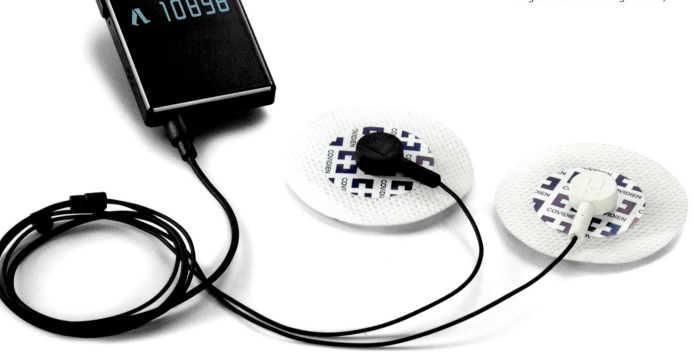

eMotion Faros
ECG Sensor

Manufacturer
Mega Electronics Ltd, Kuopio, Finland
In-house design
Mega Electronics Design Team
Design
Tactical Technologies KPH Nordic Inc.
(Lasse Kettunen), Kuopio, Finland
Web
www.megaemg.com
www.tactech.fi

The eMotion Faros electrocardiogram sensor is very small, weighs merely 13 grams and is operated using a single button. The data may either be stored on the device or transferred to the Internet via Bluetooth. The measuring tool can be attached to the body using electrodes, a cable set or a chest strap. The device is available in translucent red, blue and yellow.

Statement by the jury
The ECG sensor impresses with its minimal size. The coloured surfaces give it a friendly and sporty look.

Der EKG-Sensor eMotion Faros ist extrem klein, wiegt nur 13 Gramm und wird über eine einzige Taste bedient. Die gemessenen Daten können entweder auf dem Gerät gespeichert oder online per Bluetooth versendet werden. Das Messinstrument kann mit Elektroden, einem Kabelset oder einem Brustgürtel am Körper befestigt werden. Es steht in den durchscheinenden Farben Rot, Blau und Gelb zur Verfügung.

Begründung der Jury
Der EKG-Sensor beeindruckt durch seine minimale Größe. Die farbigen Oberflächen sorgen für einen sympathisch-sportlichen Auftritt.

QardioCore
ECG Monitor

Manufacturer
Qardio, San Francisco, USA
In-house design
Web
www.getqarcio.com

The QardioCore ECG monitor allows for continuous and medically precise measurement of cardiac functions, body temperature, activity levels, stress monitoring and other data. The system is attached to the torso without the need for patches, gels, and wires, using a simple strap. It works with a smartphone or tablet and uses advanced dry-sensor technology and bespoke design to fit a modern lifestyle.

Statement by the jury
The ends of the QardioCore are rounded in opposite directions and unite to form an elegant S-shaped line, giving it the look of an accessory rather than a medical device.

Der EKG-Monitor QardioCore erlaubt eine kontinuierliche und medizinisch exakte Messung von Herzfunktionen, Körpertemperatur, Aktivitätsniveau, Stressmonitoring und weiterer Daten. Das System wird ohne Pflaster, Gels und Drähte lediglich mithilfe eines einfachen Gurtes am Körper befestigt. Es lässt sich über ein Smartphone oder Tablet bedienen und nutzt eine fortschrittliche Trockensensor-Technologie sowie ein ansprechendes Design, das einem modernen Lebensstil gerecht wird.

Begründung der Jury
Die gegenläufig gerundeten Enden des QardioCore vereinen sich zu einer eleganten S-Linie und lassen ihn eher wie ein Accessoire wirken denn als Medizingerät.

ezFit
Health Tracker
Vitalmessgerät

Manufacturer
Eclat Textile Co., Ltd.,
New Taipei, Taiwan
In-house design
Kevin Chen, Stephanie Sung, Lydia Lin
Design
Taiwan Textile Research Institute
(Chienlung Shen, Fenling Chen),
New Taipei, Taiwan
Nova Design Co., Ltd.,
New Taipei, Taiwan
Web
www.eclat.com.tw
www.ttri.org.tw
www.e-novadesign.com
Honourable Mention

The ezFit health tracker features various functions for monitoring health status, including electrocardiograms, electroneuromyography and respiratory monitoring. The device is woven into several layers of viscose and conductive fibres and may be worn under tight clothing without any additional fasteners. Furthermore, it can communicate wirelessly with a smartphone or smartwatch via Bluetooth.

Statement by the jury
With its flat form and textile quality, this health tracker is particularly comfortable to wear.

Das Vitalmessgerät ezFit verfügt über verschiedene Funktionen, um den Gesundheitszustand zu erfassen, u. a. Elektrokardiographie, Elektroneurographie und Atemüberwachung. Das Gerät ist in mehrere Schichten aus Viskose und leitfähigen Fasern eingewebt und wird ohne zusätzliche Befestigungsmittel unter enger Kleidung getragen. Des Weiteren kann das Gerät über Bluetooth mit einem Smartphone oder einer Smartwatch kommunizieren.

Begründung der Jury
Das Vitalmessgerät bietet durch seine flache Form und textile Beschaffenheit einen ausgesprochen hohen Tragekomfort.

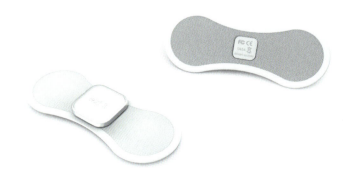

VISUSCOUT® 100
Handheld Fundus Camera
Tragbare Funduskamera

Manufacturer
Carl Zeiss Meditec AG, Jena, Germany
Design
designaffairs GmbH, Erlangen, Germany
Web
www.meditec.zeiss.com
www.designaffairs.com

The Visuscout 100 handheld fundus camera is easy to transport and suitable for any medical practice thanks to its compact size. Due to its powerful battery there are no annoying cables. Nine internal fixation LEDs facilitate the capture of peripheral images of the retina and help align the patient correctly. Images and videos can be immediately transferred to a PC or mobile device via a Wi-Fi connection.

Statement by the jury
The housing of the fundus camera sets fascinating accents with its black-and-white contrasts, while the bottom provides a very stable contact surface.

Die tragbare Funduskamera Visuscout 100 ist aufgrund ihrer kompakten Größe leicht zu transportieren und für jede Praxisumgebung geeignet. Der leistungsstarke Akku macht störende Kabel überflüssig. Neun interne Fixierungs-LEDs erleichtern die Aufnahme von Bildern der Netzhautperipherie und helfen bei der optimalen Ausrichtung des Patienten. Über Wi-Fi können die Bilder und Videos sofort auf einen PC oder ein mobiles Endgerät übertragen werden.

Begründung der Jury
Das Gehäuse der Funduskamera setzt durch die Schwarz-Weiß-Kontraste spannende Akzente, während die Unterseite eine ausgesprochen stabile Auflagefläche bildet.

MagniLink ZIP
Video Magnifier
Bildschirmlesegerät

Manufacturer
LVI Low Vision International AB, Växjö, Sweden
In-house design
Joachim Schill
Design
Myra Industriell Design AB, Stockholm, Sweden
Web
www.lvi.se
www.myra.se

The MagniLink Zip is a fully functional video magnifier, which can be folded when not in use or for transport. The system is available with a 13" or 17" monitor that is connected to a full HD reading/distance camera. The easily accessible rotating knobs and buttons facilitate intuitive operation. The device can also be used with a battery and an adjustable reading table.

Statement by the jury
The MagniLink Zip has all the functions of a stationary reading device plus the unlimited flexibility of a mobile system.

Das MagniLink Zip ist ein vollwertiges Bildschirmlesegerät, das zusammengeklappt werden kann, wenn es gerade nicht benötigt wird oder transportiert werden soll. Das System ist lieferbar mit einem 13"- oder einem 17"-Monitor, der jeweils mit einer Full-HD-Lese-/Distanzkamera gekoppelt ist. Leicht zugängliche Drehknöpfe und Tasten ermöglichen eine intuitive Bedienung. Das Gerät kann mit einem Akku und einem feststellbaren Lesetisch genutzt werden.

Begründung der Jury
Das MagniLink Zip bietet die volle Funktionalität eines stationären Lesegeräts, ergänzt um die uneingeschränkte Flexibilität eines mobilen Systems.

Candy 5 HD 2
Electronic Magnifier
Elektronische Lupe

Manufacturer
HIMS International Corporation,
Daejeon, South Korea
Design
BornPartners Design & Branding,
Daejeon, South Korea
Web
www.himsintl.com
www.bornpartners.com

The Candy 5 HD 2 electronic magnifier features four control buttons in different colours, which can also be easily distinguished by visually impaired individuals. The 5" LCD display has an anti-glare function that absorbs reflections of light. The continuous 22x zoom, the five different colour modes and the four levels of brightness can all be adjusted to the visual requirements of each user. The adjustable handle and the reading stand ensure a comfortable reading process.

Die elektronische Lupe Candy 5 HD 2 besitzt vier unterschiedlich farbige Bedientasten, die auch für Menschen mit einer Sehbehinderung gut voneinander zu unterscheiden sind. Ein 5"-LCD-Display mit Blendschutz absorbiert Lichtreflektionen. Der stufenlose 22-fache Zoom, fünf verschiedene Farbmodi sowie vier unterschiedliche Helligkeitsstufen können zudem auf die individuellen Sehbedürfnisse des Anwenders abgestimmt werden. Der verstellbare Handgriff und das Lesestativ sorgen für einen angenehmen Lesekomfort.

Statement by the jury
The sculpted surfaces and colourful highlights have a visual and tactile appeal, while also giving the magnifier a playful character.

Begründung der Jury
Die ausmodellierten Flächen und Farbmarkierungen sind nicht nur haptisch und visuell gut erfahrbar, sie verleihen der Lupe auch einen spielerischen Charakter.

LINKAGE EF
Hearing Aid
Hörgerät

Manufacturer
SK telecom, Seoul, South Korea
In-house design
Web
www.sktelecom.com

Linkage EF is an intelligent hearing aid, which is suited for people with hearing loss and can also - when the hearing aid function is turned off - be used as a conventional Bluetooth headset. Fine acoustic adjustment can be accomplished using an app. The fashionable design aims to increase the wearer's self-confidence and achieve a more open approach to the use of hearing aids.

Statement by the jury
The Linkage EF hearing aid is characterised by its smooth, seamless form, which is further emphasised by its monochrome colour scheme and looks very stylish.

Linkage EF ist ein intelligentes Hörgerät, das für schwerhörige Menschen geeignet ist, aber auch – wenn die Hörgerätfunktion ausgeschaltet ist – als ganz normales Bluetooth-Headset genutzt werden kann. Die akustische Feinjustierung erfolgt über eine App. Das modische Design zielt darauf ab, das Selbstbewusstsein des Trägers zu stärken und einen offeneren Umgang mit Hörhilfen zu erreichen.

Begründung der Jury
Das Hörgerät Linkage EF ist geprägt von seiner nahtlosen Form, die von der monochromen Farbgebung noch betont wird und sehr stilvoll aussieht.

Moxi Fit
Hearing Aid
Hörgerät

Manufacturer
Unitron, Kitchener, Canada
In-house design
Design
AWOL Company, Calabasas, USA
Web
www.unitron.com
www.awolcompany.com

The Moxi Fit hearing aid combines style with intuitive functionality. A sculpted form fits the ear's natural anatomy and a smooth luminescent finish impresses in the hand. The button controlling volume and programme selection can be operated effortlessly even by those with dexterity challenges. The battery door is easily opened, and colour markers allow for a clear differentiation between right and left. Moreover, the hearing aid is available in a variety of pleasing colours.

Statement by the jury
Softly curved forms and a discreet sheen lend this hearing aid a classically elegant aesthetic appearance.

Das Hörgerät Moxi Fit vereint Stil mit Funktionalität. Die skulpturale Form passt sich der natürlichen Anatomie des Ohres an und das sanft lumineszierende Finish besticht in der Hand. Auch bei eingeschränkter Fingerfertigkeit ist der Taster für die Lautstärke und Programmauswahl problemlos zu bedienen. Die Batterielade ist einfach zu öffnen. Durch Farbmarkierungen lassen sich rechts und links gut unterscheiden. Das Hörgerät ist in vielen ansprechenden Farben erhältlich.

Begründung der Jury
Die sanft geschwungene Form und der dezente Schimmer verleihen dem Hörgerät eine klassisch-elegante Ästhetik.

saga R
Hearing Aid
Hörgerät

Manufacturer
audifon GmbH & Co. KG,
Kölleda, Germany
Design
gobrecht ID (Thomas Gobrecht),
Gelnhausen, Germany
Web
www.audifon.com
www.gobrechtdesign.de
Honourable Mention

The saga R hearing aid is suitable for all kinds of hearing loss and automatically adjusts to the acoustics of the individual situation. It features complex technology for signal processing and is characterised by its dynamically curved lines and elaborately lacquered housing, which comes in many different colours. The aesthetic form of the hearing aid corresponds to the ear's sensitive anatomy.

Statement by the jury
The textures of the different materials used in this hearing aid reflect light to varying degrees, thus fostering an elegant look.

Das Hörgerät saga R ist für alle Arten von Hörverlusten geeignet und passt sich automatisch der jeweiligen akustischen Situation an. Es verfügt über komplexe Techniken zur Signalverarbeitung und zeichnet sich durch dynamisch geschwungene Linien und ein aufwendig lackiertes Gehäuse aus, das in vielen verschiedenen Farben zur Verfügung steht. Die ästhetische Form des Hörgeräts entspricht der sensiblen Anatomie des Ohres.

Begründung der Jury
Die verschiedenen Materialstrukturen des Hörgeräts reflektieren das Licht unterschiedlich stark und erzeugen dadurch eine anmutige Eleganz.

Beltone Boost
Hearing Aid
Hörgerät

Manufacturer
Beltone, Glenview, USA
In-house design
Web
www.beltone.com

The Beltone Boost hearing aid was developed for children and adults with profound hearing loss. It can be connected with iOS devices such as the iPhone, enabling phone calls and music or video chats to be transferred directly from the smartphone to the hearing aid without requiring any intermediary devices. Moreover, the cordless connection improves the understanding of speech, particularly during phone conversations and in loud environments.

Statement by the jury
Thanks to technical innovation and a remarkably small design, the Beltone Boost is very discreet and also comfortable to wear.

Das Hörgerät Beltone Boost wurde für Kinder und Erwachsene entwickelt, die von einem hochgradigen Hörverlust betroffen sind. Es lässt sich mit iOS-Geräten wie dem iPhone koppeln, sodass Anrufe, Musik oder Videochats vom Smartphone direkt und ohne Zusatzgeräte auf das Hörgerät übertragen werden können. Darüber hinaus verbessert die Schnurlosverbindung das Sprachverstehen insbesondere bei Telefongesprächen und bei lauten Umgebungsgeräuschen.

Begründung der Jury
Technische Innovation und eine besonders kleine Bauform machen Beltone Boost zu einem Hörgerät, das sehr diskret und komfortabel zu tragen ist.

Cochlear Baha 5 Sound Processor
Cochlear Baha 5 Soundprozessor

Manufacturer
Cochlear Bone Anchored Solutions AB, Mölnlycke, Sweden
Design
Attention (Henrik Jeppesen, Martin Pråme), Copenhagen, Denmark
Web
www.cochlear.com
www.attention-group.com

The Cochlear Baha 5 Sound Processor was developed for individuals who rely on a bone conduction system to hear and communicate. It is equipped with a powerful and efficient transducer that is almost 50 per cent smaller than the previous generation. The sound processor is a Made for iPhone hearing device and can be controlled by a Smart App. Its small size and range of hair-matching colours make it a very discreet device.

Statement by the jury
This sound processor is a highly successful translation of modern technology into an exceptionally compact design.

Der Cochlear Baha 5 Soundprozessor wurde für Menschen entwickelt, die auf ein Knochenleitungsimplantat angewiesen sind, um kommunizieren zu können. Ausgestattet ist er mit einem leistungsfähigen Wandler, der fast halb so groß ist wie bei dem Vorgängermodell. Der Soundprozessor ist „Made for iPhone"-fähig und kann über eine Smart App gesteuert werden. Aufgrund der kleinen Größe und einer großen Farbauswahl, die sich auf die Haarfarbe abstimmen lässt, ist das Gerät diskret zu tragen.

Begründung der Jury
Bei dem Soundprozessor ist es auf beeindruckende Weise gelungen, moderne Technik in eine außergewöhnlich kompakte Bauform zu überführen.

3M ESPE dental face shield
3M ESPE zahnärztlicher Gesichtsschutzschild

Manufacturer
3M, Shanghai, China
In-house design
Annie Yang
Web
www.3m.com
Honourable Mention

The 3M ESPE dental face shield protects the dentist from flying particles and liquid splash during treatment. The anti-fog and anti-reflection coating ensures clear visibility. The frame made from polycarbonate and thermoplastic elastomer makes it pleasantly light to wear and adjustable to different head sizes. The upper half of the shield has an arched shape to provide space for practitioners wearing glasses.

Statement by the jury
With its narrow and light frame, this face shield has a very unobtrusive look and therefore does not have an interfering effect on the dentist or the patient.

Der Gesichtsschutzschild 3M ESPE schützt den Zahnarzt während der Behandlung vor umherfliegenden Partikeln und Flüssigkeiten. Die reflexions- und beschlagfreie Beschichtung sorgt für eine stets klare Sicht. Der Rahmen aus Polycarbonat und thermoplastischem Elastomer erzeugt ein angenehm leichtes Tragegefühl und kann an verschiedene Kopfgrößen angepasst werden. Die obere Hälfte des Schildes ist stärker gewölbt, um Brillenträgern genügend Platz zu bieten.

Begründung der Jury
Der Gesichtsschutzschild wirkt durch den schmalen und hellen Rahmen sehr unauffällig, sodass es in keiner Weise den Arzt oder Patienten stört.

D1 pure
Dental Milling Machine
Dental-Fräsmaschine

Manufacturer
DATRON AG, Mühltal, Germany
In-house design
Frank Wesp
Web
www.datron.de

Thanks to its compact size, the five-axis dental milling machine called D1 pure requires very little space. It was developed for the manufacturing of dental restorations made from soft materials in small laboratories and dental practices. The integrated 9" multitouch display merges with the smooth acrylic surface. The strong slope of the housing's side contour improves access to the interior and touchscreen use. The pronounced contrasts in the housing's colour scheme direct the focus towards the control unit.

Die 5-achsige Dental-Fräsmaschine D1 pure benötigt durch ihre kompakten Abmessungen nur eine geringe Stellfläche. Sie wurde speziell für die Fertigung von Zahnrestaurationen aus weichen Materialien in kleinen Labors und Zahnarztpraxen entwickelt. Das integrierte 9"-Multitouch-Display verschmilzt mit der glatten Acryl-oberfläche. Die starke Abschrägung der seitlichen Gehäusekontur verbessert die Zugänglichkeit zum Innenraum und die Bedienung des Touchscreens. Durch hohe Kontraste in der Gehäusefarbgebung wird der Fokus auf die Bedieneinheiten gelenkt.

Tyscor VS 2
Radial Suction Machine
Radialabsaugmaschine

Manufacturer
Dürr Dental AG, Bietigheim-Bissingen, Germany
Design
formstudio merkle park (Ulrich Merkle), Stuttgart, Germany
Web
www.duerrdental.com
www.formstudio.com

The Tyscor VS 2 for dental practices is lightweight, quiet and energy-saving. The sound-absorbing and bacteria-inhibiting foam housing has only half the weight of the previous model. During the production process, the material itself requires only a very low amount of energy. The environmentally friendly radial suction motor saves up to 50 per cent of energy and also enables a noticeably space-saving construction. The device is monitored via a network and thus requires no additional control elements. There are new freedoms in designing the exterior.

Die Absaugmaschine Tyscor VS 2 für Zahnarztpraxen ist leicht, leise und energiesparend. Das schallabsorbierende, bakterienhemmende Schaumgehäuse ist nur noch halb so schwer wie beim bisherigen Gerät. Das Material selbst erfordert bereits in der Herstellung einen geringen Energieaufwand. Der umweltfreundliche Radialsaugmotor erreicht bis zu 50 Prozent Energieeinsparung und ermöglicht darüber hinaus eine deutlich platzsparende Bauweise. Das Gerät wird netzwerkgebunden überwacht und benötigt daher keine eigenen Bedienelemente. An den Außenflächen ergeben sich neue Freiheiten der Gestaltung.

Statement by the jury
Thanks to the selection of materials, the Tyscor VS 2 is surprisingly lightweight, and the foam housing also gives rise to a special aesthetic.

Begründung der Jury
Die Tyscor VS 2 ist durch die innovative Materialwahl nicht nur überraschend leicht, sondern aus dem Schaumgehäuse ergibt sich auch eine besondere Ästhetik.

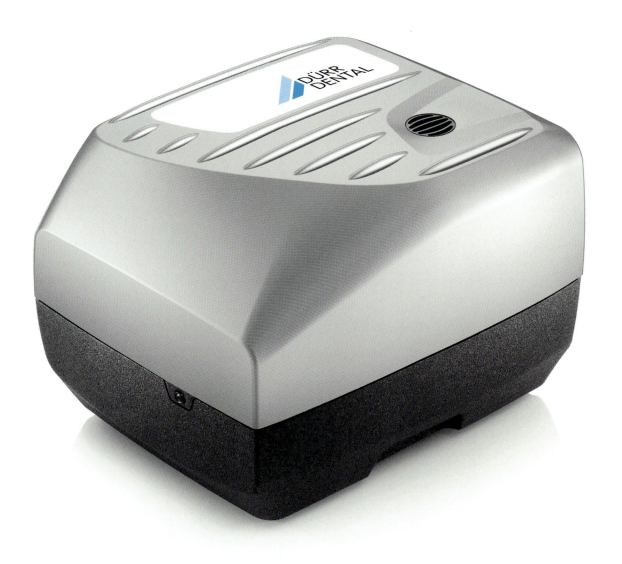

inLab MC X5
Dental Milling and Grinding Unit
Dentale Fräs- und Schleifeinheit

Manufacturer
Sirona Denta Systems GmbH, Bensheim, Germany
Design
Puls Produktdesign (Andreas Ries, Torsten Richter),
Darmstadt, Germany
Web
www.sirona.com
www.puls-design.de

The inLab MC X5 milling and grinding unit is used in dental laboratories to produce aesthetically sophisticated restorations. The five-axis machine offers high flexibility, because it can quickly switch between different materials such as zirconium oxide, polymers and ceramics. Moreover, an automatic change from dry to wet processing is possible while working on a single workpiece. The high-quality design of the chamber ensures easy access for maintenance and is therefore easy to clean.

Die Fräs- und Schleifeinheit inLab MC X5 wird in zahntechnischen Laboren zur Herstellung ästhetisch anspruchsvoller Restaurationen eingesetzt. Die 5-Achs-Maschine bietet eine hohe Flexibilität, da sie zwischen verschiedenen Materialien wie Zirkonoxid, Kunststoff oder Keramik zügig wechseln kann. Darüber hinaus ist während der Bearbeitung eines Werkstücks ein automatischer Wechsel von Trocken- auf Nassfertigung möglich. Die hochwertige Innenraumkonstruktion ist gut zugänglich und dadurch schnell und einfach zu warten und zu reinigen.

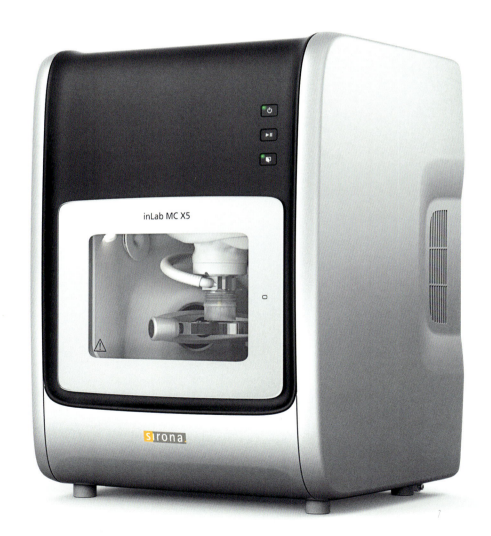

HAAKE
Viscotester iQ Rheometer

Manufacturer
Thermo Fisher Scientific, Karlsruhe, Germany
Design
Indeed Innovation GmbH,
Hamburg, Germany
Web
www.thermoscientific.com/VTiQ
www.indeed-innovation.com

The HAAKE Viscotester iQ Rheometer measures and analyses the rheological properties of a variety of different samples – ranging from low-viscous liquids to stiff pastes. The intuitive user interface increases its efficiency and minimises the risk of operator errors. Its compact size and small footprint enable mobile use within the laboratory. Thanks to its modular design, the unit can be configured to meet individual requirements.

Statement by the jury
This rheometer fascinates with its very delicate and accentuated shape. The operating field faces the user and features a deliberately simple design.

Das HAAKE Viscotester iQ Rheometer misst und analysiert die Fließeigenschaften einer Vielzahl von Proben – von Flüssigkeiten mit geringer Viskosität bis hin zu dicken Pasten. Die intuitive Bedienoberfläche erhöht die Effizienz und minimiert das Risiko einer Fehlbedienung. Die kompakten Abmessungen und die geringe Standfläche ermöglichen eine mobile Anwendung im Labor. Durch das modulare Design kann das Gerät individuell angepasst werden.

Begründung der Jury
Das Rheometer besticht durch seine sehr feine und akzentuierte Form. Das Bedienfeld ist dem Anwender zugewandt und bewusst einfach gehalten.

Roller 10 digital
Roller Shaker
Rollenschüttler

Manufacturer
IKA-Werke GmbH & Co. KG,
Staufen, Germany
In-house design
Web
www.ika.com

The Roller 10 digital roller shaker has been developed for shaking solid and liquid substances in containers and bottles with a variety of diameters and sizes. The transparent side panels that enclose the working area adjust to the position of the outermost rollers, preventing individual containers from falling off the device. The panels are removable and therefore quick and easy to clean.

Statement by the jury
The design of this roller shaker is highly functional and simple, making it very intuitive and safe to use.

Mit dem Rollenschüttler Roller 10 digital können feste und flüssige Substanzen in Gefäßen und Flaschen mit unterschiedlichen Durchmessern und Größen geschüttelt werden. Die transparenten Seitenschilder, die den Arbeitsbereich umgeben, passen sich stets an die Lage der äußersten Rollen an und verhindern so, dass einzelne Gefäße vom Gerät fallen. Die Schilder sind abnehmbar und können somit schnell und einfach gereinigt werden.

Begründung der Jury
Die Gestaltung des Rollenschüttlers ist ausgesprochen funktional und einfach, sodass er sehr intuitiv und sicher zu bedienen ist.

Trayster digital
Overhead Shaker
Überkopfschüttler

Manufacturer
IKA-Werke GmbH & Co. KG,
Staufen, Germany
In-house design
Web
www.ika.com

The Trayster digital is an overhead shaker for separating and mixing powder and liquid samples. Thanks to its range of attachments, which can be combined as required, the device is compatible with the container sizes commonly used in laboratories. All attachments may be mounted or removed swiftly thanks to its quick-release catch, which also makes the device easier to clean after use. If required, all attachments can be installed permanently.

Statement by the jury
This overhead shaker captivates with its flowing transition from the control area to the tray. Its open design ensures excellent access to the samples.

Der Trayster digital ist ein Überkopfschüttler zum Trennen und Mischen von pulverförmigen und flüssigen Proben. Durch unterschiedliche Aufsätze, die frei kombinierbar sind, deckt er die gängigen Gefäßgrößen im Labor ab. Alle Aufsätze können durch einen Schnappverschluss schnell montiert oder demontiert werden. Dies erleichtert auch die Reinigung des Geräts nach der Anwendung. Bei Bedarf können alle Aufsätze dauerhaft fixiert werden.

Begründung der Jury
Der Überkopfschüttler gefällt durch den fließenden Übergang vom Bedien- zum Spannbereich. Die offene Form bietet einen hervorragenden Zugriff auf die Proben.

Luxeo 6i
Stereo Microscope
Stereomikroskop

Manufacturer
Labo America, Fremont, USA
In-house design
Neeraj Jain
Web
www.laboamerica.com

The Luxeo 6i stereo microscope features a fully integrated illumination system, giving the base an unusually compact design. The stand can be extended with an easy-to-operate lifting mechanism to accommodate large objects. The coaxial arrangement and the gooseneck double LED light enable the illumination of all kinds of complex surfaces. The in-base, dark- and bright-field arrangement can be easily controlled using a lever.

Statement by the jury
This stereo microscope is expressive in form and colour. The flowing lines and strict geometric design yield a fascinating contrast.

Das Stereomikroskop Luxeo 6i verfügt über ein komplett integriertes Beleuchtungssystem. Dadurch fällt der Standfuß ungewöhnlich kompakt aus. Der Ständer kann mit einem leicht zu bedienenden Hubmechanismus verlängert werden, um auch große Objekte zu platzieren. Der Koaxial-Aufbau und das Schwanenhals-Doppel-LED-Licht ermöglichen die Ausleuchtung aller Arten von komplexen Oberflächen. Die Dunkel- und Hellfeld-Anordnung in der Basis lässt sich einfach per Hebel betätigen.

Begründung der Jury
Das Stereomikroskop ist expressiv in Form und Farbe. Aus den fließenden Linien und dem streng geometrischen Aufbau ergibt sich ein spannender Kontrast.

MALDI Biotyper Pilot
Workstation for Sample Preparation
Arbeitsstation für die Probenpräparation

Manufacturer
Bruker Dalton k GmbH, Bremen, Germany
In-house design
Design
TO DO Design GmbH & Co. KG (Tiago Faria), Bremen, Germany
Web
www.bruker.com
www.todo-design.de

The Maldi Biotyper Pilot assists in the preparation of samples in microbiology laboratories. The mini-projector displays cross hairs on the sample carrier and therefore indicates the exact position where the sample is to be placed. The device is operated without a keyboard, as the assignment of the sample to the sample carrier is solely carried out through barcode scanning. The cross hairs automatically jump to the next free position.

Statement by the jury
The Maldi Biotyper Pilot captivates with its minimalist design language, which clearly identifies the different areas of function.

Der Maldi Biotyper Pilot unterstützt die Präparation von Proben im mikrobiologischen Labor. Dazu projiziert der Mini-Beamer ein Fadenkreuz auf den Probenträger und zeigt dadurch die genaue Position an, wohin die Probe übertragen werden soll. Für die Bedienung ist keine Tastatur notwendig. Die Zuordnung der Proben zum Probenträger geschieht allein durch Scannen von Barcodes, wobei das Fadenkreuz stets auf die nächste freie Position springt.

Begründung der Jury
Der Maldi Biotyper Pilot besticht durch seine reduzierte Formensprache, bei der die Funktionsbereiche eine klare Zuweisung erfahren.

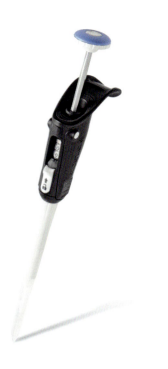

MICROMAN E
Positive Displacement Pipette
Direktverdrängungspipette

Manufacturer
Gilson SAS, Villiers-le-Bel, France
In-house design
Web
www.gilson.com

The Microman E positive displacement pipette prevents a layer of air from forming between the pipette and the sample. Thus, problematic liquids such as very thick or slightly volatile solutions can be pipetted safely and precisely. The slim, ergonomic handle features a volume-control button, a wide finger-rest and a large push-button. Its patented QuickSnap system makes it as intuitive to use as a regular pipette.

Statement by the jury
Thanks to its extremely slim form and dynamically curved lines, this positive displacement pipette rests comfortably in the hand.

Die Direktverdrängungspipette Microman E verhindert, dass eine Luftschicht zwischen Pipette und Probe entsteht. Dadurch lassen sich problematische Flüssigkeiten wie sehr zähe oder leicht flüchtige Lösungen sicher und präzise pipettieren. Der schmale ergonomische Handgriff verfügt über einen Volumen-Kontroll-Druckknopf, einen breiten Fingerabstand und einen großen Druckknopf. Das patentierte QuickSnap-System erlaubt eine ebenso intuitive Handhabung wie bei einer herkömmlichen Pipette.

Begründung der Jury
Die Direktverdrängungspipette liegt dank ihrer überaus schlanken Form und der dynamisch geschwungenen Linienführung hervorragend in der Hand.

General Purpose Water Baths
Mehrzweck-Wasserbäder

Manufacturer
PolyScience, Niles, USA
In-house design
Philip Preston, Tom Wallbaum
Design
Cesaroni Design, Glenview, USA
Web
www.polyscience.com
www.cesaroni.com
Honourable Mention

The digital General Purpose Water Baths offer a volume range from 2–28 litres. The laboratory devices feature an integrated timer and preprogrammed temperature set-points. The gabled, polycarbonate cover accommodates glassware of varying heights and tilts away when loading and unloading samples. It also allows the condensate to drain neatly back into the bath.

Statement by the jury
The General Purpose Water Baths impress with the high functionality provided by their covers. Their slanted control panels facilitate ease of use.

Diese digitalen Mehrzweck-Wasserbäder bieten ein Fassungsvermögen von 2–28 Litern. Die Laborgeräte verfügen über einen integrierten Timer und voreinstellbare Temperatursollwerte. Die giebelförmige Abdeckung aus Polycarbonat erlaubt die Verwendung von Gläsern mit unterschiedlichen Höhen. Zum Be- und Entladen der Proben wird die Abdeckung aus dem Weg gekippt, außerdem ermöglicht sie ein sauberes Abfließen des Kondensats zurück in das Bad.

Begründung der Jury
Die Wasserbäder überzeugen durch die hohe Funktionalität der Abdeckung. Das angeschrägte Bedienfeld vereinfacht die Benutzung.

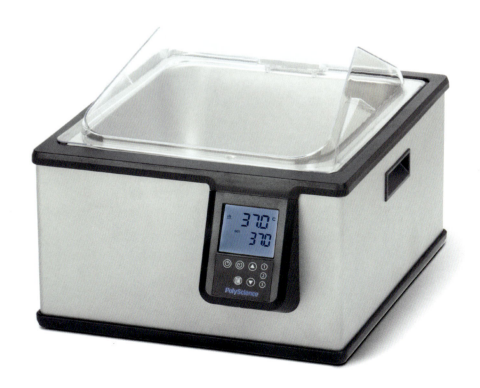

PG 85 Series
Washer-Disinfectors
Reinigungs- und
Desinfektionsautomaten

Manufacturer
Miele & Cie. KG, Gütersloh, Germany
In-house design
Web
www.miele.de

The PG 85 series washer-disinfectors enable automatic cleaning of sensitive items in medical and laboratory environments. A large range of baskets and accessories offer flexible adaptation to individual loading requirements. The smooth housing surfaces and the stainless-steel operating panel are easy to clean. The well-accessible salt container in the door and the AutoClose function ensure a fluent workflow.

Statement by the jury
With its straight lines, the design of the PG 85 series radiates robustness and strength. The monochrome colour scheme supports its industrial look.

Die Reinigungs- und Desinfektionsautomaten der Serie PG 85 ermöglichen die maschinelle Aufbereitung von sensiblem Spülgut im Medizin- und Laborbereich. Eine große Auswahl an Körben und Zubehör bietet eine flexible Anpassung an die jeweilige Beladungssituation. Die glatten Geräteoberflächen und das Bedienfeld aus Edelstahl sind gut zu reinigen. Das leicht zugängliche Salzgefäß in der Tür und die AutoClose-Funktion erleichtern die Arbeitsabläufe.

Begründung der Jury
Die geradlinige Bauweise der Serie PG 85 strotzt vor Robustheit und Stärke. Die monochrome Farbgebung unterstützt ihre industrielle Anmutung.

GETINGE GSS67H
Steriliser
Sterilisator

Manufacturer
Getinge Infection Control AB,
Getinge, Sweden
In-house design
Cecilia Anderberg, Per Englesson
Design
Etteplan (Tomas Stringdahl),
Halmstad, Sweden
Web
www.getinge.com
www.etteplan.com

The Getinge GSS67H steriliser is suitable for a wide range of applications related to the sterilisation of equipment in hospitals. The large chamber, which is supplemented by a special sliding-door design, enables a greater load capacity than found in comparable sterilisers. The smooth, polished stainless-steel surfaces and the flush mounted door make the front easy to clean. The control panel is made of Corian, a non-porous material, so stains cannot penetrate its surface.

Statement by the jury
The ratio of footprint to volume is excellent in this steriliser. The materials employed meet high requirements with regard to cleanliness and hygiene.

Der Dampfsterilisator Getinge GSS67H bietet vielseitige Einsatzmöglichkeiten zur Aufbereitung von Sterilgut im Krankenhaus. Die große Kammer, die durch eine spezielle Schiebetürkonstruktion ergänzt wird, erlaubt höhere Beladekapazitäten als bei vergleichbaren Sterilisatoren. Die glatten Oberflächen aus poliertem Edelstahl und die flächenbündige Tür ermöglichen eine einfache Reinigung der Front. Das Bedienpanel besteht aus Corian, einem festen und geschlossenen Mineralwerkstoff, sodass Verunreinigungen der Oberfläche nichts anhaben können.

Begründung der Jury
Das Verhältnis von Stellfläche und Fassungsvermögen ist bei diesem Sterilisator vortrefflich gestaltet. Die Materialien erfüllen hohe Ansprüche an Sauberkeit und Hygiene.

PSPIX
Imaging Plate Scanner for Dental Radiology
Speicherfolienscanner für die Dentalradiologie

Manufacturer
Sopro, Acteon Group, La Ciotat, France
Design
Axena Design and Engineering, Bailly, France
Web
www.acteongroup.com
www.axena.fr

The PSPIX imaging plate scanner can be set up directly at the place of treatment due to its compact size. Therefore, the dentist does not need to move to another room to scan X-ray images. The large colour touchscreen makes the device intuitive, fast and easy to use. The imaging plates are kept in hygienic sleeves to protect them from light and contamination. The removable parts can be cleaned in a thermal disinfector or optionally in an autoclave.

Der Speicherfolienscanner PSPIX kann aufgrund seiner kompakten Größe direkt am Behandlungsplatz aufgestellt werden. Dadurch muss der Zahnarzt nicht in einen anderen Raum gehen, um die Röntgenbilder zu scannen. Der große Farb-Touchscreen ermöglicht eine intuitive, schnelle und einfache Bedienung. Die Speicherfolien werden in Hygienehüllen vor Licht und Verunreinigungen geschützt. Die abnehmbaren Teile können in einem Thermodesinfektor oder optional in einem Autoklaven gereinigt werden.

Statement by the jury
Flowing lines and glossy effects characterise the PSPIX, giving it a premium and elaborate appearance.

Begründung der Jury
Fließende Linien und Hochglanzeffekte charakterisieren den PSPIX, der dadurch ein sehr wertiges und extravagantes Erscheinungsbild erhält.

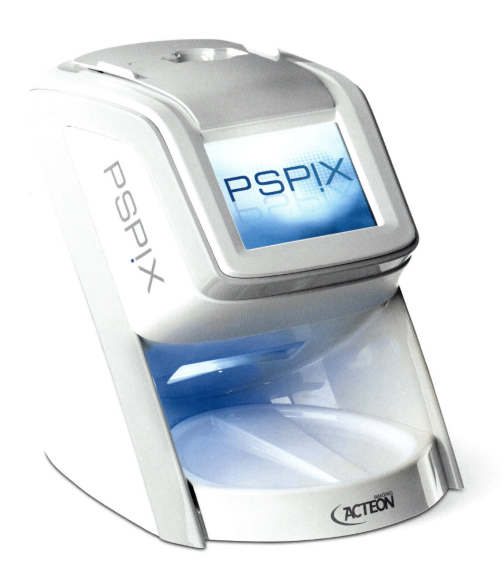

Molecular Diagnosis System
System für die Molekulardiagnostik

Manufacturer
CapitalBio Corporation, Beijing, China
Design
Tsinghua University,
Healthcare Design Innovation Lab (Chao Zhao),
Beijing, China
Web
www.capitalbio.com
www.ad.tsinghua.edu.cn

The Molecular Diagnosis System is an automated solution for the execution and analysis of molecular biological tests. It is used in clinical settings and research laboratories to detect changes in DNA and thus identify illnesses at an early stage. The laboratory system consists of a high-throughput scanner for imaging and data analysis of gene arrays, a device for the isothermal amplification of DNA molecules, and a workstation that identifies specific DNA sequences using a dynamic hybridisation method.

Dieses Laborsystem bildet eine automatisierte Komplettlösung zur Durchführung und Analyse molekularbiologischer Tests. Es wird in der Klinik und in der Forschung eingesetzt, um Veränderungen in der Erbsubstanz nachzuweisen und dadurch Erkrankungen früh zu entdecken. Das System besteht aus einem Hochdurchsatz-Scanner für die Bildgebung und Datenanalyse von Gen-Arrays; einem Gerät zur isothermischen Vervielfältigung von DNA-Molekülen; einer Arbeitsstation, die spezifische DNA-Sequenzen durch ein dynamisches Hybridisierungsverfahren nachweist.

Statement by the jury
The Molecular Diagnosis System surprises with its asymmetrical shapes, which are framed by turquoise-coloured side panels and give the devices a personality of their own.

Begründung der Jury
Das Laborsystem überrascht mit asymmetrischen Formen, die von türkisfarbenen Seitenwänden eingerahmt werden und den Geräten Persönlichkeit verleihen.

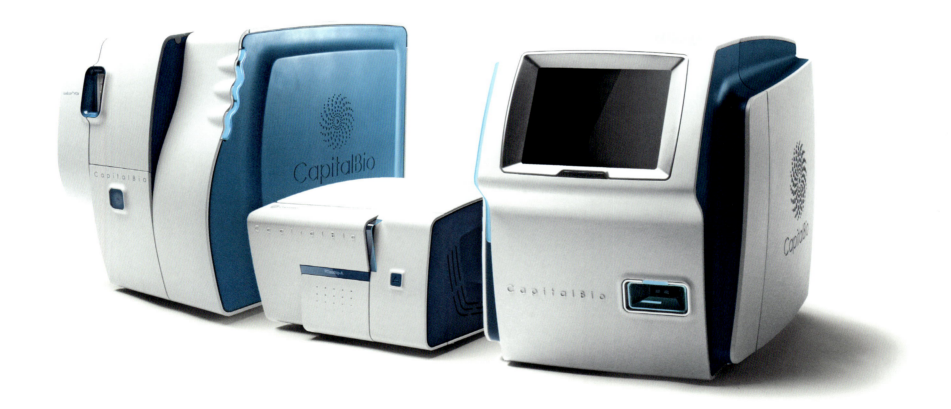

Rowa ProLog®
Automatic Storage System
Vollautomatisches
Einlagerungssystem

Manufacturer
CareFusion Germany 326 GmbH,
Kelberg, Germany
Design
Henssler und Schultheiss
Fullservice Productdesign GmbH
(Martin Schultheiss),
Schwäbisch Gmünd, Germany
Web
www.carefusion.com/rowa
www.henssler-schultheiss.de

ProLog automatically stores large quantities of pharmaceutical products in a connected order-picking system. It empties the contents of large wholesale boxes into the storage flap, and the system sorts the packages automatically. Made predominately of sheet metal, the casing components structure the system in a striking manner, while the aluminium surfaces contrast with the technical look. The hopper is length-adjustable, and the control module can be positioned in three different ways.

ProLog erledigt vollautomatisch die Einlagerung großer Mengen Arzneimittel-packungen in ein angeschlossenes Kommissioniersystem. Die Großhandelskisten werden durch die Einlagerungsklappe entleert, und das System sortiert die Packungen selbständig. Die größtenteils in Blech gefertigten Gehäuseteile sorgen für eine markante Strukturierung der Anlage, die Aluminiumflächen bilden einen Kontrast zu der technischen Anmutung. Der Schüttwagen ist längenverstellbar und das Bedienmodul auf drei verschiedene Arten positionierbar.

FXS Combi
FXS Kombi
Filling and Closing Machine
Füll- und Verschließmaschine

Manufacturer
Robert Bosch GmbH, Packaging Technology,
Crailsheim, Germany
In-house design
Web
www.boschpackaging.com
Honourable Mention

The FXS Combi filling and closing machine is designed for pharmaceutical filling operations. According to the output and the container, the machine with either two or five positions processes pre-sterilised syringes, vials and cartridges at low and medium output rates. The integrated capping station for vials and cartridges enables a space-saving adaptation to the production environment and high pharmaceutical safety, since the containers are sealed safely immediately after filling. The parts for the different packaging types can be exchanged quickly.

Die Füll- und Verschließmaschine FXS Kombi ist eine Anlage für die pharmazeutische Abfüllung. Je nach Ausbringung und Behältnis verarbeitet die zwei- bzw. fünfstellige Maschine vorsterilisierte Spritzen, Vials und Karpulen im unteren bis mittleren Ausbringungsbereich. Die integrierte Bördelstation für Vials und Karpulen ermöglicht eine platzsparende Anpassung an die Produktionsumgebung und hohe pharmazeutische Sicherheit, da das Behältnis unmittelbar nach der Füllung sicher verschlossen wird. Die Formatteile für die verschiedenen Packmittel sind schnell auswechselbar.

ViVAA
Biodynamic Room Lighting
Biodynamische
Raumbeleuchtung

Manufacturer
Derungs Licht AG, Gossau, Switzerland
In-house design
Design
zeug design gmbh, Salzburg, Austria
Web
www.derungslicht.com
www.zeug.at

Vivaa bathes corridors, receptions and waiting areas in natural light, which simulates the 24-hour daylight cycle. The cold-tone light in the morning has an activating effect, while the warm-tone light starting in the afternoon prepares for relaxing sleep. The seamless luminaire made from polished stainless steel is slim and appears to hover in the room. The high proportion of indirect light, the LED module and an anti-glare technology all ensure that the illumination is homogeneous and free of both glare and shadows.

Statement by the jury
Form and function harmonise in this lighting system: it simulates the diurnal rhythm but is also visually reminiscent of the sun and moon.

Vivaa taucht Flure, Empfangs- und Warte-bereiche in ein natürliches Licht, das den 24-Stunden-Tageslichtverlauf simuliert. Das Kalttonlicht am Morgen aktiviert, während das Warmtonlicht ab dem Nach-mittag auf einen erholsamen Schlaf vor-bereitet. Der nahtlose Leuchtkörper aus geschliffenem Edelstahl ist schlank und scheint im Raum zu schweben. Der hohe Anteil an indirektem Licht, das LED-Modul und eine Entblendungstechnik sorgen für eine homogene, blend- und schattenfreie Ausleuchtung.

Begründung der Jury
Form und Funktion stehen bei der Leuchte im Einklang: Sie ahmt nicht nur den Tages-rhythmus nach, sondern erinnert auch optisch an die Sonne und den Mond.

Adaptive Healing Room
Adaptives Patientenzimmer

Manufacturer
Royal Philips, Eindhoven, Netherlands
In-house design
Web
www.philips.com

The Adaptive Healing Room increases well-being and supports the healing process. The amount of stimuli provided by the room can be gradually adjusted. This is beneficial particularly for stroke patients whose brain activity has been impacted. There are three programmes available that foster stimulus in different ways, designed to respectively vary the amount of artificial light, the atmos-phere at night and the intensity of the screen's brightness.

Statement by the jury
The holistic design concept of the heal-ing room creates visual spaciousness and clarity, which have an immediate, calming effect.

Dieses Patientenzimmer steigert das Wohlbefinden und unterstützt beim Gene-sungsprozess. Das Reizangebot, das von dem Raum ausgeht, lässt sich graduell anpassen. Davon profitieren insbesondere Schlaganfallpatienten, deren Gehirn-aktivitäten beeinträchtigt sind. Es stehen drei Programme zur Auswahl, die auf unterschiedliche Weise stimulieren, indem jeweils die Menge des künstlichen Lichts, die Atmosphäre in der Nacht sowie die Intensität der Bildschirmhelligkeit variiert werden kann.

Begründung der Jury
Das ganzheitliche Gestaltungskonzept des Patientenzimmers schafft optische Weite und Klarheit, was eine unmittelbar beruhi-gende Wirkung ausübt.

XICOO AVS-01
Vital Signs Monitor
Vitalzeichenmonitor

Manufacturer
Guangzhou Shirui Electronics Co., Ltd.,
Guangzhou, China
In-house design
Wang Wukun, Xiao Dongyi
Web
www.cvte.com

The XICOO AVS-01 portable vital signs monitor has been designed for professional use in medical environments and covers all clinically relevant parameters in one device. It measures blood pressure, oxygen saturation, pulse frequency and temperature. If values are critical, the alarm function is activated. The vital data can be transferred wirelessly to a network to share with doctors or store digitally.

Statement by the jury
The haptic qualities of the vital signs monitor – rounded edges, smooth surfaces and soft-grip keys – foster a favourable working experience.

Der tragbare Vitalzeichenmonitor XICOO AVS-01 ist für den professionellen Einsatz in medizinischen Einrichtungen konzipiert und deckt alle klinisch relevanten Parameter in einem Gerät ab. Er misst Blutdruck, Sauerstoffsättigung, Pulsfrequenz und Temperatur. Bei kritischen Werten schlägt die Alarmfunktion an. Die Vitaldaten können kabellos an ein Netzwerk übermittelt werden, um sie mit dem Arzt zu teilen und papierlos zu speichern.

Begründung der Jury
Die haptischen Merkmale des Vitalzeichenmonitors – gerundete Kanten, glatte Oberflächen und Softgrip-Tasten – machen die Arbeit damit zu einem angenehmen Erlebnis.

Ascom Myco
Smartphone for Nurses
Smartphone für Pflegekräfte

Manufacturer
Ascom Wireless Solutions,
Gothenburg, Sweden
In-house design
Ellen Hartelius, Linnea Fogelmark
Design
Semcon (Patric Svensson),
Gothenburg, Sweden
Web
www.ascommyco.com
www.ascom.com/ws
www.semcor.com
Honourable Mention

The Ascom Myco is a purpose-built smartphone for health care, allowing caregivers to access the information they need at any given time. The top display provides notifications of alerts, which combined with colour, sound and vibration enables caregivers to instantly understand the situation. The ergonomic design of the smartphone is impact-resistant and easy to clean. The carrying clip evenly distributes the weight, thus providing a firm grip even on thin fabrics.

Statement by the jury
The Ascom Myco combines the familiar design of a smartphone with innovative additional functions for everyday tasks in the hospital.

Das Ascom Myco ist ein spezialgefertigtes Smartphone für das Gesundheitswesen, das der Pflegekraft zu jeder Zeit und an jedem Ort die Informationen liefert, die sie gerade braucht. Das Top-Display zeigt Alarme an, die zusammen mit Farb-, Ton- und Vibrationssignalen dabei helfen, die Situation zu erfassen. Das Smartphone ist ergonomisch gestaltet, stoßfest und gut zu reinigen. Der Trageclip sorgt für eine gleichmäßige Gewichtsverteilung, sodass das Gerät selbst an dünnem Stoff hält.

Begründung der Jury
Das Ascom Myco kombiniert das dem Anwender vertraute Design eines Smartphones mit innovativen Zusatzfunktionen für den Krankenhausalltag.

sentida 7-i
Care Bed
Pflegebett

Manufacturer
wissner-bosserhoff GmbH,
Wickede, Germany
In-house design
Web
www.wi-bo.de

The sentida 7-i intelligent care bed features sophisticated safety sensor technology, which can be connected to in-house emergency call systems. The sensor technology monitors whether the resident is in bed or not, as well as if the side guards and brakes are locked. The nursing staff can control all functions via a 7" touchscreen. In addition, the bed is equipped with a humidity sensor mat and a weighing system. The under-bed light provides orientation at night. A pull-out handle helps residents get out of bed by themselves.

Das intelligente Pflegebett sentida 7-i verfügt über eine ausgeklügelte Sicherheitssensorik, die mit dem Hausrufsystem vernetzt werden kann. Die Sensorik überprüft, ob der Bewohner im Bett liegt oder es verlassen hat und ob die Seitensicherungen und Bremsen verriegelt sind. Das Pflegepersonal kann alle Funktionen über den 7"-Touchscreen steuern. Zudem ist das Bett mit einer Nässe-Sensormatte und einer Körperwaage ausgestattet. Die Unterbettbeleuchtung bietet auch nachts Orientierung. Ein ausziehbarer Griff hilft dem Bewohner, selbständig aufzustehen.

Venta
Care Bed
Pflegebett

Manufacturer
Stiegelmeyer Pflegemöbel GmbH & Co. KG,
Herford, Germany
Design
Joh. Stiegelmeyer GmbH & Co. KG
(Martin Bansmann), Herford, Germany
Web
www.stiegelmeyer.com
Honourable Mention

The Venta care bed features soft textile covers that are simply pulled over the head- and footboards. There are a variety of colours, patterns and fabrics to choose from, thanks to which the bed blends in with any interior design. Matching side panels in an upholstered version are also available. The side protection smoothly retracts into the bed when not in use.

Statement by the jury
With its many individual personalisation options, this care bed fosters an inviting atmosphere and thus enhances well-being.

Das Pflegebett Venta verfügt über textile Softcover, die einfach über Kopf- und Fußteil gezogen werden. Zur Auswahl stehen eine Vielfalt an Farben, Mustern und Stoffen, mit denen sich das Bett an jeden Geschmack und jedes Ambiente anpassen lässt. Passend dazu werden die Seitenblenden in einer gepolsterten Variante angeboten. Die Seitensicherung lässt sich mühelos im Bett versenken, wenn sie nicht gebraucht wird.

Begründung der Jury
Das Pflegebett schafft durch seine individuellen Gestaltungsmöglichkeiten eine einladende Atmosphäre und fördert dadurch das Wohlbefinden.

Seba
Raising Aid
Aufsetzhilfe

Manufacturer
ArjoHuntleigh, Malmö, Sweden
In-house design
ArjoHuntleigh R&D
Design
Veryday, Bromma, Sweden
Web
www.arjohuntleigh.com
www.veryday.com

The Seba raising aid facilitates the safe and comfortable raising of patients with restricted mobility from a supine position to sitting at the edge of a bed. The aid reduces the load on a caregiver's back by up to 80 per cent, since the patient's weight is used as a lever. In addition, it drives patient compliance by promoting comfort, security and participation during the positioning process.

Statement by the jury
This raising aid features a construction that is both simple and practical, significantly reducing the physical load on caregivers.

Die Aufsetzhilfe Seba ermöglicht es, Patienten mit eingeschränkter Mobilität aus einer liegenden Position sicher und komfortabel aufzurichten und auf die Bettkante zu setzen. Das Hilfsmittel reduziert Rückenbelastungen des Pflegepersonals um bis zu 80 Prozent, da das Gewicht des Pflegebedürftigen als Hebel genutzt wird. Darüber hinaus fördert es die Patienten-Compliance durch erhöhten Komfort, mehr Sicherheit und seinem Mitwirken während des Positionierungsablaufes.

Begründung der Jury
Die Aufsetzhilfe stellt eine ebenso simple wie praktische Konstruktion dar, die eine deutliche körperliche Entlastung für die Pflegekraft bedeutet.

JWX-2
Power Assist Unit for Wheelchairs
Elektrische Antriebshilfe für Rollstühle

Manufacturer
Yamaha Motor Co., Ltd.,
Shizuoka, Japan

In-house design
Design Center, IM Business Unit
(Masanori Yonemitsu,
Masashi Nomura)

Web
www.yamaha-motor.co.jp

reddot award 2015
best of the best

Free mobility

Wheelchair users often need help from others to help them to get around. Steps, a kerb or even a gradient can quickly become an insuperable obstacle in every-day life that they cannot get over without assistance. The JWX-2 power assist unit addresses this problem with a very well thought-out concept. It is extremely easy to install on an existing manual wheelchair and fits onto both of the chair's wheels in a way that is consistent both from a visual and functional point of view. It also gives it a sporty, agile appearance. Installation is simple, intuitive and self-evident. In this way, a manual wheelchair can quickly and easily be turned into one that is electrically driven. The advantage is that the user does not, in principle, need to change his or her method of operating the wheelchair. The electric power assist unit makes it much easier to get going, climb gradients and ride over surfaces that show a lot of resistance. As the abilities of every wheelchair user vary, the JW Smart Tune software allows for adjustment of the controls to match the strength of the arms. With its excellent design, which takes into account the needs of wheelchair users, this power assist unit extends their mobility, maintaining the users' functional capacity, and gives them the chance to move around more freely.

Freie Mobilität

Rollstuhlfahrer sind für ihre Mobilität oftmals auf fremde Hilfe angewiesen. Stufen, Bordsteine oder auch Steigungen können im Alltag rasch zu unüber-windlichen Hindernissen werden, die sie alleine nicht bewältigen können. Mit einem sehr gut ausgearbeiteten Konzept widmet sich die Gestaltung des elektrischen Zusatzantriebs JWX-2 dieser Problematik. Ausgespro-chen unkompliziert kann diese Antriebshilfe nach-träglich an manuellen Rollstühlen befestigt werden. Sie integriert sich auf formal wie funktional überaus schlüssige Weise in die Räder des Rollstuhles und strahlt eine Anmutung sportlicher Agilität aus. Die Montage ist einfach, intuitiv und selbsterklärend. Der vorhandene manuelle Rollstuhl wandelt sich damit schnell und problemlos in einen elektrisch betriebe-nen. Ein wichtiger Vorteil ist, dass der Rollstuhlfahrer dabei prinzipiell seine ihm vertraute Bedienweise nicht ändern muss. Der elektrische Zusatzantrieb erleichtert ihm jedoch das An-, Bergauf- sowie das Fahren über Flächen mit hohem Fahrwiderstand erheblich. Da die Konstitution eines jeden Rollstuhlfahrers anders ist, ermöglicht die JW Smart Tune-Software eine individu-elle Anpassung an die Armstärke des Nutzers. Mit einer ausgezeichnet die Situation von Rollstuhlfahrern nach-vollziehenden Gestaltung wird mit dieser Antriebshilfe deren Mobilität erweitert – ihre funktionelle Kapazität bleibt erhalten und sie können sich viel freier bewegen.

Statement by the jury

This is an impressively simple and well thought-through concept. The motor of the JWX-2 power assist unit is integrated into the wheels in a very coherent way, thus offering the greatest possible level of as-sistance to the wheelchair user. The auxiliary-power function perfectly combines simple technology, good design and effective support. Well-engineered functionality transforms a manual wheelchair into a powerful, electrically operated one.

Begründung der Jury

Eine beeindruckend einfache und durchdachte funktio-nale Lösung. Bei der elektrischen Antriebshilfe JWX-2 wird der Motor sehr schlüssig in die Räder integriert und ermöglicht so eine maximale Unterstützung für den Rollstuhlfahrer. Dieser Zusatzantrieb bietet die perfekte Kombination aus einfacher Technik, guter Gestaltung und unterstützender Wirkung. Seine aus-gereifte Funktionalität verwandelt den manuell betrie-benen Rollstuhl unkompliziert in eine leistungsfähige elektrische Version.

Designer portrait
See page 64
Siehe Seite 64

Action 5 / MyOn HC
Wheelchair
Rollstuhl

Manufacturer
Invacare France Operations SAS, Fondettes, France
In-house design
Aurélie Lauret, Jean-Michel Roncin, Maël Robert, Wilfrid Da Cunha
Web
www.invacare.com

The Action 5 / MyOn HC wheelchair features a folding mechanism with an automatic locking system, which provides ease of use and high rigidity. In addition, the back can be folded forward. With the innovative seat-rail profile, continuous adjustments may be made discreetly. A wide range of frame colours and three shades for components allow the wheelchair to be personalised according to the user's individual tastes.

Der Rollstuhl Action 5 / MyOn HC verfügt über ein Faltsystem mit automatischer Einrastung, das eine leichte Handhabung und hohe Starrheit gewährleistet. Zudem kann der Rücken vorwärts gefaltet werden. Mit dem neuartigen Sitzschienenprofil lassen sich stufenlose Anpassungen vornehmen. Durch eine Vielzahl von Rahmenfarben und drei Komponentenfarben kann der Rollstuhl nach dem persönlichen Geschmack des Nutzers gestaltet werden.

Statement by the jury
This wheelchair convinces with its comprehensive functionality. The appealing mix of materials and colours fosters an individual look.

Begründung der Jury
Der Rollstuhl überzeugt durch seine umfassende Funktionalität. Ein ansprechender Mix aus Materialien und Farben sorgt für ein individuelles Erscheinungsbild.

OA Kneetrac

Knee Brace

Kniegelenkorthese

Manufacturer
Changeui Medical Co., Ltd.,
Seoul, South Korea
In-house design
Sang Moon Jung
Web
www.diskdr.co.kr

The OA Kneetrac knee brace is designed to alleviate pain in individuals suffering from moderate to severe osteoarthritis of the knee joint. It provides traction treatment, which increases the gap between two bone joints. Thus, it reduces body weight on the worn-out joint. The H-shaped frame design provides a stable structure. The frame is made of Duralumin, a metal that is light but strong. The surface of the frame is anodised using a special technique.

Statement by the jury
The frame with its metal appearance gives this knee brace a clear structure while simultaneously conveying an appealing, technical look.

Die Orthese OA Kneetrac wird bei einer mittleren bis schweren Arthrose des Kniegelenks eingesetzt, um Schmerzen zu lindern. Sie übt Zug auf das Knie aus, was den Gelenkspalt vergrößert und das Körpergewicht, das auf dem verschlissenen Gelenk lastet, verringert. Das H-förmige Rahmendesign sorgt für eine stabile Konstruktion. Der Rahmen ist aus Duralumin gefertigt, einem leichten, aber stabilen Metall. Die Rahmenoberflächen wurden in einem speziellen Verfahren eloxiert.

Begründung der Jury
Die Rahmenstruktur in Metalloptik verleiht der Kniegelenkorthese eine klare Gliederung und erzielt zugleich eine ansprechende technische Anmutung.

3R62 Pheon

Prosthetic Knee Joint

Prothesenkniegelenk

Manufacturer
Otto Bock HealthCare GmbH,
Duderstadt, Germany
In-house design
Christian Noack
Web
www.ottobock.com

The 3R62 Pheon polycentric prosthetic knee joint is especially suited for prosthesis wearers with low mobility, since it provides the corresponding functionality for swing and stance phase control. At the same time, the prosthesis supports the therapeutic goals of restoring the ability to stand and regaining moderate walking ability. The integrated lock can be activated or deactivated.

Statement by the jury
The high functionality of this prosthetic joint is reflected in its simplicity. The interior is transparent and very easy to access.

Das polyzentrische Prothesenkniegelenk 3R62 Pheon ist besonders für Prothesenträger mit niedriger Mobilität geeignet, da es die entsprechende Funktionalität für die Schwungphasensteuerung und Standphasensicherung bereitstellt. Zugleich dient die Prothese dem therapeutischen Zweck, die Stehfähigkeit wiederherzustellen und eine moderate Gehfähigkeit zurückzuerlangen. Eine integrierte Sperre kann optional aktiviert oder deaktiviert werden.

Begründung der Jury
Die hohe Funktionalität des Prothesengelenks spiegelt sich in ihrer Schlichtheit wider. Das Innenleben ist freigelegt und sehr gut zugänglich.

Genium Protector

Protector for Prosthetic Knee Joint

Protektor für Prothesenkniegelenk

Manufacturer
Otto Bock HealthCare GmbH,
Duderstadt, Germany
Design
KISKA GmbH, Anif (Salzburg), Austria
Web
www.ottobock.com
www.kiska.com

The Genium Protector shields the Genium Prosthetic Knee Joint from impact in everyday life. Its champagne-coloured, shimmering surface conveys a premium impression. In the calf area the protective cover evinces a natural shape. An innovative foot cuff consisting of flexible fabric and contoured plastic conceals the technical connection to the prosthetic foot while simultaneously accommodating movement dynamics in the ankle area.

Statement by the jury
This protective cover is accurately modelled on the structure of the human knee. Thanks to its natural form and soft lines, it does not appear bulky under clothing.

Der Genium Protector schützt das Genium Prothesenkniegelenk vor Stößen im alltäglichen Gebrauch. Seine champagnerfarben schimmernde Oberfläche erzeugt einen edlen Eindruck. Im Wadenbereich erzeugt der Protector ein natürlich wirkendes Volumen. Eine innovative Fußmanschette aus flexiblem Textil und formgebendem Kunststoff verdeckt die technische Anbindung an den Prothesenfuß und trägt gleichzeitig zur Bewegungsdynamik im Knöchelbereich bei.

Begründung der Jury
Der Protector ist dem Skelettbau des Knies genau nachempfunden. Dank der natürlichen Form und sanften Linienführung trägt er auch unter Kleidung nicht auf.

1C10 Terion
Prosthetic Foot
Prothesenfuß

Manufacturer
Otto Bock HealthCare GmbH,
Duderstadt, Germany
In-house design
Jeff Friesen, Oleg Pianykh
Web
www.ottobock.com

The 1C10 Terion prosthetic foot has been developed for walking on different surfaces at a variety of speeds. It has a minimalist form with an anatomically shaped elastic heel, which facilitates the natural rolling movement of the foot. In the process, the spring element made from carbon releases stored energy. Moreover, the design focuses on robustness and also on simple handling for the orthopaedic technician.

Statement by the jury
The 1C10 Terion captivates with its minimalist design language. The integrated carbon spring is very light and strong at the same time.

Der Prothesenfuß 1C10 Terion eignet sich zum Gehen auf unterschiedlichen Untergründen und in verschiedenen Geschwindigkeiten. Seine minimalistische Form mit anatomisch geformter, elastischer Ferse erlaubt eine natürliche Abrollbewegung. Dabei gibt das Federelement aus Carbon die gespeicherte Energie wieder ab. Des Weiteren wurde beim Design auf Robustheit sowie auf eine einfache Handhabung für den Orthopädietechniker Wert gelegt.

Begründung der Jury
Der 1C10 Terion begeistert durch seine reduzierte Formensprache. Die integrierte Carbon-Feder ist überaus leicht und zugleich belastbar.

OPPO Carver Wrist Brace
OPPO Carver Handgelenkstütze

Manufacturer
OPPO Medical Corporation, Taipei, Taiwan
In-house design
Ming-Jhih Wu, Chien-Min Fang
Web
www.oppomedical.com

The Oppo Carver wrist brace provides excellent fixation of the wrist without limiting the thumb's freedom of movement. The contact surface between the wrist brace and the palm of the hand has been minimised to prevent skin irritation. The silicone surface makes the brace comfortable to wear. The interior support structure made of aluminium is adjustable and thus fits any wrist and required functional position.

Statement by the jury
This wrist brace is a successful combination of select materials, high ergonomic quality and an open design.

Die Oppo Carver Handgelenkstütze bietet eine sehr gute Fixierung des Handgelenks, ohne die Bewegungsfreiheit des Daumens einzuschränken. Die Kontaktfläche zwischen Handgelenkstütze und Handfläche ist minimiert, um Hautirritationen vorzubeugen. Die Oberfläche aus Silikon sorgt für ein angenehmes Tragegefühl. Das eingearbeitete Stützgestell aus Aluminium ist verstellbar und kann an das individuelle Handgelenk und die notwendige Funktionsstellung angepasst werden.

Begründung der Jury
Die Handgelenkstütze stellt eine gelungene Verbindung aus ausgesuchten Materialien, hoher Ergonomie und einer offenen Konstruktion dar.

Dynamics Plus
Elbow Support
Dynamics Plus
Ellenbogenbandage

Manufacturer
Ofa Bamberg GmbH, Bamberg, Germany
In-house design
Sandra Abels
Design
jojorama (Joachim Möllmann),
Hannover, Germany
Web
www.ofa.de
www.jojorama.de

The Dynamics Plus Elbow Support features a large Arthroflex comfort zone in the sensitive area of the crook of the arm. It is made from a special knitted fabric that is highly stretchable and fits like a second skin. The Plus pads at the outside and inside of the forearm relieve the entheses at the elbow and have a gentle massaging effect, while the silicone material absorbs irritations in the forearm that occur as a result of movements.

Statement by the jury
The Dynamics Plus Elbow Support impresses with its effective support function and high elasticity, which enables a very good fit.

Die Dynamics Plus Ellenbogenbandage verfügt im sensiblen Bereich der Armbeuge über eine großflächige Arthroflex-Komfortzone. Dabei handelt es sich um ein dehnfähiges Spezialgestrick, das sich wie eine zweite Haut anpasst. Die Plus-Pelotten außen und innen am Unterarm entlasten die Sehnenansätze am Ellenbogen und haben einen sanften Massageeffekt, während das Silikonmaterial die Weiterleitung von Bewegungsreizen im Unterarm dämpft.

Begründung der Jury
Die Dynamics Plus Ellenbogenbandage beeindruckt durch ihre effektive Stützfunktion und hohe Elastizität, die eine sehr gute Passform ermöglicht.

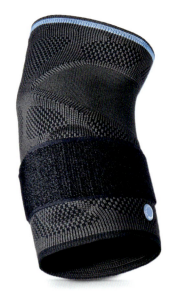

JuzoFlex® Genu Xtra STYLE
Knee Support
Kniebandage

Manufacturer
Julius Zorn GmbH, Aichach, Germany
In-house design
Siegfried Lechner
Web
www.juzo.ccm

The JuzoFlex Genu Xtra STYLE knee support is available in six different neon colour combinations. The anatomically shaped patella ring made from polyurethane is very light and therefore does not slip out of place easily. Above the patella ring is an area with vertical elasticity, and there is a comfort zone in the hollow of the knee. The support is made of a breathable, skin-friendly fabric, into which additional spiral rods have been integrated for a functional fit.

Statement by the jury
Thanks to its loud colours, this knitted knee support with its high-quality textile craftsmanship looks less like a therapeutic aid and more like a fashionable accessory.

Die Kniebandage JuzoFlex Genu Xtra STYLE gibt es in sechs verschiedenen Neon-Farbkombinationen. Der anatomisch geformte Patellaring aus Polyurethan ist sehr leicht und verrutscht daher kaum. Oberhalb des Patellarings liegt ein längselastischer Dehnungsbereich, in der Kniekehle befindet sich eine Komfortzone. Die Bandage besteht aus atmungsaktivem und hautfreundlichem Komfort-Gestrick, in das zusätzlich Spiralstäbe für einen funktionsgerechten Sitz eingearbeitet sind.

Begründung der Jury
Die auf hohem Niveau gefertigte Kniebandage sieht dank ihrer knalligen Farben weniger wie ein Therapiehilfsmittel, sondern wie ein modisches Accessoire aus.

PROMASTER
Support Series
Bandagen-Serie

Manufacturer
THUASNE Deutschland GmbH, Burgwedel, Germany
Design
f/p design GmbH, Munich, Germany
Web
www.thuasne.de
www.fp-design-gmbh.de

The Promaster support series features pads made with a specially developed FlexAir structure. The pads optimally adapt to the shape of the body. Additional silicone pimples enhance the massage effect and promote blood circulation to accelerate the healing process. Moreover, a breathable knitted fabric in intelligent 3D structure ensures a pleasantly firm fit, while the soft, wavy compression seam prevents the support from becoming too constricting. It is easy to put on thanks to sturdy grip surfaces.

Die Bandagen-Serie Promaster besitzt Pelotten mit einer speziell entwickelten FlexAir-Struktur, welche sich der Körperform optimal anpassen. Zusätzliche Noppen verstärken den Massageeffekt und fördern die Durchblutung zur Beschleunigung des Heilungsprozesses. Darüber hinaus sorgt das atmungsaktive Gestrick in intelligenter 3D-Struktur für einen angenehm festen Sitz, während der weiche und wellenförmige Kompressionsübergang verhindert, dass die Bandagen einschnüren. Stabile Griffflächen erleichtern das Anziehen.

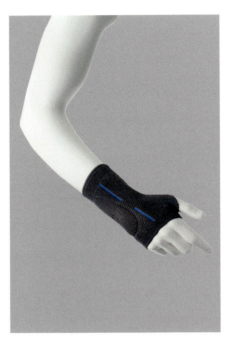
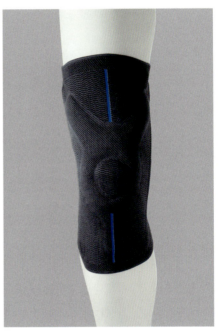
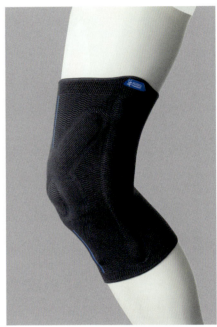
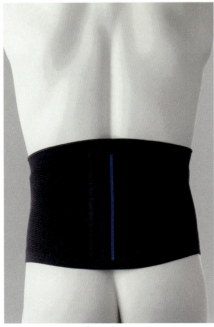
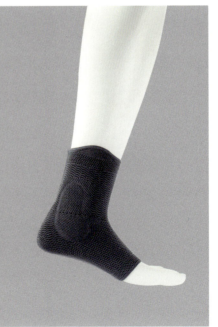
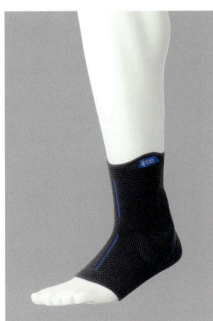

Healing Tattoo
Heilungstattoo
Medical Plaster
Medizinisches Pflaster

Manufacturer
DCB Development Center for Biotechnology,
New Taipei, Taiwan
Design
Bonnsu, Taipei, Taiwan
Web
www.dcb.org.tw
www.bonnsu.com

These temporary healing tattoos were developed especially for children and teenagers suffering from arthritis. They contain a natural, anti-inflammatory herbal extract, which is absorbed through the skin. While medications often have an effect on the whole body, here the extract is only applied to the targeted areas by way of the tattoos. The patterns resemble woven bands, bracelets or jewellery and can be combined to create different styles.

Statement by the jury
With their playful and fashionable designs, these healing tattoos are strongly geared to the needs of young people.

Die temporären Heilungstattoos wurden speziell für Kinder und Jugendliche mit Arthritis entwickelt. Sie enthalten ein natürliches entzündungshemmendes Kräuterextrakt, das über die Haut aufgenommen wird. Während Medikamente oft auf den gesamten Körper wirken, wird das Extrakt mit den Tattoos gezielt auf die betroffenen Bereiche aufgetragen. Die Motive ähneln geflochtenen Bändern, Armbändern oder Schmuck und lassen sich zu verschiedenen Mustern kombinieren.

Begründung der Jury
Mit ihren spielerischen und trendigen Designs sind die Heilungstattoos in hohem Maße auf die Bedürfnisse von jungen Menschen abgestimmt.

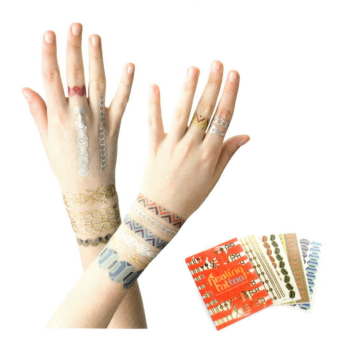

Anti-Stroke Treatment Helmet
Schlaganfall-Therapie-Helm

Manufacturer
Zhejiang Xinfeng Medical Apparatus Co., Ltd.,
Shangyu, China
Design
HUMTA design
(Yunfang Fan, Gangxiong Chen),
Kunshan, China
Web
www.zjxinfeng.com
www.humtaid.com

This helmet for the treatment of stroke patients has been developed to help prevent further cerebral infarcts. To this end, the device emits special microcurrent impulses in areas where traditional acupuncture points are located. The helmet can be adjusted to fit different head sizes. The sliders and the needle electrodes are also adjustable to ensure optimal positioning over the acupuncture points.

Statement by the jury
This helmet convinces with its open form. It provides a clear view of the treatment area and makes the design appear less bulky.

Dieser Helm dient zur Behandlung von Schlaganfallpatienten und soll dabei helfen, weitere Hirninfarkte zu verhindern. Dazu sendet das Gerät an den Stellen, an denen sich traditionelle Akupunkturpunkte befinden, spezielle Mikrostrom-Impulse aus. Der Helm kann an verschiedene Kopfgrößen angepasst werden. Die Schieber und Nadelelektroden sind ebenfalls verstellbar, um eine optimale Ausrichtung auf die Akupunkturpunkte zu erreichen.

Begründung der Jury
Der Helm überzeugt durch seine offene Form. Dadurch bleibt die Sicht auf das Behandlungsfeld frei, und die Konstruktion wirkt weniger massiv.

Valedo
Digital Back Coach
Digitaler Rückentrainer

Manufacturer
Hocoma, Volketswil, Switzerland
In-house design
Design
NOSE Design (Christian Harbeke),
Zürich, Switzerland
BLYSS (Christian Werler),
Zürich, Switzerland
Web
www.valedotherapy.com
www.hocoma.com
www.nose.ch
www.blyss.ch
Honourable Mention

The Valedo digital back coach has been specially developed for the long-term promotion of back health. The movements of the torso and pelvis are registered by two wireless sensors attached to the body, and then transmitted to a smartphone or tablet. These movement signals are analysed and provide real-time feedback to users in a game-like form as to whether they are moving correctly while performing exercises.

Statement by the jury
This therapeutic device for the back pleases with the compact design of its sensors, which stay attached to the body and can hardly be felt by the wearer.

Der digitale Rückentrainer Valedo wurde speziell für die langfristige Förderung der Rückengesundheit entwickelt. Die Rumpf- und Beckenbewegungen werden mit zwei drahtlosen Sensoren, die am Körper angebracht werden, registriert und auf ein Smartphone oder ein Tablet übertragen. Diese Bewegungssignale werden analysiert und liefern in einem spielerischen Umfeld Echtzeit-Feedback darüber, ob sich der Benutzer, während er die Übungen ausführt, richtig bewegt.

Begründung der Jury
Das medizinische Trainingsgerät besticht durch die kompakte Gestaltung der Sensoren, die gut am Körper haften und kaum zu spüren sind.

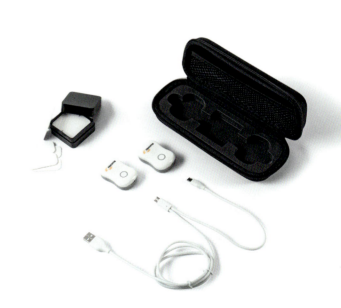

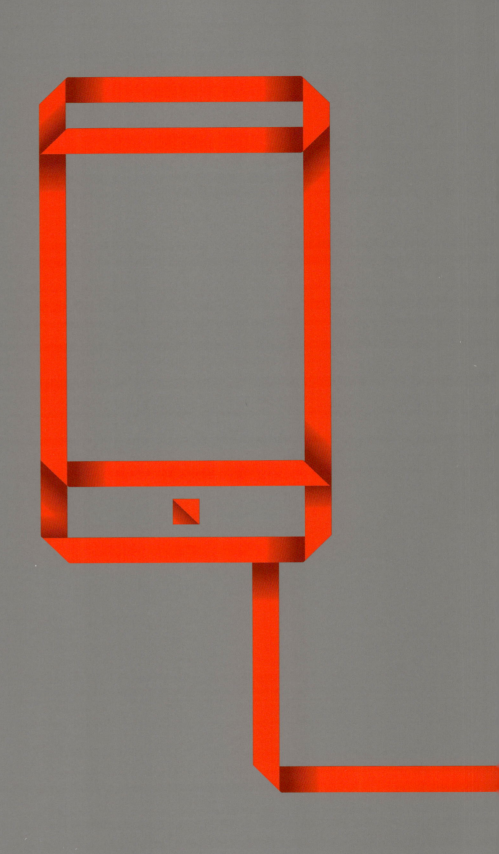

2-way radio sets	Funkgeräte
Charging stations	Headsets
Conference technology	Konferenztechnik
Headsets	Ladestationen
Mobile phones	Mobiltelefone
Power banks	Power Banks
Routers	Router
Smartphones	Smartphones
Telephones and telephone	Telefone und
systems	Telefonanlagen

Communication
Kommunikation

Xperia™ E3
Smartphone

Manufacturer
Sony Mobile Communications Inc.,
Tokyo, Japan

In-house design
Sony Mobile Communications Inc.

Web
www.sonymobile.com

reddot award 2015
best of the best

Slim elegance

Smartphones are at the centre of everyday reality in which they have become constant companions. Since smartphones are continuously in use, they have to be robust and easy to operate. The development of the Xperia E3 focuses particularly on reliability in use and thus on an equally high standard of user-friendliness. The aim was to combine design and quality into one product that perfectly adapts to the lifestyle of its users. Furthermore, this smartphone has emerged by taking on the OmniBalance design concept with the aim of offering balance and symmetry in all angles. The result is a device with a well-balanced appearance and a form that impresses with purist self-sufficiency. The matte finish of the frame lends it both a slim profile and a pleasant tactile feel. The phone rests comfortably in the hand, thanks to its rounded side edges that also lend it a high degree of sturdiness. The Xperia E3 can be operated with just one hand, which is of great benefit in many everyday situations, and even offers users an innovative photo app. Also outstanding are the processing speed and the high degree of durability since the soft yet robust edges ensure that it does not dent and thus keeps looking new for a longer period of time. With its slim and symmetric design it fascinates users anew every single day.

Schlanke Eleganz

Das Smartphone steht im Mittelpunkt einer Lebenswelt, in der es zum allgegenwärtigen Begleiter geworden ist. Da es im Alltag viel beansprucht wird, muss es robust und einfach bedienbar sein. Bei der Entwicklung des Xperia E3 standen besonders die Verlässlichkeit im Gebrauch und eine damit einhergehende hohe Nutzerfreundlichkeit im Mittelpunkt. Design und Qualität sollten sich in einem Produkt vereinen, das sich in das Leben des Nutzers perfekt einfügt. Dieses Smartphone entstand zudem nach der Maxime des OmniBalance-Designs mit dem Ziel, eine in alle Richtungen ausgewogene Symmetrie zu erreichen. Das Ergebnis ist ein Gerät, welches sehr ausgewogen wirkt und dessen puristische Form durch ihre Geschlossenheit besticht. Die matte Lackierung des Rahmens verleiht ihm eine schlanke Silhouette und macht es zugleich haptisch angenehm. Es liegt sehr gut austariert in der Hand, wobei die Gestaltung mit abgerundeten Seitenprofilen ihm eine hohe Stabilität gibt. Das Xperia E3 lässt sich leicht mit nur einer Hand bedienen, was im Alltag viele Vorteile hat, und bietet dem Nutzer darüber hinaus innovative Foto-Apps. Bemerkenswert sind auch seine Geschwindigkeit und das hohe Maß an Langlebigkeit, da seine weichen und zugleich robusten Kanten dafür sorgen, dass es lange wie neu aussieht und keine Dellen entstehen. Mit seiner schlanken und symmetrischen Gestaltung begeistert es dabei seine Nutzer immer wieder aufs Neue.

Statement by the jury

Designed with balanced proportions and a soft appearance, the Xperia E3 emotionalises the user. The phone impresses with its symmetry and rests highly pleasant in the hand. Thanks to its rounded side edges, it is easy to operate with just one hand. The convincing design embodies a perfect balance of material, finish and colour. This smartphone thereby is oriented towards durability and pioneering in this regard as well.

Begründung der Jury

Gestaltet mit ausgewogenen und weich anmutenden Proportionen, emotionalisiert das Xperia E3 den Nutzer. Es beeindruckt mit seiner Symmetrie und liegt überaus angenehm in der Hand. Durch die abgerundeten Seitenprofile lässt es sich problemlos mit nur einer Hand bedienen. Seiner Gestaltung gelingt auf perfekte Weise die Balance von Farbe, Finish und Material. Dieses Smartphone ist dabei auf Langlebigkeit ausgerichtet und auch in dieser Hinsicht zukunftsweisend.

Designer portrait
See page 66
Siehe Seite 66

Xperia Z3
Smartphone

Manufacturer
Sony Mobile Communications Inc.,
Tokyo, Japan
In-house design
Web
www.sonymobile.com

The ultra-flat, rounded aluminium frame and the clear symmetry of the Xperia Z3 catch the eye immediately. A high-quality on/off switch of aluminium and the robust tempered glass panels give it an elegant and upmarket appearance. The surface proves to have a pleasant feel, the smartphone is also very slim and light and thus fits ergonomically well in the hand. It guarantees reliable intuitive operability.

Statement by the jury
The design of the Xperia Z3 convinces due to the unmistakeable slim silhouette. High-quality materials and a good haptic emphasise its sophisticated functionality.

Der ultraflache abgerundete Aluminiumrahmen und die klare Symmetrie des Xperia Z3 fallen sofort ins Auge. Ein qualitätvoller, aus Aluminium gefertigter Ein-/Ausschalter und die widerstandsfähigen Hartglasflächen verleihen ihm ein elegantes wie auch hochwertiges Erscheinungsbild. Die Oberfläche beweist eine angenehme Haptik, zudem ist das Smartphone sehr schmal und leicht und passt sich so ergonomisch gut der Hand an. Es verspricht zuverlässige intuitive Bedienbarkeit.

Begründung der Jury
Gestalterisch überzeugt das Xperia Z3 durch die unverkennbar schlanke Silhouette. Hochwertige Materialien und gute Haptik unterstreichen seine ausgereifte Funktionalität.

Xperia Z3 Compact
Smartphone

Manufacturer
Sony Mobile Communications Inc.,
Tokyo, Japan
In-house design
Web
www.sonymobile.com

The Xperia Z3 Compact is shaped according to the OmniBalance design principle, creating a symmetrical appearance in any position by means of rounded edges and smooth surfaces. A high-gloss frame of tempered mineral glass on both sides gives the smartphone an upmarket, timeless impression and also prevents scratching. Weighing only 129 grams and with a screen diagonal of 4.6" it lies pleasantly in the hand.

Statement by the jury
With its construction of tempered mineral glass at the front and the back, the Xperia Z3 Compact enthrals with both its durability and its aesthetic design.

Das Xperia Z3 Compact ist nach dem Prinzip des OmniBalance-Designs geformt, das mit abgerundeten Kanten und glatten Oberflächen eine symmetrische Optik in jeder Position schafft. Ein hochglänzender Rahmen aus gehärtetem Mineralglas auf beiden Seiten verleiht dem Smartphone eine hochwertige, zeitlose Anmutung und verhindert zudem das Entstehen von Kratzspuren. Mit einem Gewicht von nur 129 Gramm und einer Bildschirmdiagonale von 4,6" liegt es bequem in der Hand.

Begründung der Jury
In seiner Ausführung aus gehärtetem Mineralglas an Vorder- wie Rückseite begeistert das Xperia Z3 Compact sowohl in Hinblick auf Langlebigkeit wie auch ästhetische Gestaltung.

Xperia T2 Ultra
Smartphone

Manufacturer
Sony Mobile Communications Inc.,
Tokyo, Japan
In-house design
Web
www.sonymobile.com

The name Xperia T2 Ultra refers to the high-resolution 6" display of this smartphone. In spite of its remarkable display size it is slim and weighs only 152 grams, making it a good companion. The materials used are above all chosen with weight-saving in mind and ensure precise engineering. The OmniBalance design creates balance from every viewing angle as well as visual symmetry.

Statement by the jury
The Xperia T2 Ultra smartphone with its generous 6" display and light weight catches the eye particularly with regard to its functional qualities.

Der Name Xperia T2 Ultra verweist auf das hochauflösende 6"-Display dieses Smartphones. Trotz seiner beachtlichen Bildschirmgröße ist es schlank und wiegt nur 152 Gramm, was es zu einem guten Begleiter macht. Das verwendete Material steht ganz im Zeichen der Gewichtsersparnis und gewährleistet zugleich präzise Fertigung. Das OmniBalance-Design schafft ein in alle Richtungen ausgewogenes Gleichgewicht sowie optische Symmetrie.

Begründung der Jury
Das Smartphone Xperia T2 Ultra fällt mit großzügigem 6"-Display bei geringem Gewicht hinsichtlich seiner Nutzungseigenschaften besonders ins Auge.

Xperia M2 Aqua
Smartphone

Manufacturer
Sony Mobile Communications Inc.,
Tokyo, Japan
In-house design
Web
www.sonymobile.com

The reduced and user-friendly design gives no indication that the device is water- and dust-proof. The matt back panel of the smartphone with its robust finish presents an exciting contrast to the camera ring which is high-gloss. The side panels in brushed aluminium and the aluminium on-switch prove to be very precise and are responsible for the high-quality impression of the product.

Statement by the jury
The Xperia M2 Aqua impresses due to its reliable protection from environmental influences such as water and dust and thus earns merit with regard to durability and user value.

Die reduzierte und auf Benutzerfreundlichkeit ausgelegte Gestaltung lässt von außen nicht erahnen, dass es sich um ein wasser- und staubdichtes Gerät handelt. Die mattierte Rückseite des Smartphones mit ihrer widerstandsfähigen Oberfläche steht in spannendem Kontrast zu dem auf Hochglanz polierten Kameraring. Die Seitenflächen in gebürsteter Aluminiumoptik und der aus Aluminium gefertigte Einschaltknopf erweisen sich als sehr präzise und machen den hochwertigen Eindruck des Produkts aus.

Begründung der Jury
Das Xperia M2 Aqua beeindruckt durch zuverlässigen Schutz vor Umwelteinflüssen wie Wasser und Staub und überzeugt daher hinsichtlich Langlebigkeit und Nutzwert.

Arrows NX F-02G
Smartphone

Manufacturer
Fujitsu Limited, Kawasaki, Japan
Design
Fujitsu Design Limited (Makoto Sawaguchi, Kentaro Yoshihashi, Takeshi Watanabe), Kawasaki, Japan
Web
www.fujitsu.com
http://jp.fujitsu.com/group/fdl

The smartphone focuses on high-resolution image quality and ease of handling. The four distinctive corners are set in metal, providing stability also by alternating between formats; the haptic is also pleasant. The device has a display resolution of 1440 × 2560 pixels at 564 ppi as well as a camera with an image resolution of 20.70 MP and a camera resolution of 2.10 MP, which qualifies it for belonging in the top league.

Statement by the jury
With regard to display and camera quality, Arrows NX F-02G ranks in the top position among smartphones.

Der Fokus liegt bei diesem Smartphone auf hochauflösender Bildqualität und guter Handhabung. Die vier ausgeprägten Kanten sind in Metall eingefasst, das verleiht Stabilität, auch beim Wechsel zwischen den Formaten, sowie eine angenehme Haptik. Technisch spielt das Gerät mit einer Displayauflösung von 1.440 x 2.560 Pixeln bei 564 ppi sowie einer Kamera mit Bildauflösung von 20,70 MP und einer Kameraauflösung von 2,10 MP in der oberen Liga.

Begründung der Jury
Arrows NX F-02G positioniert sich mit beachtlichen Display- und Kameraqualitäten im Spitzenfeld unter den Smartphones.

G Flex2
Smartphone

Manufacturer
LG Electronics Inc., Seoul, South Korea
In-house design
Seunghoon Yu, Sangmin Park
Web
www.lg.com

The smartphone G Flex 2 is shaped in an ergonomic, gentle curve with a radius of 500 mm as opposed to conventional telephones and fits pleasantly around the face. It also assures a secure hold and reduces the distance between mouth and microphone. The curvature is facilitated by an innovative POLED-Display. The matt metal back cover with metallic or red hairline structure provides a good feel.

Statement by the jury
The gentle curvature of the Smartphone G Flex 2 not only looks good – the excellent handling and enhanced talk convenience are convincing.

Das Smartphone G Flex 2 beschreibt eine ergonomisch sanfte Krümmung, die mit einem Radius von 500 mm im Gegensatz zu herkömmlichen Telefonen angenehm die Gesichtsform umspielt und zudem für sicheren Griff sorgt. Die Distanz zwischen Mund und Mikrofon verringert sich ebenfalls. Ermöglicht wird die gebogene Formgebung durch ein neuartiges POLED-Display. Die mattierte Metallrückseite mit Haarlinienstruktur in Metallic oder Rot bietet eine gute Haptik.

Begründung der Jury
Die sanfte Krümmung des Smartphones G Flex 2 sieht nicht nur gut aus – auch das hervorragende Handling und der gute Sprachkomfort überzeugen.

iPhone 6
Smartphone

Manufacturer
Apple, Cupertino, USA
In-house design
Web
www.apple.com

The case of the iPhone 6 is of anodised aluminium. Its special unibody design with rounded edges forms a visually seamless unit with the glass surface of the 4.7" Retina HD display. Both its surface qualities and the soft forms create a pleasant feel which is further supported by the very thin case, 6.9 mm in thickness. Weighing only 129 grams, the smartphone is also fairly lightweight. The high-performance mobile device with advanced technology is available in metallic gold, silver and space grey.

Das Gehäuse des iPhone 6 ist aus eloxiertem Aluminium gefertigt. Sein spezielles Unibody-Design mit abgerundeten Kanten bildet eine optisch nahtlose Einheit mit der Glasoberfläche des 4,7"-Retina-HD-Displays. Seine Oberflächenbeschaffenheit und die weichen Formen erzeugen eine angenehme Haptik, die von dem mit 6,9 mm Stärke sehr dünnen Gehäuse unterstützt wird. Mit nur 129 Gramm ist das Smartphone zudem ziemlich leicht. Das leistungsstarke Mobilgerät mit fortschrittlicher Technologie ist in den Metalltönen Gold, Silber und Spacegrau erhältlich.

iPhone 6 Plus
Smartphone

Manufacturer
Apple, Cupertino, USA
In-house design
Web
www.apple.com

With dimensions of approximately 158 × 78 mm and a 5.5" display, the iPhone 6 Plus achieves almost the viewing quality of a tablet. The high-quality Retina HD-display with a contrast ratio of 1,300:1 provides a high level of convenience. The standard apps have been adapted to the display size and offer more menu options in horizontal format. The case is of anodised aluminium and has rounded edges, which creates an apparently seamless unit with the glass surface of the display. In spite of the largely dimensioned display, the smartphone is only 7.1 mm thick and weighs 172 grams.

Mit Maßen von rund 158 × 78 mm und einer Bildschirmgröße von 5,5" kommt das iPhone 6 Plus nahe an den Betrachtungskomfort eines Tablets heran. Das hochwertige Retina-HD-Display weist mit 1.300:1 einen sehr hohen Kontrastwert auf. Die Standard-Apps sind an die Bildschirmgröße angepasst und zeigen im Querformat mehr Menüoptionen an. Das Gehäuse aus eloxiertem Aluminium bildet mit seinen abgerundeten Kanten eine optisch nahtlose Einheit mit der Glasoberfläche des Displays. Trotz großer Bildschirmmaße ist das Smartphone nur 7,1 mm stark bei einem Gewicht von 172 Gramm.

Statement by the jury
The 5.5" Retina HD display of the iPhone 6 Plus impresses with a convincing viewing convenience. The noble, high-quality appearance enthrals from a technical and from an aesthetic point of view.

Begründung der Jury
Das 5,5"-Retina-HD-Display des iPhone 6 Plus beeindruckt mit überzeugendem Bildschirmkomfort. Das edle wie hochwertige Erscheinungsbild begeistert in technischer Hinsicht wie in seiner Ästhetik.

Aquos Crystal
Smartphone

Manufacturer
Sharp Corporation, Hiroshima, Japan
In-house design
Keiichi Koyama, Philippe Poulin
Web
www.sharp.co.jp

An almost frameless appearance is the brand feature of Aquos Crystal. Thanks to innovative technology, the screen goes right to the edge of the phone. The user has the impression he is handling nothing but the display itself and can authentically capture the three-dimensional surroundings. The frameless design makes a new touchscreen interface possible which has an innovative gesture control. A screenshot can be made with a horizontal swipe in the top corner of the display. Rounded corners and a noble finish make operation simple and pleasant.

Ein nahezu rahmenloses Erscheinungsbild ist das Markenzeichen von Aquos Crystal. Dank innovativer Technologie füllt der Bildschirm fast die kompletten Abmessungen des Gehäuses aus. Der Nutzer gewinnt den Eindruck, nur noch mit dem reinen Display selbst zu hantieren und damit die dreidimensionale Umgebung unverfälscht einfangen zu können. Die rahmenlose Gestaltung ermöglicht eine neue Touchscreen-Oberfläche mit innovativer Gestensteuerung. Ein Screenshot wird mittels horizontaler Wischbewegung in der oberen Ecke des Displays ausgelöst. Abgerundete Kanten und eine edle Oberfläche machen die Handhabung einfach und angenehm.

Statement by the jury
Aquos Crystal finds new dimensions in design: the user seems to hold nothing but the pure display in the hand. Innovative gesture display makes operation straightforward.

Begründung der Jury
Aquos Crystal findet gestalterisch zu neuen Dimensionen: Nichts als das reine Display scheint der Nutzer in Händen zu halten. Innovative Gestensteuerung macht die Bedienung denkbar einfach.

LG AKA
Smartphone

Manufacturer
LG Electronics Inc., Seoul, South Korea
In-house design
Bo-Ra Choi, Dong-Eun Kim
Web
www.lg.com

With this smartphone the user can express his own personality. Together with the matching case – optional colours are white, yellow, pink and navy blue – and a personalised ringing tone, four individually selectable pairs of eyes in an interactive display at the top permit to generate a kind of avatar. When the battery is low, for instance, the colour of the eyes changes. To use the display, the housing cover is simply attached to the other side.

Statement by the jury
The Smartphone LG AKA is pleasing due to its high emotional content. Members of the certainly younger target group will create their own individual, graphically well-solved characters.

Mit diesem Smartphone kann der Benutzer seine eigene Persönlichkeit zum Ausdruck bringen. Bei inaktivem Display erscheinen am oberen Rand vier individuell auswählbare Augenpaare, die zusammen mit passender Hülle – zur Wahl stehen die Farben Weiß, Gelb, Pink und Marineblau – und persönlichem Klingelton eine Art Avatar bilden. Bei kritischem Batteriestatus etwa ändern die Augen ihre Farbe. Die Gehäuseabdeckung wird bei Gebrauch des Displays einfach umgedreht aufgesteckt.

Begründung der Jury
Das Smartphone LG AKA gefällt wegen seines hohen emotionalen Gehalts. Eine zweifelsohne jüngere Zielgruppe kann ihre individuellen, grafisch gut gelösten Charaktere kreieren.

emporiaSMART
Smartphone

Manufacturer
emporia Telecom, Linz, Austria
Design
Mango Design, Braunschweig, Germany
Web
www.emporia.at
www.mangodesign.de

This smartphone is designed for the needs of the elderly. The touchscreen offers a clear menu structure and is easily readable. If required, a standard keypad cover can be fitted over it. This means still being able to dial numbers as usual with the familiar keypad without having to sacrifice the convenience of a smartphone. It is also possible to dial using a pen stylus. The most popular apps have been optimized for intuitive handling.

Statement by the jury
emporiaSMART succeeds in convincing elderly people to lose their inhibitions towards modern smartphones in a charming way by not depriving them of the familiar keyboard.

Dieses Smartphone ist auf die Bedürfnisse älterer Personen ausgerichtet. Der Touchscreen bietet eine klare Menüstruktur und gute Lesbarkeit. Darüber wird bei Bedarf ein herkömmliches Tastencover geklappt. Somit muss nicht auf den Komfort eines Smartphones verzichtet werden bei gleichzeitiger Verwendung der gewohnten Tastatur. Auch eine Eingabe mittels Stift ist möglich. Die am häufigsten genutzten Apps wurden für ein intuitives Handling optimiert.

Begründung der Jury
Dem emporiaSMART gelingt es auf charmante Weise, älteren Menschen die Scheu vor modernen Smartphones zu nehmen, da sie nicht auf die gewohnte Tastatur verzichten müssen.

P'9983
Smartphone

Manufacturer
BlackBerry, Waterloo, Canada
Design
Porsche Design Studio,
Zell am See, Austria
Web
www.blackberry.com
www.porsche-design.com

When choosing the materials for the high-performance P'9983 smartphone, the focus was on excellent quality; for example, sapphire glass is used for the camera lens, forged stainless steel for the logo and the case. A special glass fibre weave was chosen for the back cover. The QWERTY keyboard consists of special keys made of an extremely robust glass-like material with a 3-D effect. The smartphone's appearance is of timeless elegance.

Statement by the jury
The design of the P'9983 smartphone is indicative of robustness. This impression is emphasised by the integration of high-quality materials.

Bei der Materialauswahl des leistungsstarken Smartphones P'9983 wurde insbesondere auf exzellente Qualität geachtet: Saphirglas für die Kameralinse, geschmiedeter Edelstahl für das Logo und für das Gehäuse sowie ein spezielles Glasfasergewebe für die Rückseite. Die QWERTZ-Tastatur verfügt über spezielle Tasten aus einem äußerst robusten glasähnlichen Material mit 3-D-Effekt. In seinem Erscheinungsbild zeigt es zeitlose Eleganz.

Begründung der Jury
Die Gestaltung des Smartphones P'9983 zeugt von Beständigkeit. Unterstrichen wird dieser Eindruck durch die Ausführung in hochwertigen Materialien.

BlackBerry Passport
Smartphone

Manufacturer
BlackBerry, Waterloo, Canada

In-house design
BlackBerry

Web
www.blackberry.com

reddot award 2015
best of the best

Form appealing to the senses

Since the BlackBerry smartphones merge an effective approach toward managing content and data with sophisticated state-of-the-art user-friendliness, these devices have become highly popular among users. The BlackBerry Passport takes its inspiration from the size of a passport and thus ties in with the notion of mobility. The user elements and connections follow a clear arrangement, delivering intuitive user guidance. The smartphone features a large-size display that has been effectively integrated into the overall form. The display provides easy navigation and management of data, as well as an amazingly clear viewing of content. Another highly user-friendly aspect is presented by the text line length, which is similar to that of a book. The innovative touch-sensitive keypad of the BlackBerry Passport offers a particularly pleasing tactile experience, ensuring efficient typing and creating the impression of a seamless blend of the digital and the physical. To enhance this experience, the typeface Slate Pro was used in the same size and colour on the physical and the on-screen keyboard. The BlackBerry Passport features a distinctive stainless steel frame structure that skilfully visualises the device's strength and durability yet at the same time a sense of lightness. Reminiscent of a passport, this smartphone successfully merges a well-balanced use of forms with a user concept that touches the senses in an inspiring way.

Form für die Sinne

Da sich bei den BlackBerry-Smartphones eine effektive Art der Datenverwaltung mit einer hochentwickelten Nutzerfreundlichkeit verbindet, haben diese Geräte viele Anhänger. Die Form des BlackBerry Passport ist inspiriert von den Maßen des Reisepasses, um so an die damit verbundene Symbolik der Mobilität anzuknüpfen. Seine Bedienelemente und Anschlüsse sind übersichtlich angeordnet und erlauben eine klare Nutzerführung. Das Smartphone ist mit einem großformatigen Display gestaltet, das schlüssig in die Gesamtform integriert wurde. Auf ihm kann leicht navigiert werden und die Inhalte lassen sich verblüffend gut darstellen. Überaus nutzerfreundlich ist dabei auch die Länge der Textzeilen, da sie mit der eines Buches vergleichbar ist. Eine besondere taktile Erfahrung bietet die innovative, berührungsempfindliche Tastatur des BlackBerry Passport. Man kann auf ihr ausgesprochen effizient schreiben und hat dabei das Gefühl, als würde das Digitale mit dem Physischen verschmelzen. Um diese Erfahrung zu intensivieren, wurde die Schriftart Slate Pro in gleicher Größe und Farbe für die physische und die Bildschirmtastatur gewählt. Das BlackBerry Passport ist mit einem markanten Edelstahlrahmen gestaltet, der gekonnt seine Solidität visualisiert und gleichzeitig Leichtigkeit vermittelt. Die vertraut anmutende, wohl proportionierte Formensprache dieses Smartphones verbindet sich überaus gelungen mit einem Nutzerkonzept, das auf inspirierende Weise die Sinne berührt.

Statement by the jury

The interface of the BlackBerry Passport delivers exciting new tactile experiences. Following an outstandingly innovative approach, its huge display is highly touch-sensitive and offers easy operation and effective navigation. Another fascinating aspect is the display's clarity, optimised particularly for apps. Inspired in shape by a passport, this smartphone impresses with a design that is perfectly elaborate right down to the very last detail. It creates a sense of familiarity and, at the same time, is highly functional.

Begründung der Jury

Das Interface des BlackBerry Passport bietet dem Nutzer aufregend neue taktile Erfahrungen. Sein großes Display ist auf bemerkenswert innovative Weise berührungsempfindlich und erlaubt ein effektives und leichtes Navigieren und Arbeiten. Faszinierend ist die gute Darstellung insbesondere für die Lesbarkeit von Apps. Formal angelehnt an einen Reisepass, beeindruckt dieses Smartphone so mit einer bis in jedes Detail perfekt durchdachten Gestaltung. Es schafft Vertrautheit und ist zugleich sehr funktional.

Designer portrait
See page 68
Siehe Seite 68

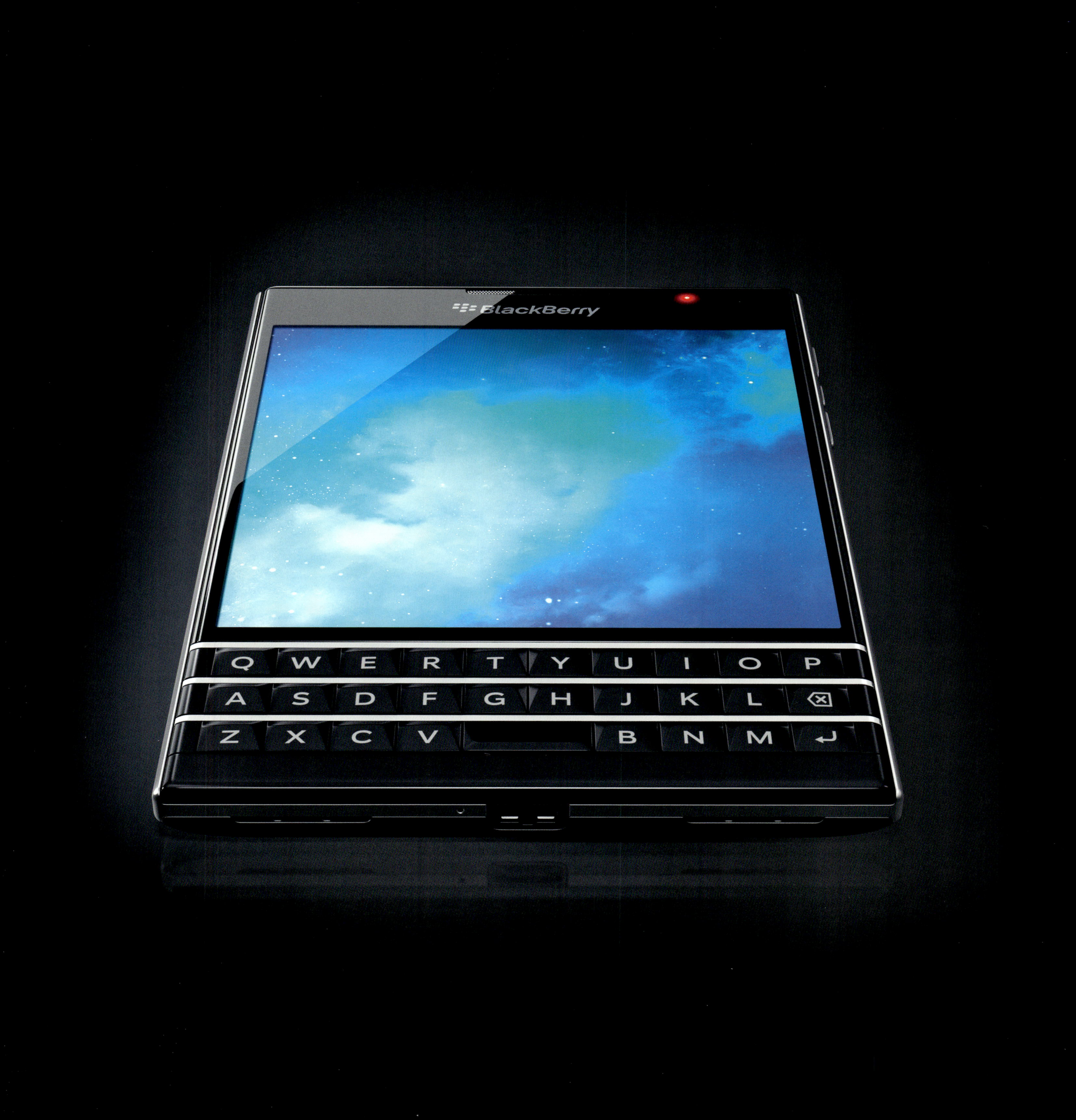
BlackBerry
Q W E R T Y U I O P
A S D F G H J K L
Z X C V B N M

Medion Life P63039
DECT phone set
DECT Telefon-Set

Manufacturer
Medion AG, Essen, Germany
In-house design
Medion Design Team
Web
www.medion.com

Medion Life P63039 is a desk telephone in duplex design and features DECT technology. The sensitive touch-keys ensure direct and pleasant operability. A number of mobile devices can be easily connected with each other via its Bluetooth function. The Eco setting and the answering machine complete the equipment package. High-gloss surfaces and elegant lines emphasise the high-quality standard.

Medion Life P63039 ist ein Tischtelefon in Twin-Ausführung mit DECT-Technologie. Die berührungsempfindlichen Touch-Tasten gewährleisten eine direkte und angenehme Bedienbarkeit. Durch seine Bluetooth-Funktion können mehrere Mobilteile bequem miteinander verbunden werden. Die Eco-Einstellung und der Anrufbeantworter runden die Ausstattung ab. Hochglänzende Oberflächen und eine elegante Linienführung unterstreichen den hochwertigen Anspruch.

Philips LINEA
Cordless Phone
Schnurloses Telefon

Manufacturer
Gibson Innovations, Hong Kong
In-house design
Web
www.gibson.com

From every viewing angle, the Philips Linea Cordless Phone presents itself as a reduced, clear pillar which separates when lifting the handset. It provides good sound qualities as well as well-conceived functions which offer a high level of convenience such as the easily readable high-contrast display and a calibrated keyboard. Weights integrated in the base of the device provide additional stability, while well adjusted magnets assure easy and secure docking.

Statement by the jury
From an aesthetic viewpoint, the Philips Linea Cordless Phone stands out clearly from conventional telephones and gains additional merit due to its well-considered user-friendly details.

Das Philips Linea Cordless Phone präsentiert sich aus jedem Blickwinkel in Gestalt einer reduziert-schlichten Säule, die sich beim Abnehmen des Hörers teilt. Es zeigt gute Klangqualitäten sowie durchdachte Funktionen, die viel Komfort bieten, etwa das gut lesbare, kontrastreiche Display oder eine kalibrierte Tastatur. Im unteren Bereich des Geräts sorgen integrierte Gewichte für zusätzliche Stabilität, gut eingestellte Magnete gewährleisten leichtes und sicheres Andocken.

Begründung der Jury
In ästhetischer Hinsicht hebt sich das Philips Linea Cordless Phone von vielen herkömmlichen Telefonen deutlich ab, besticht aber auch mit gut durchdachten nutzerfreundlichen Details.

Philips Luceo M6
Cordless Phone
Schnurloses Telefon

Manufacturer
Gibson Innovations, Hong Kong
In-house design
Web
www.gibson.com

The slim, bi-convex form of the cordless telephone is immediately eye-catching. The 360-degree design of high-quality construction is available in black-and-white, red-and-black or all-black. A display on the outside indicates the time or data of a caller. The large display on the inside facilitates multi-line telephone directory entries; the keys are calibrated precisely. The handset can be replaced accurately and can be charged from either side.

Statement by the jury
Its modern and unusual design makes the Philips Luceo M6 Cordless Phone a noble accessory in any living environment, which also convinces through its performance characteristics.

Die schlanke, bikonvexe Formgebung des schnurlosen Telefons fällt sofort ins Auge. Das 360-Grad-Design in hochwertiger Ausführung ist in den Farben Schwarz-Weiß, Rot-Schwarz oder Vollschwarz erhältlich. Eine Anzeige an der Außenseite gibt die Uhrzeit oder Daten eines Anrufers an. Das große Display im Inneren ermöglicht mehrzeilige Telefonbucheinträge, die Tastatur ist exakt kalibriert. Der Hörer lässt sich präzise auflegen und kann beidseitig aufgeladen werden.

Begründung der Jury
Seine moderne und ungewöhnliche Gestaltung macht das Philips Luceo M6 Cordless Phone zum edlen Accessoire in jeder Wohnumgebung und überzeugt überdies durch seine Gebrauchseigenschaften.

The HTC Dot View Case
Protective Sleeve
Schutzhülle

Manufacturer
HTC Corporation, Smart Phone, New Taipei, Taiwan
In-house design
Web
www.htc.com

A smartphone protected by the HTC Dot View Case eliminates the need to flip open the cover when you want to use it. The curved back of the case is adapted to the phone's shape and is particularly slim and light as it is made using polycarbonate. That also makes it very durable. As soon as the silicone flip cover is opened, it activates the smartphone thanks to built-in magnets. A grid of small viewing holes makes it possible to always see the weather and time.

Geschützt vom HTC Dot View Case muss bei Gebrauch des Smartphones kein Cover aufgeklappt werden. Mit einer gebogenen Rückseite passt es sich in seiner Form dem Telefon gut an und ist dank der Verwendung von Polycarbonat besonders dünn und leicht. Dennoch bietet es viel Stabilität und aktiviert anhand des vorderen Silikon-Flip-Covers mit eingebauten Magneten das Smartphone, sobald dieses geöffnet ist. Durch ein Raster mit Sichtlöchern sind Wetterinformationen oder die Uhr stets gut sichtbar.

Mesh Case
Case for iPhone 6 Plus
Hülle für iPhone 6 Plus

Manufacturer
DAQ, AndMesh, Tokyo, Japan
In-house design
Web
www.andmesh.com

Mesh Case is a convenient case for an iPhone 6 Plus made with a Japanese elastomer which protects it from damage. As the material is harder than that used in soft cases and softer than a hard case, this model offers elasticity and durability at the same time. The case is virtually shatter-proof and also resistant to dirt. With the slightly rounded back of this case the iPhone fits perfectly into the palm of the hand. The Mesh Case is available in eight different colours.

Statement by the jury
This case for the iPhone 6 Plus combines the pleasant haptic and flexibility of a soft case with the stability of a hard case.

Mesh Case eignet sich als Hülle für das iPhone 6 Plus und schützt dieses mit speziellem japanischem Elastomer vor Beschädigungen. Da das Material härter ist als das eines Softcases und weicher als das eines Hardcases, steht diese Ausführung für Formbarkeit und Stabilität zugleich. Das Produkt ist so gut wie bruchsicher und zudem resistent gegen Schmutz. Dank der leicht gewölbten Rückseite liegt das iPhone in der Hülle gut in der Hand. Erhältlich ist Mesh Case in acht verschiedenen Farben.

Begründung der Jury
Diese Hülle für das iPhone 6 Plus vereint die angenehme Haptik und Flexibilität eines Softcases mit der Stabilität eines Hardcases.

ICON Sleeve with Tensaerlite
Apple MacBook and iPad Protective Sleeve
Schutzhülle für Apple MacBook und iPad

Manufacturer
Incase Designs, Chino, USA
In-house design
Allen Min Choi, Evan Hyun Hong
Web
www.goincase.com

This remarkably thin sleeve is very robust and offers Apple products superior protection. The special material used in the frame was chosen for its ability to absorb impacts – just like the Neoprene used in the casing. The shock-absorbing effect of the frame protects all the edges of the device and thus makes it less sensitive to damage if dropped.

Statement by the jury
Thanks to its particularly hybrid construction, ICON Sleeve with Tensaerlite is both extremely thin and robust. It thus offers reliable protection for iPad and MacBook.

Die auffallend dünne Hülle ist strapazierfähig und bietet Apple-Produkten einen hohen Schutz. Für das Material des Rahmens wurden spezielle Rohstoffe gewählt, die mit ihren Eigenschaften die Einwirkungen von außen abfangen und sich – ebenso wie das für die Verkleidung verwendete Neopren – durch besondere Leichtigkeit auszeichnen. Die stoßdämpfende Wirkung des Rahmens schützt die Außenkanten des Gerätes und macht es so weniger empfindlich bei Stürzen.

Begründung der Jury
Dank eines besonderen hybriden Aufbaus ist ICON Sleeve with Tensaerlite zugleich extrem dünn und sehr strapazierfähig. So bietet es zuverlässigen Schutz für iPad und MacBook.

Power Stack Duo
Heavy-duty Batteries
Hochleistungsakkus

Manufacturer
Nexiom Company Limited, Hong Kong
In-house design
Vincent Lau, Lei Zheng
Web
www.nexiom.cc

Power Stack Duo in friendly orange and black colouring is a three-part set consisting of heavy-duty batteries. They can be stacked together, providing one multiply-reinforced battery with power reserves for high demands. When charged as a stack, the Top-Is-The-King principle applies: as the element on top of the stack has priority in charging, the operational component can be simply removed and used. Two cables with Lightning USB and one with micro USB connectors are integrated. Each element offers individual capacity.

Power Stack Duo in freundlichem Orange mit Schwarz ist ein dreiteiliges Set aus Hochleistungsakkus. Sie können zusammengesteckt werden und ergeben so einen mehrfach verstärkten Akku mit Stromreserven für gehobene Ansprüche. Werden sie in einem Stapel geladen, gilt das Top-is-the-King-Prinzip: Das jeweils zuoberst liegende Element hat Priorität beim Ladevorgang, somit kann der einsatzbereite Teil immer einfach abgenommen und verwendet werden. Zwei Kabel mit Lightning- und ein Mikro-USB-Anschluss sind integriert. Jedes Element bietet individuelle Kapazität.

Xtorm Power Bank Air 6000
Mobile Charging Unit
Mobiles Ladegerät

Manufacturer
A-Solar b.v., Houten, Netherlands
In-house design
Web
www.xtorm.eu

Xtorm Power Bank Air 6000 is a mobile unit which allows to charge two mobile devices at the same time via USB interfaces. As the maximum charging power is 2.5 amps, an average smartphone can be charged within one hour. The lithium-ion battery of the unit has a capacity of 6,000 mAh. The haptically pleasant, anthracite-coloured, rubberised material assures impact resistance; a discreet, blue LED indicates the battery status.

Statement by the jury
This Xtorm Power Bank Air 6000 mobile charging unit convinces as much due to its high power capacity as to its attractively reduced appearance.

Xtorm Power Bank Air 6000 ist ein mobiles Ladegerät. Bis zu zwei Mobilgeräte können über USB-Schnittstellen gleichzeitig aufgeladen werden. Die maximale Ladeleistung beträgt 2,5 Ampere, ein durchschnittliches Smartphone ist damit in unter einer Stunde geladen. Sein Lithium-Ionen-Akku verfügt über eine Kapazität von 6.000 mAh. Die haptisch angenehme Ausführung aus gummiertem anthrazitfarbenen Material sorgt für Stoßfestigkeit, den Batteriestatus kommuniziert eine dezente blaue LED-Anzeige.

Begründung der Jury
Das mobile Ladegerät Xtorm Power Bank Air 6000 überzeugt durch seine hohe Ladekapazität ebenso wie durch ein reduziertanmutiges Auftreten.

AP005/006
External Powerbanks
Externe Akkus

Manufacturer
Huawei Device Co., Ltd., Shenzhen, China
In-house design
Yuan Ding, Shuo Zhang
Web
http://consumer.huawei.com

The external powerbanks are an addition to smartphones of the Huawei P series. They feature a plain and elegant design and a tailor-made, discreet range of colours. Ultrathin at only 8 or 9.3 mm, they offer a good balance between mobility and performance. The polycarbonate case inhibits dints and scratches; tests have shown that the base withstands up to 100,000 abrasions thanks to its metal oxidation processing. High-performance soft lithium polymer batteries assure reliability.

Statement by the jury
The elegantly designed external powerbanks of the AP005/006 series are a convincingly efficient addition to Huawei smartphones.

Die externen Akkus ergänzen die Smartphones der Huawei-P-Serie. Sie zeigen eine schlicht-elegante Gestaltung und eine maßgeschneiderte, dezente Farbpalette. Mit nur 8 bzw. 9,3 mm sind sie sehr dünn und bieten eine gute Balance zwischen Mobilität und Leistung. Das Gehäuse aus Polycarbonat ist unempfindlich gegen Beulen oder Kratzer. Der Boden wurde in einem Metalloxidationsverfahren behandelt und hält im Belastungstest bis zu 100.000 Abrieben stand. Hochwertige Soft-Lithium-Polymer-Batterien sorgen für Zuverlässigkeit.

Begründung der Jury
Die elegant gestalteten externen Akkus der Serie AP005/006 sind als Zubehör der Huawei-Smartphones überzeugend effizient.

Power Solution Metallic Digital Charger
Portable Power Charger
Portables Ladegerät

Manufacturer
Lifetrons Switzerland AG, Dicken, Switzerland
In-house design
Lifetrons Switzerland Design Team
Web
www.lifetrons.ch

The charger in metallic look has an impressive capacity of 8,000 mAh and two USB ports. Its patent digital display and shiny exterior impart a high-quality, robust impression. With the one-touch charging function, two devices can be charged at the same time, quickly and efficiently. An integrated safety control prevents surge, overcharge and short circuits.

Statement by the jury
With a capacity of 8,000 mAh the portable power charger proves efficient. A metallic finish and the discreet digital display emphasise its high quality.

Das Ladegerät im Metallic-Look verfügt über eine eindrucksvolle Kapazität von 8.000 mAh und zwei USB-Ausgänge. Eine patentierte digitale Anzeige und eine glänzende metallene Oberfläche sorgen für einen hochwertigen, robusten Eindruck. Mit der One-Touch-Aufladefunktion können schnell und effizient zwei Geräte gleichzeitig aufgeladen werden. Die integrierte Sicherheitssteuerung verhindert Überspannung, Überladung und Kurzschlüsse.

Begründung der Jury
Mit einer Kapazität von 8.000 mAh beweist das portable Ladegerät Effizienz. Eine metallene Oberfläche und die dezente Digitalanzeige unterstreichen seine Hochwertigkeit.

Power Solution XL Metallic Digital Charger
Portable Power Charger
Portables Ladegerät

Manufacturer
Lifetrons Switzerland AG, Dicken, Switzerland
In-house design
Lifetrons Switzerland Design Team
Web
www.lifetrons.ch

The charger impresses with a capacity of 13,000 mAh. From an aesthetic viewpoint, the consistent hexahedron form is eye-catching. Compact dimensions emphasise the noble impression and allow the device to lie snugly in the hand. The patent digital display convinces in its discreet simplicity. Thanks to its one-touch charging control, two devices can be charged reliably at the same time.

Statement by the jury
With its impressive capacity of 13,000 mAh this portable power charger ranks among the leaders in its class.

Dieses Ladegerät beeindruckt mit einer Kapazität von 13.000 mAh. Unter ästhetischem Gesichtspunkt betrachtet fällt besonders die konsequente Hexaederform auf. Die kompakten Maße unterstreichen die edle Anmutung und lassen das Gerät gut in der Hand liegen. Die patentierte Digitalanzeige überzeugt in ihrer zurückhaltenden Schlichtheit. Dank One-Touch-Aufladefunktion werden zuverlässig zwei Geräte gleichzeitig geladen.

Begründung der Jury
Mit einer imponierenden Kapazität von 13.000 mAh reiht sich dieses portable Ladegerät unter den Spitzenreitern seiner Klasse ein.

Just Mobile TopGum
Portable Power Charger
Portables Ladegerät

Manufacturer
Just Mobile Ltd., Taichung, Taiwan
In-house design
Nils Gustafsson, Erich Huang
Web
www.just-mobile.com

Just Mobile TopGum with housing of high-quality aluminium is a power source for smartphones, tablets and other devices with USB connectors. With a charging capacity of 6,000 mAh the device can fully charge a smartphone up to three times. Thanks to two integrated Lightning cables, iPhones and iPads can be charged simultaneously; furthermore, a fast-charge USB port is provided. A magnetic charging station facilitates wireless charging in an elegant vertical position.

Just Mobile TopGum mit einem Gehäuse aus hochwertigem Aluminium versorgt Smartphones, Tablets und andere Geräte mit USB-Anschluss. Mit einer Ladekapazität von 6.000 mAh lädt er ein Smartphone bis zu drei Mal komplett wieder auf. Dank zweier integrierter Lightning-Kabel können zeitgleich iPhone und iPad geladen werden, zudem ist ein Fast-Charge-USB-Ausgang vorhanden. Eine magnetische Ladestation ermöglicht kabelloses Aufladen in elegant aufrechter Position.

Statement by the jury
With Just Mobile TopGum, mobile phones and tablets remain reliably charged when travelling. Apart from its distinctive functionality, it gains prominence due to a stylish form.

Begründung der Jury
Mit Just Mobile TopGum bleiben Mobiltelefon und Tablet unterwegs zuverlässig aufgeladen. Neben seiner ausgeprägten Funktionalität fällt er durch eine stilvolle Formgebung auf.

Electroluminescent Charge Cable
Elektrolumineszierendes Ladekabel

Manufacturer
Pilot Electronics, City of Industry, USA
In-house design
Calvin Wang, Yin Zheng Kai, Hai Tran, Lesley Leong,
EL Shenzen, Joseph Dong
Web
www.pilotelectronic.com

This charging and data cable transmits an electroluminescent light which indicates the charging status of the device. If the device's battery power capacity is low, the light flow moves fast; when the device is almost fully charged it moves more slowly and finally shuts off when the battery is fully charged. There are many types of electroluminescent colours for a unique experience.

Dieses Ladekabel sendet einen elektrolumineszenten Lichtimpuls, durch den der aktuelle Ladestatus eines Endgerätes zu erkennen ist. Zu Beginn des Ladevorgangs bewegt sich der Lichtfluss schnell, bei nahezu vollständig geladenem Gerät wird er langsamer und erlischt schließlich bei vollem Akku. Unterschiedliche Farbgebungen machen jedes Kabel zu einem individuellen Erlebnis.

Statement by the jury
The flowing light impulse indicates the charge status of a device even from far away and makes the often troublesome observation of the display obsolete: an innovative solution which exhibits a playful approach.

Begründung der Jury
Der fließende Lichtimpuls zeigt schon von Ferne den Ladestatus eines Geräts an und macht den oft mühsamen Blick auf das Display obsolet: eine innovative Lösung, die von einer spielerischen Herangehensweise zeugt.

Docomo Portable AC Adapter 01
Charger
Ladegerät

Manufacturer
NTT docomo, Inc., Tokyo, Japan
In-house design
Web
www.nttdocomo.co.jp

For this charger, two possible modes of use have been considered: space-saving usage at home or at work and practical use when travelling. The extremely small dimensions allow uncomplicated operation on the desk or in the proximity of a power socket without occupying space unnecessarily. When on the move, the silicone attachment is very useful for easily rolling up the cable; it is available in blue, red or grey, giving it a friendly, individual effect. The micro-USB plug is designed with a curvature, so that it is easy to tell the difference between top and bottom.

Bei diesem Ladegerät wurde an zwei Nutzungsmöglichkeiten gedacht – die platzsparende Verwendung zu Hause bzw. am Arbeitsplatz sowie den praktischen Gebrauch unterwegs. Die extrem kleinen Maße erlauben die unkomplizierte Anwendung auf dem Schreibtisch oder in der Nähe jeder Steckdose, ohne unnötig Platz zu verbrauchen. Unterwegs macht sich das Zusatzteil aus Silikon zum einfachen Aufrollen des Kabels bezahlt, das in den Farben Blau, Rot oder Grau eine freundliche individuelle Note ermöglicht. Der Micro-USB-Stecker ist mit einer Wölbung so gestaltet, dass Ober- und Unterseite leicht zu unterscheiden sind.

Statement by the jury
The charger convinces with clear, compact dimensions and especially through its innovative idea of an aid for simply rolling up the cable. This means no more tangled cables in the pocket.

Begründung der Jury
Das Ladegerät überzeugt mit klaren, kompakten Maßen und insbesondere die Idee, das Kabel einfach aufrollen zu können. So gibt es keinen lästigen Kabelsalat mehr in der Tasche.

Aircard 340U
USB Modem

Manufacturer
NETGEAR, Inc., San Jose, USA
In-house design
James Hathway
Design
Alloy Ltd. (Jim Blyth, Matt Plested), Farnham, Great Britain
Web
www.netgear.com
www.thealloy.com

The AirCard 340U is an ultra-light, hinged USB modem which is designed to enable access to 4G/LTE mobile networks. The device features an integrated LCD display which provides at-a-glance information on data usage, network and signal strength. The crafted dual axis hinge and USB connection enable easy access to adjacent PC connections when in use, and creates a slim profile when stored. The AirCard 340U provides 100Mbps download and 50 Mbps upload speeds to multiple devices, and offers roaming in over 200 countries.

Die Aircard 340U ist ein ultraleichtes, klappbares USB-Modem für den Zugang zu 4G-/LTE-Netzwerken. Ein integrierter LCD-Bildschirm informiert auf einen Blick über Datennutzung, Netzwerk und Signalstärke. Das hochwertig verarbeitete zweigelenkige Scharnier sowie der USB-Anschluss stellen bei Gebrauch einen einfachen Anschluss an den PC sicher, für die Aufbewahrung wird es zu einer schlanken Form zusammengeklappt. Die Aircard 340U ermöglicht einen Download von bis zu 100 Mbps und eine Uploadleistung von bis zu 50 Mbps für mehrere Geräte sowie Roaming in über 200 Ländern.

Statement by the jury
Thanks to an innovative, integrated LCD display, the Aircard 340U makes key data on connection strength visible, without sacrificing space on the laptop monitor.

Begründung der Jury
Dank innovativem, integriertem LCD-Bildschirm macht die Aircard 340U Eckdaten zur Verbindungsstärke sichtbar, ohne dafür Platz auf dem Laptop-Monitor zu verschwenden.

Share Stick
Travel-Router

Manufacturer
Trend Power Limited, Hong Kong
In-house design
Ho Yeung Lam
Web
www.powertrend.me

The Travel-Router works multifunctionally: as portable wireless router and as 3G to Wi-Fi converter it facilitates simultaneous internet access for up to ten devices. Via USB and SD connections data are transferred without occupying device memory space. Furthermore the Share Stick with 5,200 mAh functions as mobile charger. The small stick is available in five friendly colours and has a pleasant, soft-touch surface.

Statement by the jury
Exhibiting an amazing multifunctionality, the Share Stick presents itself as a well-conceived travel companion. Its fresh appearance is also pleasing.

Der Travel-Router arbeitet multifunktional: Als Portable Wireless Router und 3G-to-Wi-Fi-Converter ermöglicht er bis zu zehn Endgeräten den zeitgleichen Zugang zum Internet. Mittels USB- und SD-Anschluss werden Daten geladen, ohne den Speicherplatz des Endgerätes zu belasten. Zusätzlich funktioniert der Share Stick mit 5.200 mAh als mobiles Ladegerät. Der kleinformatige Stick wird in fünf freundlichen Farben angeboten und weist eine angenehme Soft-Touch-Oberfläche auf.

Begründung der Jury
Der Share Stick zeigt eine erstaunliche Multifunktionalität, die ihn zum wohldurchdachten Reisebegleiter macht. Zugleich erfreut sein frisches Erscheinungsbild.

L–01G
Mobile Wi-Fi Router
Mobiler Wi-Fi-Router

Manufacturer
LG Electronics, Seoul, South Korea
In-house design
Tomoyuki Akutsu
Web
www.lg.com

With the mobile Wi-Fi router it is possible to connect several devices to the Internet when travelling. The 3" touchscreen is, like that of smartphones, easy to operate. A Wi-Fi connection and other settings are uncomplicated. By means of the charging function mobile devices can also be charged. An energy-saving function controls power consumption when sending data.

Statement by the jury
The Mobile Wi-Fi Router L-01G facilitates access to the Internet while travelling without any problems. The well-conceived 3" touchscreen provides convenient control.

Mit dem mobilen Wi-Fi-Router kann man unterwegs mit mehreren Geräten gleichzeitig über einen Zugang ins Internet gelangen. Der 3"-Touchscreen ist wie von Smartphones gewohnt einfach zu bedienen. Eine Wi-Fi-Verbindung und weitere Einstellungen sind unkompliziert eingerichtet. Mit der Ladefunktion können mobile Geräte auch aufgeladen werden; eine Energiesparfunktion kontrolliert den Stromverbrauch jeder Datenübertragung.

Begründung der Jury
Der Mobile Wi-Fi-Router L-01G macht die Internetnutzung auch unterwegs problemlos möglich. Der durchdacht gestaltete 3"-Touchscreen sorgt für unkomplizierte Handhabung.

ZU100
Smart Router

Manufacturer
Shenzhen Rapoo Technology Co., Ltd., Shenzhen, China
Design
Shenzhen Zivoo Technology Co., Ltd., Shenzhen, China
Web
www.zivoo.com

With its high-quality outer appearance the router sets itself clearly apart from conventional devices. The speed of wired broadband connections is up to 1 GB. A standard USB port provides connection, for example, of a flash drive or card reader; the 1 TB hard drive offers sufficient memory capacity. The cubic housing with elegant colouration is optically convincing and displays a pleasant surface structure.

Statement by the jury
Contrary to conventional routers, this sophsticated product integrates also in environments with high aesthetic demands.

Der Router hebt sich in seiner edlen Erscheinungsform deutlich von herkömmlichen Geräten ab. Die Geschwindigkeit der kabelgebundenen Breitbandverbindung beträgt bis zu 1 GB. Ein Standard-USB-Port ermöglicht z. B. den Anschluss eines Flash-Laufwerks oder Kartenlesers, die integrierte 1-TB-Festplatte bietet genügend Speicherplatz. Das kubische Gehäuse mit eleganter Farbgebung überzeugt optisch und zeigt eine angenehme Oberflächenstruktur.

Begründung der Jury
Anders als herkömmliche Router integriert sich dieses ausgefeilt gestaltete Produkt auch in Umgebungen mit hohen ästhetischen Ansprüchen.

N300 Smart
Wi-Fi Extender

Manufacturer
Edimax Technology Co., Ltd., New Taipei, Taiwan
In-house design
Edimax Design Center
Web
www.edimax.com

The Wi-Fi Extender assures good Wi-Fi reception in remote corners of a household and offers family-friendly, additional functions via an App: for instance decoupled online accesses can be installed. Internet access for the younger family members can be regulated to times or their online sessions documented. A night mode deactivates Wi-Fi and LED display and assures undisturbed sleep. The discreet blue LED is visually attractive.

Der Wi-Fi-Extender sorgt auch in weiten Winkeln eines Haushaltes für einen guten Wi-Fi-Empfang und bietet über eine App familienfreundliche Zusatzfunktionen. So lassen sich z. B. unterschiedliche Wi-Fi-Netzwerke einrichten. Der Internetzugang jüngerer Familienmitglieder kann zeitlich reglementiert und deren Onlinezeit dokumentiert werden. Ein Nachtmodus deaktiviert Wi-Fi und LED-Anzeige und sichert ungestörte Nachtruhe. Optisch ansprechend ist dazu der dezente blaue LED-Ring.

Statement by the jury
The Wi-Fi Extender limits online access to individual, terminal household appliances – a family-friendly device for undisturbed learning and rest times.

Begründung der Jury
Der Wi-Fi-Extender begrenzt den Onlinezugang einzelner Endgeräte eines Haushalts – eine familienfreundliche Einrichtung für störungsfreie Lern- und Ruhezeiten.

Smart BT-Plug
Bluetooth-enabled Plug-in Socket
Bluetoothfähige Zwischensteckdose

Manufacturer
Gunitech Corp., Hsinchu County, Taiwan
In-house design
Kelly Chen
Web
www.gunitech.com

Smart BT-Plug is a Bluetooth-enabled plug-in socket which considerably eases operation of electrically driven appliances in households: sockets and USB connectors can be remote-controlled. Appliances can be switched on and off wirelessly via mobile phone, tablet or other mobile devices; a timer or standby mode can be activated or deactivated. Several outlets can be controlled simultaneously and automatic energy consumption can be monitored.

Smart BT-Plug ist eine bluetoothfähige Zwischensteckdose, die das Bedienen strombetriebener Geräte im Haushalt erheblich erleichtert: Steckdosen und USB-Anschlüsse lassen sich ferngesteuert bedienen. Kabellos werden Geräte per Mobiltelefon, Tablet oder andere mobile Instrumente an- und ausgeschaltet, ein Timer oder Ruhemodus aktiviert bzw. deaktiviert. Es können mehrere Ausgänge gleichzeitig gesteuert und auch eine automatische Stromverbrauchskontrolle vorgenommen werden.

Statement by the jury
With the Smart BT-Plug, it is possible to control household electrical connections from the couch.

Begründung der Jury
Mit Smart BT-Plug wird die Kontrolle über die Stromanschlüsse des Haushalts vom Sofa aus möglich.

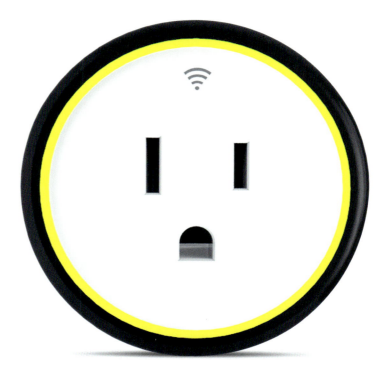

smanos K1 SmartHome DIY Kit
Smart Home Solution
Smart-Home-Lösung

Manufacturer
Chuango Security Technology Corporation, Shenzhen, China
In-house design
Chen Li, Shifeng Xu, Zhongning Wu
Web
www.chuango.com
www.smanos.com

The smanos K1 SmartHome DIY Kit allows property owners to access the electrical system in the house at any time via wireless network and to make use of intuitive remote control or on-site direct control. Status of light switches and electrical sockets, locks and cameras, alarm systems and environmental sensors can be retrieved, set and monitored by smartphone. Up to 50 wireless Smart Home elements can be integrated in the house automation and controlled without technical knowledge, making it usable even for children and the elderly.

Der smanos K1 SmartHome DIY Kit ermöglicht es dem Immobilienbesitzer, sich durch drahtlose Vernetzung jederzeit Zugriff auf die Elektrik im Haus zu verschaffen und eine intuitive Bedienung aus der Ferne oder direkt vor Ort zu nutzen. Lichtschalter und Steckdosen, Schlösser und Kameras, Alarmanlagen und Umgebungssensoren können abgerufen, eingestellt und mit dem Smartphone überprüft werden. Bis zu 50 drahtlose Smart-Home-Elemente lassen sich in die Hausautomation integrieren und sind ohne technische Kenntnisse intuitiv steuerbar, was eine Nutzung auch für Kinder und ältere Menschen möglich macht.

Statement by the jury
smanos K1 SmartHome DIY Kit efficiently provides the facilities of a modern security service and assures a cosy atmosphere in the home.

Begründung der Jury
Der smanos K1 SmartHome DIY Kit bietet effizient den Service eines modernen Sicherheitsdienstes und sorgt für anheimelnde Atmosphäre zu Hause.

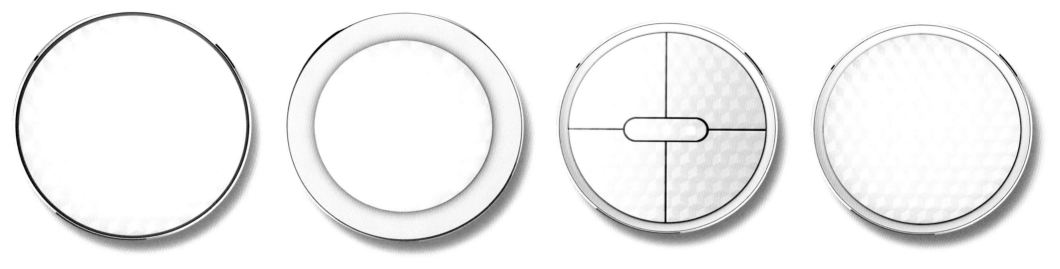

SwiftCam M3s
Handheld Stabilizing Gimbal for Smartphones
Handheld-Gimbal für Smartphones

Manufacturer
SwiftCam Technologies Group Company Limited, Hong Kong
In-house design
Jacky Cheung, Kenneth Choi
Web
www.swiftcam.com

SwiftCam M3s is an innovative, handheld gimbal for smartphones. Thanks to its pressure clip mechanism, models like iPhone 5, iPhone 6 Plus or Samsung Galaxy Note 4 can be easily connected. Technically, the device offers three-axis stabilisation, a 360-degree horizontal pan, panning and tilting with joystick control as well as standard camera pivoting. By upgrading it with an app, the gimbal will always remain at state of the art. An ergonomic grip supports handling; a magnetic locking system facilitates storage and transport.

SwiftCam M3s ist ein innovatives Handheld-Gimbal für Smartphones. Über einen Druck- und Klemmmechanismus werden Modelle wie das iPhone 5, iPhone 6 Plus oder das Samsung Galaxy Note 4 ganz einfach angeschlossen. Technisch bietet das Gerät eine Drei-Achsen-Stabilisierung, einen 360-Grad-Horizontal-Schwenk, Schwenken und Neigen per Joystick-Steuerung sowie die Einstellung typischer Kameraschwenks. Durch Upgrade mittels App bleibt das Gimbal stets auf dem Stand der Technik. Ein ergonomischer Griff unterstützt die Handhabung, ein Magnetverschluss erleichtert Lagerung und Transport.

Statement by the jury
The SwiftCam M3s is compatible with many standard smartphones and impresses with this versatility for a wide range of periphery equipment.

Begründung der Jury
Das SwiftCam M3s ist mit zahlreichen gängigen Smartphones kompatibel und beeindruckt mit dieser Offenheit für eine breite Produktperipherie.

Cardboard
Flat-pack Cardboard Virtual Reality Viewer
Virtual-Reality-Erweiterung für Smartphones

Manufacturer
Google, Mountain View, USA
In-house design
Web
www.google.com

You can also find this product in
Dieses Produkt finden Sie auch in
Doing
Page 420
Seite 420

Cardboard is a smartphone accessory that facilitates experiencing virtual reality in an uncomplicated and cost-effective manner. It is die-cut from a single sheet of cardboard and dispatched as a flat envelope. The user provides input via a sliding magnet on the side of the box. The magnetic sensor of the smartphone recognises and processes its movement. An ordinary rubber band prevents the phone from sliding out.

Statement by the jury
Cardboard makes virtual reality come true for everyone who owns a smartphone. The die-cut cardboard solution is affordable and quite simple.

Cardboard ist eine Erweiterung für Smartphones, die das Erleben von Virtual Reality unkompliziert und kostengünstig möglich macht. Es wird aus einem einzigen Stück Karton gestanzt und als flacher Umschlag versandt. Der Benutzer kann den betrachteten Film einfach über einen Magneten an der Seite des Kartongehäuses steuern. Der Magnetsensor des Mobiltelefons erkennt und verarbeitet diese Bewegung. Ein herkömmliches Gummiband verhindert das Herausrutschen des Telefons.

Begründung der Jury
Cardboard macht Virtual Reality für praktisch jeden Besitzer eines Smartphones erlebbar. Die Lösung aus gestanztem Karton ist erschwinglich und ganz einfach.

Breeze
Breathalyser
Alkomat

Manufacturer
Breathometer Inc., San Francisco, USA
Design
NewDealDesign LLC, San Francisco, USA
Web
www.breathometer.com
Honourable Mention

The Breeze breathalyser is very light, weighing only 30 gm, and because of its size fits easily in the pocket. The user blows into the mouthpiece and an electro-chemical fuel cell breath sensor with Fluid Dynamics (patent applied for) sends the blood alcohol level readings immediately to an App. The Breathometer App can be used to order an Uber, local cab, call a friend or even find a hotel nearby. For the user's own reference the App also records the previously measured values.

Statement by the jury
The alcohol level can be discreetly measured anywhere, since Breeze weighs only 30 gm and fits in every pocket. An App makes it easier to react correctly.

Der Alkomat Breeze ist mit 30 Gramm sehr leicht und findet dank seiner geringen Größe in jeder Tasche Platz. Der Benutzer bläst in das Mundstück, ein elektrochemisch arbeitender Atemsensor mit zum Patent angemeldeter Fluiddynamik sendet die Blutalkoholwerte sofort an eine App. Über diese kann bei Bedarf ein Uber-Dienst oder Taxi bestellt, ein Kontakt angerufen oder sogar ein nahegelegenes Hotel gefunden werden. Zur Eigenkontrolle dokumentiert die App zudem die zuletzt gemessenen Werte.

Begründung der Jury
Der Alkoholspiegel kann überall diskret getestet werden, denn Breeze wiegt nur 30 Gramm und passt in jede Tasche. Eine App macht es leichter, gegebenenfalls richtig zu reagieren.

Charger for Warming Hands
Akku und Handwärmer

Guangdong Pisen Electronics Co., Ltd., Shenzhen, China
In-house design
Web
www.pisen.com.cn
Honourable Mention

This two-in-one product is both a hand warmer and a power supply. When connected to a mobile end device, the appliance provides constant warmth. An integrated temperature sensor switches it off automatically when a temperature of 60 degrees centigrade is reached. When required, the device functions also as a portable battery and reliable charger, for example, for a mobile phone. The exterior is of anodised aluminium, has a pleasant feel and is crafted with laser engraving.

Statement by the jury
Whoever is out and about in the open in the cold season needs two things: warm hands and sufficient battery power. The device offers a practical solution for both.

Dieses 2-in-1-Produkt ist Handwärmer und zusätzliches Stromaggregat in einem. An ein mobiles Endgerät angeschlossen, gibt das Gerät konstante Wärme ab. Ein integrierter Temperaturfühler schaltet es ab einer Oberflächentemperatur von 60 Grad Celsius automatisch aus. Bei Bedarf funktioniert das Objekt auch als portabler Akku und zuverlässiges Ladegerät, z. B. für ein Mobiltelefon. Die Oberfläche besteht aus haptisch angenehmem eloxiertem und mit Lasergravur bearbeitetem Aluminium.

Begründung der Jury
Wer in der kalten Jahreszeit viel im Freien unterwegs ist, braucht zweierlei: warme Hände und ausreichend Akku. Das Gerät bietet für beides eine praktische Lösung.

Cisco TelePresence IX5000
Video Conferencing System
Videokonferenzsystem

Manufacturer
Cisco Systems, San Jose, USA

In-house design
Cisco Systems

Web
www.cisco.com

reddot award 2015
best of the best

Experiencing communication

Since video conferencing systems greatly enhance communication, they have become common workplace tools. Thanks to its convincing concept, the Cisco TelePresence IX5000 delivers the best possible communication experience for such fully integrated, audio, video and data conferencing systems. At first glance it fascinates with its design featuring perfectly balanced elements. The optimally placed cameras and life-sized screens are rounded out by innovative, seamlessly integrated loudspeakers. The presentation cables and system controls are elegantly integrated into the table surface and hidden within the uni-cast aluminium legs. The centre of communication is marked by an aesthetically pleasing, triple LCD display arc, which is framed by a seamless textile panel to create an impressive panoramic visual effect. The conference system also provides a unique way of responding to incoming calls. When a call comes in, the textile panel is, almost magically, illuminated by hidden LED lighting panels ensuring optimal lighting for video broadcast. Consistent and natural audio capture is guaranteed by signature aluminium microphones integrated into the table. Thus, based on a fundamental reinterpretation, a novel conferencing system has emerged with an elegant design tailored to create a highly natural communication experience that will fascinate all who use it.

Erlebnis Kommunikation

Da sie die Kommunikation erheblich erleichtern, sind Videokonferenzsysteme heute allgemein üblich. Das Cisco TelePresence IX5000 bietet dank seines überzeugenden Konzepts das bestmögliche Kommunikationserlebnis eines solchen vollintegrierten Systems für Audio, Video und Daten. Auf den ersten Blick begeistert es durch seine perfekt aufeinander abgestimmten Elemente. Optimal platzierte Kameras und lebensgroße Bildschirme werden ergänzt durch innovative, schlüssig integrierte Lautsprecher. Die Präsentationskabel und die Systemsteuerung sind stilvoll und fließend in den Tisch eingelassen und in den Aluminiumtischbeinen verborgen. Im Mittelpunkt der Kommunikation steht eine ästhetisch ansprechende, gebogene 3-fach-LCD-Bildschirmfläche. Diese bietet einen beeindruckenden visuellen Panoramaeffekt, der durch die Einrahmung mit einem nahtlosen Stoffpanel noch erhöht wird. Außergewöhnlich ist bei diesem Konferenzsystem, wie es auf einen eingehenden Anruf reagiert. Auf geradezu magische Weise aktiviert sich dann eine hinter Textil verborgene LED-Beleuchtung, die optimale Beleuchtungsverhältnisse für die visuelle Kommunikation gewährleistet. In den Tisch eingelassene Mikrofone in charakteristischer Aluminium-Optik stellen eine hochwertige und natürliche Aufnahme der Sprache sicher. Auf der Basis einer grundlegenden Neuinterpretation entstand so ein Videokonferenzsystem, das alle, die es benutzen, durch seine Eleganz wie auch die Natürlichkeit der mit ihm möglichen Kommunikation fasziniert.

Statement by the jury

The Cisco TelePresence IX5000 fully integrated conferencing system convinces with its consistently iconic design. It showcases an elegant appearance and offers conference participants an entirely new conferencing experience. Comfort pairs with details that are manifested in an outstandingly intelligent way. Particularly worth mentioning is how incoming calls activate the LED lighting hidden behind textile to perfectly illuminate the conferencing environment.

Begründung der Jury

Das Cisco TelePresence IX5000 besticht als vollintegriertes Konferenzsystem durch seine schlüssige ikonische Gestaltung. Es hat eine sehr stilvolle Anmutung und ermöglicht den Konferenzteilnehmern ein völlig neues Kommunikationserlebnis. Komfort verbindet sich mit ausgesprochen intelligent gelösten Details. Bemerkenswert ist vor allem, wie sich bei Anrufen eine hinter Textil verborgene LED-Beleuchtung einschaltet und die Kommunikationssituation perfekt ausleuchtet.

Designer portrait
See page 70
Siehe Seite 70

Cisco DX70
Desktop Collaboration Device
Multifunktionales Desktop-
Gerät

Manufacturer
Cisco Systems, San Jose, USA
In-house design
Web
www.cisco.com

Cisco DX70 is a device developed for video communication and combines desktop sharing and other functionalities. It features an aesthetic composition of a straightforward touchscreen and an integrated speaker; viewed from the front, the camera on top and the rear support stand form one unit. The flexible tilt angle provides ergonomic convenience. A cover at the back facilitates discreet cable management.

Statement by the jury
Cisco DX70 fulfills high demands of video communication due to its well-conceived technical details and optically appealing appearance.

Cisco DX70 ist ein für die Videokommunikation entwickeltes Endgerät, das auch Desktop-Sharing und weitere Funktionalitäten in sich vereint. Ästhetisch gelöst ist die Komposition aus schnörkellosem Touchscreen und integriertem Lautsprecher, frontal betrachtet bilden die aufgesetzte Kamera und der hintere Stützfuß eine Einheit. Der flexible Neigungswinkel sorgt für ergonomischen Komfort. Eine Abdeckung an der Rückseite ermöglicht unauffälliges Kabelmanagement.

Begründung der Jury
Cisco DX70 erfüllt mit durchdachten technischen Details und optisch ansprechendem Auftreten hohe Anforderungen bei der Videokommunikation.

Dolby Conference Phone
Dolby-Konferenztelefon

Manufacturer
Dolby Laboratories, San Francisco, USA
In-house design
Lucas Saule, Peter Michaelian
Design
frog, San Francisco, USA
Web
www.dolby.com
www.frogdesign.com
Honourable Mention

The Dolby Conference Phone is provided with multiple microphones and high-performance software. All speakers can be clearly heard and listeners can differentiate distinctly between the points in the room from where each voice emanates. The round, perforated metal cover with its concave centre ensures complete detection of all contributions in the room and also issues a cue for participation in the discussion. The user interface with touch screen is intuitively operated. A clearly visible bright light ring displays the call status.

Statement by the jury
The sound is so convincing as if the discussion partners were in the same room – a functionally fully developed instrument for efficient discussion management.

Das Dolby-Konferenztelefon verfügt über mehrere Mikrofone und eine leistungsstarke Software. Alle Sprecher sind deutlich hörbar, und Zuhörer können klar unterscheiden, von welchem Punkt im Raum die jeweilige Stimme kommt. Die runde, perforierte Metallhülle mit konkaver Mitte stellt die vollständige Erfassung sämtlicher Beiträge im Raum sicher und lädt zudem zur Diskussionsbeteiligung ein. Das Bedienfeld mit Touchscreen ist intuitiv erfassbar. Ein gut sichtbarer heller Lichtring zeigt den Anrufstatus an.

Begründung der Jury
Der Klang ist so überzeugend, als befänden sich die Gesprächspartner im selben Raum – ein funktional ausgereiftes Instrument für effizientes Diskussionsmanagement.

DICENTIS
Extended Conference Device
Erweitertes Konferenzgerät

Manufacturer
Bosch Sicherheitssysteme GmbH,
Grasbrunn, Germany
Design
TEAMS Design GmbH, Esslingen, Germany
Web
www.boschsecurity.com
www.teamsdesign.com

Dicentis in its extended version provides an elegant, capacitive 4.3" touchscreen, a discreet, high-directive microphone and an integrated NFC reader for user recognition. The touchscreen can be given a digital company logo. Each device can be used either by participants in the discussion or by the chairperson. With its high-quality materials and noble design it is well adaptable for any conference.

Statement by the jury
In its extended version Dicentis convinces above all by its highly functional as well as intuitively operated touchscreen.

Dicentis in seiner erweiterten Ausführung verfügt über einen eleganten kapazitiven 4,3"-Touchscreen, ein diskretes hochdirektives Mikrofon und einen integrierten NFC-Reader zur Benutzeridentifikation. Der Touchscreen kann mit einem digitalen Firmenlogo versehen werden. Jedes Gerät lässt sich entweder als Diskussions- oder als Vorsitzendeneinheit nutzen. Hochwertige Materialien und edle Gestaltung passen sich gut in den Rahmen jeder Konferenz ein.

Begründung der Jury
In seiner erweiterten Ausführung überzeugt Dicentis vor allem durch den hoch funktionalen wie intuitiv erfassbaren Touchscreen.

DICENTIS
Conference Device
Konferenzgerät

Manufacturer
Bosch Sicherheitssysteme GmbH,
Grasbrunn, Germany
Design
TEAMS Design GmbH, Esslingen, Germany
Web
www.boschsecurity.com
www.teamsdesign.com

The conference device Dicentis facilitates interference-free and highly flexible wireless conferences. The discussion unit is used by the individual user or the chairperson. Via the wireless access point, participants can speak, register a request to speak and listen to the speaker. A timeless design with high-quality materials emphasises the useful functionality of the device.

Statement by the jury
The Dicentis conference device provides high functionality and reliable operability, creatively emphasised by a timeless appearance.

Das Konferenzgerät Dicentis ermöglicht störungsfreie und hoch flexible drahtlose Konferenzen. Die Diskussionseinheit wird vom Einzelbenutzer oder Vorsitzenden verwendet. Über den Wireless Access Point können Teilnehmer sprechen, eine Wortmeldung registrieren und dem aktiven Sprecher zuhören. Eine zeitlose Gestaltung in hochwertigen Materialien unterstreicht die seriöse Funktionalität des Geräts.

Begründung der Jury
Das Konferenzgerät Dicentis weist eine hohe Funktionalität und zuverlässige Bedienbarkeit auf, gestalterisch unterstrichen von seinem zeitlosen Erscheinungsbild.

Orbis
Conference System
Konferenzsystem

Manufacturer
beyerdynamic GmbH & Co. KG,
Heilbronn, Germany
Design
Matthias Leipholz, Karlsruhe, Germany
Web
www.beyerdynamic.de
www.matthiasleipholz.de

The Orbis conference system scores points with a clear, formal language. The flat microphone units with separate speaker module as well as the small dimensioned central controls are reduced to essentials, their functions defined by form, position and size. Installation of the system in high-quality interiors is simplified by a particularly flat construction and rounded corners. Thanks to a standardised XLR port, different types of microphones can be connected.

Statement by the jury
Due to reduced lines the Orbis conference system integrates even into an upscale ambience. Its controls are operable intuitively.

Das Konferenzsystem Orbis fällt durch eine klare Formensprache auf. Die flachen Einbausprechstellen mit separatem Lautsprechermodul wie auch die kleinformatigen zentralen Steuereinheiten sind auf das Wesentliche reduziert und definieren anhand von Form, Position und Größe ihre Funktionen. Der Einbau des Systems in hochwertige Interieurs wird vereinfacht durch eine besonders niedrige Bauhöhe und abgerundete Ecken. Mittels standardisierter XLR-Steckverbindung können unterschiedliche Mikrofontypen angeschlossen werden.

Begründung der Jury
Mit seiner reduzierten Linienführung integriert sich das Konferenzsystem Orbis auch in gehobeneres Ambiente. Seine Bedienelemente sind intuitiv erfassbar.

Catchbox 2.4
Wireless Microphone
Funkmikrofon

Manufacturer
Trick Technologies Oy, Espoo, Finland
In-house design
Pyry Taanila, Timo Kauppila, Mikelis Studers
Web
www.getcatchbox.com

Catchbox 2.4 is a soft, wireless microphone designed for audiences. The device is covered with a soft cube-shaped cover and can thus be thrown back and forth between the various speakers. This promotes playful interaction and makes awkwardness and time-wasting obsolete. A special magnetic mechanism secures the technical components inside. In order to avoid unwanted noise when being thrown or dropped, the device continuously monitors each movement and switches the microphone on and off.

Das weiche, schnurlose Mikrofon Catchbox 2.4 wurde für große Gesprächsrunden entworfen. Das Gerät ist mit einem würfelförmigen Soft-Mantel umgeben und kann so zwischen den einzelnen Sprechern einer Gruppe hin und her geworfen werden. Das unterstützt die spielerische Interaktion und macht umständliches und zeitraubendes Herumreichen obsolet. Ein spezieller Magnetmechanismus sichert die technischen Komponenten im Inneren. Um unerwünschte Geräusche beim Werfen oder Fallenlassen zu vermeiden, misst das Gerät kontinuierlich jede Bewegung und schaltet das Mikrofon automatisch ein und aus.

Statement by the jury
A wireless microphone that people can throw to each another – adoption of this original and well-conceived design idea will break the ice in every discussion round.

Begründung der Jury
Ein Funkmikrofon, das man einander zuwerfen kann – diese originelle und durchdacht umgesetzte Gestaltungsidee bricht in jeder Diskussionsrunde das Eis.

1MORE Chat Bear
Communication Device
Kommunikationsmittel

Manufacturer
1more Inc., Shenzhen, China
In-house design
Web
www.1more.com
Honourable Mention

1MORE Chat Bear eases communication between parent and child when they are in different places. The hi-fi microphone and speaker located in the toy produce a very life-like sound. By means of an app, content such as songs or stories can be sent. The friendly appearance of the bear, which is made of easily cleaned PU leather, creates a trusting atmosphere.

1MORE Chat Bear erleichtert die Kommunikation zwischen Eltern und Kind, wenn diese sich an unterschiedlichen Orten befinden. Hi-Fi-Mikrofon und Lautsprecher erzeugen einen sehr lebensechten Klang. Über eine App lassen sich auch Inhalte wie Lieder oder Geschichten senden. Das freundliche Erscheinungsbild des Bären aus pflegeleichtem PU-Leder schafft für das Kind eine vertrauensvolle Atmosphäre.

Statement by the jury
Sometimes children must spend time without their minders, but 1MORE Chat Bear brings them nearer to each other.

Begründung der Jury
Manchmal müssen Kinder etwas Zeit ohne ihre Vertrauenspersonen verbringen, aber der 1MORE Chat Bear bringt beide einander ein wenig näher.

ConferenceCam Connect
Video Conference System
Videokonferenzsystem

Manufacturer
Logitech, Newark, California, USA
In-house design
Logitech Design
Design
MNML, Chicago, USA
Web
www.logitech.com
www.mnml.com

The ConferenceCam Connect designed for up to six people is quickly installed, offers a generous 90 degree field of view with digital panning and mechanical tilt function as well as 4x digital zoom in full HD and ZEISS optics with autofocus. The mobile speakerphone features Bluetooth wireless technology, NFC and USB. Thanks to good surround sound as well as echo and noise suppression, users can communicate within a 3.6-metre range without problem. For mobile workplaces, this cost-effective solution is operated via Wi-Fi or rechargeable battery.

Die ConferenceCam Connect für bis zu sechs Personen ist rasch installiert, bietet ein großzügiges 90-Grad-Blickfeld mit digitaler Schwenk- und mechanischer Kippfunktion sowie digitalem 4-fach-Zoom in Full HD, zudem eine ZEISS-Optik mit Autofokus. Die mobile Freisprecheinrichtung ist mit Bluetooth-Technologie, NFC und USB ausgestattet. Dank gutem Rundumklang sowie Echo- und Rauschunterdrückung können Nutzer in einem Umkreis von 3,6 Metern problemlos miteinander kommunizieren. Für den mobilen Arbeitsplatz wird diese preisgünstige Lösung über Wi-Fi bzw. mit Akku betrieben.

Statement by the jury
The ConferenceCam Connect proves to be functionally impressive and an uncomplicated video conference solution for groups as well as for mobile use.

Begründung der Jury
Die ConferenceCam Connect erweist sich als funktional beeindruckende, zudem unkomplizierte Videokonferenz-Lösung für Gruppen und im mobilen Einsatz.

Highfive
Video Conference System
Videokonferenzsystem

Manufacturer
Highfive, Redwood City, USA
Design
Whipsaw Inc, San Jose, USA
Web
https://highfive.com
www.whipsaw.com

Highfive is an all-in-one video conferencing device and cloud service. The technical unit includes a camera in a cubic housing, a microphone in the side wings as well as an integrated computer processor. The device rests securely on top of a flat screen monitor. Up to eight people can join a Highfive video call from the conference room, their computer or a mobile device. When used with a conference room device, every external participant sees the room completely in HD and can hear those present in the room from up to nine metres away from the microphones in the device.

Highfive ist ein All-in-One-Gerät für Videokonferenzen mit Cloud-Service. Die technische Vorrichtung umfasst eine Kamera in einem kubischen Gehäuse, ein Mikrofon in den Seitenarmen sowie einen integrierten Computerprozessor. Die Einheit wird an der Oberseite eines Bildschirms sicher fixiert. Bis zu acht Personen können an einem Gespräch im Konferenzraum via Computer oder mobilem Gerät teilnehmen. Jeder externe Teilnehmer sieht den Raum vollständig in HD und hört Gesprächsbeiträge, die aus bis zu neun Metern Distanz vom Mikrofon erfolgen.

Statement by the jury
With Highfive and its well-conceived cloud service every company monitor screen can be adapted for professional video conferences – a functional as well as an innovative solution.

Begründung der Jury
Mit Highfive und seinem durchdachten Cloud-Service wird jeder Bildschirm für die professionelle Videokonferenz nutzbar – eine funktionale wie innovative Lösung.

Jabra Stealth
Headset

Manufacturer
GN Netcom A/S, Jabra, Ballerup, Denmark
Design
Johan Birger, Höllviken, Sweden
Web
www.jabra.com
www.johanbirger.com

The Bluetooth headset Jabra Stealth with in-ear-design is very small, measuring only 6.5 × 1.5 × 2.4 cm and with a weight of 7.9 grams so light that it sits perfectly balanced in the ear. You get Noise Blackout with two microphones and digital signal processors which identify and filter out disturbing background noises, conversely enhancing voice for calls to produce HD voice quality. Via the voice control button, users can activate Siri and Google Now with a simple click.

Statement by the jury
The Jabra Stealth facilitates well interference-suppressed communication in public places and is comfortable to carry. It thus scores points for functionality and well-conceived ergonomics.

Das Bluetooth-Headset Jabra Stealth im In-Ear-Design ist mit nur 6,5 x 1,5 x 2,4 cm sehr klein und mit 7,9 Gramm so leicht, dass es gut ausbalanciert am Ohr sitzt. Noise Blackout mit zwei Mikrofonen und digitale Signalprozessoren erkennen und filtern störende Hintergrundgeräusche und verstärken im Gegenzug die Stimme für Gespräche in HD-Voice-Qualität. Über die Sprachsteuerungstaste aktivieren Nutzer Siri und Google Now ganz einfach per Tastendruck.

Begründung der Jury
Jabra Stealth ermöglicht gut entstörte Kommunikation im öffentlichen Raum bei hohem Tragekomfort und überzeugt so mit Funktionalität und durchdachter Ergonomie.

Jabra Storm
Headset

Manufacturer
GN Netcom A/S, Jabra, Ballerup, Denmark
Design
Johan Birger, Höllviken, Sweden
Web
www.jabra.com
www.johanbirger.com

The Behind-the-Ear Bluetooth-headset with its battery installed directly in the microphone housing and a high degree of wearing comfort offers up to ten hours talk time and ten days in standby. Wind-Noise-Blackout technology assures clear talk in HD voice quality. A multifunctional button controls volume; announcements inform on connection and battery status. Up to eight mobile end devices can be connected and two connections held simultaneously.

Statement by the jury
Well-conceived user-friendliness and a high level of wearing comfort make the headset a valuable tool for everyone who makes extensive use of mobile telecommunication.

Das Behind-the-Ear-Bluetooth-Headset mit direkt im Lautsprechergehäuse verbautem Akku und hohem Tragekomfort bietet bis zu zehn Stunden Gesprächs- und zehn Tage Stand-by-Zeit. Die Wind-Noise-Blackout-Technologie sorgt für klare Gespräche in HD-Voice-Qualität. Eine Multifunktionstaste regelt die Lautstärke, Sprachansagen informieren zum Verbindungs- und Batteriestatus. Es können bis zu acht mobile Endgeräte gekoppelt und zwei Verbindungen gleichzeitig gehalten werden.

Begründung der Jury
Durchdachte Anwenderfreundlichkeit und hoher Tragekomfort machen das Headset zu einem wertvollen Tool für alle, die viel mobil kommunizieren.

EncorePro HW510/520
Contact Center Headset
Callcenter-Headset

Manufacturer
Plantronics, Inc., Santa Cruz, USA
In-house design
Web
www.plantronics.com

The EncorePro HW510/520 cabled headset was conceived for professional use. Thanks to the concentration on relevant components and the use of high-quality materials, it is light in weight. The ergonomic headband is adjustable on six axes and, together with the upholstered ear cushions, ensures comfortable wearing for hours. An intuitive optical as well as clearly audible feedback system with good haptic qualities supports the optimal microphone positioning.

Statement by the jury
Thanks to its light weight, ergonomic construction and intuitive operability, EncorePro HW510/520 proves to be a functionally and formally efficient work tool.

Das schnurgebundene Headset EncorePro HW510/520 ist für den professionellen Einsatz konzipiert. Dank der Konzentration auf relevante Komponenten und Verarbeitung hochwertiger Materialien hat es nur ein geringes Gewicht. Der ergonomische, über sechs Achsen verstellbare Kopfbügel sowie die Polsterung der Ohrkissen sorgen für stundenlangen bequemen Sitz. Ein intuitives optisches, haptisches wie auch hörbares Feedbacksystem unterstützt die optimale Mikrofonpositionierung.

Begründung der Jury
EncorePro HW510/520 erweist sich dank geringen Gewichts, ergonomischer Bauweise und intuitiver Bedienbarkeit als funktional wie formal effizientes Arbeitsinstrument.

Tone Infinim
Headset

Manufacturer
LG Electronics Inc., Seoul, South Korea
In-house design
Jonghak Lee, Sangmin Park
Web
www.lg.com

The Tone Infinim is an ergonomically formed headset which is worn behind the neck. The integrated earbuds can be retracted and stored together with the cable in the headset interior. The element is connected to the mobile phone via Bluetooth and thus offers a number of additional services such as playback of text messages or a name alert. The practical jog-switch function facilitates simple and intuitive control.

Das Tone Infinim ist ein ergonomisch geformtes Bügel-Headset, das um den Nacken gelegt wird. Die integrierten In-Ear-Kopfhörer können eingezogen und samt Kabel im Inneren des Headsets verstaut werden. Über Bluetooth ist das Element mit dem Mobiltelefon verbunden und bietet so zahlreiche Zusatzservices wie die Wiedergabe von Textnachrichten oder einen Name-Alert. Die praktische Jog-Switch-Funktion ermöglicht eine einfache und intuitive Bedienung.

Statement by the jury
Tone Infinim assures the wearer full movement of the head. It also scores points through its acoustic qualities.

Begründung der Jury
Tone Infinim sichert dem Träger volle Bewegungsfreiheit des Kopfes. Zudem punktet es durch seine akustischen Qualitäten.

Ceecoach
Communication System
Kommunikationssystem

Manufacturer
peiker acustic GmbH & Co. KG,
Friedrichsdorf, Germany
Design
Stroschein Design
(Prof. Sebastian Stroschein), Berlin, Germany
Web
www.ceecoach.de
www.stroschein.de

With Ceecoach, up to six participants can communicate by means of Bluetooth technology over a distance of up to 600 metres without depending on smartphones or mobile radio networks. The device can be controlled simply and intuitively without even looking at it. Due to its wave-shaped design it can be pleasantly held and easily operated even when wearing bulky gloves.

Statement by the jury
Simple handling and functionality make Ceecoach a suitable communication system for use in the open air when it is necessary to concentrate on events nearby.

Mit Ceecoach können bis zu sechs Teilnehmer über eine Distanz von maximal 600 Metern via Bluetooth-Technologie kommunizieren, ohne dabei von Smartphones oder Mobilfunknetzen abhängig zu sein. Selbst ohne den Blick auf das Gerät zu richten, kann man es einfach und intuitiv bedienen. Durch die geschwungene Formgebung kann es zuverlässig gehalten und auch mit derberen Handschuhen gut genutzt werden.

Begründung der Jury
Einfache Handhabung und Funktionalität machen Ceecoach zum geeigneten Kommunikationssystem, wenn man im Freien auf das Geschehen um sich herum konzentriert bleiben muss.

MOTOTRBO SL300
Two-Way Radio
Funkgerät

Manufacturer
Motorola Solutions, Schaumburg, USA
In-house design
Web
www.motorolasolutions.com

The two-way radio MOTOTRBO SL300 weighs about half of what a standard device does. Its digital display is activated as soon as the device is used and deactivates automatically as soon as it is put aside. The channel is selected not with a button but with a rocker-switch, which ensures simple and intuitive one-handed use. Due to its stub antenna, rounded edges and robust construction it fits in any pocket.

Statement by the jury
An intuitively operable rocker-switch, light weight and compactness make the two-way radio a functional and ergonomic device.

Das Funkgerät MOTOTRBO SL300 wiegt nur etwa halb so viel wie ein herkömmliches Funkgerät. Sein digitales Display wird aktiviert, sobald das Gerät benutzt wird, anschließend deaktiviert es sich selbsttätig wieder. Der Kanal wird nicht mittels Knopf, sondern durch einen Wippschalter angewählt und ist damit einfach und intuitiv im Einhandbetrieb zu nutzen. Dank einer stumpfen Antenne, gerundeter Kanten und der robusten Konstruktion findet es in jeder Tasche Platz.

Begründung der Jury
Das Funkgerät beweist seine funktionalen und ergonomischen Qualitäten durch einen intuitiv bedienbaren Wippschalter, sein geringes Gewicht und seine Kompaktheit.

Kelio visio X7
Time Management System
Zeitmanagementsystem

Manufacturer
Bodet software, Cholet, France
In-house design
Web
www.bodet-software.com

Kelio visio X7 is an efficient time management system for companies. Even for firms and institutions with a very high personnel count, attendance times, external appointments, breaks and holiday times can be determined at the press of a button and analysed in bundled form. In its plain design in a reserved anthracite colour, the device integrates well in any environment; the screen background can be selected individually by the employees.

Statement by the jury
The Kelio visio X7 time management System impresses through its functionality and convincing self-explanatory features.

Kelio visio X7 ist ein effizientes Zeitmanagementsystem für den Officebereich. Anwesenheiten, auswärtige Termine, Pausen- und Urlaubszeiten sind für Institutionen auch mit sehr hohem Personalaufkommen auf Knopfdruck einlesbar und können gebündelt analysiert werden. In seiner schlichten Gestaltung in dezentem Anthrazit passt sich das Gerät gut in jede Umgebung ein, der Bildschirmhintergrund kann von den Mitarbeitern individuell gewählt werden.

Begründung der Jury
Das Zeitmanagementsystem Kelio visio X7 besticht mit seiner Funktionalität und den überzeugenden Selbsterklärungsqualitäten.

EF Pro
LTE Smart Communication Device
LTE-Mobilfunkgerät

Manufacturer
Bluebird Inc., Seoul, South Korea
In-house design
Bluebird Design Group, Seoul, South Korea
Web
www.mypidion.com

EF Pro is a stylish LTE smart communication device, for instance for restaurants, department stores or hotels of the up-market class. The volume, a push-to-talk function and an emergency call button are controlled by a graphic user interface. To provide high performance, antenna and speakers are located in the top section of the device. Integrated NFC and RFID chips facilitate easy connectivity with other devices.

Statement by the jury
An elegant impression, high functionality and good connectivity with the product peripheral equipment make the EF Pro a suitable communication device for the premium business class.

EF Pro ist ein stilvolles LTE-Mobilfunkgerät etwa für Restaurants, Kaufhäuser oder Hotels des gehobenen Standards. Über eine grafische Benutzeroberfläche werden die Lautstärke, eine Push-to-talk-Funktion sowie eine Notrufmöglichkeit gesteuert. Zugunsten einer hohen Leistung sind Antenne und Lautsprecher im oberen Bereich des Geräts untergebracht. Integrierte NFC- und RFID-Chips ermöglichen eine einfache Verbindung mit anderen Geräten.

Begründung der Jury
Eine elegante Anmutung, hohe Funktionalität und gute Einbindung in die Produktperipherie machen das EF Pro zum geeigneten Mobilfunkgerät für den Premium-Business-bereich.

NCR Orderman7
Mobile POS Solution
Mobile POS-Lösung

Manufacturer
Orderman GmbH, NCR Corporation, Salzburg, Austria
Design
tomasini formung (Bernd Tomasini), Salzburg, Austria
Web
www.orderman.com
www.ncr.com
www.tomasini.com

The mobile POS solution NCR Orderman7 has been designed for use in restaurants. The device is operated via multi-touch display, usable with the fingers or pen as well as via programmable hardware keys. It features various radio standards, is watertight, rugged and has a battery life of up to 18 hours. The battery is quickly changed at the press of a button. The 5" display, which is readable in sunlight, is located in an ergonomic metal and plastic casing and can be controlled with one hand.

Statement by the jury
Service staff operate the NCR Orderman7 as they are used to from smartphones, intuitively and faultlessly, thanks to a functional and ergonomically consistent design.

Die mobile POS-Lösung NCR Orderman7 ist für den Einsatz in der Gastronomie konzipiert. Das Gerät wird via Multitouch-Display mit Fingern oder Stift sowie über programmierbare Hardware-Tasten bedient. Es beherrscht verschiedene Funkstandards, ist wasserdicht, bruchfest und läuft bis zu 18 Stunden. Der Akku lässt sich rasch per Knopfdruck wechseln. Das sonnentaugliche 5"-Display sitzt in einem ergonomischen, mit einer Hand bedienbaren Metall-Kunst-stoff-Gehäuse.

Begründung der Jury
Servicekräfte bedienen den NCR Orderman7 wie vom Smartphone gewohnt intuitiv und fehlerfrei – dank funktional wie ergonomisch schlüssiger Gestaltung.

LEX L10
Communication Device for Task Forces
Kommunikationsmittel für Einsatzkräfte

Manufacturer
Motorola Solutions, Schaumburg, USA
In-house design
Web
www.motorolasolutions.com

The LEX L10 is a mission-critical communication device for first responders that runs the Security Enhanced Android OS as a decisive asset. It can communicate in covert mode with radios via Bluetooth. A non-slip surface allows one-handed operation. The device incorporates two forward-directional speakers, a tri-microphone with noise suppressor and echo cancellation as well as a push-to-talk button. It is watertight, conforming to IP67, and withstands drops from up to 1.21 metres.

Statement by the jury
The LEX L10 proves to be a reliable communication device for security forces and gains merit by good integration in the product periphery and a high level of functionality.

Das LEX L10 ist ein missionskritisches Kommunikationsmittel für Helfer vor Ort, das über Security Enhanced Android OS verfügt und einen entscheidenden Beitrag zur sicheren Kommunikation leistet. Es kann im Verdecktmodus über Bluetooth mit Funkgeräten kommunizieren. Eine rutschfeste Oberfläche erlaubt einhändige Bedienung. Das Gerät weist zwei nach vorn gerichtete Lautsprecher, ein Dreifachmikrofon mit Geräusch- und Echounterdrückung sowie einen Push-to-talk-Bedienknopf auf. Zudem ist es wasserdicht nach IP67 und hält Stürzen aus bis zu 1,21 Metern Höhe stand.

Begründung der Jury
Das LEX L10 erweist sich als zuverlässiges Kommunikationsmittel für Einsatzkräfte und überzeugt durch gute Integration in die Produktperipherie und hohe Funktionalität.

VoyagerECK
Portable Secure Communications System
Tragbares sicheres Kommunikationssystem

Manufacturer
Klas Telecom, Dublin, Ireland
In-house design
Frank Murray, Mark Ryan
Design
Dolmen (Christopher Murphy, James Ryan), Dublin, Ireland
Web
www.klastelecom.com
www.dolmen.ie

VoyagerECK facilitates worldwide fast access to secure networks. The two modular units, handset and encryption device, are particularly small due to innovative designs and modern materials which can be densely packed. An outer shell of ultra-thin carbon fibre is extremely impact-resistant and space-saving. Heat is dissipated via the heat sink made of high-precision engineered aluminium. Removable caps at each end guarantee simple access to the communication equipment.

VoyagerECK ermöglicht weltweit schnellen Zugriff auf sichere Netzwerke. Die zwei modularen Einheiten, Mobilteil und Verschlüsselungsgerät, sind dank innovativen Designs und moderner Materialien, die dicht bestückt werden können, besonders klein. Eine Außenhülle aus ultradünner Karbonfaser, die extrem hohen Belastungen standhält, spart zusätzlich Platz. Hitze wird durch den Kühlkörper aus hoch präzisem Aluminium abgeführt. Abnehmbare Deckel an beiden Enden erlauben den leichten Zugriff auf das Kommunikationsequipment.

Statement by the jury
First responders and other institutions in crisis zones must be able to rely on communication equipment – the highly functional VoyagerECK fulfils this demand convincingly.

Begründung der Jury
Helfer vor Ort und andere Institutionen in Krisengebieten müssen sich auf ihr Kommunikationsmittel verlassen können – das hoch funktionale VoyagerECK erfüllt diesen Anspruch überzeugend.

Philips Voice Tracer DVT2500
Digital Dictation Device
Digitales Diktiergerät

Manufacturer
Speech Processing Solutions GmbH,
Vienna, Austria
In-house design
Web
www.speech.com

Simple form, precise contours and rounded surfaces ensure ergonomic handling of the dictation device. The compact body is divided into convex surfaces which emphasise the functional areas such as microphone, display and buttons. The dark, matte case, which contrasts with the metallic surface, creates a high-quality appearance. A light sensor regulates the background lighting of the display in order to minimise power consumption.

Statement by the jury
The Philips Voice Tracer DVT2500 lies pleasantly in the hand. Its functions are simply and understandably arranged – it also optically upgrades every workplace.

Einfache Formgebung, präzise Konturen und gerundete Oberflächen gewährleisten eine ergonomische Handhabung des Diktiergerätes. Der kompakte Körper ist in konvexe Oberflächen unterteilt, die Funktionsbereiche wie Mikrofon, Display und Tasten betonen. Das mit den metallischen Oberflächen kontrastierende dunkle, matte Gehäuse sorgt für ein hochwertiges Aussehen. Ein Lichtsensor regelt die Hintergrundbeleuchtung des Displays, um den Stromverbrauch zu minimieren.

Begründung der Jury
Der Philips Voice Tracer DVT2500 liegt gut in der Hand, seine Funktionsbereiche sind einfach und verständlich angeordnet – zudem wertet es jeden Arbeitsplatz optisch auf.

HERO
Tracking System
Ortungssystem

Manufacturer
HERO GmbH, Düsseldorf, Germany
In-house design
Web
www.gps-hero.de

The tracking system operates using a conventional SIM card. In an emergency the clearly visible and easily felt SOS button connects automatically with up to three assigned numbers. The device is equipped with a fall sensor as well as a geo-fence system which signals departure from a predefined zone. Control is enabled via an app or web portal. The device is ergonomically simple to operate and is available in three friendly colours as well as discreet black.

Statement by the jury
Elderly persons and persons in need of help as well as children who travel unaccompanied involve a great deal of responsibility – HERO provides efficient and functional support.

Das Ortungssystem wird mit einer handelsüblichen SIM-Karte betrieben. Im Notfall verbindet der deutlich sicht- und tastbare SOS-Knopf automatisch mit bis zu drei Nummern. Weiter ist das Gerät mit einem Sturzsensor ausgestattet sowie mit einem Geo-Fence-System, welches das Verlassen vordefinierter Zonen meldet. Die Steuerung erfolgt via App bzw. Webportal. Das ergonomisch einfach bedienbare Gerät ist in drei freundlichen Farben sowie in dezentem Schwarz erhältlich.

Begründung der Jury
Ältere und hilfsbedürftige Personen sowie Kinder, die sich ohne Begleitung bewegen, stellen eine große Herausforderung für die Verantwortlichen dar – Hero bietet hier effiziente und funktionale Unterstützung.

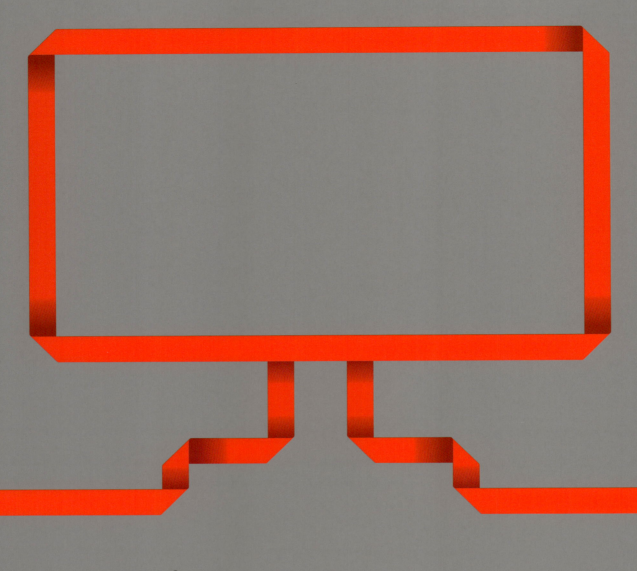

Computers
and information
technology
Computer und
Informationstechnik

Dell UltraSharp 34 Monitor

Manufacturer
Dell Inc., Round Rock, Texas, USA

In-house design
Experience Design Group

Web
www.dell.com

reddot award 2015
best of the best

Enhanced panorama

The computer monitor is the most important device for the visualisation of data and information, in front of which we spend a lot of time every day. Since home offices and living areas are becoming more and more alike, the monitor is a substantial element of the room interior. The Dell UltraSharp 34 Monitor impresses with an elegant new design that becomes an eye-catcher in any room. The innovative, curved 34" monitor stands out with a wide panoramic field of view, offering users a fascinating viewing experience. In addition, it features an excellent WQHD resolution of 3440 x 2440 pixels, an aspect ratio of 21:9 and HDMI 2.0. The size and curvature design of the screen are specifically selected based on human factors research and investigations to ensure that the screen covers both the primary field-of-view as well as the secondary peripheral vision and thus achieves ideal visual ergonomic performance. With its aesthetics and excellent image quality, the monitor fascinates private users. It is however just as interesting for dedicated gamers and commercial users who no longer have to use two monitors, because it creates an expanded, more immersive viewing environment with its panoramic view. Thus, the Dell UltraSharp 34 Monitor combines a path-breaking new design with user-friendly versatility – its ergonomics changes viewing habits.

Erweitertes Panorama

Der Computermonitor ist das wichtigste Darstellungsmedium für Daten und Informationen, vor dem wir tagtäglich viel Zeit verbringen. Da sich zudem das Home-Office und der Wohnbereich immer mehr einander angleichen, ist der Monitor auch ein maßgeblicher Bestandteil des Ambiente. Der Dell UltraSharp 34 Monitor beeindruckt durch eine elegante neue Formensprache, mit der er im Raum die Blicke auf sich zieht. Der innovative, gewölbte 34"-Monitor besticht durch ein weites Panorama-Sichtfeld, welches dem Nutzer ein faszinierendes Erlebnis bietet. Er verfügt darüber hinaus über eine ausgezeichnete WQHD-Auflösung von 3440 x 2440 Pixel, ein Seitenverhältnis von 21:9 und HDMI 2.0. Bei der Gestaltung dieses Monitors wurden das Format und die Wölbung des Bildschirms auf der Basis ergonomischer Forschungen und Untersuchungen so gewählt, dass er sowohl das unmittelbare als auch das periphere Blickfeld abdeckt und deshalb eine optimale visuelle Darstellung ermöglicht. Der Monitor begeistert durch seine Ästhetik und exzellente Bildqualität den privaten Anwender. Er ist jedoch ebenso interessant für anspruchsvolle Gamer und auch kommerzielle Anwender, die bislang zwei Bildschirme für ihre Arbeit benötigten, da er durch sein Panorama-Sichtfeld eine erweiterte, umfassende Bildumgebung schafft. Der Dell UltraSharp 34 Monitor verbindet so eine wegweisend neue Formensprache mit einer nutzerfreundlichen Vielseitigkeit – seine Ergonomie verändert auch die visuellen Gewohnheiten.

Statement by the jury

The Dell UltraSharp 34 Monitor presents an extremely successful integration of form and function. With its innovative curvature, the purist 34" monitor offers an ultra-wide panoramic view and thus optimises visual ergonomics. The computer monitor is hereby redefined in an exciting way. It is user-friendly and opens up new perspectives for both private and professional users. A further impressive feature is the elegant combination of materials and colours.

Begründung der Jury

Eine überaus gelungene Integration von Form und Funktion stellt der Dell UltraSharp 34 Monitor dar. Der puristisch gestaltete 34"-Monitor bietet durch seine innovative Wölbung des Bildschirms ein Panorama-Sichtfeld und damit eine optimierte visuelle Ergonomie. Der Computermonitor wird hier auf spannende Weise neu definiert. Er ist dabei sehr nutzerfreundlich und eröffnet privaten wie auch professionellen Anwendern neue Perspektiven. Beeindruckend ist zudem die stilvolle Kombination der verwendeten Materialien und Farben.

Designer portrait
See page 72
Siehe Seite 72

RL Series
Monitor

Manufacturer
BenQ Corporation, Taipei, Taiwan
In-house design
Web
www.benq.com

The RL series consists of professional monitors for console players, featuring a very short response time so that games are displayed smoothly and without blur. With a hot key beneath the display, the gamer can comfortably switch between inputs. The monitor stand is designed at an angle and, with the triangular base covered in non-slip TPU coating, offers convenient storage for console or keyboard.

Die RL-Serie umfasst professionelle Monitore für Konsolenspieler. Ihre Reaktionszeit ist äußerst gering, sodass Spiele flüssig und ohne Schlieren dargestellt werden. Mit einer Schnelltaste unterhalb des Displays kann der Gamer bequem zwischen den Eingängen wechseln. Der Ständer des Monitors ist gewinkelt und bietet mit dem dreieckigen Fuß mit rutschfester TPU-Beschichtung eine praktische Ablage für Konsole oder Tastatur.

Statement by the jury
This monitor series is consistently geared to the needs of console players and offers, in addition to optimised image quality, useful equipment details.

Begründung der Jury
Die Monitorserie ist konsequent auf die Bedürfnisse von Konsolenspielern ausgelegt und bietet neben einer optimierten Bildqualität nützliche Ausstattungsmerkmale.

SW Series
Monitor

Manufacturer
BenQ Corporation, Taipei, Taiwan
In-house design
Web
www.benq.com

The SW monitor series works with the Adobe RGB colour space for natural colour reproduction on the screen. The high QHD resolution facilitates image processing, even strongly magnified image details are displayed with high picture sharpness. The monitor offers a large number of placement and positioning options: the user can adjust height, pivot, tilt and vertical rotation according to his or her personal needs. A position marker memorises the preferred settings.

Statement by the jury
The SW series is ideal for photographers, as the monitors display high image quality and leave nothing to be desired with regard to positioning.

Die Monitorserie SW arbeitet mit dem Adobe-RGB-Farbraum und stellt Farben auf dem Bildschirm naturgetreu dar. Die hohe QHD-Auflösung erleichtert die Bildbearbeitung, da selbst stark vergrößerte Bildausschnitte scharf dargestellt werden. Die Möglichkeiten, den Monitor zu verstellen, sind vielfältig: Höhe, Drehpunkt, Neigung und vertikale Ausrichtung kann der Nutzer an seine Bedürfnisse anpassen. Ein Positionsmarker merkt sich die gewünschte Einstellung.

Begründung der Jury
Die SW-Serie ist für Fotografen ideal, denn die Monitore stellen Aufnahmen in hoher Qualität dar und lassen bei der Positionierung keine Wünsche offen.

EW Series
Monitor

Manufacturer
BenQ Corporation, Taipei, Taiwan
In-house design
Web
www.benq.com

The frame of the monitors from the EW series is only 0.8 mm thin. When two monitors are placed next to each other in a dual-monitor configuration, the two screens appear to seamlessly flow together as one. Work space is thus doubled, and no irritating side edges impair the user experience. With their slim frame, the monitors also render a more slender overall impression and fit better on the desktop.

Statement by the jury
As individual screens, the monitors have a pleasantly slim appearance, and in a dual-monitor configuration they enable a comfortable working situation.

Monitore der EW-Serie haben einen nur 0,8 mm dünnen Rand. Werden zwei Monitore im Rahmen einer Dual-Monitor-Konfiguration nebeneinandergestellt, ergibt sich ein Bild, das nahtlos ineinander überzugehen scheint. Der Arbeitsbereich wird dadurch verdoppelt, ohne dass störende Ränder das Nutzererlebnis beeinträchtigen. Mit ihren dünnen Rahmen sind die Monitore auch insgesamt schlanker und finden leicht Platz auf einem Schreibtisch.

Begründung der Jury
Als Einzelbildschirme wirken die Monitore angenehm schlank, in einer Dual-Monitor-Konfiguration ermöglichen sie bequemes Arbeiten.

ThinkVision X24
Monitor

Manufacturer
Lenovo, Morrisville, North Carolina, USA
In-house design
Web
www.lenovo.com

This monitor with a 23.8" IPS display is merely 7.5 mm thick and thus one of the slimmest devices of its kind. The active viewing area nearly meets the frame, which is only 6 mm wide. The screen rests on a stand with a minimalist, elegant design. It consists of a chrome-plated stem leading into a circular black base. The simple red cable clip attached to the shaft creates a vibrant colour accent.

Statement by the jury
The ThinkVision X24 monitor has a characteristic stand deriving its appeal from the elegant shaft and colour-accentuated round base.

Dieser Monitor mit einem 23,8"-IPS-Display ist nur 7,5 mm dick und damit eines der dünnsten Geräte dieser Art. Der aktive Sichtbereich erstreckt sich fast bis an den lediglich 6 mm breiten Rahmen. Der Bildschirm ruht auf einem minimalistischen, eleganten Ständer. Dieser besteht aus einem verchromten Schaft, der in eine kreisförmige schwarze Basis mündet. Der einfache, am Schaft angebrachte rote Kabelhalter sorgt für einen farblichen Akzent.

Begründung der Jury
Der Monitor ThinkVision X24 besitzt einen charakteristischen Ständer, der seinen Reiz aus dem eleganten Schaft und dem farblich abgesetzten runden Fuß bezieht.

ThinkVision Pro2840m
Monitor

Manufacturer
Lenovo, Morrisville, North Carolina, USA
In-house design
Web
www.lenovo.com

The ThinkVision Pro2840m is a 28" 4K monitor for professional users that places emphasis on colour accuracy and high resolution. Its reduced design with a slim bezel directs the focus of attention exclusively to the screen. The touch controls below the screen are only illuminated when activated to avoid irritating effects during work. The monitor stand is adjustable to any desired position and also provides intelligent cable management.

Statement by the jury
The monitor impresses with its minimalist design and elegant base, offering a wide variety of adjustment options.

Der ThinkVision Pro2840m ist ein 28"-4K-Monitor für professionelle Anwender, die Wert auf Farbtreue und eine hohe Auflösung legen. Seine reduzierte Gestaltung mit einem dünnen Rahmen lenkt die Aufmerksamkeit ganz auf den Bildschirm. Die Touch-Bedientasten unterhalb des Bildschirms leuchten nur bei Gebrauch, damit sie bei der Arbeit nicht stören. Der Ständer erlaubt das Ausrichten des Monitors in jede Position und ein intelligentes Kabelmanagement.

Begründung der Jury
Der Monitor beeindruckt mit seiner minimalistischen Gestaltung und der elegant wirkenden Halterung, die zahlreiche Einstellmöglichkeiten bietet.

34UC97
Curved Monitor

Manufacturer
LG Electronics Inc., Seoul, South Korea
In-house design
Gangho Woo, Jaeneung Jung
Web
www.lg.com

This 34" monitor in 21:9 UltraWide format creates an intensive viewing experience for the user thanks to its curved screen. The design deliberately does without any decorative elements so as to engender minimalist aesthetics. The metal stand design is likewise slim and reserved, emphasising the lightness of the monitor. The curved rear in a champagne hue gives it an elegant look.

Statement by the jury
With its graceful and straightforward design, the monitor blends well into any interior. The viewer's attention stays completely focused on the screen.

Der 34"-Monitor im 21:9-UltraWide-Format verschafft dem Anwender durch seine gekrümmte Bildschirmform ein intensives Erlebnis. Um eine minimalistische Ästhetik zu erzielen, wurde auf schmückende Elemente verzichtet. Auch der Metallständer ist schmal und zurückhaltend gestaltet und unterstreicht dadurch die Leichtigkeit des Monitors. Der gewölbte Rücken in Champagner-Farbe verleiht ihm eine elegante Anmutung.

Begründung der Jury
Da der Monitor anmutig und geradlinig gestaltet ist, fügt er sich in jedes Interieur ein. Die Aufmerksamkeit des Betrachters bleibt dabei ganz auf dem Bildschirm.

Blackmagic SmartView 4K
Ultra HD Broadcast Monitor

Manufacturer
Blackmagic Design Pty Ltd, Melbourne, Australia
In-house design
Web
www.blackmagicdesign.com

The SmartView 4K is a professional broadcast monitor designed for different video production environments. It features 12 Gbit/s SDI connections for the high-speed decoding of 4K signals. The Ultra HD display is enhanced by an intuitive backlit control panel facilitating expedient access to technical data like colour balance and image noise. The precision CNC-machined aluminium front panel offers the rigidity required for rack mounting and secure anchoring of the robust interior chassis. The monitor can be flexibly installed on walls, desks or articulated arms.

Der SmartView 4K ist ein für verschiedene Videoproduktionsumgebungen konzipierter Profi-Broadcastmonitor. Er besitzt 12-Gbit/s-fähige SDI-Anschlüsse für die Hochgeschwindigkeitsentschlüsselung von 4K-Signalen. Das Display in Ultra-HD-Auflösung wird durch eine intuitive Tastatur mit Hintergrundbeleuchtung abgerundet und gewährt den schnellen Zugriff auf technische Daten wie Farbbalance und Bildrauschen. Die CNC-präzisionsgefräste Aluminiumfrontblende bietet die für den Rackeinbau nötige Formfestigkeit und verankert das robuste Innenchassis. Der Monitor kann flexibel an Wänden, auf Tischen oder an Gelenkarm-Halterungen installiert werden.

Statement by the jury
The SmartView 4K offers a well-conceived design for professional users and can be implemented efficiently even under difficult conditions.

Begründung der Jury
Der SmartView 4K bietet eine durchdachte Gestaltung für professionelle Nutzer und lässt sich auch unter schwierigen Bedingungen effizient einsetzen.

Revo One RL85
Personal Computer

Manufacturer
Acer Incorporated, New Taipei, Taiwan
In-house design
Web
www.acer.com

The Revo One RL85 is a compact PC for the home environment with a large number of entertainment functions. As such, it can serve as external hard drive, network-attached storage or multi-media device. The design was kept as minimalist and compact as possible so that the PC may be unobtrusively placed anywhere. With its rounded shape, it also renders a friendly, untechnical impression. The indicator lights were designed to avoid irritating effects in dim lighting conditions.

Der Revo One RL85 ist ein kompakter PC für den Heimbereich mit zahlreichen Unterhaltungsfunktionen. So dient er beispielsweise als externe Festplatte, Netzwerkspeicher oder Multimediagerät. Er wurde so minimalistisch und kompakt wie möglich gestaltet, damit er überall unauffällig aufgestellt werden kann. Mit seiner rundlichen Form wirkt er zudem freundlich und wenig technisch. Die Leuchtanzeige wurde so angebracht, dass sie auch bei Dunkelheit nicht stört.

Statement by the jury
The Revo One has a very congenial appearance, requires only little space and offers comprehensive functionality, thus making it ideally suited to home settings.

Begründung der Jury
Der Revo One wirkt sehr sympathisch, benötigt wenig Platz und bietet eine umfassende Funktionalität. Damit eignet er sich hervorragend für den Heimbereich.

Kano
Computer Kit
Computer-Bausatz

Manufacturer
Kano, London, Great Britain
Design
Kano and Map, London, Great Britain
Web
www.kano.me
www.mapprojectoffice.com

Kano is an inexpensive and open computer construction kit. Children as young as six have used it to build computers, speakers, games, radios and Minecraft worlds. The kit was developed with the input of young people, artists and teachers from all over the world, so that children of all ages can discover technology for themselves. The packaging is recyclable and fits in most mailboxes, thus reducing shipping costs.

Kano ist ein preiswerter und offener Computer-Bausatz. Kinder ab sechs Jahren haben damit bereits Computer, Lautsprecher, Spiele, Minecraft-Welten und Radios zusammengebaut. Entwickelt wurde der Bausatz mit jungen Leuten, Künstlern und Lehrern aus der ganzen Welt, damit Kinder jeden Alters Technologie für sich entdecken können. Die Verpackung ist wiederverwendbar und passt in die meisten Briefkästen. Dadurch werden Transportkosten reduziert.

Statement by the jury
Kano promotes creativity, teaches children new technologies in a playful way, and is produced in an environmentally friendly manner.

Begründung der Jury
Kano fördert die Kreativität, bringt Kindern auf spielerische Weise neue Technologien näher und ist umweltbewusst produziert.

Core V51
Midi Tower Chassis
Midi-Tower-Gehäuse

Manufacturer
Thermaltake Technology Co., Ltd., Taipei, Taiwan
In-house design
Web
www.thermaltakecorp.com

The Core V51 midi tower chassis is designed to house the latest E-ATX, ATX and Micro ATX motherboards. Its large interior offers ample space for water-cooling systems and ventilators. Conventional high-end GPU cards can be comfortably accommodated as well. As there are no tools needed for installation, the assembly of components is quickly accomplished. The perforated top panel allows for efficient heat dissipation and guarantees optimum ventilation of the interior.

Das Midi-Tower-Gehäuse Core V51 ist für aktuelle Motherboards mit E-ATX-, ATX- und Micro-ATX-Formfaktor geeignet. Sein geräumiges Inneres bietet viel Platz für Wasserkühlungsanlagen und Lüfter. Auch gängige High-End-GPU-Karten lassen sich bequem unterbringen. Da für die Montage kein Werkzeug benötigt wird, geht der Einbau der Komponenten schnell vonstatten. Die perforierte Oberseite leitet die Wärme effizient ab und sorgt für die optimale Belüftung des Inneren.

Statement by the jury
This chassis offers outstanding features for gamers in particular. Components can be easily added, and heat management is efficient and well conceived.

Begründung der Jury
Das Gehäuse bietet gerade für Spieler hervorragende Eigenschaften. Komponenten lassen sich leicht einbauen, das Wärmemanagement ist effizient und durchdacht.

Silent Base 800
PC Chassis
PC-Gehäuse

Manufacturer
Listan GmbH & Co. KG, be quiet!, Glinde, Germany
Design
Brandis Industrial Design (Michael Brandis), Nuremberg, Germany
Web
www.bequiet.com
www.brandis-design.com

The Silent Base 800 is designed for silent operation and optimal cooling performance. The air ducts in the upper and lower covers precisely control the airflow, providing silent air circulation within the unit. The three fans are decoupled from the casing with rubber elements, and the hard drives are fitted onto silicone rails so as to reduce vibrations. No tools are needed to install the drives in the chassis. Intelligent cable routing and easily accessible air filters contribute to a high degree of user-friendliness.

Das Silent Base 800 ist auf einen leisen Betrieb bei optimaler Kühlung ausgelegt. Zu diesem Zweck sorgen die Luftkanäle des Gehäusedeckels und -bodens dafür, dass die Luft geräuschlos zirkulieren kann. Die drei Lüfter sind durch Gummielemente vom Gehäuse entkoppelt, und auch die Festplatten lagern auf Silikonschienen, um Vibrationen zu reduzieren. Die Laufwerke werden ohne Werkzeug im Gehäuse installiert, eine intelligente Kabelführung und leicht zugängliche Luftfilter tragen zur Benutzerfreundlichkeit bei.

Statement by the jury
This chassis convinces with its efficient airflow management, thus reducing noise emission to a minimum while in operation.

Begründung der Jury
Das Gehäuse überzeugt mit seinem effizienten Luftstrommanagement, das die Geräuschentwicklung im Betrieb auf ein Minimum reduziert.

ThinkStation P900
Personal Computer

Manufacturer
Lenovo, Morrisville, North Carolina, USA
In-house design
Web
www.lenovo.com

The ThinkStation P900 presents an innovative interior layout. Key component access is easy, intuitive and tool-less. The precisely formed steel enclosure protects the control panel and drive bays, and it also includes integral lift handles. Large front intakes and rear exits allow efficient cooling of internal components. The computer epitomises thoughtful design, clever engineering and careful attention to detail.

Statement by the jury
The ThinkStation P900 impresses with its functional design, which offers high flexibility with regard to configuration.

Die ThinkStation P900 ist durch ein innovatives internes Layout gekennzeichnet. Wichtige Komponenten sind ohne Werkzeug einfach und intuitiv zu erreichen. Das präzise ausgeformte Stahlgehäuse schützt die Steuerungseinheit und Laufwerksschächte und beherbergt integrierte Griffe. Große Lufteinlässe an Vorder- und Rückseite gewährleisten eine effektive Kühlung der Komponenten. Der Computer verkörpert ein durchdachtes Design, eine ausgeklügelte Technik und Sorgfalt für Details.

Begründung der Jury
Die ThinkStation P900 beeindruckt durch ihre funktionale Gestaltung, die bei der Konfigurierung hohe Flexibilität bietet.

KLEVV Cras
Computer Memory Module
Computer-Speichermodul

Manufacturer
Essencore Limited, Hong Kong
Design
PlayLab (Seo-Ryong Kwak),
Seoul, South Korea
Web
www.essencore.com
www.kodasdesign.com

The Klevv Cras DDR4 storage modules have a perforated structure that enlarges the overall surface for better heat dissipation. This enhances the performance and operational stability of the modules, which is of particular advantage for high-end gaming PCs. The modules convince with their stylish design, especially with the enclosures featuring transparent side panels: at the upper edge, LEDs give off a deep blue light, and the modules themselves are fully covered by an elongated fine metal band.

Die Klevv Cras DDR4-Speichermodule besitzen eine gelochte Struktur, welche die Gesamtoberfläche vergrößert, damit die Wärme besser abstrahlen kann. Dadurch sind die Module leistungsfähiger und laufen stabiler, was besonders für High-End-Gaming-PCs vorteilhaft ist. In Gehäusen mit durchsichtigen Seitenteilen überzeugen die Bausteine mit ihrem stylischen Design: Am oberen Rand leuchten LEDs in Tiefblau, über die Module selbst zieht sich der Länge nach ein feines Metallband.

Statement by the jury
The storage modules are ideally suited for gamers who want to equip their PC with attractive, high-performance components.

Begründung der Jury
Die Speichermodule eignen sich hervorragend für Gamer, die ihren PC mit attraktiven, leistungsfähigen Komponenten bestücken möchten.

T-Station 27ST67
PC

Manufacturer
LG Electronics Inc., Seoul, South Korea
In-house design
Gangho Woo, Jaewon Lim
Web
www.lg.com

The T-Station combines the functionality of a PC with a 27" television. When the rear cover is detached, PC upgrades are simple to accomplish. All ports and cables are located behind this cover to maintain an overall clean appearance. Positioned between the transparent stand and screen, the speaker can be varied with regard to colour and pattern.

Statement by the jury
Screen, speaker and stand of the T-Station harmoniously blend into a single line, giving rise to a sleek silhouette.

Die T-Station vereint die Funktionalitäten eines PCs und eines 27"-Fernsehers. Wird die Klappe auf der Rückseite abgenommen, können PC-Upgrades einfach vorgenommen werden. Alle Anschlüsse und Kabel befinden sich hinter dieser Klappe, wodurch der aufgeräumte Gesamteindruck erhalten bleibt. Zwischen dem Bildschirm und dem durchsichtigen Ständer befindet sich der Lautsprecher, der in Farbe und Muster variiert werden kann.

Begründung der Jury
Bei der T-Station gehen Bildschirm, Lautsprecher und Ständer in einer Linie harmonisch ineinander über, wodurch der Bildschirm eine schlanke Silhouette zeigt.

Aspire V Nitro Series
Notebook

Manufacturer
Acer Incorporated, New Taipei, Taiwan
In-house design
Web
www.acer.com

The Aspire V Nitro notebook series was designed for gamers and multimedia users. It impresses with its clear line management, defined edges and slim profile. The fine nanostructure on the top cover conveys elegance, offers a pleasant tactile appeal and ensures a better grip. Premium materials and the keyboard backlit in red also underline its high performance.

Statement by the jury
These notebooks show a dynamic design and colour scheme that perfectly harmonise with the high-performance hardware equipment.

Die Notebook-Serie Aspire V Nitro wurde besonders für Gamer und Multimedia-Nutzer entwickelt. Sie besticht durch ihre klare Linienführung, definierte Kanten und das flache Profil. Die feine Nanostruktur auf dem Gehäusedeckel verleiht den Notebooks Eleganz, bietet eine angenehme Haptik und sorgt zudem für einen besseren Grip. Hochwertige Materialien und die Tastatur mit roter Hintergrundbeleuchtung unterstreichen ihre Leistungsfähigkeit.

Begründung der Jury
Diese Notebooks zeigen eine dynamische Gestaltung und Farbgebung, die bestens mit der leistungsfähigen Hardware-Ausstattung harmoniert.

Dell Latitude Education Solution
Notebooks and Mobile Computing Cart
Notebooks und PC-Wagen

Manufacturer
Dell Inc., Round Rock, Texas, USA
In-house design
Experience Design Group
Web
www.dell.com

The notebooks of the Latitude 13 Education series feature a 13.3" HD display with an optional touchscreen. The rubberised display bezel and bottom case absorb shocks and thus enhance drop protection. The matching mobile computing cart is accessible from two sides and offers simple access to as many as 30 computers. The cart, which is secured by lockable steel doors, can also charge the notebooks and provide software updates.

Statement by the jury
The notebooks and the mobile computing cart are coordinated in a user-friendly way and are thus perfectly suited to use in educational facilities.

Die Notebooks der Serie Latitude 13 Education besitzen ein 13,3"-HD-Display mit optionalem Touchscreen. Die gummierte Einfassung von Bildschirm und Unterschale absorbiert Stöße und verbessert so den Fallschutz. Der dazu passende mobile PC-Wagen ist von zwei Seiten zugänglich und bietet einfachen Zugriff auf bis zu 30 Computer. Der Wagen kann die Notebooks laden und Software-Updates aufspielen. Gesichert ist er durch abschließbare Stahltüren.

Begründung der Jury
Notebooks und PC-Wagen sind nutzerfreundlich aufeinander abgestimmt und dadurch bestens für den Einsatz in Bildungseinrichtungen geeignet.

ThinkPad Yoga 14
Convertible Notebook

Manufacturer
Lenovo, Morrisville, North Carolina, USA
In-house design
Web
www.lenovo.com

The ThinkPad Yoga 14 is a four-in-one notebook with 14" full HD display. It weighs merely 1.9 kg and offers a high degree of mobility. The magnesium alloy frame and scratch-resistant display render the device sufficiently robust for professional use. A special feature is the keyboard locking mechanism in tablet mode, giving the user a better grip on the tablet without accidentally pressing a key.

Statement by the jury
With its well-conceived keyboard locking mechanism, the ThinkPad Yoga 14 can be securely held in the hand and operated in tablet mode.

Das ThinkPad Yoga 14 ist ein 4-in-1-Notebook mit 14"-Full-HD-Display. Es wiegt lediglich 1,9 kg und bietet damit eine hohe Mobilität. Der Rahmen aus einer Magnesiumlegierung und das kratzresistente Display machen das Gerät robust genug für die berufliche Anwendung. Als Besonderheit arretiert das Notebook im Tablet-Modus die Tastatur, damit der Anwender das Tablet besser fassen kann, ohne versehentlich eine Taste zu drücken.

Begründung der Jury
Dank der durchdachten Tastenarretierung lässt sich das ThinkPad Yoga 14 im Tablet-Modus sicher in der Hand halten und bedienen.

ThinkPad X1 Carbon
Notebook

Manufacturer
Lenovo, Morrisville, North Carolina, USA
In-house design
Web
www.lenovo.com

The ThinkPad X1 Carbon is very lightweight yet does not compromise durability. The hinges have been enhanced to allow seamless conversion from the typical clamshell to a complete lay-flat orientation. Using carbon fibre on the top cover and light magnesium alloys for the main body, the slim form factor is lightweight and strong. The keyboard and clickpad have been refined to improve productivity and usability.

Statement by the jury
The ThinkPad X1 Carbon impressively shows how elegance, slim design and robustness can go hand in hand.

Das ThinkPad X1 Carbon ist sehr leicht und dennoch widerstandsfähig. Die Scharniere wurden verstärkt, um den nahtlosen Übergang von einer typischen aufgeklappten zu einer vollständig flachen Orientierung zu ermöglichen. Für die Abdeckung wurde Carbon verwendet, für das Gehäuse eine Magnesiumlegierung. Dadurch ist das Gerät bei einer schlanken Form leicht und stabil. Die Tastatur und das Clickpad wurden verbessert, um Produktivität und Benutzerfreundlichkeit zu erhöhen.

Begründung der Jury
Das ThinkPad X1 Carbon beweist eindrucksvoll, wie Eleganz, eine schlanke Form und Robustheit Hand in Hand gehen können.

Curved AIO PC 29V950
Curved All-in-One PC

Manufacturer
LG Electronics Inc., Seoul, South Korea
In-house design
GRO, Seung Mo Kang, Kang Ho Woo
Web
www.lg.com

The body of this all-in-one PC is situated as far as possible towards the back of the stand, thus fostering a slimmer look while emphasising the screen. With its UltraWide 21:9 format, the monitor is well suited to countless application areas. The filigree design of the stand and the curved screen lend the PC the aesthetic appeal of an art object.

Statement by the jury
This PC impresses with its balanced design and iconic, ring-shaped stand, which overweigh its technical character.

Der Körper dieses All-in-one-PCs ist am Ständer so weit wie möglich nach hinten gesetzt, um ihn schlanker aussehen zu lassen und den Bildschirm in den Vordergrund zu rücken. Mit seinem UltraWide-Format von 21:9 ist der Bildschirm für zahlreiche Anwendungen geeignet. Der filigran gestaltete Fuß und der geschwungene Bildschirm verleihen dem PC die Ästhetik eines Kunstobjekts.

Begründung der Jury
Der PC besticht durch seine ausbalancierte Gestaltung und den eigenständigen ringförmigen Standfuß. Sein technischer Charakter tritt dadurch in den Hintergrund.

Horizon 2s
All-in-One PC

Manufacturer
Lenovo (Beijing) Ltd., Beijing, China
In-house design
Innovation Design Centre
Web
www.lenovo.com

The Horizon 2s has a 19.5" display and a magnesium-aluminium alloy unibody. With a weight of 2.45 kg, the device is easily transported. A special function is that the computer starts automatically as soon as the display is placed horizontally. This positioning allows multiple users to operate the device simultaneously per multitouch, for instance when playing games or in educational settings. As needed, a foldable stand at the backside holds the display in an upright position.

Statement by the jury
Thanks to its unibody casing, the Horizon 2s is light and robust. With its expanded multitouch functionality, it is particularly suitable for groups.

Der Horizon 2s besitzt ein 19,5"-Display und ein Unibody-Gehäuse aus einer Magnesium-Aluminium-Legierung. Mit einem Gewicht von 2,45 kg ist der Computer leicht zu transportieren. Eine besondere Funktion ist, dass er automatisch startet, sobald das Display flach hingelegt wird. Dann kann er von mehreren Anwendern per Multitouch gleichzeitig genutzt werden, z. B. für Spiele oder Bildungsangebote. Bei Bedarf hält ein aufklappbarer Ständer an der Rückseite das Display in aufrechter Position.

Begründung der Jury
Der Horizon 2s ist dank seines Unibody-Gehäuses leicht und robust. Mit der erweiterten Multitouch-Funktionalität eignet er sich besonders gut für Gruppen.

Aspire U5-620
All-in-One PC

Manufacturer
Acer Incorporated, New Taipei, Taiwan
In-house design
Web
www.acer.com

With its slim chassis and silver frame, the Aspire U5-620 all-in-one PC renders an elegant and modern impression. The screen is supported by an aluminium alloy stand attached via a hinge so that the user can position the screen at any desired angle – from nearly vertical to completely flat. The front-facing speakers are located beneath the 23" screen and are protected by a metal mesh surface.

Der All-in-one-PC Aspire U5-620 wirkt mit dem flachen Gehäuse und silbernen Rahmen elegant und modern. Der Bildschirm wird mit einem Aluminium-Standfuß aufgerichtet, der über ein Scharnier befestigt ist. Dadurch kann der Benutzer den Bildschirm in jedem Winkel positionieren – von nahezu senkrecht bis flach. Unterhalb des 23"-Bildschirms befinden sich die nach vorne gerichteten Lautsprecher. Sie werden von einem Metallgitter geschützt.

Statement by the jury
Thanks to the clever stand solution, the screen of the Aspire U5-620 can be adjusted to individual preferences. This enhances comfort and offers more flexibility for the user.

Begründung der Jury
Der Bildschirm des Aspire U5-620 kann dank der cleveren Standlösung individuell ausgerichtet werden. Das ist komfortabel und bietet dem Benutzer Flexibilität.

Transformer Book Chi Series
Convertible Notebook

Manufacturer
ASUSTek Computer Inc., Taipei, Taiwan
In-house design
Asus Design Center
Web
www.asus.com

Chi is a series of Windows notebooks with a detachable display that can be used as a Windows tablet. They are very slim and thus offer a special mobile experience. The intelligent magnetic fastener between display and keyboard allows the user to effortlessly detach the display and reinsert it for smooth and easy handling. The metallic surface conveys plain elegance and draws attention to the slim device.

Statement by the jury
The Chi series offers very slim and lightweight devices with the full functionality of both a notebook and a tablet, making them highly versatile in mobile settings.

Chi ist eine Serie von Windows-Notebooks mit abnehmbarem Display, die als Windows-Tablet genutzt werden können. Sie sind sehr dünn und ermöglichen dadurch eine besondere mobile Erfahrung. Der intelligente Magnetverschluss zwischen Display und Tastatur erlaubt es, das Display mühelos abzunehmen und wieder einzustecken, sodass das flüssige Handling gewährleistet ist. Die metallische Oberfläche vermittelt schlichte Eleganz und setzt das sehr schlanke Gerät gut in Szene.

Begründung der Jury
Die Chi-Serie bietet sehr dünne und leichte Geräte mit allen Funktionen entweder eines Notebooks oder eines Tablets. Das macht sie zum mobilen Multitalent.

Portégé Z20t
Convertible Notebook

Manufacturer
Toshiba Corporation, Personal &
Client Solution Company, Tokyo, Japan
In-house design
Toshiba Corporation Design Center
(Toshio Takada, Yasushi Fukuoka)
Web
www.toshiba.eu
www.toshiba.co.jp/design/en

The Portégé Z20t is a 12.5" business notebook with a display that can be detached and used as a tablet. The magnesium case is robust and very compact, thanks to the well-considered layout. As such, the relatively large RGB and LAN ports are located in the lower part of the case, while smaller ports are accommodated at the side of the screen. In notebook mode, the palm rest can be tilted to the desired angle.

Statement by the jury
The Portégé Z20t is a reliable two-in-one device for business people. It has a durable case and facilitates comfortable working.

Das Portégé Z20t ist ein 12,5"-Business-Notebook, dessen Display abgetrennt und als Tablet verwendet werden kann. Das Magnesiumgehäuse ist robust und dank des durchdachten Layouts überaus kompakt. So befinden sich die vergleichsweise großen RGB- und LAN-Schnittstellen im unteren Gehäuse, während kleinere Schnittstellen seitlich am Bildschirm untergebracht sind. Im Notebook-Modus kann die Handauflage in den jeweils gewünschten Winkel geneigt werden.

Begründung der Jury
Das Portégé Z20t ist ein zuverlässiges 2-in-1-Gerät für Geschäftsleute. Es besitzt ein widerstandsfähiges Gehäuse und ermöglicht bequemes Arbeiten.

Transformer Book T Series
Notebook

Manufacturer
ASUSTek Computer Inc., Taipei, Taiwan
In-house design
Asus Design Center
Web
www.asus.com

Representatives of this minimalist, elegant notebook series can be used either as a Windows notebook or as an Android-based tablet, yet still only require one processor. The functional hinge design allows for quick conversion from notebook to tablet. The brushed texture on the cover creates appealing light and shadow effects, while the gently rounded edges accentuate mobile freedom and convey comfortable handling.

Statement by the jury
With just a flick of the wrist, the notebooks in this series turn into a tablet. They offer the full functionality of a PC and are convenient to carry.

Vertreter dieser minimalistischen, eleganten Notebookserie können als Windows-Notebook oder als Android-basiertes Tablet verwendet werden und benötigen trotzdem nur einen Prozessor. Das funktionale Scharnier-Design ermöglicht dabei die schnelle Verwandlung von Notebook zu Tablet. Die gebürstete Textur des Deckels erzeugt reizvolle Licht- und Schatteneffekte, während die sanft gerundeten Kanten die mobile Freiheit betonen und eine angenehme Handhabung vermitteln.

Begründung der Jury
Die Notebooks dieser Serie werden mit nur einem Handgriff zum Tablet. Sie bieten die volle Funktionalität eines PCs und lassen sich leicht mitnehmen.

HP Elite x2 1011
Convertible Notebook

Manufacturer
Hewlett-Packard, Houston, USA
In-house design
Web
www.hp.com

Elite x2 1011 is a tablet and notebook in one unit. A support is harmoniously integrated into the detachable display and makes it possible to position the tablet for comfortable touch input. High-grade materials such as anodised aluminium harmonise with the clear design and convey a sense of exclusivity. Thanks to the fully fledged keyboard and stable hinge, the device proves to be a reliable tool for everyday use in a professional environment.

Statement by the jury
Elite x2 1011 is a well-conceived two-in-one notebook. The select materials render it elegant and sturdy.

Elite x2 1011 ist Tablet und Notebook in einem. Ein Ständer ist harmonisch in das abnehmbare Display integriert und gestattet es, das Tablet für eine bequeme Toucheingabe aufzustellen. Hochwertige Materialien wie eloxiertes Aluminium harmonieren mit der klaren Gestaltung und vermitteln ein Gefühl von Exklusivität. Dank der vollwertigen Tastatur und dem stabilen Scharnier erweist es sich im täglichen Gebrauch als zuverlässiges Arbeitsgerät.

Begründung der Jury
Elite x2 1011 ist ein durchdacht gestaltetes 2-in-1-Notebook. Die ausgewählten Materialien machen es elegant und widerstandsfähig.

HP Pro x2 612
Convertible Notebook

Manufacturer
Hewlett-Packard, Houston, USA
In-house design
Web
www.hp.com

The Pro x2 612 is an elegant notebook with a detachable display. It features a particularly easy and comfortable screen-keyboard connection, providing a seamless transition between tablet and notebook modes. The design process placed emphasis on the device showing pleasant tapering curves from any viewing angle. This makes the notebook compact and invites the user to engage and work with the device.

Statement by the jury
This two-in-one notebook with its curved design is a veritable pleasure to hold and use on the go.

Das Pro x2 612 ist ein elegantes Notebook mit einem abnehmbaren Display. Der Bildschirm lässt sich besonders leicht mit der Tastatur verbinden, was den nahtlosen Übergang vom Tablet- in den Notebook-Modus ermöglicht. Bei der Gestaltung wurde Wert darauf gelegt, dass das Gerät aus jedem Blickwinkel angenehme Kurven zeigt. Dadurch wirkt es kompakt und lädt geradezu dazu ein, es in die Hand zu nehmen, um damit zu arbeiten.

Begründung der Jury
Dieses 2-in-1-Notebook wird mit seiner geschwungenen Linienführung zu einem wahren Handschmeichler, den man gerne mitnimmt.

CB30-B
Notebook

Manufacturer
Toshiba Corporation, Personal &
Client Solution Company, Tokyo, Japan
In-house design
Toshiba Corporation Design Center
(Yasushi Fukuoka)
Web
www.toshiba.eu
www.toshiba.co.jp/design/en

This 13.3" notebook with the Chrome operating system is optimised for the use of cloud services. With its long battery life of nine hours and Wi-Fi connectivity, it enables Internet access at any time. The fine surface texture is pleasant to the touch, encouraging the user to take the notebook along and use it in various locations. In addition, the notebook features a compact, high-quality sound system.

Statement by the jury
The CB30-B convinces with a timeless design. Thanks to its extended battery life and comprehensive Internet connectivity, a high degree of mobility is provided as well.

Dieses 13,3"-Notebook mit dem Betriebssystem Chrome ist für die Nutzung von Cloud-Diensten optimiert. Mit seiner langen Batterielaufzeit von neun Stunden und Wi-Fi-Konnektivität ermöglicht es jederzeit den Zugriff auf das Internet. Die feine Textur der Gehäuseoberfläche bietet eine angenehme Haptik, sodass der Anwender das Gerät gerne mit sich trägt und an verschiedenen Orten verwendet. Zudem verfügt das Notebook über ein hochwertiges Soundsystem, das kompakt untergebracht ist.

Begründung der Jury
Das CB30-B überzeugt mit seiner zeitlosen Gestaltung. Dank langer Akkulaufzeit und umfassender Internetanbindung bietet es zudem eine hohe Mobilität.

Dell XPS 13
Notebook

Manufacturer
Dell Inc., Round Rock, Texas, USA
In-house design
Experience Design Group
Web
www.dell.com

The XPS 13 was developed for mobile use. Its screen has a bezel that is only 5.2 mm wide, creating a nearly frameless effect. This type of construction gives the 13" notebook the enclosure size of an 11" notebook. The outer protective covers are made of anodised aluminium, and the palm rest with its soft-touch surface offers a pleasant tactile feel. With the backlit keyboard, the notebook is easy to use even in dim lighting conditions.

Statement by the jury
With its slim bezel, the XPS 13 is very compact for a device of its size. The seamless design renders a timeless and elegant impression.

Das XPS 13 wurde für den mobilen Einsatz entwickelt. Sein Bildschirm ist mit einer nur 5,2 mm breiten Umrandung eingefasst und wirkt nahezu rahmenlos. Zudem führt diese Bauweise dazu, dass das Notebook mit seinen 13" lediglich so groß ist wie ein 11"-Notebook. Die äußeren Schutzdeckel bestehen aus eloxiertem Aluminium, die Handauflage mit Softtouch-Oberfläche bietet eine angenehme Haptik. Durch die beleuchtete Tastatur lässt es sich auch im Dunkeln gut bedienen.

Begründung der Jury
Das XPS 13 ist dank seines schmalen Rahmens für seine Größe sehr kompakt. Mit seiner nahtlosen Gestaltung wirkt es zeitlos und elegant.

Zenbook UX501
Notebook

Manufacturer
ASUSTek Computer Inc., Taipei, Taiwan
In-house design
Asus Design Center
Web
www.asus.com

The Zenbook UX501 is a high-performance notebook displaying a slender design. The aluminium body gives this very thin notebook stability for mobile use and also renders a timeless, elegant impression. The 4K display presents sharp images with vivid colours even at a slanted viewing angle. High-quality speakers, together with the brilliant display, ensure a comprehensive multimedia experience.

Statement by the jury
A slim shape and brushed aluminium body underscore the high performance of this notebook.

Das Zenbook UX501 ist ein leistungsfähiges Notebook mit einer schlanken Gestaltung. Das Gehäuse aus Aluminium verleiht dem sehr dünnen Notebook Stabilität für den mobilen Einsatz und wirkt überdies zeitlos und elegant. Das 4K-Display stellt Inhalte auch bei einem schrägen Betrachtungswinkel scharf und mit lebhaften Farben dar. Hochwertige Lautsprecher bieten zusammen mit dem brillanten Display ein umfassendes Multimedia-Erlebnis.

Begründung der Jury
Die schlanke Form und das Gehäuse aus gebürstetem Aluminium unterstreichen die hohe Leistungsfähigkeit dieses Notebooks.

G751
Notebook

Manufacturer
ASUSTek Computer Inc., Taipei, Taiwan
In-house design
Asus Design Center (Mary Hsi)
Web
www.asus.com

The G751 is a high-performance notebook for professional computer gamers. Inspired by a sports car, it combines flowing lines and surfaces with an iconic design, creating a powerful and dynamic appearance. The soft-touch coating of the display reduces reflections and improves recognisability of display content. The keys are illuminated and can be seen well even under dim lighting conditions.

Statement by the jury
The notebook is conceived for high-performance gaming, a fact that is underscored by its sporty design. This fosters a high degree of independence as well.

Das G751 ist leistungsstarkes Notebook für professionelle Computerspieler. Inspiriert von einem Sportwagen, kombiniert es fließende Linien und Flächen mit einer ikonischen Gestaltung und wirkt dadurch kraftvoll und dynamisch. Die Softtouch-Beschichtung des Displays reduziert Reflexionen, wodurch sich Bildschirminhalte besser erkennen lassen. Die Tasten sind beleuchtet und dadurch auch bei Dämmerlicht gut zu erkennen.

Begründung der Jury
Das Notebook ist für hohe Belastungen beim Spielen ausgelegt. Seine sportliche Gestaltung unterstreicht dies und verleiht ihm darüber hinaus Eigenständigkeit.

Y40
Notebook

Manufacturer
Lenovo (Beijing) Ltd., Beijing, China
In-house design
Innovation Design Centre
Web
www.lenovo.com

The Y40 gaming notebook shows a design inspired by a sports car: the keys are framed in red, while the speakers are enclosed in red metal mesh. The chamfered edges at the top and bottom allude to the airflow running from the bottom of the device towards the back. The robust casing is made of carbon fibre, rendering the device durable, preventing fingerprints and offering good haptic properties.

Statement by the jury
Carefully applied colour accents and slanted lines lend this gaming notebook an unobtrusive, sporty character.

Das Gaming-Notebook Y40 zeigt ein von einem Sportwagen inspiriertes Design: Die Tasten sind rot umrandet, die Lautsprecher mit einem roten Metallgitter versehen. Die abgeschrägten Kanten an Ober- und Unterseite deuten den Luftstrom an, der sich von der Unterseite des Geräts über die Rückseite zieht. Das robuste Gehäuse besteht aus Carbon. Dies macht das Gerät widerstandsfähig, verhindert Fingerabdrücke und bietet eine gute Haptik.

Begründung der Jury
Sorgfältig dosierte Farbakzente und schräge Linienführung verleihen dem Gaming-Notebook einen unaufdringlichen, sportlichen Charakter.

EeeBook Series
Netbook

Manufacturer
ASUSTek Computer Inc., Taipei, Taiwan
In-house design
Asus Design Center
Web
www.asus.com

These netbooks with an 11.6" HD display are very compact and weigh less than 1 kg. With a quad-core processor and fully fledged keyboard, they enable smooth and fluent working with 12 hours of battery life. The design is reduced to the bare essentials so that the notebooks render a solid impression with a comfortably smooth feel. The series is available in four classic colours to suit users' varying preferences.

Statement by the jury
The netbooks of the EeeBook series render a favourable impression on the go. They are lightweight, compact and enrich the market with contemporary colours.

Diese Netbooks mit 11,6"-HD-Display sind sehr kompakt und wiegen weniger als 1 kg. Mit ihrem Vierkernprozessor und der vollwertigen Tastatur ermöglichen sie flüssiges Arbeiten bei einer Akkulaufzeit von zwölf Stunden. Die Gestaltung ist auf das Nötigste reduziert, sodass die Netbooks solide erscheinen und sich angenehm glatt anfühlen. Die Serie bietet vier klassische Farben zur Auswahl, die den unterschiedlichen Vorlieben der Anwender entsprechen.

Begründung der Jury
Die Netbooks der EeeBook-Serie machen unterwegs einen guten Eindruck. Sie sind leicht, kompakt und bereichern den Markt mit aktuellen Farben.

HP Slatebook
Notebook

Manufacturer
Hewlett-Packard, Houston, USA
In-house design
Design
Native Design, London, Great Britain
Web
www.hp.com
www.native.com

The Slatebook is a notebook that runs on the Android operating system. Apps familiar from smartphones or tablets can be controlled via the 14" touchscreen or fully fledged keyboard. An Nvidia Tegra 4 processor provides the required speed for an optimised graphics presentation. The notebook is 16 mm thin and weighs merely 1.7 kg. The device with its contemporary design is available in a range of vibrant colours.

Statement by the jury
With the Slatebook, Android apps can be used more comfortably. Thanks to the wide variety of colours, models are available to suit any taste.

Das Slatebook ist ein Notebook, das mit dem Betriebssystem Android läuft. Die bisher von Smartphone oder Tablet bekannten Apps werden über den 14"-Touchscreen oder über die vollwertige Tastatur gesteuert. Ein Nvidia-Tegra-4-Prozessor sorgt dabei für die nötige Schnelligkeit und eine optimierte Grafikdarstellung. Das Notebook ist knapp 16 mm dünn und wiegt lediglich 1,7 kg. Das zeitgemäß gestaltete Gerät ist in verschiedenen leuchtenden Farben erhältlich.

Begründung der Jury
Mit dem Slatebook können Android-Apps komfortabler genutzt werden. Dank der unterschiedlichen Farbvarianten ist für jeden Geschmack ein Modell verfügbar.

Aspire Switch 12 SW5-271
Convertible Notebook

Manufacturer
Acer Incorporated, New Taipei, Taiwan
In-house design
Web
www.acer.com

The Aspire Switch 12 is a versatile five-in-one notebook with detachable wireless keyboard. It can for instance be used as a notebook or tablet – with or without keyboard – and the display may be situated in different positions, providing an optimised viewing angle for each respective application. With the magnetic snap hinge mechanism, transition between different operation modes is quite simple.

Statement by the jury
The Aspire Switch 12 is a versatile device catering to nearly any requirement and thus offers extensive freedom for mobile computing.

Das Aspire Switch 12 ist ein vielseitiges 5-in-1-Notebook mit abnehmbarer Funktastatur. Es kann z. B. als Notebook oder Tablet verwendet werden – mit oder ohne Tastatur – und das Display lässt sich in unterschiedlichen Positionen platzieren, die passend zur Anwendung immer die optimale Betrachtung ermöglichen. Der Übergang zwischen den einzelnen Modi gestaltet sich dank des magnetischen Schnappmechanismus denkbar einfach.

Begründung der Jury
Das Aspire Switch 12 ist ein vielseitiges Gerät für nahezu jede Anforderung und bietet beim mobilen Computing sämtliche Freiheiten.

TravelMate P645
Notebook

Manufacturer
Acer Incorporated, New Taipei, Taiwan
In-house design
Web
www.acer.com

The TravelMate P645 was designed with business people in mind. Its carbon-fibre top cover, reinforced hinges and body fashioned from a magnesium-aluminium alloy all render the notebook robust and durable in everyday use. The hard drive is shock- and impact-resistant and thus protected against loss of data, and it also has integrated anti-theft software. The technical equipment, with a fast i5 processor and HD graphics, enables an efficient working experience.

Statement by the jury
The TravelMate P645 is perfectly suited to the business field. Particularly convincing features are the notebook's durable workmanship and clever security functions.

Das TravelMate P645 wurde für Geschäftsleute konzipiert. Seine Abdeckung aus Kohlefaser, die verstärkten Scharniere und das Gehäuse aus einer Magnesium-Aluminium-Legierung machen das Notebook robust und unempfindlich im täglichen Gebrauch. Die Festplatte ist gegen Stöße und damit gegen Datenverlust geschützt, ebenso ist eine Anti-Diebstahl-Software integriert. Die technische Ausstattung mit schnellem i5-Prozessor und HD-Grafik ermöglicht effizientes Arbeiten.

Begründung der Jury
Das TravelMate P645 ist bestens für den Business-Bereich ausgestattet. Besonders die robuste Beschaffenheit und die intelligenten Sicherheitsfunktionen des Notebooks überzeugen.

Aspire R 13
Convertible Notebook

Manufacturer
Acer Incorporated, New Taipei, Taiwan
In-house design
Web
www.acer.com

The Aspire R 13 is characterised by an intelligent construction with clever functions for improved operation during work and leisure time. Thanks to the innovative display attachment, switching between different modes is quite simple, for the display can easily be swivelled from notebook to tablet position. In addition, the notebook is equipped with high-grade components like Gorilla Glass and a WQHD display.

Statement by the jury
With the swivel mechanism, this notebook can be quickly converted into a tablet, yet it remains a functional unit.

Das Aspire R 13 zeichnet sich durch eine intelligente Bauweise mit cleveren Funktionen für eine bessere Bedienung bei der Arbeit und in der Freizeit aus. Dank der innovativen Befestigung des Displays ist der Wechsel zwischen verschiedenen Modi ganz einfach, denn der Bildschirm lässt sich leicht von der Notebook- in eine Tablet-Position schwenken. Ausgestattet ist das Notebook darüber hinaus mit hochwertigen Komponenten wie Gorilla-Glas und WQHD-Display.

Begründung der Jury
Durch den Schwenkmechanismus wird das Notebook schnell in ein Tablet verwandelt, bleibt aber als funktionale Einheit bestehen.

ASUSPRO Series
Notebook

Manufacturer
ASUSTek Computer Inc., Taipei, Taiwan
In-house design
Asus Design Center
Web
www.asus.com

This series consists of notebooks designed for small and medium businesses. The reliable, high-performance devices have a reduced, clear design and thus correlate well with the elegant appearance of business people. With their robust casings, shock-proof hard drives and splash-resistant keyboards, the devices master virtually any stressful situation in everyday professional life. The notebooks are easy to operate, enabling the user to focus exclusively on the task at hand.

Statement by the jury
With a reduced design and a wide variety of functions, these devices meet all workplace requirements.

Diese Serie umfasst Notebooks, die für kleine und mittlere Unternehmen konzipiert wurden. Die leistungsfähigen und zuverlässigen Geräte sind reduziert und klar gestaltet und passen damit gut zur eleganten Erscheinung von Geschäftsleuten. Mit ihren robusten Gehäusen, stoßfesten Festplatten und spritzwassergeschützten Tastaturen verkraften sie problemlos die Belastungen im Arbeitsalltag. Sie sind einfach zu handhaben, damit der Anwender sich ganz auf seine Aufgaben konzentrieren kann.

Begründung der Jury
Die zurückhaltend gestalteten Geräte bieten mit ihren zahlreichen Funktionen alles, was am Arbeitsplatz benötigt wird.

HP Omen
Gaming Notebook

Manufacturer
Hewlett-Packard, Houston, USA
In-house design
Design
Native Design, London, Great Britain
Web
www.hp.com
www.native.com

The Omen gaming notebook is equipped with high-performance hardware such as the fourth-generation i7 processor, a GeForce GTX graphic card and an SSD drive to guarantee the speed needed for gaming. A large number of I/O ports on the backside offer comprehensive connection options. The dynamic design with red keys and LED accents against the deep black body accentuate the high performance of this notebook.

Statement by the jury
Hardware equipment and design are optimally coordinated, resulting in an attractive product for demanding gamers.

Das Gaming-Notebook Omen ist mit leistungsfähiger Hardware wie einem i7-Prozessor der vierten Generation, einer GeForce-GTX-Grafikkarte und einem SSD-Laufwerk ausgestattet, um die für das Gaming gewünschte Schnelligkeit zu garantieren. Zahlreiche I/O-Ports auf der Rückseite bieten umfangreiche Anschlussmöglichkeiten. Die dynamische Gestaltung mit roten Tasten und LED-Akzenten auf dem tiefschwarzen Gehäuse unterstreicht die hohe Performance des Notebooks.

Begründung der Jury
Hardware-Ausstattung und Design sind optimal aufeinander abgestimmt und bilden ein gelungenes Produkt für anspruchsvolle Gamer.

HP Spectre x360
Convertible Notebook

Manufacturer
Hewlett-Packard, Houston, USA
In-house design
Design
Native Design, London, Great Britain
Web
www.hp.com
www.native.com

The Spectre x360 is a precision-crafted portable device that can be converted from a notebook into a tablet. Its chassis has been machined from aluminium, giving the product an elegant appearance and structural stability at the same time. The 360-degree hinge enables fast switching between the various application areas, such as notebook and tablet modes.

Statement by the jury
Thanks to its hinge, the Spectre x360 offers a high degree of flexibility and, with its solid aluminium chassis, renders a valuable impression.

Das Spectre x360 ist ein präzise gefertigtes, mobiles Gerät, das sich von einem Notebook in ein Tablet verwandeln lässt. Sein Gehäuse wurde aus Aluminium hergestellt, was dem Produkt eine elegante Anmutung verleiht und gleichzeitig für seine Stabilität sorgt. Das 360-Grad-Scharnier gestattet den schnellen Wechsel zwischen den unterschiedlichen Anwendungen, wie Notebook- oder Tablet-Modus.

Begründung der Jury
Das Spectre x360 bietet dank seines Scharniers eine hohe Flexibilität bei der Nutzung und wirkt mit seinem massiven Aluminiumgehäuse überaus hochwertig.

HP Envy x2 Family
Convertible Notebook

Manufacturer
Hewlett-Packard, Houston, USA
In-house design
Web
www.hp.com

The Envy x2 notebook family was designed to meet the diverse demands of modern computer users with regard to entertainment, education, work and surfing. Thanks to the aluminium chassis, the devices are lightweight and stable, thus heightening ease of transport. The screen has a practical stand attached to the reverse side, which can be adjusted to any desired angle in both notebook and tablet modes.

Statement by the jury
These two-in-one notebooks captivate with their elegant design. The cleverly integrated stand renders the use of the devices even more flexible.

Die Notebook-Familie Envy x2 wurde entwickelt, um den vielfältigen Ansprüchen moderner Computernutzer hinsichtlich Unterhaltung, Bildung, Arbeit und Surfen zu entsprechen. Die Geräte sind dank ihres Aluminiumgehäuses leicht und stabil, lassen sich also bequem transportieren. Der Bildschirm besitzt auf der Rückseite einen praktischen Ständer, der sowohl im Notebook- als auch im Tablet-Modus in einem beliebigen Winkel eingestellt werden kann.

Begründung der Jury
Die 2-in-1-Notebooks bestechen durch ihre elegante Gestaltung. Der clever integrierte Ständer macht die Nutzung der Geräte noch flexibler.

HP Pavilion 360x 11
Convertible Notebook

Manufacturer
Hewlett-Packard, Houston, USA
In-house design
Web
www.hp.com

The conception of the Pavilion 360x 11 was focused on modern manufacturing techniques, new materials, fresh colours and tasteful finishes. The result is a lightweight and robust notebook for young users offering comprehensive functionality at an affordable price. The hinge construction enables flexible, need-based use in both working and entertainment contexts. The notebook is available in three appealing colours.

Statement by the jury
The Pavilion 360x 11 is an attractive two-in-one notebook for a young target group. It is lightweight, durable and fun to use.

Bei der Konzeption des Pavilion 360x 11 standen moderne Herstellungsverfahren, neue Materialien, frische Farben und geschmackvolle Oberflächen im Vordergrund. Das Ergebnis ist ein leichtes und robustes Notebook für junge Anwender, das alle Funktionen bietet und dennoch bezahlbar bleibt. Dank des Scharniers ist die Nutzung flexibel, so kann das Gerät ganz nach Bedarf für Arbeit oder Unterhaltung genutzt werden. Das Notebook ist in drei ansprechenden Farben verfügbar.

Begründung der Jury
Das Pavilion 360x 11 ist ein attraktives 2-in-1-Notebook für eine junge Zielgruppe. Es ist leicht, widerstandsfähig und vermittelt Spaß bei der Nutzung.

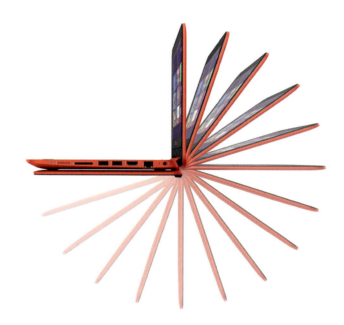

HP EliteBook 1020
Notebook

Manufacturer
Hewlett-Packard, Houston, USA
In-house design
Web
www.hp.com

Designed for business users, the Elite-Book 1020 with a thickness of merely 15.7 mm offers a high degree of mobility. The magnesium-aluminium chassis makes the notebook so stable that it performs tasks reliably even under demanding conditions. It is nearly silent in operation, offers comprehensive connectivity and features a backlit keyboard for good visibility of the keys even under poor lighting conditions.

Statement by the jury
A very slim silhouette, strongly tapered to the front, gives this notebook an elegant appearance, accentuating its premium quality.

Das für Business-Anwender gedachte Elite-Book 1020 bietet mit seiner Dicke von nur 15,7 mm ein hohes Maß an Mobilität. Das Magnesium-Aluminium-Gehäuse macht das Notebook so stabil, dass es auch unter anspruchsvollen Bedingungen zuverlässig seinen Dienst tut. Es arbeitet nahezu geräuschlos, bietet eine umfassende Konnektivität, und ein Keyboard mit Hintergrundbeleuchtung sorgt für gute Sichtbarkeit der Tasten bei schlechten Lichtverhältnissen.

Begründung der Jury
Die sehr schlanke, sich nach vorne stark verjüngende Silhouette verleiht dem Notebook eine elegante Anmutung, die seine Wertigkeit unterstreicht.

HP Chromebook 14
Notebook

Manufacturer
Hewlett-Packard, Houston, USA
In-house design
Design
Native Design, London, Great Britain
Web
www.hp.com
www.native.com

The Chromebook 14 was designed with the idea to create a lightweight, mobile device with vivid colours and 3G access. It features the Google Chrome operating system and has access to applications on the Internet where data can be stored as well. It is equipped with an HD webcam and various connector ports such as USB, HDMI, and audio. The device weighs just slightly more than 1.5 kg and is thus convenient to carry.

Statement by the jury
The Chromebook 14 offers all the important functions and inspires with its fresh colour scheme, which emphasises the lightness of the device.

Das Chromebook 14 wurde mit dem Gedanken entwickelt, ein leichtes, mobiles Gerät mit frischen Farben und 3G-Zugang zu kreieren. Es läuft auf Google Chrome und greift auf Apps im Internet zu, wo auch Daten abgelegt werden können. Ausgestattet ist es mit einer HD-Webcam und diversen Anschlüssen wie USB, HDMI und Audio. Es wiegt nur etwas mehr als 1,5 kg und ist daher leicht zu transportieren.

Begründung der Jury
Das Chromebook 14 bietet alle wichtigen Funktionen und begeistert mit seiner frischen Farbgebung, die die Leichtigkeit des Geräts unterstreicht.

TabBook Duo
Tablet with Keyboard
Tablet mit Tastatur

Manufacturer
LG Electronics Inc., Seoul, South Korea
In-house design
Gangho Woo, Seungdon Lee
Web
www.lg.com

The TabBook Duo is a tablet consisting of a main body and detachable keyboard. The two components are connected using a magnetic mount, which eliminates the need for connectors or screws. The keyboard can be used as protective cover on the tablet or fixed at its rear, serving as a base. Moreover, the tablet features a stand, allowing it to be set up in any position or viewing angle.

Statement by the jury
Tablet and keyboard form a harmonious unit so that the TabBook Duo may be operated, depending on the application, by touchscreen or keyboard.

Das TabBook Duo ist ein Tablet, das aus einem Hauptkörper und einer abnehmbaren Tastatur besteht. Beide Komponenten werden mit einer magnetischen Halterung verbunden, sodass keine Stecker oder Schrauben notwendig sind. Die Tastatur kann als schützender Deckel auf dem Tablet oder auf seiner Rückseite als Unterlage fixiert werden. Darüber hinaus ist das Tablet mit einer Stütze versehen, sodass es in jeder Position und jedem Betrachtungswinkel aufgestellt werden kann.

Begründung der Jury
Tablet und Tastatur bilden eine harmonische Einheit. So kann das TabBook Duo je nach Anwendung per Touchscreen oder Tastatur bedient werden.

Aspire Switch
10 E SW3-013
Convertible Notebook

Manufacturer
Acer Incorporated, New Taipei, Taiwan
In-house design
Web
www.acer.com

The Aspire Switch 10 E SW3-013 is equipped with a magnetic snap hinge for the display, so that users can swiftly change between tablet and notebook modes. The keys of the chiclet keyboard are arranged with sufficient space in between and offer reliable tactile feedback when typing. The coating for both the cover and the underside is non-slip and available in striking colours to match different lifestyles.

Statement by the jury
With its detachable display, this notebook is suitable for both work and leisure. It has a modern design and offers extended battery runtime.

Das Aspire Switch 10 E SW3-013 ist mit einem magnetischen Steckplatz für das Display ausgestattet, damit Anwender leicht zwischen Tablet- und Notebook-Modus wechseln können. Die Tasten der Chiclet-Tastatur sind in ausreichendem Abstand angeordnet und bieten eine zuverlässige taktile Rückmeldung beim Tippen. Der Überzug für Abdeckung und Unterseite ist griffig und in auffälligen Farben verfügbar, die zu unterschiedlichen Lifestyles passen.

Begründung der Jury
Das Notebook eignet sich mit seinem abnehmbaren Display für Arbeit und Freizeit gleichermaßen. Es ist modern gestaltet und bietet eine lange Akkulaufzeit.

Aspire Switch
11 V SW5-173
Convertible Notebook

Manufacturer
Acer Incorporated, New Taipei, Taiwan
In-house design
Web
www.acer.com

The Aspire Switch 11 V SW5-173 features an 11" display for clear graphical display. When detached from the magnetic snap hinge, the display can be used like a tablet. A unique feature is the crossed-line stripe pattern on the reverse side of the display. It underlines the classic elegance of the notebook and generates a sense of familiarity. In contrast, the keyboard is characterised by functional clarity.

Statement by the jury
With the uniquely textured reverse side of the display, this two-in-one notebook is a stylish, elegant companion in any situation.

Das Aspire Switch 11 V SW5-173 bietet ein 11"-Display für die klare Grafikwiedergabe. Wird das Display aus der magnetischen Halterung genommen, kann es wie ein Tablet genutzt werden. Erwähnenswert ist das Karomuster auf der Rückseite des Displays. Es unterstreicht die klassische Eleganz des Notebooks und erzeugt ein Gefühl der Vertrautheit. Im Gegensatz dazu ist die Tastatur von einer funktionalen Klarheit geprägt.

Begründung der Jury
Dieses 2-in-1-Notebook ist dank der eigenständig gemusterten Display-Rückseite ein stilvoller, eleganter Begleiter in jeder Situation.

Dell Latitude Rugged
Extreme Notebooks

Manufacturer
Dell Inc., Round Rock, Texas, USA
In-house design
Experience Design Group
Web
www.dell.com

The Latitude Rugged Extreme notebooks are available in 12" and 14" versions. Both models have a solid magnesium-alloy frame and are made of shock-absorbing polymer materials to be able to withstand hazards such as dust, moisture, falls, vibrations and extreme temperatures, as well as other harsh conditions. The hinge construction enables seamless switching between notebook and tablet modes.

Statement by the jury
These robust devices with their two-in-one functionality bring more flexibility to those fields of application posing high challenges to the material.

Die Notebooks Latitude Rugged Extreme sind als 12"- und 14"-Ausführung erhältlich. Beide Modelle besitzen einen Rahmen aus einer festen Magnesiumlegierung und stoßdämpfenden Polymerwerkstoffen, der sie widerstandsfähig gegen Gefährdungen wie Staub, Feuchtigkeit, Stürze und Erschütterungen, extreme Temperaturen und andere schwierige Bedingungen macht. Ihre Gelenkkonstruktion ermöglicht einen stufenlosen Wechsel zwischen Notebook- und Tablet-Modus.

Begründung der Jury
Diese robusten Geräte bringen mit ihrer 2-in-1-Funktionalität mehr Flexibilität in Bereiche, die hohe Anforderungen an das Material stellen.

ThinkPad 10 / ThinkPad Helix and Acessories
Tablet and Accessories
Tablet und Zubehör

Manufacturer
Lenovo, Morrisville, North Carolina, USA
In-house design
Web
www.lenovo.com

The ThinkPad 10 is a versatile tablet that can be combined with comprehensive accessories. It is equipped with a 2 MP webcam for video conferencing and an 8 MP camera with autofocus, flash and video function. The device can be connected to a keyboard or a docking station. In addition, a cover is available to protect the tablet and serve as a stand when shooting images with the camera. The ThinkPad Helix with a larger 11.6" screen offers similar versatility.

Statement by the jury
The tablet convinces with its functional diversity. With its accessories, the application range is expanded in a sensible way.

Das ThinkPad 10 ist ein vielseitiges Tablet, das mit umfangreichem Zubehör kombiniert werden kann. Ausgestattet ist es mit einer 2-MP-Webcam für Videokonferenzen und einer 8-MP-Kamera mit Autofokus, Blitzlicht und Videofunktion. Das Gerät kann mit einer Tastatur und einer Dockingstation verbunden werden. Zusätzlich ist ein Cover erhältlich, welches das Tablet schützt und bei Aufnahmen mit der Kamera als Ständer hilfreich ist. Das ThinkPad Helix mit einem größerem 11,6''-Bildschirm ist ebenso vielseitig.

Begründung der Jury
Das Tablet überzeugt mit seiner Funktionsvielfalt. Durch sein Zubehör wird der Anwendungsbereich sinnvoll erweitert.

Yoga II Pro
Tablet

Manufacturer
Lenovo (Beijing) Ltd., Beijing, China
In-house design
Innovation Design Centre
Web
www.lenovo.com

The 13" Yoga II Pro tablet can be adjusted to any desired angle with the integrated stand: positioned upright, it serves as screen, such as for watching videos. Flat on the table or slightly angled, the touchscreen is convenient to use. The stand also functions as a wall mount, enabling the tablet to be hung anywhere for watching videos or listening to music.

Statement by the jury
The stand is a clever design feature. It turns the tablet into a versatile device that is comfortable to use in any situation.

Das 13"-Tablet Yoga II Pro wird mit dem integrierten Ständer je nach persönlicher Vorliebe in einem beliebigen Winkel ausgerichtet: Aufgestellt dient es als Bildschirm, um beispielsweise Videos zu betrachten. Wird es flach oder leicht geneigt auf den Tisch gelegt, kann der Touchscreen bequem bedient werden. Der Ständer ist gleichzeitig Wandhalterung, um es überall aufzuhängen und Videos zu sehen oder Musik zu hören.

Begründung der Jury
Der Ständer ist ein cleveres Gestaltungsmerkmal. Er macht aus dem Tablet ein vielseitiges Gerät, das in jeder Situation bequem genutzt werden kann.

HP Pro Slate 8
Tablet

Manufacturer
Hewlett-Packard, Houston, USA
In-house design
Web
www.hp.com

The Pro Slate 8 is a slim Android tablet with a high-resolution 8" display in 4:3 format. The functional device was designed for business people and, with its diamond-cut, anodised aluminium housing and front-facing speakers, it renders a solid and elegant impression. The design of the reverse side harmonises well with accessories such as cases, covers and keyboards. Moreover, it has a pleasant tactile feel and rests well in the hand.

Statement by the jury
This tablet renders a valuable impression. It is lightweight, flat and durable, and thus particularly well suited for business professionals.

Das Pro Slate 8 ist ein schlankes Android-Tablet mit einem hochauflösenden 8"-Display im 4:3-Format. Das funktionale Gerät wurde für Geschäftsleute konzipiert und wirkt mit seinem diamantgeschliffenen Aluminiumgehäuse und den nach vorne gerichteten Lautsprechern solide und elegant. Die Rückseite wurde so gestaltet, dass sie mit Zubehör wie Hüllen, Abdeckungen und Tastaturen gut harmoniert. Zudem lässt sie sich angenehm anfassen und liegt sicher in der Hand.

Begründung der Jury
Dieses Tablet macht einen wertigen Eindruck. Es ist leicht, flach und widerstandsfähig und damit besonders gut für Geschäftsleute geeignet.

HP Pro Slate 12
Tablet

Manufacturer
Hewlett-Packard, Houston, USA
In-house design
Web
www.hp.com

The 12" tablet Pro Slate 12 has a 4:3 HD display with durable Concore Glass. Of interest to business people are the integrated hardware security functions, which can be managed via the software already installed. The tablet runs on Android and thus offers access to a large number of apps. The stylus provided with the tablet ensures comfortable and flexible use.

Statement by the jury
The lightweight and elegant tablet impresses with its large screen, displaying content in a reader-friendly way.

Das 12"-Taolet Pro Slate 12 besitzt ein 4:3-HD-Display, das aus widerstandsfähigem Concore-Glas besteht. Für Geschäftsleute interessant sind die integrierten Hardware-Sicherheitsfunktionen, die sich über bereits installierte Software verwalten lassen. Das Tablet läuft mit Android und bietet damit Zugang zu einer Vielzahl an Apps. Der mitgelieferte Eingabestift macht die Bedienung komfortabel und flexibel.

Begründung der Jury
Das leichte und elegante Tablet beeindruckt durch seinen großen Bildschirm, der Inhalte lesefreundlich darstellt.

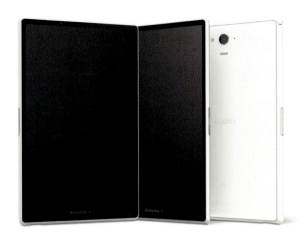

SH-06F
Tablet

Manufacturer
Sharp Corporation, Osaka, Japan
In-house design
Web
www.sharp.co.jp

This tablet has a high-definition 7" IGZO display for sharp and brilliant presentation of content. The tablet is compact and one of the lightest devices of its kind at only 233 grams. The honeycomb form guarantees a safe grip and prevents accidental slipping. Moreover, the device is waterproof and includes a built-in receiver that enables use as mobile phone.

Statement by the jury
This tablet is slim and, thanks to its special coating, can be held securely, which is of particular advantage when used as a phone.

Dieses Tablet bietet ein 7" großes, hochauflösendes IGZO-Display, das Inhalte gestochen scharf und brillant darstellt. Das Tablet ist kompakt und mit 233 Gramm eines der leichtesten Geräte dieser Art. Durch die Wabenform ist es rutschfest und liegt sicher in der Hand. Das Gerät ist zudem wasserdicht und besitzt einen eingebauten Empfänger, damit es auch als Telefon verwendet werden kann.

Begründung der Jury
Dieses Tablet ist schlank und dank seiner besonderen Beschichtung sicher zu halten, was besonders beim Telefonieren von Vorteil ist.

Fonepad Series FE375/FE380
Tablet

Manufacturer
ASUSTek Computer Inc., Taipei, Taiwan
In-house design
Asus Design Center
Web
www.asus.com

The Fonepads FE375 and FE380 are 7" and 8" tablets with phone functionality. Thanks to the soft-touch surface, the slim tablets can be securely held in the hand. There is room for two SIM cards, for instance one for data transmission and another for making phone calls. The tablets are designed to address the needs of users throughout the world.

Statement by the jury
These tablets facilitate both comfortable surfing and phone calls. The solution with two SIM cards offers flexibility and enables better control.

Die Fonepads FE375 und FE380 sind 7"-und 8"-Tablets mit Telefonfunktion. Die dünnen Tablets lassen sich dank der Softtouch-Oberfläche sicher festhalten. Sie können mit zwei SIM-Karten verwendet werden, von denen eine beispielsweise für den Datenverkehr und die andere für das Telefonieren genutzt werden kann. Die Tablets sind so gestaltet, dass sie den Ansprüchen von Anwendern aus aller Welt genügen.

Begründung der Jury
Mit diesen Tablets lässt es sich bequem surfen und telefonieren. Die Lösung mit zwei SIM-Karten bietet Flexibilität und ermöglicht eine bessere Kontrolle.

Xperia™ Z3 Tablet Compact

Manufacturer
Sony Mobile Communications Inc., Tokyo, Japan
In-house design
Web
www.sonymobile.com

The Xperia Z3 Tablet Compact is waterproof and features an 8" full HD display. It is very thin at 6.4 mm and weighs less than 260 grams, allowing it to be carried along anywhere. Its reverse side is made of solid fibreglass material, making this tablet robust and durable for everyday use. The sound workmanship impresses with its stainless-steel corners, slim frame and surfaces that are pleasant to the touch.

Statement by the jury
This elegant, featherlight tablet is made from durable materials and is thus able to master any situation.

Das Xperia Z3 Tablet Compact ist wasserdicht und verfügt über einen 8"-Full-HD-Bildschirm. Es ist lediglich 6,4 mm dünn und wiegt weniger als 260 Gramm, lässt sich also leicht überallhin mitnehmen. Die Rückseite ist aus festem Glasfasermaterial gefertigt und macht das Tablet robust für die Anwendung im Alltag. Die solide Verarbeitung besticht mit aus Edelstahl gefertigten Ecken, einem schlanken Rahmen und angenehm in der Hand liegenden Oberflächen.

Begründung der Jury
Dieses elegante, federleichte Tablet ist aus widerstandsfähigen Materialien gefertigt und kann damit in jeder Situation bestehen.

A740
All-in-One PC

Manufacturer
Lenovo (Beijing) Ltd., Beijing, China
In-house design
Innovation Design Centre
Web
www.lenovo.com

The A740 is an all-in-one PC with a 27" 4K touchscreen. The display is 4.5 mm thin and encased in an aluminium unibody, which gives the PC a high degree of rigidity despite its slim design. The variable stand offers the option to comfortably adjust the display to the desired position, whether at a 45 degree angle for ergonomic typing or horizontal placement for shared use by more than one person.

Statement by the jury
Thanks to its large, slim display, the A740 is convenient to operate in various settings. With its aluminium body, it renders a high-grade and elegant impression.

Der A740 ist ein All-in-one-PC mit 27"-4K-Touchscreen. Das nur 4,5 mm dicke Display wird von einem Unibody-Gehäuse aus Aluminium umschlossen. Dadurch besitzt der PC trotz seiner schlanken Bauweise eine hohe Steifigkeit. Der variable Ständer bietet die Möglichkeit, das Display bequem auf die gewünschte Position einzustellen, sei es in einer 45-Grad-Position für das ergonomische Tippen oder flach hingelegt für die gemeinsame Nutzung durch mehrere Personen.

Begründung der Jury
Der A740 ist dank seines großen, flachen Displays in verschiedenen Kontexten komfortabel zu bedienen. Mit seinem Aluminiumgehäuse wirkt er hochwertig und elegant.

Dell Inspiron 2-in-1 Family
Convertible Notebook

Manufacturer
Dell Inc., Round Rock, Texas, USA
In-house design
Experience Design Group
Web
www.dell.com

The Inspiron two-in-one PCs are available in 11" and 13" versions. Both devices are characterised by an elegant, frameless display and an innovative hinge design. The hinge is rotatable by 360 degrees to switch between the various modi, such as notebook or tablet. The 11" version weighs 1.39 kg, is only 19.4 mm thick and can thus be effortlessly carried. The 13" version offers full HD resolution as well as an integrated stylus.

Statement by the jury
These two-in-one PCs combine reliability with an appealing design. Thanks to the hinge, the user can quickly select the most comfortable position.

Die Inspiron 2-in-1-PCs sind als 11"- und 13"-Version verfügbar. Beide Geräte zeichnen sich durch einen eleganten, rahmenlosen Bildschirm und ein innovatives Scharnier aus. Das Scharnier lässt sich um 360 Grad drehen, so lässt sich problemlos zwischen den verschiedenen Modi wechseln, wie z. B. Notebook und Tablet. Die 11"-Version wiegt 1,39 kg und ist nur 19,4 mm dünn, lässt sich also leicht tragen. Die 13"-Version bietet Full-HD-Auflösung sowie einen integrierten Stift.

Begründung der Jury
Die 2-in-1-PCs vereinen Zuverlässigkeit mit ansprechendem Design. Dank des Scharniers kann der Anwender schnell die für ihn angenehmste Position wählen.

Dell Venue 8 7000 Series
Tablet PC

Manufacturer
Dell Inc., Round Rock, Texas, USA
In-house design
Experience Design Group
Design
Intel Design Team,
Santa Clara, California, USA
Web
www.dell.com
www.intel.com

The Venue 8 7000 Series Android tablet is a mere 6 mm thin, thanks to a compact circuit-board design and thoughtful component layout. The rear housing is fashioned from solid aluminium, making it very durable despite the slender design. There is room for three cameras, including a 3D camera. The stereo speakers are facing forward and deliver clear sound when chatting or watching videos.

Statement by the jury
Thanks to the aluminium body, this tablet with its high-grade workmanship and outstanding equipment is both thin and sturdy.

Das Android-Tablet Venue 8 7000 Series ist dank eines kompakten Leiterplattendesigns und einer gut durchdachten Anordnung der Komponenten lediglich 6 mm dünn. Der hintere Teil des Gehäuses besteht aus massivem Aluminium, sodass es trotz seiner geringen Dicke sehr widerstandsfähig ist. Es bietet Platz für drei Kameras, darunter eine 3D-Kamera. Die Stereolautsprecher sind nach vorne gerichtet und liefern einen klaren Klang bei Videos und Chat.

Begründung der Jury
Dank des Aluminiumgehäuses ist dieses hochwertig verarbeitete Tablet mit ausgezeichneter Ausstattung gleichermaßen dünn wie stabil.

Cintiq 27QHD touch
Pen Display
Stift-Display

Manufacturer
Wacom Co. Ltd., Tokyo, Japan
Design
dingfest | design (Volker Hübner), Erkrath, Germany
Web
www.wacom.com
www.dingfest.de

This pen display features a frameless 27" screen with a resolution of 2,560 x 1,440 pixels and a wide viewing angle, facilitating work on large-format content with rich details. Content is realistically rendered with 1.07 billion colours and a 97 per cent Adobe RGB colour gamut. In combination with a pressure-sensitive pen or using multitouch gestures, users can operate the display swiftly and intuitively, which is of particular advantage for creative work. Robust, integrated support legs allow for comfortable placement in two positions.

Das Stift-Display verfügt über ein randloses 27"-Display mit einer Auflösung von 2.560 x 1.440 Pixeln und einem weiten Blickwinkel, was das Arbeiten an großformatigen Inhalten mit vielen Details erleichtert. Inhalte werden mit 1,07 Milliarden Farben und 97 Prozent des Adobe-RGB-Farbraums lebensecht dargestellt. In Verbindung mit einem druckempfindlichen Stift oder per Multitouch-Gesten können Anwender das Display intuitiv und schnell verwenden, was besonders beim kreativen Arbeiten von Vorteil ist. Für eine bequeme Position sorgen robuste integrierte Stützfüße, die sich in zwei Stellungen arretieren lassen.

Intuos Creative Stylus 2
Active Stylus for iPads
Aktiver Stylus für iPads

Manufacturer
Wacom Co. Ltd., Tokyo, Japan
In-house design
Naoya Nishizawa
Web
www.wacom.com

The Intuos Creative Stylus 2 has a fine solid tip for increased visibility and precise strokes while drawing on an iPad. Thanks to 2,048 pressure levels, the stylus senses the lightest movements, just like a real pen or brush on paper or canvas. It comes with a practical case protecting the device and also offering space for the USB charging cable and replacement nibs. The stylus thus becomes an ideal pocket-sized companion for designers on the go.

Der Intuos Creative Stylus 2 verfügt über eine feine, feste Spitze, die beim Zeichnen auf einem iPad eine gute Sicht auf die Arbeit und eine präzise Strichführung ermöglicht. Dank der 2.048 Druckstufen reagiert der Stylus sensibel auf kleine Berührungen, ganz wie beim echten Zeichnen oder Malen auf Papier und Leinwand. Er wird in einem praktischen Etui ausgeliefert, das das Gerät schützt und zudem Platz für das USB-Ladekabel und Ersatzspitzen bietet. So wird der Stylus zum idealen Begleiter für den mobilen Designer.

Statement by the jury
Thanks to its high sensitivity, the Intuos Creative Stylus 2 offers a realistic drawing experience. With its selection of materials, it renders an elegant and high-grade impression.

Begründung der Jury
Dank seiner hohen Empfindlichkeit ermöglicht der Intuos Creative Stylus 2 eine realistische Zeichenerfahrung. Durch die Materialauswahl wirkt er elegant und hochwertig.

ASUSPRO Ultra
Docking Station

Manufacturer
ASUSTek Computer Inc., Taipei, Taiwan
In-house design
Asus Design Center
Web
www.asus.com

This docking station for notebooks was specifically designed for businesses in order to facilitate a more flexible and efficient working experience. Vertically mounted notebooks are securely held to prevent dislodging even if accidentally bumped. After turning it on, the system is immediately available and can be expanded to include monitors and other peripheral devices, thanks to a large number of connectors. In this way, the docking station offers a complete office environment.

Diese Dockingstation für Notebooks wurde speziell für Unternehmen entwickelt, um das Arbeiten flexibler und effizienter zu gestalten. Sie ist so konzipiert, dass vertikal eingesteckte Notebooks auch bei versehentlichen Stößen sicher gehalten werden. Das System ist sofort nach dem Einschalten verfügbar und kann dank zahlreicher Anschlüsse mit Monitoren und weiteren Peripheriegeräten erweitert werden. Dadurch bietet die Dockingstation eine vollständige Office-Umgebung.

CAP1750
Access Point

Manufacturer
Edimax Technology Co., Ltd.,
New Taipei, Taiwan
In-house design
Edimax Design Center
Web
www.edimax.com

The CAP1750 is a high-performance, ceiling-mounted access point with wireless connectivity designed for large business environments. With its plain, modern design, the device stands out against other industrial solutions. Recessed into the front of the housing are eye-catching LEDs, situated for quick indication of device status. The access point is easy to handle and offers a wide operating range.

Statement by the jury
The design of this access point is so discreet that it stays completely inconspicuous when mounted next to other ceiling units like lamps or smoke detectors.

CAP1750 ist ein leistungsstarker kabelloser Access Point für die Montage an der Decke. Das Gerät wurde für große Unternehmen entwickelt und hebt sich durch seine schlichte, moderne Gestaltung von anderen industriellen Lösungen ab. Als Blickfang sind auf der Vorderseite LEDs in das Gehäuse eingelassen, an denen sich der Gerätestatus ablesen lässt. Das Gerät ist einfach zu handhaben und bietet eine große Reichweite.

Begründung der Jury
Dieser Access Point ist so zurückhaltend gestaltet, dass er neben anderen Deckengeräten wie Lampen oder Rauchmeldern vollkommen unauffällig bleibt.

SR 1420
Outdoor Wi-Fi Antenna
Wi-Fi-Antenne für den Außenbereich

Manufacturer
Vivint, Lehi, USA
In-house design
Vivint Innovation Center
Web
www.vivint.com

The SR 1420 antenna is installed on rooftops to enable Internet connectivity via Wi-Fi outdoors. With its minimalist design, the unit blends unobtrusively with other rooftop installations. Design thinking helped add functionality, increased durability, reduced antenna cost and cut install time in half.

Statement by the jury
This Wi-Fi antenna harmonises excellently with the surrounding environment. With its bright, smooth surface and rounded edges, it renders a very friendly impression.

Die Antenne SR 1420 wird auf Dächern installiert, um den Internet-Empfang per Wi-Fi im Freien zu ermöglichen. Ihre Gestaltung ist minimalistisch, damit sie sich unauffällig neben anderen Dachinstallationen einfügt. Durch den ganzheitlichen Designprozess wurde die Funktionalität erweitert, die Haltbarkeit erhöht, die Montagezeit halbiert und die Herstellungskosten reduziert.

Begründung der Jury
Diese Wi-Fi-Antenne harmoniert vorzüglich mit ihrer Umgebung und wirkt mit der hellen, glatten Oberfläche und den gerundeten Kanten überaus freundlich.

Blackmagic ATEM
Live Production Switchers
Live-Produktionsmischer

Manufacturer
Blackmagic Design Pty Ltd, Melbourne, Australia
In-house design
Web
www.blackmagicdesign.com

The ATEM series live production switchers are equipped with functions for demanding Ultra HD workflows as employed in the production of news, sports events and TV features. The well-conceived product architecture is geared to high-pressure live broadcast situations. Moreover, the compact switchers are ideal for use in small spaces. As the control interfaces are software-based, a laptop can be used instead of a bulky switching panel. Precision CNC-machined front panels offer consistent and intuitive workflows for error-free switching and remote camera control.

Die Live-Produktionsmischer der ATEM-Serie sind mit Funktionen für anspruchsvolle Ultra-HD-Workflows ausgestattet, wie sie besonders bei Nachrichten, Sportevents und TV-Produktionen bestehen. Die wohlüberlegte Gestaltung des Produkts ist auf den stressigen Live-Broadcast-Betrieb abgestimmt, zudem können die kompakten Mischer ideal auf engem Raum eingesetzt werden. Da die Bedienoberflächen softwarebasiert sind, benötigt der Anwender keine sperrige Konsole, sondern kann seinen Laptop benutzen. CNC-präzisionsgefräste Frontblenden gestatten konsequente und intuitive Workflows für makelloses Mischen und die Fernsteuerung von Kameras.

Statement by the jury
The production switchers are compact and do without an extra switching panel, making them mobile and cost-effective.

Begründung der Jury
Die Produktionsmischer sind kompakt und kommen ohne eigene Konsole aus. Dadurch sind sie mobil und kostengünstig.

ThinkPad Stack
Computer Accessory
Computer-Zubehör

Manufacturer
Lenovo, Morrisville, North Carolina, USA
In-house design
Web
www.lenovo.com

The ThinkPad Stack is a modular system of computing equipment for mobile users. The four modules – wireless router, hard drive, speaker and battery – can be combined and stacked at the user's discretion. Matching notches and integrated magnets solidly lock the modules together so they stay in position. The clear lines and subtle, rounded corners facilitate transport and lend the product a friendly appearance.

Der ThinkPad Stack ist ein modular aufgebautes System von Computerzubehör für mobile Anwender. Die vier Module – kabelloser Router, Festplatte, Lautsprecher und Akku – lassen sich nach Belieben kombinieren und stapeln. Dabei sorgen passende Aussparungen und integrierte Magnete dafür, dass sie fest in ihrer Position bleiben. Die sauberen Linien und leicht gerundeten Kanten sind angenehm beim Transport und verleihen dem Produkt eine sympathische Anmutung.

Statement by the jury
The ThinkPad Stack assembles intelligently coordinated devices within a small space, thus offering a wide variety of functions for professional mobile users.

Begründung der Jury
Der ThinkPad Stack versammelt auf kleinstem Raum intelligent aufeinander abgestimmte Geräte. Damit bietet es professionellen, mobilen Anwendern eine Vielzahl an Funktionen.

ProSAFE® Click Switches™
Gigabit Ethernet Switches
Gigabit-Netzwerk-Switches

Manufacturer
NETGEAR, Inc., San Jose, USA
In-house design
John Ramones, Rose Hu, Ed Kalubiran
Design
Enlisted Design (Beau Oyler, Julian Bagirov,
Charles Bates, Jeff Tung, Jared Aller), Oakland, USA
Web
www.netgear.com
www.enlisteddesign.com

The ProSAFE Click Switches feature a bracket mounting system for easy installation and repair. Since they do not require a fan, operation is completely silent. With their nano suction feet, the switches may be attached to suitable surfaces without any tools. Additional features are the integrated USB charging port, a multidirectional power cable and the narrow ventilation slots for minimal dust penetration. The design of the housing is slender and inconspicuous, so that the switches cut an attractive figure not only in a server cabinet, but on a conference table as well.

Die ProSAFE Click Switches besitzen ein Click-Montagesystem für die einfache Installation und Reparatur. Da sie keinen Lüfter benötigen, arbeiten sie geräuschlos. Mit ihren Nano-Saugfüßen können sie ohne Werkzeug an geeigneten Flächen angebracht werden. Weitere Merkmale sind der integrierte USB-Anschluss, ein multi-direktionales Netzkabel und schmale Lüftungsschlitze, durch die kaum Staub dringen kann. Das Gehäuse ist schlank und unauffällig gestaltet, sodass die Switches nicht nur in einem Serverschrank, sondern auch auf einem Konferenztisch eine gute Figur machen.

Statement by the jury
The ProSAFE Click Switches are easy to install and, thanks to their discreet look, can be integrated into any environment.

Begründung der Jury
Die ProSAFE Click Switches lassen sich leicht installieren und können dank ihrer zurückhaltenden Gestaltung in jede Umgebung integriert werden.

Dell X1008/X1008P
Network Switches
Netzwerk-Switches

Manufacturer
Dell Inc., Round Rock, Texas, USA
In-house design
Experience Design Group
Web
www.dell.com

The X1008 and X1008P network switches are designed for use in small and medium-sized businesses, accommodating both tabletop operation and mounting under a desk or on a wall. They have fan-less aluminium enclosures, are very compact and also stackable. Both versions include eight 1 GbE ports and enable Power over Ethernet (PoE), thus supplying the switches themselves and also any auxiliary devices with power and data.

Statement by the jury
The network switches have an inconspicuous design and require only little space, allowing for flexible positioning or mounting.

Die Netzwerk-Switches X1008 und X1008P sind für den Tischbetrieb wie auch für die Montage unter dem Tisch und an der Wand konzipiert und eignen sich für kleine und mittlere Unternehmen. Sie besitzen lüfterlose Aluminiumgehäuse, sind sehr kompakt und lassen sich zudem stapeln. Beide verfügen über acht 1-GbE-Ports und ermöglichen die Stromversorgung über Ethernet (PoE), sodass die Switches selbst sowie Hilfsgeräte mit Daten und Strom versorgt werden.

Begründung der Jury
Die Netzwerk-Switches sind unauffällig gestaltet und benötigen nur wenig Platz. Dadurch können sie flexibel aufgestellt oder montiert werden.

MPH2™
Managed Rack Power Distribution Unit
Steuerbare Stromversorgungseinheit

Manufacturer
Knürr GmbH, Emerson Network Power, Arnstorf, Germany
In-house design
Christian Stepputat
Web
www.emersonnetworkpower.com

The MPH2 is a power distribution unit for data centres with remote monitoring and control functions, as well as options for controlling environmental conditions such as rack temperature and air humidity. It was designed for high operation temperatures and can thus be used under difficult conditions as well. Its design is functional and ergonomic, and both physical and electrical installation is easily accomplished.

Statement by the jury
This power distribution unit accommodates many functions in minimal space and can also be easily integrated into existing networks.

MPH2 ist eine Stromversorgungseinheit für Rechenzentren. Sie bietet Fernüberwachungs- und Steuerfunktionen sowie Optionen für die Kontrolle von Umgebungsbedingungen wie Racktemperatur oder Luftfeuchtigkeit. Sie wurde für hohe Betriebstemperaturen ausgelegt und ist deshalb auch unter schwierigen Bedingungen einsetzbar. Ihre Gestaltung ist funktional und ergonomisch, die physische wie elektrische Installation geht einfach vonstatten.

Begründung der Jury
Die Stromversorgungseinheit bringt auf engstem Raum viele Funktionen unter und lässt sich auch leicht in bestehende Netzwerke integrieren.

Wireless Mouse M320
Kabellose Maus

Manufacturer
Logitech, Newark, California, USA
In-house design
Logitech Design
Design
Design Partners, Bray, Ireland
Web
www.logitech.com
www.designpartners.com

The Wireless Mouse M320 combines functionality, aesthetics and comfort. With its asymmetrical shape and the textured, soft rubber surface, it fits comfortably in the hand for long-lasting use. The cursor can be controlled with precision on a wide variety of surfaces, and the wide wheel facilitates smooth scrolling. The small USB receiver provides a reliable wireless 2.4 GHz connection. As the mouse is energy-efficient, a single battery will last for two years.

Die Wireless Mouse M320 verbindet Funktionalität, Ästhetik und Komfort. Mit ihrer asymmetrischen Form und der strukturierten, weich gummierten Oberfläche liegt sie bequem in der Hand und ermöglicht dadurch eine lange Nutzung. Der Cursor lässt sich präzise auf einer Vielzahl von unterschiedlichen Oberflächen steuern, zusätzlich ermöglicht das breite Scrollrad flüssige Bildläufe. Der kleine USB-Empfänger bietet eine zuverlässige kabellose 2,4-GHz-Verbindung. Da die Maus energiesparend arbeitet, reicht eine Batterie für zwei Jahre.

Statement by the jury
With its smooth structures, the mouse has a clean aesthetic appearance. The generously laid out contact surfaces are appealing to the touch and promise a high degree of comfort.

Begründung der Jury
Die Maus zeigt mit ihren glatten Strukturen eine aufgeräumte Ästhetik. Die großzügigen, haptisch ansprechenden Auflageflächen versprechen einen hohen Komfort.

Ventus Laser
Gaming Mouse
Gaming-Maus

Manufacturer
Thermaltake Technology Co., Ltd., Taipei, Taiwan
In-house design
Web
www.thermaltakecorp.com

With its symmetrical construction, this gaming mouse is suited for both right- and left-handers. As such, it features a thumb button on either side, which can be activated or disabled through the mouse software. The laser sensor offers a resolution of up to 5,700 DPI and enables precise navigation. The honeycomb cut-out is an appealing design feature, keeping the user's hand cool and dry when operating the mouse over an extended period of time.

Die Gaming-Maus eignet sich dank ihres symmetrischen Aufbaus für Rechts- und für Linkshänder. So hat sie beispielsweise eine Daumentaste auf jeder Seite, die jeweils nicht benötigte Taste lässt sich per Treiber deaktivieren. Der Laser bietet eine Auflösung von bis zu 5.700 DPI und ermöglicht das präzise Navigieren. Die wabenförmige Öffnung ist ein ansprechendes Gestaltungsmerkmal und lässt Luft an die Handfläche, was bei längerer Benutzung angenehm ist.

Statement by the jury
Ventus Laser offers ergonomic features for right- and left-handers in a sporty design with a black body and red accents.

Begründung der Jury
Ventus Laser bietet Ergonomie für Rechts- wie für Linkshänder in einem sportlichen Design mit schwarzem Gehäuse und roten Akzenten.

Aorus Thunder M7
MMO Gaming Mouse
MMO-Gaming-Maus

Manufacturer
Gigabyte Technology Co., Ltd., Taipei, Taiwan
In-house design
Wanyun Chou
Web
www.gigabyte.com

The intuitive macro key arrangement of the Thunder M7 gaming mouse was designed to facilitate the quick and accurate actuation of commands, which is of vital importance in the gaming field. Moreover, the ergonomic arrangement offers comprehensive comfort during extended gaming sessions. With a design inspired by a sports car and straightforward line management, the mouse heightens the gaming experience. The replica of a V8 engine with a transparent cover is a particularly attractive visual accent. Its gaming laser sensor supports a resolution of up to 8,200 dpi and a tracking speed of 150 ips.

Das intuitive Makrotastenlayout der Gaming-Maus Thunder M7 wurde so konstruiert, dass jede Taste im richtigen Moment präzise betätigt werden kann. Dies ist gerade im Gaming-Bereich von Vorteil, wo exakte, schnelle Befehle unabdingbar sind. Zusätzlich bietet die ergonomische Anordnung auch bei längeren Spielesessions einen umfassenden Komfort. Mit ihrem von einem Sportwagen inspirierten Design und der schlichten Linienführung unterstreicht die Maus das Spieleerlebnis. Besonders die Nachbildung eines V8-Motors mit einer transparenten Haube ist ein attraktiver optischer Akzent. Ihr Gaming-Lasersensor unterstützt eine Auflösung von bis zu 8.200 dpi und 150 ips Erkennungsgeschwindigkeit.

Statement by the jury
This gaming mouse conveys a very sporty appearance. It offers a sophisticated key layout, guaranteeing the precision and speed so important for gamers.

Begründung der Jury
Diese Gaming-Maus wirkt überaus sportlich. Sie bietet ein ausgefeiltes Tastenlayout, das die für Gamer so wichtige Präzision und Schnelligkeit gewährleistet.

K480
Bluetooth Multi-Device Keyboard
Bluetooth-Tastatur

Manufacturer
Logitech, Newark, California, USA
In-house design
Logitech Design
Design
Feiz Design, Amsterdam, Netherlands
Web
www.logitech.com
www.feizdesign.com

The Bluetooth Multi-Device Keyboard K480 can be used with computers, tablets and smartphones. Up to three devices may be wirelessly connected at the same time. The user can effortlessly switch between these devices with the Easy-Switch dial. Moreover, the keyboard offers the familiar arrangement of keys and the shortcuts for Windows, Mac and Chrome computers and/or Android and iOS mobile devices. The integrated cradle holds multiple smartphone or tablet devices, so that the user always has an overview of their displays.

Das Bluetooth Multi-Device Keyboard K480 ist eine Tastatur für Computer, Tablet und Smartphone. Bis zu drei Geräte lassen sich gleichzeitig kabellos verbinden. Über einen Easy-Switch-Schalter kann der Anwender dann zwischen diesen Geräten unkompliziert wechseln. Darüber hinaus bietet die Tastatur die vertraute Tastenanordnung und Shortcuts für Windows-, Mac- oder Chrome-Computer bzw. für Android- und iOS-Mobilgeräte. In die integrierte Halterung für Smartphones und Tablets können mehrere Geräte gleichzeitig gestellt werden, sodass deren Displays immer im Blick bleiben.

Statement by the jury
With this keyboard, switching between multiple devices is very simple, which makes working with more than one mobile device simultaneously very convenient.

Begründung der Jury
Das Wechseln zwischen den Geräten ist mit dieser Tastatur sehr einfach. Sie ermöglicht es, sehr bequem mit mehreren Mobilgeräten zu arbeiten.

Universal Mobile Keyboard
Tastatur

Manufacturer
Microsoft Asia Center for Hardware, Shenzhen, China
In-house design
MACH Design Team
Web
www.microsoft.com

The Universal Mobile Keyboard was designed for the highest possible mobility. It offers a protective cover and a practical stand for tablets and smartphones. Up to three devices with Windows, iOS or Android operating systems can be paired with the keyboard via Bluetooth. This allows the user to easily switch between the devices. The keyboard automatically turns on as soon as the cover is opened. At the same time, the Bluetooth connection is established with the paired devices so that the user can start typing at once.

Das Universal Mobile Keyboard wurde im Hinblick auf eine größtmögliche Mobilität konzipiert. Es bietet eine schützende Abdeckung und eine praktische Halterung für Tablets oder Smartphones. Bis zu drei Geräte mit Windows-, iOS- oder Android-Betriebssystem können per Bluetooth mit der Tastatur gekoppelt werden. Dies ermöglicht dem Anwender, ganz einfach zwischen diesen Geräten zu wechseln. Mit dem Öffnen der Abdeckung schaltet sich die Tastatur ein. Gleichzeitig wird die Bluetooth-Verbindung mit gekoppelten Geräten hergestellt, sodass der Anwender sofort mit dem Tippen beginnen kann.

Statement by the jury
This keyboard considerably facilitates working with smartphones or tablets. Connecting the devices to the keyboard is child's play.

Begründung der Jury
Die Tastatur macht das Arbeiten mit Smartphones oder Tablets wesentlich komfortabler. Die Anbindung der Geräte ist dabei kinderleicht.

Universal Mobile Keyboard
Tastatur

Manufacturer
Microsoft Corporation,
Redmond, Washington, USA
In-house design
Web
www.microsoft.de/hardware

The Universal Mobile Keyboard was designed for use with Windows tablets, iPads, iPhones and Android devices. It can be paired with up to three devices with different operating systems, allowing the user to easily switch back and forth between them. The keyboard has a protective cover and a built-in stand that can hold a tablet or smartphone. A single battery charge provides power for up to six months.

Statement by the jury
The integrated stand is practical and protects the keys at the same time, turning the keyboard and paired device into a single unit.

Das Universal Mobile Keyboard wurde für den Einsatz mit Windows-Tablets, iPads, iPhones und Android-Geräten entwickelt. Es lässt sich mit bis zu drei Geräten mit unterschiedlichen Betriebssystemen koppeln, der Anwender kann dann problemlos zwischen diesen Geräten wechseln. Die Tastatur besitzt eine Schutzhülle und eine Halterung, in die ein Tablet oder Smartphone hineingestellt werden kann. Eine Akkuladung reicht bis zu sechs Monate.

Begründung der Jury
Die integrierte Halterung ist praktisch und schützt gleichzeitig die Tasten. Dank ihr werden Tastatur und gekoppeltes Gerät zu einer Einheit.

Poseidon Z Forged
Mechanical Gaming Keyboard
Mechanische Gaming-Tastatur

Manufacturer
Thermaltake Technology Co., Ltd.,
Taipei, Taiwan
In-house design
Web
www.thermaltakecorp.com

The Poseidon Z Forged gaming keyboard provides 114 keys with red and blue illumination. Additionally, it features ten macro keys and different profiles so that a total of 150 macros can be programmed for normal and gamer modes. Illumination brightness is adjustable in five steps. With its aluminium upper surface and high-grade mechanical keys, the keyboard is constructed for a long life cycle.

Statement by the jury
This keyboard renders a solid impression and gives the user many options for customising the keys according to individual preferences.

Die Gaming-Tastatur Poseidon Z Forged bietet 114 Tasten mit roter und blauer Beleuchtung. Zusätzlich verfügt sie über zehn Makrotasten und verschiedene Profile, sodass sich insgesamt 150 Makros für den normalen und für den Spielemodus programmieren lassen. Auch die Helligkeit der Tastenbeleuchtung kann in fünf Stufen angepasst werden. Die Tastatur ist mit ihrer Oberseite aus Aluminium und den hochwertigen mechanischen Tasten auf eine lange Lebensdauer ausgelegt.

Begründung der Jury
Die Tastatur macht einen soliden Eindruck und bietet dem Anwender zahlreiche Möglichkeiten, Tasten individuell zu belegen.

Keys-To-Go
Keyboard
Tastatur

Manufacturer
Logitech, Newark, California, USA
In-house design
Logitech Design
Web
www.logitech.com

The Keys-To-Go portable mobile keyboard was specifically designed for the iPad and any smart cover. It features a comfortable key arrangement and well-spaced keys, as well as iOS shortcuts for easy typing. Thanks to its small size, the keyboard is convenient to carry with an iPad or iPad mini. The water-repellant and washable FabricSkin surface protects the keyboard against drops and enables easy cleaning. The rechargeable lithium-ion polymer battery provides up to three months of runtime, and Bluetooth connectivity provides wireless freedom anywhere.

Die tragbare Tastatur Keys-To-Go wurde speziell als Ergänzung für das iPad und jedes beliebige Cover entwickelt. Die Tasten sind komfortabel, mit angenehm großen Abständen angeordnet, iOS-Shortcuts erleichtern das Tippen. Dank ihrer geringen Größe lässt sich die Tastatur bequem zusammen mit einem iPad oder iPad mini transportieren. Die wasserabweisende und abwaschbare FabricSkin-Oberfläche schützt gegen Spritzwasser und kann leicht gereinigt werden. Der Lithium-Polymer-Akku bietet bis zu drei Monate Laufzeit, und die Bluetooth-Verbindung bietet kabellose Freiheit an jedem Ort.

Statement by the jury
With its rounded edges, this keyboard harmonises very well with an iPad. As it is very handy, it facilitates productivity on the go.

Begründung der Jury
Diese Tastatur harmoniert mit ihren gerundeten Kanten sehr gut mit einem iPad. Da sie sehr handlich ist, erleichtert sie das Arbeiten unterwegs.

Aorus Thunder K7
Mechanical Gaming Keyboard
Mechanische Gaming-Tastatur

Manufacturer
Gigabyte Technology Co., Ltd., Taipei, Taiwan
In-house design
Wanyun Chou
Web
www.gigabyte.com

This modular mechanical gaming keyboard with magnetic coupling offers gamers a high degree of flexibility. The detachable macro keypad can be used all by itself or else connected from the side. Without this keypad module, the keyboard is especially handy and compact. The two keyboard modules, along with an ergonomic, full-size wrist rest, are connected magnetically, enabling fast reconfiguration and withstanding even rigorous operating conditions. Volume and illumination can be adjusted in no time at all with the wheel controllers.

Die modulare, mechanische Gaming-Tastatur mit magnetischer Kopplung bietet Gamern eine hohe Flexibilität. Der abnehmbare Makrotastenblock kann unabhängig genutzt oder an die Seiten angekoppelt werden. Ohne dieses Modul wird die Tastatur besonders handlich und kompakt. Die beiden Tastaturmodule und eine ergonomische Handgelenkauflage in voller Größe werden magnetisch gekoppelt, erlauben die schnelle Neukonfiguration und halten auch intensive Belastungen aus. Lautstärke und Beleuchtung lassen sich mit Drehreglern im Handumdrehen anpassen.

Statement by the jury
The Thunder K7 keyboard is compact, flexibly adaptable and thus perfectly suitable for computer gamers. The angular design renders a robust and sporty impression.

Begründung der Jury
Die Tastatur Thunder K7 ist kompakt, flexibel anpassbar und damit für Computerspieler bestens geeignet. Die kantige Gestaltung wirkt robust und sportlich.

G910 Orion Spark
RGB Mechanical Gaming Keyboard
Mechanische RGB-Gaming-Tastatur

Manufacturer
Logitech, Newark, California, USA
Design
Design Partners, Bray, Ireland
Web
www.logitech.com
www.designpartners.com

The G910 Orion Spark is a fast mechanical gaming keyboard with exclusive Romer-G mechanical switches. Up to 25 per cent faster actuation and reliable anti-ghosting protection provide the decisive edge in battles were every millisecond matters. The intelligent illumination of each key can be customised from a palette of more than 16.8 million colours, helping gamers to execute commands more swiftly. The key arrangement has been optimised with respect to speed and accessibility, and two interchangeable palm rests ensure high comfort.

Die G910 Orion Spark ist eine schnelle mechanische Gaming-Tastatur mit exklusiven mechanischen Romer-G-Schaltern. Eine um 25 Prozent schnellere Tastenreaktion und eine verlässliche Anti-Ghosting-Funktion bieten einen entscheidenden Vorsprung in Kämpfen, bei denen jede Millisekunde zählt. Die Beleuchtung jeder Taste lässt sich auf der Grundlage von über 16,8 Millionen Farben anpassen, damit der Anwender Gaming-Befehle schneller ausführen kann. Die Anordnung der Tasten wurde in Hinblick auf Geschwindigkeit und Erreichbarkeit optimiert, zwei austauschbare Handballenauflagen sorgen für einen hohen Komfort.

Statement by the jury
The G910 Orion Spark is optimally geared to gamers' requirements. Thanks to the tailorable keyboard illumination, each gamer maintains a strong overview.

Begründung der Jury
Die G910 Orion Spark ist optimal auf die Erfordernisse von Gamern ausgelegt. Dank der anpassbaren Tastenbeleuchtung behält der Spieler den Überblick.

HD-PZNU3
Portable External Hard Drive
Tragbare externe Festplatte

Manufacturer
Buffalo Inc., Nagoya, Japan
In-house design
Gakuto Takahashi
Web
www.buffalotech.com

The HD-PZNU3 is a portable hard drive with secure protection from data leakage, shock, water and dust. It has a robust enclosure with internal shock absorbers protecting the drive from accidental drops of up to two metres. Thanks to the USB cable, the drive is always ready for use. Data are securely stored on the drive and can be unlocked via smart card or NFC-compatible smartphones.

Statement by the jury
This hard drive impresses with its functional design and guarantees outstanding data protection even in extreme environments.

Die tragbare Festplatte HD-PZNU3 ist vcr Datenverlust, Erschütterungen, Wasser und Staub geschützt. Sie zeichnet sich durch ein robustes Gehäuse aus und besitzt zudem im Inneren Dämpfer, die Stürze aus einer Höhe von bis zu zwei Metern absorbieren. Da das USB-Kabel eingebaut ist, ist es immer griffbereit. Daten werden auf der Festplatte geschützt abgelegt und mittels Chipkarte oder NFC-kompatiblem Smartphone wieder freigegeben.

Begründung der Jury
Diese Festplatte gefällt mit ihrer funktionalen Gestaltung und gewährleistet selbst in extremen Umgebungen einen hervorragenden Schutz der Daten.

Portable SSD T1
External SSD
Externe SSD

Manufacturer
Samsung Electronics GmbH,
Schwalbach, Germany
In-house design
Web
www.samsung.com

The Portable SSD T1 is a representative of the new sort of external storage product capable of storing data considerably faster than their conventional counterparts; even large amounts of data may be accommodated. Its compact and slim housing is made of black metal and decorated with a characteristic pattern. Data is protected against unauthorised access through 256-bit AES hardware encryption. The SSD is available with storage capacities of 250 GB, 500 GB and 1 TB.

Statement by the jury
The Portable SSD T1 is an elegant, very mobile data storage solution that considerably facilitates the process of managing data on the go.

Die Portable SSD T1 ist ein Vertreter der neuen Laufwerkstypen, die Daten deutlich schneller speichern als ihre konventionellen Pendants. Dadurch ist sie besonders für große Datenmengen geeignet. Sie besitzt ein kompaktes und schlankes Gehäuse aus schwarzem Metall, das mit einem charakteristischen Muster verziert ist. Daten sind durch die AES-Hardware-Verschlüsselung mit 256 Bit gegen Fremdzugriff geschützt. Erhältlich ist die SSD mit den Kapazitäten 250 GB, 500 GB und 1 TB.

Begründung der Jury
Die Portable SSD T1 ist ein eleganter, sehr mobiler Datenspeicher, der das Datenmanagement unterwegs erheblich erleichtert.

SE-208GB
External Optical Disc Drive
Externes optisches Disc-Laufwerk

Manufacturer
TSST Korea, Suwon, South Korea
In-house design
Jaesik Park, Kyel Ko
Web
www.tsstodd.com

The SE-208GB external optical disc drive burns both CDs and DVDs and reads all conventional CD and DVD formats. It offers a high degree of mobility and user-friendliness, as the eject button and the LED status indicator are located at the top and thus easily accessible. The device has a thickness of only 1.4 cm and a weight of merely 331 grams, enabling it to be easily stowed away and for instance carried in a bag together with a laptop. Available in blue, red, black and white, it places importance on the user's individuality.

Das externe optische Disc-Laufwerk SE-208GB brennt CDs und DVDs und liest alle gängigen CD- und DVD-Formate. Es bietet ein hohes Maß an Mobilität und Nutzerfreundlichkeit, denn Auswurfknopf und LED-Statusanzeige befinden sich auf der Oberseite und sind damit leicht zugänglich. Das Gerät ist nur 1,4 cm dick und mit 331 Gramm sehr leicht, wodurch es sich beispielsweise neben einem Laptop gut in der Tasche verstauen und tragen lässt. Verfügbar in Blau, Rot, Schwarz und Weiß, betont es die Individualität des Nutzers.

Statement by the jury
This disc drive is hardly thicker than a CD and, with colours ranging from striking to discreet, caters to the tastes of various users.

Begründung der Jury
Das Laufwerk ist nur wenig dicker als eine CD und entspricht mit seinen unterschiedlichen Farben von auffallend bis dezent dem Geschmack vieler Nutzer.

Seven
External Hard Drive
Externe Festplatte

Manufacturer
Seagate, Cupertino, USA
In-house design
Web
www.seagate.com

Seven is a very slim external hard drive whose design was inspired by internal 2.5" hard drives. It pays homage to the roots of computer storage and simultaneously embodies the advances with respect to innovation of storage. The stainless-steel enclosure was machined using the deep-draw process and is merely 7 mm thin, yet durable. The hard drive includes USB 3 connectivity and provides 500 GB of memory.

Statement by the jury
The Seven hard drive elevates the classic storage medium to a new technical and visual level in an impressive way.

Seven ist eine sehr schlanke externe Festplatte, deren Formgebung von internen 2,5"-Festplatten inspiriert wurde. Sie ist eine Hommage an die Wurzeln von Computerspeichern und verkörpert gleichzeitig den Fortschritt in Sachen Speicherinnovation. Das im Tiefziehverfahren hergestellte Edelstahlgehäuse ist lediglich 7 mm dünn, dabei jedoch stabil. Die Festplatte ist mit einer USB-3-Schnittstelle ausgestattet und bietet 500 GB Speicherplatz.

Begründung der Jury
Die Festplatte Seven hebt ein klassisches Speichermedium in beeindruckender Weise auf ein neues technisches und optisches Niveau.

iKlips
Flash Drive
Flash-Laufwerk

Manufacturer
Adam Elements International Co., Ltd., New Taipei, Taiwan
In-house design
Hung-Yi Lin
Web
www.adamelements.com

iKlips is a flash drive with lightning and USB interfaces. It allows simple data transfer between iPhone, iPad, Mac and PC. The drive achieves storage of up to 256 GB, thus offering enough capacity for multimedia files as well. It can be attached to clothing or other objects with a clip. A metal frame prevents accidental bending of the device, and elastic silicone caps protect the connectors in case the drive is dropped.

Statement by the jury
iKlips is a versatile data-storage medium that can be easily taken along anywhere. It is robust and, thanks to its clip, cannot be lost.

iKlips ist ein Flash-Laufwerk mit Lightning- und USB-Schnittstelle. Es erlaubt den einfachen Transfer von Dateien zwischen iPhone, iPad, Mac und PC. Das Laufwerk hat bis zu 256 GB Speicher und bietet damit auch für Multimedia-Dateien genügend Kapazität. Mit einem Clip wird es an der Kleidung oder an anderen Gegenständen befestigt. Dabei verhindert ein Metallrahmen, dass es sich verbiegt, elastische Kappen aus Silikon schützen die Stecker, wenn es hinfällt.

Begründung der Jury
iKlips ist ein vielseitiger Datenspeicher, der sich leicht überallhin mitnehmen lässt. Er ist robust und kann dank seines Clips nicht verloren gehen.

DTSE9 G2
USB Flash Drive
USB-Stick

Manufacturer
Kingston Technology Company, Inc., Fountain Valley, USA
Design
Emamidesign (Arman Emami), Berlin, Germany
Web
www.kingston.com
www.emamidesign.de

With its large opening and tapered end, this USB flash drive may be effortlessly attached to a key ring or karabiner and is easily detached as well. Its design is reduced to the bare essentials and does without unnecessary details. The ergonomic grip identifies the top and bottom during the process of inserting the drive into a USB slot. During the design phase, the geometry of the standardised metal frames of dock connectors was used for the entire casing, so that the flash drive forms a seamless unit. The metal casing provides protection from scratches.

Der USB-Stick kann durch die große Öffnung und das schmalere Ende ganz einfach z. B. an einem Schlüsselbund oder Karabiner angebracht und wieder abgenommen werden. Seine Form ist auf das Wesentliche reduziert und verzichtet auf unnötige Details. Der ergonomische Griffbereich erleichtert die Zuordnung von Ober- und Unterseite während des Einsteckens in den USB-Slot. Bei der Gestaltung wurde die Geometrie von genormten Metallrahmen des Dock-Connectors für das gesamte Gehäuse übernommen, sodass der USB-Stick eine nahtlose Einheit bildet. Das Metallgehäuse bietet Schutz vor Kratzern.

HP Color LaserJet Enterprise M553
Printer
Drucker

Manufacturer
Hewlett-Packard, Boise, USA

Design
Hewlett-Packard In-House Team
Native Design

Web
www.hp.com
www.native.com

reddot award 2015
best of the best

Elegant icon

Ever since computers found their way into work environments, the aesthetics of printers have also been adjusted to fit daily office routines. Both devices often constitute a formal unity. However, the HP Color LaserJet Enterprise M553 is a powerful colour laser printer with a purist appearance that also allows it to stand on its own. It impresses with clear radii and well-balanced proportions. Since even the vents are concealed, nothing distracts from its pure form. This printer was designed for modern office environments, into which it blends seamlessly, yielding an almost iconic look. An innovative feature is the combination of intelligent, media-oriented technology and a specially developed toner for extremely low energy consumption and lower overall costs. It also achieves very good print results in work groups, and even prints two-sided documents with the same speed as one page. Moreover, this printer offers highly user-friendly operation, featuring a colour combination that enhances interaction during printing tasks. Ergonomically shaped side-pull handles make opening and closing it easy. The printer merges traditional, familiar functionality with an elegant progressive appearance to present an image of an overall harmonious whole, designed for modern office environments.

Elegante Ikone

Seitdem der Computer Einzug in die Arbeitswelt gehalten hat, wurden auch die Drucker in ihrer Ästhetik dem Büroalltag angepasst. Beide Geräte bilden oft eine formale Einheit. Der HP Color LaserJet Enterprise M553 ist ein leistungsfähiger Farblaserdrucker, der durch seine puristische Anmutung aber auch gut für sich alleine stehen kann. Er beeindruckt mit klaren Radien und ausgewogenen Proportionen. Da auch die Lüftungsschlitze versteckt gestaltet sind, stört nichts die pure Form. Konzipiert wurde dieser Drucker für die moderne Büroumgebung, in die er sich perfekt einfügt und dort eine geradezu ikonische Ausstrahlung gewinnt. Innovativ ist seine Kombination aus intelligenter, medien-sensitiver Technologie und speziell entwickeltem Toner, um einen extrem niedrigen Energieverbrauch zu erzielen und die Gesamtkosten deutlich zu senken. Auch bei Arbeitsgruppen liefert er dabei sehr gute Druckergebnisse und kann sogar zweiseitige Dokumente mit der gleichen Geschwindigkeit wie eine Seite drucken. Dieser Drucker bietet zudem ein nutzerfreundliches Bedienkonzept. So ist er gestaltet mit einer Farbgebung, die die Interaktion bei der Bedienung erleichtert. Ergonomisch geformte Handgriffe ermöglichen es, ihn unkompliziert zu öffnen und wieder zu schließen. Traditionell gewohnte Funktionalität verbindet sich bei diesem Drucker mit einer progressiv anmutenden Eleganz zu einem überaus harmonischen Gesamteindruck für das moderne Büroumfeld.

Statement by the jury

With its perfectly implemented purism, the HP Color LaserJet Enterprise M553 exudes an elegant appearance. Each detail is consistently honed, featuring a carefully crafted technology inside that also fascinates. Based on a sophisticated, intelligent printing technique and a specially developed toner, this user-friendly printer delivers excellent printing results with very low energy consumption.

Begründung der Jury

Mit seinem perfekt ausgeführten Purismus verbreitet der Farblaserdrucker HP Color LaserJet Enterprise M553 eine elegante Anmutung. Jedes Detail ist stimmig ausgearbeitet, wobei auch sein sorgfältig gestaltetes Innenleben fasziniert. Auf der Basis einer ausgereiften, intelligenten Drucktechnik und eines speziell entwickelten Toners bietet dieser benutzerfreundliche Drucker exzellente Druckergebnisse sowie einen sehr niedrigen Energieverbrauch.

Designer portrait
See page 74
Siehe Seite 74

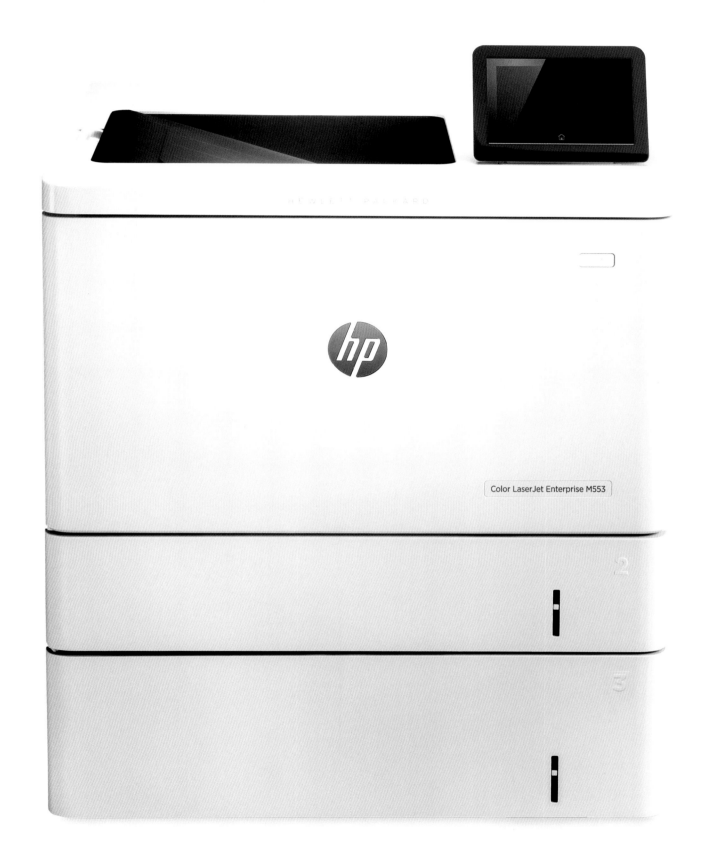

HP Color LaserJet Pro MFP M277
Printer
Drucker

Manufacturer
Hewlett-Packard, Boise, USA
In-house design
Web
www.hp.com

This elegant multifunction printer is compact and, with its reduced form, blends seamlessly into any modern work environment. Its well-conceived colour contrasts and subtle curves invite the user to interact with the device. When the printer wakes up from sleep mode, it prints very fast as compared to other printers in its class. The scanner cover closes slowly, making the scanning experience even more convenient.

Statement by the jury
Contoured line management and reduced colour design give this laser printer a progressive aesthetic appearance that enriches any environment.

Dieser elegante Multifunktionsdrucker ist kompakt und passt mit seiner reduzierten Form in jede moderne Arbeitsumgebung. Seine durchdacht gesetzten Farbkontraste und die feinen Kurven laden den Benutzer zur Interaktion ein. Aus dem Ruhemodus heraus druckt er im Vergleich zu anderen Geräten seiner Klasse sehr schnell. Die Scannerabdeckung schließt angenehm langsam, was das Scannen noch komfortabler macht.

Begründung der Jury
Die runde Linienführung und die reduzierte Farbgebung verleihen diesem Laserdrucker eine fortschrittliche Ästhetik, die jede Umgebung bereichert.

Leitz Icon
Labelling System
Etikettiersystem

Manufacturer
Esselte Leitz GmbH & Co KG, Stuttgart, Germany
Design
Whipsaw Inc, San Jose, USA
Web
www.leitz.com/icon
www.whipsaw.com

The Leitz Icon is a compact, portable label printer with accompanying cartridge in a reduced design for professional and private users. It prints adhesive and non-adhesive labels for a wide range of applications such as letters and parcels, as well as labelling for office, household and storage use. Featuring Wi-Fi connectivity and a rechargeable battery, the printer can be operated virtually anywhere via a smartphone or tablet. The label cartridges are made of compostable paper pulp, which simultaneously serves as packaging. Thanks to thermal printing, the system does not require ink or toner.

Statement by the jury
This label printer is so compact that it can be used precisely where it is needed, thus giving the user a high degree of freedom.

Leitz Icon ist ein kompakter, tragbarer Etikettendrucker mit Kassette in reduzierter Formgebung für den professionellen und privaten Nutzer. Er druckt Etiketten und Einsteckschilder für eine Vielzahl von Anwendungen, wie Briefe und Pakete sowie Beschriftungen in Büro, Haushalt und Lager. Da er mit Wi-Fi ausgestattet ist und auch per Akku betrieben werden kann, ist er überall einsatzbereit. Bedient wird der Drucker über Smartphone oder Tablet. Die Etikettenkartuschen sind aus kompostierbarem Pappmaché gefertigt, das gleichzeitig als Verpackung dient. Dank Thermodruck ist weder Tinte noch Toner nötig.

Begründung der Jury
Dieser Etikettendrucker ist so kompakt, dass er genau dort eingesetzt werden kann, wo er benötigt wird. Das gibt dem Nutzer viele Freiheiten.

HP Envy 5660
e-All-in-One
Printer
Drucker

Manufacturer
Hewlett-Packard, San Diego, USA
In-house design
Justin Francke, David Leong
Web
www.hp.com

The Envy 5660 e-All-in-One printer was designed for the home environment. It prints photos and text in high quality from smartphones and tablets. In addition to functions like duplex printing and flatbed scanning, the printer features dedicated photo and media trays. As the colour touchscreen is slightly slanted and located in central position, it can be operated comfortably. The scanner lid has a soft-touch surface with a pleasant tactile appeal.

Statement by the jury
With its shape slightly tapering to the front and round contours, this printer has a friendly appearance that blends well into the home environment.

Der Drucker Envy 5660 e-All-in-One wurde für den Heimbereich entwickelt. Er druckt Fotos und Text in hoher Qualität von Smartphones oder Tablets. Neben Funktionen wie Duplexdruck und Flachbettscannen besitzt er spezielle Zufuhrschächte für Fotopapier und Medien. Da der Farb-Touchscreen leicht angeschrägt in zentraler Position liegt, kann er bequem bedient werden. Die Abdeckung des Scanners mit ihrer Softtouch-Oberfläche fasst sich angenehm an.

Begründung der Jury
Mit seiner sich nach vorne leicht verjüngenden Form und den runden Konturen macht der Drucker einen freundlichen Eindruck und fügt sich gut in den Heimbereich ein.

HP Officejet 8040
e-All-in-One
Printer
Drucker

Manufacturer
Hewlett-Packard, San Diego, USA
In-house design
Justin Francke, David Leong
Web
www.hp.com

The Officejet 8040 e-All-in-One is a full-featured inkjet printer for the home office. Data is automatically stored and managed in the cloud or in the 1 TB onboard storage. Documents may be printed from mobile devices, and the automated photo tray allows for unattended printing. Additional functions such as duplex printing, fax and automatic scanning enhance productivity during work.

Statement by the jury
The printer supports home-office work with intelligent functions like the large integrated storage solution and access to the cloud.

Der Officejet 8040 e-All-in-One ist ein voll ausgestatteter Tintenstrahldrucker für das Homeoffice. Daten können automatisch in der Cloud oder im 1 TB großen Onboard-Speicher abgelegt und verwaltet werden. Dokumente lassen sich von Mobilgeräten drucken, und die automatische Fotopapierzuführung ermöglicht auch den unbeaufsichtigten Druck. Weitere Funktionen wie Duplexdruck, Fax und automatisches Scannen erhöhen die Produktivität bei der Arbeit.

Begründung der Jury
Der Drucker unterstützt die Arbeit im Homeoffice mit sinnvollen Funktionen wie beispielsweise dem großen integrierten Speicher und dem Zugriff auf die Cloud.

HP Officejet Pro 8620
e-All-in-One
Printer
Drucker

Manufacturer
Hewlett-Packard, Singapore
In-house design
IPS Global Experience Design
Web
www.hp.com

The Officejet Pro 8620 e-All-in-One is an inkjet printer with the additional capacity for faxing, scanning, and copying. Smartphones, tablets and notebooks can be connected via Wi-Fi or NFC, and the HP ePrint app also allows mobile printing of documents. With the 4.3" touchscreen, the user can directly access web content and also manage pending printing requests. The device has a minimalist design and thus blends into any environment.

Statement by the jury
This user-friendly multifunction printer offers outstanding network connectivity, thus enabling comfortable printing from anywhere.

Der Officejet Pro 8620 e-All-in-One ist ein Tintenstrahldrucker mit zusätzlichen Fax-, Scan- und Kopierfunktion. Smartphone, Tablet oder Notebook werden per Wi-Fi oder NFC angebunden, und auch von unterwegs lassen sich mit der App HP ePrint Dokumente ausdrucken. Über den 4,3"-Touchscreen kann der Anwender auf das Internet zugreifen sowie die anstehenden Druckaufträge verwalten. Das Gerät ist minimalistisch gestaltet und fügt sich in jedes Umfeld ein.

Begründung der Jury
Dieser benutzerfreundliche Multifunktionsdrucker bietet eine ausgezeichnete Netzwerkanbindung und ermöglicht dadurch das bequeme Drucken von überall.

Maxify MB5350 Series
Inkjet Business Printer
Business-Tintenstrahldrucker

Manufacturer
Canon Inc., Tokyo, Japan
In-house design
Yasunori Senshiki, Takahiro Yamamoto,
Ayako Kurita, Rikiya Saito, Nobuo Komiya
Web
www.canon.com

The inkjet printers of the Maxify MB5350 series are suited for small businesses and home offices. They are space-saving and, with their symmetrical shape, blend into all kinds of office interiors. The devices, which connect via LAN or Wi-Fi, offer a wide range of functionality, including printing, scanning, copying, and faxing. Clearly arranged buttons and a touchscreen simplify operation.

Statement by the jury
The Maxify MB5350 series offers user-friendly, compact printers optimally suited for small offices, thanks to their well-conceived, reduced design.

Die Tintenstrahldrucker der Serie Maxify MB5350 eignen sich für kleinere Büros und das Home-Office. Sie nehmen wenig Platz in Anspruch und passen mit ihrer symmetrischen Form in alle Arten von Büros. Die Geräte können drucken, scannen, kopieren und faxen und werden per LAN oder Wi-Fi angesteuert. Das übersichtliche Tastenfeld mit Touchscreen sorgt für eine einfache Bedienung.

Begründung der Jury
Die Serie Maxify MB5350 bietet nutzerfreundliche, kompakte Drucker, die dank ihrer durchdachten, zurückhaltenden Gestaltung bestens für kleinere Büros geeignet sind.

instax Share SP-1
Smartphone Printer
Smartphone-Drucker

Manufacturer
FUJIFILM Corporation, Tokyo, Japan
In-house design
Makoto Isozaki
Web
www.fujifilm.com

With the instax Share SP-1 printer, pictures taken by a smartphone or digital camera are printed with ease onto instant film via Wi-Fi. When using the related smartphone app, different templates are available, such as for printing the time, place, weather, temperature and serial numbers. The printer has an upright shape so that it can be placed in the middle of a user group to share photos directly. LEDs in the upper area indicate the hardware status of the printer.

Statement by the jury
This smartphone printer has a discreet and friendly design, is convenient to carry and ideally suited to use in a group.

Mit dem Drucker instax Share SP-1 lassen sich via Wi-Fi Bilder von Smartphone oder Digitalkamera auf einen Sofortfilm drucken. Bei der Anwendung mit einem Smartphone bietet die dazugehörige App verschiedene Vorlagen, um beispielsweise Zeit, Ort, Wetter, Temperatur oder Seriennummern mitzudrucken. Der Drucker hat eine aufrechte Form, damit er in der Mitte einer Benutzergruppe platziert werden kann, um Fotos direkt zu teilen. LEDs im oberen Bereich zeigen den Status an.

Begründung der Jury
Dieser zurückhaltend und freundlich gestaltete Smartphone-Drucker lässt sich einfach mitnehmen und ideal in einer Gruppe verwenden.

ScanSnap iX100
Document Scanner
Dokumentenscanner

Manufacturer
PFU Limited, Ishikawa, Japan
In-house design
Kiichi Taniho, Shigeru Yonemura
Design
Toshi Satoji Design (Takashi Kirimoto, Toshiki Satoji), Milan, Italy
Web
www.pfu.fujitsu.com
www.toshisatojidesign.com

The ScanSnap iX100 is a portable document scanner with Wi-Fi connectivity and a built-in rechargeable battery. Documents are scanned wirelessly to smartphones, tablets or computers and can be processed, stored or shared via cloud services. The scanner weighs merely 400 grams and needs only 5.2 seconds to scan a DIN A4 page in colour. The battery is charged via USB and allows up to 260 A4 scans with a single charge. In this way, the scanner makes work more productive and facilitates information exchange.

Der ScanSnap iX100 ist ein mobiler Dokumentenscanner mit Wi-Fi-Konnektivität und eingebautem Akku. Unterlagen werden kabellos direkt auf Smartphones, Tablets oder Computer gescannt und können bearbeitet, gespeichert oder über Cloud-Dienste geteilt werden. Der Scanner wiegt nur 400 Gramm und benötigt zum Scannen einer farbigen DIN-A4-Seite lediglich 5,2 Sekunden. Der Akku wird über USB aufgeladen und erlaubt bis zu 260 A4-Scans mit einer Ladung. Damit macht der Scanner das Arbeiten produktiver und erleichtert den Austausch von Informationen.

Statement by the jury
Thanks to its handy format and reduced weight, this Wi-Fi scanner is easy to transport and suitable for use in any given location.

Begründung der Jury
Der Wi-Fi-Scanner lässt sich dank seines handlichen Formats und geringen Gewichts gut mitnehmen und ist überall einsetzbar.

Blackmagic Cintel
Film Scanner

Manufacturer
Blackmagic Design Pty Ltd,
Melbourne, Australia

In-house design
Blackmagic Design Pty Ltd

Web
www.blackmagicdesign.com

Minimalism to perfection

Motion picture film scanners are rarely affordable for filmmakers and institutions, which means limitations regarding their work. The Blackmagic Cintel represents a professional scanning solution for this target group at a very reasonable price. It provides creative individuals and institutions, such as museums, with the ability to digitise fragile archival footage. Independent filmmakers may now harness the timeless image quality of film, confident of drastically reduced production costs and lead times. Precisely tuned to eliminate unwanted resonance from digital servo motors and paired with high-end image stabilising software, Cintel ensures gentle film handling and accurate, high-quality real-time Ultra HD scans. This film scanner is light and easy to use, thus also eliminating the need for expert technicians required for the assembly and operation of bulky, complex devices. All details are designed with meticulous attention and meet the demands of a highly stable, lightweight structure. The resulting minimal aesthetic of Cintel is another impressive feature, which lends the film scanner an artistic and sculptural appearance. It thus also blends in with refined interior spaces, enriching them with its elegance.

Minimalismus in Perfektion

Für Filmschaffende und Institutionen sind Filmabtaster oftmals kaum erschwinglich, was Einschränkungen für ihre Arbeit bedeutet. Der Blackmagic Cintel stellt ein für diesen Bereich professionelles, sehr kostengünstiges Gerät dar. Kreativen und auch Einrichtungen wie etwa Museen bietet er dadurch die Möglichkeit, ihr empfindliches Archivmaterial entsprechend zu digitalisieren. Independent-Filmemacher können die zeitlose Filmbildqualität nutzen und so ihre Produktionskosten und Anlaufzeiten drastisch reduzieren. Mittels Unterdrückung der störenden Resonanz von den Digitalservos und des Einsatzes einer High-End-Bildstabilisierungssoftware sorgt der Cintel dabei für eine schonende Handhabung des Filmmaterials. Er liefert akkurate, hochqualitative Echtzeit-Scans in UHD-Auflösung. Dieser Filmscanner ist leichtgewichtig und bedienungsfreundlich konzipiert, weshalb das für eine Montage und den Betrieb komplexer, aufwendiger Geräte nötige technische Fachpersonal nicht gebraucht wird. Alle seine Details sind sehr sorgfältig ausgeführt und entsprechen den Erfordernissen einer formfesten, leichten Konstruktion. Überaus beeindruckend ist die daraus resultierende minimalistische Ästhetik des Cintel, die diesem Filmscanner eine künstlerische und skulpturale Anmutung gibt. Er kann sich damit auch in edle Innenraumkonzepte einfügen und bereichert diese mit seiner Eleganz.

Statement by the jury

The Blackmagic Cintel film scanner gives the impression of an exceedingly appealing art work. The design stands out with a minimalist form language that also skilfully emphasises the functional details. This highly affordable device offers filmmakers as well as institutions an extensive range of possibilities from screenings to the digitalisation of archival footage. For all this, it provides exceptionally high-quality real-time scans in Ultra HD.

Begründung der Jury

Der Filmscanner Blackmagic Cintel erweckt den Eindruck eines überaus ästhetischen Kunstwerkes. Die Gestaltung glänzt mit einer minimalistischen Formensprache, die auch die funktionalen Vorteile gekonnt zur Geltung bringt. Filmschaffenden wie auch Filminstitutionen bietet dieses sehr kostengünstige Gerät weitreichende Möglichkeiten, von der Vorführung bis hin zur Digitalisierung von Archivmaterial. Dafür liefert es außergewöhnlich hochwertige Echtzeit-Scans in UHD-Auflösung.

Designer portrait
See page 76
Siehe Seite 76

Cintel **Scanner**

Blackmagicdesign

MODAT-531
Industrial PDA
Industrieller PDA

Manufacturer
IEI Integration Corp., New Taipei, Taiwan
In-house design
Li Po Han
Web
www.ieiworld.com

The MODAT-531 is an Androidbased industrial digital assistant for professional use. It has a barcode scanner and NFC connectivity, along with conventional smartphone functions like Wi-Fi, Bluetooth, GPS, camera, microphone and speaker. The PDA meets the IP67 standard and is thus protected against dust and water. The device is also impact-resistant up to a drop height of 1.2 metres and can withstand temperature fluctuations ranging from –10 to +50 degrees centigrade. The device is robust, lightweight and compact, allowing it to be integrated into various working fields.

Statement by the jury
With regard to equipment for the industrial field, this PDA leaves nothing to be desired. Moreover, it is sturdy and displays a pleasing, contemporary design.

Der MODAT-531 ist ein Android-basierter industrieller PDA für den professionellen Einsatz. Er besitzt einen Barcode-Scanner und NFC, hinzu kommen die Smartphone-Funktionen Wi-Fi, Bluetooth, GPS, Kamera, Mikrofon und Lautsprecher. Der PDA erfüllt den IP67-Standard und ist somit gegen Staub und Wasser geschützt. Zudem ist er stoßfest bis zu einer Fallhöhe von 1,20 Metern und hält Temperaturschwankungen zwischen –10 und +50 Grad Celsius stand. Das Gerät ist robust, leicht und kompakt und kann in verschiedenen Arbeitsbereichen eingesetzt werden.

Begründung der Jury
Dieser PDA lässt in puncto Ausstattung für den industriellen Bereich keine Wünsche offen. Dazu ist er robust und zeigt ein zeitgemäßes, gefälliges Design.

Valueline Generation 2
Industrial PC
Industrie-PC

Manufacturer
Phoenix Contact GmbH & Co. KG,
Blomberg, Germany
Design
Tecform, Müller-Witt GbR (Gerwin Müller), Kiel, Germany
Web
www.phoenixcontact.com
www.tecform-design.de

Industrial PCs with touchscreens are increasingly becoming the central point of interaction between man and machine. The new generation of Valueline panel PCs presents complex system processes smoothly and in high detail, as modern Intel i processors together with flexible working memory extension allow the devices to be precisely adapted to the given requirements. The housing rear has been optimised for quick maintenance access, with all essential components within reach. The IP65-protected front with multitouch functionality has a clear and transparent design.

Industrie-PCs mit Touchscreen werden zunehmend zum zentralen Interaktionspunkt zwischen Mensch und Maschine. Die neue Generation der Valueline-Panel-PCs stellt komplexe Anlagenprozesse detailliert und flüssig dar, denn aktuelle Intel-i-Prozessoren sowie eine flexible Arbeitsspeichererweiterung erlauben, dass die Geräte passgenau zugeschnitten werden. Die Gehäuserückseite wurde für einen schnellen Zugriff für die Wartung optimiert, alle wichtigen Komponenten sind leicht zugänglich. Die IP65-geschützte Front mit Multitouch ist klar und übersichtlich gestaltet.

Statement by the jury
This user-friendly industry PC simplifies process monitoring and controlling to a significant degree.

Begründung der Jury
Dieser anwenderfreundliche Industrie-PC vereinfacht die Überwachung und Steuerung von Prozessen erheblich.

Vicon Cara
Motion Capture Headset

Manufacturer
Vicon, Oxford, Great Britain
Design
Curventa (Ian Murison, Adrian Bennett,
Chris Small, Tom Owen),
London, Great Britain
Web
www.vicon.com
www.curventa.com

The Vicon Cara is a portable motion capture headset featuring four adjustable miniature HD cameras, which register an actor's motion and convert it into natural 3D computer graphics. The headset weighs only 1.2 kg and is comfortably worn without restricting the actor's movement. This allows both the actors and the director to fully focus on the performance.

Statement by the jury
The Vicon Cara enables low-cost entry into the professional motion capture field, which is of particular benefit to smaller film and animation studios.

Vicon Cara ist ein tragbares Motion-Capture-Headset. Vier verstellbare HD-Miniaturkameras erfassen die Bewegungen des Akteurs und wandeln diese naturgetreu in 3D-Computergrafiken um. Das Headset wiegt lediglich 1,2 kg und ist leicht und angenehm zu tragen, ohne die Bewegungen des Schauspielers zu beeinflussen. Dadurch können sich Schauspieler und Regisseur ganz auf die Darstellung konzentrieren.

Begründung der Jury
Vicon Cara ermöglicht den kostengünstigen Einstieg in die professionelle Motion-Capture-Technik. Davon profitieren besonders kleine Film- und Animationsstudios.

tanJack Bluetooth
TAN Generator

Manufacturer
Reiner Kartengeräte GmbH & Co. KG,
Furtwangen, Germany
Design
triften design studio, Hamburg, Germany
Web
www.reiner-sct.com
www.triften.com

tanJack Bluetooth is a TAN generator for use in online banking. It connects to a smartphone or tablet via Bluetooth and to a PC via the USB slot. The device also supports the optical transmission or manual generation of TANs. It meets the online banking security standards of the German banking industry and may also be used with banking apps that support the generation of TANs via Bluetooth LE.

Statement by the jury
tanJack Bluetooth gives the user the choice between different TAN transmission methods. The clearly structured surface facilitates operation.

tanJack Bluetooth ist ein TAN-Generator für die Verwendung beim Online-Banking. Er lässt sich per Bluetooth mit einem Smartphone oder Tablet verbinden und über USB mit einem PC. Zudem unterstützt er die optische Übertragung oder das manuelle Generieren von TANs. Er erfüllt die Sicherheitsstandards der Deutschen Kreditwirtschaft beim Online-Banking und kann auch mit Banking-Apps verwendet werden, die das Generieren von TANs mit Bluetooth LE unterstützen.

Begründung der Jury
tanJack Bluetooth lässt dem Anwender die Wahl zwischen den verschiedensten TAN-Übermittlungsverfahren. Die klar strukturierte Oberfläche macht die Bedienung einfach.

B12i (L12i)
Charging Cabinet
Ladewagen

Manufacturer
AVer Information Inc., New Taipei, Taiwan
In-house design
Web
www.aver.com

This charging cabinet allows up to 12 iPads, tablets, notebooks and Chromebooks to be charged simultaneously. It offers easy expandability by plugging a second charging cabinet into the first and stacking one on top of the other, so that users can charge up to 24 devices. It also features a clever cable management system, making it an ideal solution for the hassle-free management of IT equipment.

Statement by the jury
The charging cabinet keeps mobile devices safe in a small space. With its rounded edges, it renders a friendly impression.

Dieser Ladewagen lädt gleichzeitig bis zu zwölf iPads, Tablets, Notebooks und Chromebooks auf. Er lässt sich leicht erweitern, indem ein zweiter Ladewagen an den ersten angeschlossen wird und beide übereinander gestellt werden. Dadurch lassen sich bis zu 24 Geräte aufladen. Der Ladewagen verfügt zudem über ein intelligentes Kabelmanagement und wird dadurch zu einer idealen Lösung für die mühelose Verwaltung von IT-Geräten.

Begründung der Jury
Der Ladewagen bewahrt mobile Geräte sicher auf kleinem Raum auf und macht mit seinen gerundeten Kanten einen freundlichen Eindruck.

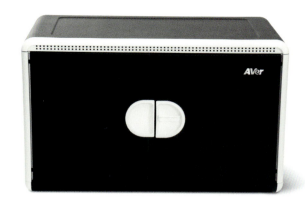

Tablet Butler
iPad Support with Theft Protection
iPad-Halter mit Diebstahlsicherung

Manufacturer
Krug & Priester, Balingen, Germany
Design
michael schad.produktdesign, Stuttgart, Germany
Web
www.krug-priester.com
www.michaelschad.de

The powder-coated Tablet Butler is designed for various iPad versions. With its purist design, the stand is an inconspicuous aid for iPads that need to be secured during use. As such, it can be positioned well in public spaces to allow for visitor interaction. Integrated theft protection prevents the iPad from being stolen, while the solid metal construction guarantees safe positioning.

Statement by the jury
The Tablet Butler is a solid, inconspicuous solution for the safe and secure placement of iPads.

Der pulverbeschichtete Tablet Butler ist für verschiedene iPad-Versionen konzipiert. Mit seinem puristischen Design ist er ein unauffälliger Helfer, wenn iPads sicher platziert werden sollen. So lässt er sich beispielsweise gut im öffentlichen Raum aufstellen, um Besuchern die Möglichkeit zur Interaktion zu geben. Der integrierte Diebstahlschutz verhindert, dass das iPad gestohlen wird, die massive Metallkonstruktion sorgt für einen sicheren Stand.

Begründung der Jury
Der Tablet Butler ist eine solide, unauffällige Lösung, wenn es darum geht, iPads für die Benutzung sicher zu platzieren.

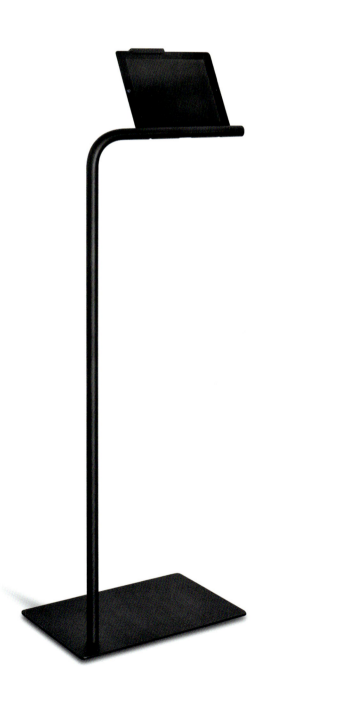

The jury 2015
International orientation and objectivity
Internationalität und Objektivität

The jurors of the Red Dot Award: Product Design
All members of the Red Dot Award: Product Design jury are appointed on the basis of independence and impartiality. They are independent designers, academics in design faculties, representatives of international design institutions, and design journalists.

The jury is international in its composition, which changes every year. These conditions assure a maximum of objectivity. The members of this year's jury are presented in alphabetical order on the following pages.

Die Juroren des Red Dot Award: Product Design
In die Jury des Red Dot Award: Product Design wird als Mitglied nur berufen, wer völlig unabhängig und unparteiisch ist. Dies sind selbstständig arbeitende Designer, Hochschullehrer der Designfakultäten, Repräsentanten internationaler Designinstitutionen und Designfachjournalisten.

Die Jury ist international besetzt und wechselt in jedem Jahr ihre Zusammensetzung. Unter diesen Voraussetzungen ist ein Höchstmaß an Objektivität gewährleistet. Auf den folgenden Seiten werden die Jurymitglieder des diesjährigen Wettbewerbs in alphabetischer Reihenfolge vorgestellt.

01

Prof.
Werner Aisslinger
Germany
Deutschland

The works of the designer Werner Aisslinger cover the spectrum of experimental and artistic approaches, including industrial design and architecture. He makes use of the latest technologies and has helped introduce new materials and techniques to the world of product design. His works are part of prestigious, permanent collections of international museums such as the Museum of Modern Art and the Metropolitan Museum of Art in New York, the Fonds National d'Art Contemporain in Paris, the Victoria & Albert Museum London, the Neue Sammlung Museum in Munich, and the Vitra Design Museum in Weil, Germany. In 2013, Werner Aisslinger opened his first solo show called "Home of the future" in the Berlin museum Haus am Waldsee. Besides numerous international awards he received the prestigious A&W Designer of the Year Award in 2014. Werner Aisslinger lives in Berlin and works for companies such as Vitra, Foscarini, Haier, Canon, 25hours Hotel Bikini and Moroso, for whom he has developed the "Hemp chair" together with BASF, which is the first ever biocomposite monobloc chair.

Die Arbeiten des Designers Werner Aisslinger umfassen das Spektrum experimenteller und künstlerischer Ansätze samt Industriedesign und Architektur. Er bedient sich modernster Technologien und hat dazu beigetragen, neue Materialien und Techniken in die Welt des Produktdesigns einzuführen. Seine Arbeiten sind Bestandteil bedeutender Museumssammlungen, darunter Museum of Modern Art und Metropolitan Museum of Art in New York, Fonds National d'Art Contemporain in Paris, Victoria & Albert Museum in London, Die Neue Sammlung in München und Vitra Design Museum in Weil. 2013 eröffnete Werner Aisslinger seine erste Solo-Show „Home of the future' im Berliner Museum „Haus am Waldsee". Neben zahlreichen internationalen Auszeichnungen wurde er 2014 als „A&W-Designer des Jahres" ausgezeichnet. Der in Berlin lebende Designer arbeitet u. a. für Unternehmen wie Vitra, Foscarini, Haier, Canon, 25hours Hotel Bikini und Moroso, für das er zusammen mit BASF den „Hemp chair", den weltweit ersten biologisch zusammengesetzten Monoblock-Stuhl, entwickelt hat.

02

"Design is one of the most
important professions of our time."
„Design ist eine der wichtigsten
Professionen unserer Zeit."

What is, in your opinion, the significance of design quality in the industries you evaluated?
The design quality of the established brands is consistently high with a tendency towards a greater focus on product design, which is overall a positive trend.

How important is design quality in the global market?
Design quality is the ultimate distinguishing criterion. Particularly in product groups in which technical evolution creates a level playing field, design becomes the only relevant distinguishing criterion for products.

Do you see a dominating trend in this year's designs?
Innovative materials and new manufacturing methods are almost the norm in ambitious product developments. What is remarkable this year, however, is the sophistication with which these topics are combined with product design.

Wie würden Sie den Stellenwert von Designqualität in den von Ihnen beurteilten Branchen einschätzen?
Die Designqualität der etablierten Marken ist gleichbleibend hoch, mit Tendenz zu einem größeren Fokus auf das Produktdesign – alles in allem ein erfreulicher Trend.

Wie wichtig ist Designqualität im globalen Markt?
Designqualität ist das Unterscheidungskriterium schlechthin – gerade in Produktgruppen, in denen sich die technische Evolution nivelliert, wird Design zum einzigen relevanten Unterscheidungsfaktor für Produkte.

Sehen Sie eine Entwicklung im Design, die sich in diesem Jahr durchsetzt?
Innovative Materialien und neue Herstellungsmethoden sind bei ambitionierten Produktentwicklungen fast schon Standard – erstaunlich in diesem Jahr ist jedoch die „Sophistication", diese Themen mit dem Produktdesign zu verknüpfen.

01

Manuel
Alvarez Fuentes
Mexico
Mexiko

Manuel Alvarez Fuentes studied industrial design at the Universidad Iberoamericana, Mexico City, where he later served as Director of the Design Department. In 1975, he also received a Master of Design from the Royal College of Art, London. He has over 40 years of experience in the fields of product design, furniture and interior design, packaging design, signage and visual communications. From 1992 to 2012, he was Director of Diseño Corporativo, a product design consultancy. Currently, he is head of innovation at Alfher Porcewol in Mexico City. He has acted as consultant for numerous companies and also as a board member of various designers' associations, including as a member of the Icsid Board of Directors, as Vice President of the National Chamber of Industry of Mexico, Querétaro, and as Director of the Innovation and Design Award, Querétaro. Furthermore, Alvarez Fuentes has been senior tutor in Industrial Design at Tecnológico de Monterrey Campus Querétaro since 2009. In 2012 and 2013, he was president of the jury of Premio Quorum, the most prestigious design competition in Mexico.

Manuel Alvarez Fuentes studierte Industriedesign an der Universidad Iberoamericana, Mexiko-Stadt, wo er später die Leitung des Fachbereichs Design übernahm. 1975 erhielt er zudem einen Master of Design vom Royal College of Art, London. Er hat über 40 Jahre Erfahrung in Produkt- und Möbelgestaltung, Interior Design, Verpackungsdesign, Leitsystemen und visueller Kommunikation. Von 1992 bis 2012 war er Direktor von Diseño Corporativo, einem Beratungsunternehmen für Produktgestaltung. Heute ist Alvarez Fuentes Head of Innovation bei Alfher Porcewol in Mexiko-Stadt. Er war als Berater für zahlreiche Unternehmen sowie als Vorstandsmitglied in verschiedenen Designverbänden tätig, z. B. im Icsid-Vorstand, als Vizepräsident der mexikanischen Industrie- und Handelskammer im Bundesstaat Querétaro und als Direktor des Innovation and Design Award, Querétaro. Seit 2009 ist Alvarez Fuentes Senior-Dozent für Industriedesign am Tecnológico de Monterrey Campus Querétaro. 2012 und 2013 war er Präsident der Jury des Premio Quorum, des angesehensten Designwettbewerbs in Mexiko.

01 ELICA MAFdi 1
Kitchen hood, limited edition,
ceiling mount, island type.
The design of the hood is based on
Elica's patented, highly effective
Evolution ventilation system.
The steel carcase finish in porcelain
enamel (Peltre) is by Alfher Porcewol.
Küchenhaube, limitierte Auflage,
Deckenmontage, Insel-Typ.
Die Gestaltung der Haube basiert auf
dem patentierten, hocheffizienten
Ventilationssystem Elica Evolution.
Das Finish aus Porzellan-Emaille
(Peltre) ist von Alfher Porcewol.

02 ELICA MAFdi 1
Kitchen hoods, limited edition,
wall mounted.
The hood includes illumination
offered also in the Twin Evolution
system. Peltre finish colours:
orange, blue and black.
Küchenhauben, limitierte Auflage,
Wandmontage.
Die Haube bietet die gleiche Be-
leuchtung wie das „Twin Evolution"-
System. Farben des Peltre-Finishs:
Orange, Blau und Schwarz.

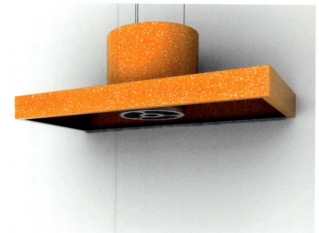

02

"My work philosophy is based on the idea that design is not only responsible for continuous innovation. Its main focus should be the users, their happiness and satisfaction."

„Meine Arbeitsphilosophie basiert auf der Idee, dass Gestaltung nicht nur für laufende Innovation verantwortlich ist. Das Haupt-augenmerk sollte vielmehr auf den Benutzern, ihrem Glück und ihrer Zufriedenheit liegen."

What are your sources of inspiration?
Living with a very talented and creative Mexican architect, my father Augusto Alvarez, was a source of inspiration at the beginning and later in my life. Subsequently, the most reliable sources came from a thorough observation of people.

What do you take special notice of when you are assessing a product as a jury member?
I search for intelligent, simple, meaningful and purposeful products; in many respects I try to differentiate originality "per se" from true innovation. I disregard those products that are only good-looking but do not contribute to a better material world.

What significance does winning an award in a design competition have for the designer?
It should mean a triggering motivation for his or her creative work, to do better every time.

Was sind Ihre Inspirationsquellen?
Das Zusammenleben mit einem sehr talentierten und kreativen mexikanischen Architekten, meinem Vater Augusto Alvarez, war eine Inspirationsquelle für mich, sowohl zu Beginn als auch später in meinem Leben. Danach ergaben sich die zuverlässigsten Inspirations-quellen aus der sorgfältigen Beobachtung von Menschen.

Worauf achten Sie besonders, wenn Sie ein Produkt als Juror bewerten?
Ich bin auf der Suche nach intelligenten, einfachen, bedeutungsvollen und zweckmäßigen Produkten. In vielerlei Hinsicht versuche ich, zwischen Originalität per se und wahrer Innovation zu unterscheiden. Ich ignoriere diejenigen Produkte, die lediglich gut aussehen und keinen Beitrag zu einer besseren materiellen Welt leisten.

Welche Bedeutung hat die Auszeichnung in einem Designwettbewerb für den Designer?
Sie sollte einen Motivationsschub für die kreative Arbeit des Designers darstellen, einen Ansporn, jedes Mal etwas besser zu machen.

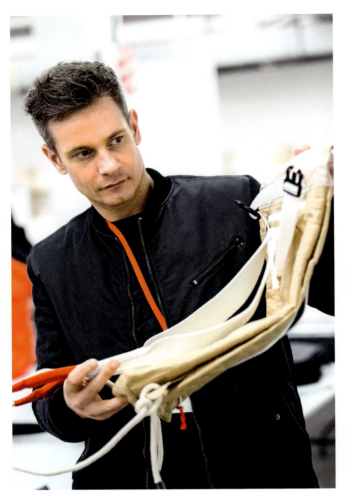

David Andersen
Denmark
Dänemark

David Andersen, born in 1978, graduated from the Glasgow School of Art and the Copenhagen Academy of Fashion and Design in 2003. In 2004, he was awarded "Best Costume Designer" and received the award "Wedding Gown of the Year" from the Royal Court Theatre in Denmark. He develops designs for ready-to-wear clothes, shoes, perfume, underwear and home wear and emerged as a fashion designer working as chief designer at Dreams by Isabell Kristensen as well as designing couture for artists and dance competitions under his own name. He debuted his collection "David Andersen" in 2007 and apart from Europe it also conquered markets in Japan and the US. In 2010 and 2011, the Danish Fashion Award nominated him in the category "Design Talent of the Year", and in 2010 as well as 2012, David Andersen received a grant from the National Art Foundation. He regularly shows his sustainable designs, which are worn by members of the Royal Family, politicians and celebrities, at couture exhibitions around the world. Furthermore, he is a guest lecturer at TEKO, Scandinavia's largest design and management college.

David Andersen, 1978 geboren, graduierte 2003 an der Glasgow School of Art und an der Copenhagen Academy of Fashion and Design. 2004 wurde er als „Best Costume Designer" sowie für das „Hochzeitskleid des Jahres" vom Royal Court Theatre in Dänemark prämiert. Er entwirft Konfektionskleidung, Schuhe, Parfüm, Unterwäsche und Heimtextilien und wurde als Modedesigner bekannt, als er als Chefdesigner bei Dreams von Isabell Kristensen sowie unter seinem Namen für Künstler und Tanzwettbewerbe arbeitete. Seine Kollektion „David Andersen", erschienen ab 2007, eroberte neben Europa Märkte in Japan und den USA. 2010 und 2011 nominierte ihn der Danish Fashion Award in der Kategorie „Design Talent of the Year" und 2010 sowie 2012 erhielt er ein Stipendium der National Art Foundation. David Andersen präsentiert seine nachhaltigen Entwürfe, die von Mitgliedern der Königsfamilie, Politikern oder Prominenten getragen werden, regelmäßig in Couture-Ausstellungen weltweit und ist zudem Gastdozent an der TEKO, der größten Hochschule für Design und Management Skandinaviens.

"My work philosophy is to be true to myself."

„Meine Arbeitsphilosophie besteht darin, mir selbst treu zu bleiben."

In your opinion, what will never be out of fashion?
Personality.

Which of the projects in your lifetime are you particular proud of?
The first dress I made for HRH Crown Princess Mary of Denmark was a cerise-coloured dress she wore at the royal wedding in Monaco. It was both thrilling and challenging. Meeting her privately was one thing, but knowing that this dress will be showcased in the press all around the world was just nerve-racking.

The Red Dot Award has been uncovering the best designs for 60 years now. Which product innovation would you like to see in the next 60 years, and why?
I hope that we as designers will own up to our responsibility for the environment through the production of our products and that the focus will shift from getting turnover to what actually happens on the journey there.

Was ist Ihrer Meinung nach nie unmodern?
Persönlichkeit.

Auf welche Ihrer bisherigen Projekte sind Sie besonders stolz?
Das erste Kleid, das ich für Ihre Königliche Hoheit Kronprinzessin Mary von Dänemark entwarf, war ein kirschrotes Kleid, das sie zur königlichen Hochzeit in Monaco trug. Dieses Projekt war sowohl aufregend als auch eine Herausforderung. Sie privat zu treffen, war eine Sache, aber zu wissen, dass dieses Kleid von der internationalen Presse weltweit zur Schau gestellt werden würde, war schlicht nervenaufreibend.

Der Red Dot Award ermittelt seit 60 Jahren die besten Gestaltungen. Welche Produktinnovation würden Sie sich für die nächsten 60 Jahre wünschen?
Ich hoffe, dass wir unserer Verantwortung als Designer für die Umwelt durch die Herstellung unserer Produkte gerecht werden und sich das Augenmerk weg von dem Ziel des Umsatzes hin zu dem verschiebt, was auf dem Weg dorthin tatsächlich passiert.

01

Prof. Martin Beeh
Germany
Deutschland

Professor Martin Beeh is a graduate in Industrial Design from the Darmstadt University of Applied Sciences in Germany and the ENSCI-Les Ateliers, Paris, and completed a postgraduate course in business administration. In 1995, he became design coordinator at Décathlon in Lille/France, in 1997 senior designer at Electrolux Industrial Design Center Nuremberg and Stockholm and furthermore became design manager at Electrolux Industrial Design Center Pordenone/Italy, in 2001. He is a laureate of several design awards as well as founder and director of the renowned student design competition "Electrolux Design Lab". In the year 2006 he became general manager of the German office of the material library Material ConneXion in Cologne. Three years later, he founded the design office beeh_innovation. He focuses on Industrial Design, Design Management, Design Thinking and Material Innovation. Martin Beeh lectured at the Folkwang University of the Arts in Essen and at the University of Applied Sciences Schwäbisch Gmünd. Since 2012, he is professor for design management at the University of Applied Sciences Ostwestfalen-Lippe in Lemgo/Germany.

Professor Martin Beeh absolvierte ein Studium in Industriedesign an der Fachhochschule Darmstadt und an der ENSCI-Les Ateliers, Paris, sowie ein Aufbaustudium der Betriebswirtschaft. 1995 wurde er Designkoordinator bei Décathlon in Lille/Frankreich, 1997 Senior Designer im Electrolux Industrial Design Center Nürnberg und Stockholm sowie 2001 Design Manager am Electrolux Industrial Design Center Pordenone/ Italien. Er ist Gewinner diverser Designpreise und gründete und leitete den renommierten Designwettbewerb für Studierende, das „Electrolux Design Lab". Im Jahr 2006 wurde er General Manager der deutschen Niederlassung der Materialbibliothek „Material ConneXion" in Köln. Drei Jahre später gründete Martin Beeh das Designbüro beeh_innovation. Seine Schwerpunkte liegen in den Bereichen Industriedesign, Design Management, Design Thinking und Materialinnovation. Martin Beeh hatte Lehraufträge an der Folkwang Universität der Künste in Essen und an der Hochschule für Gestaltung Schwäbisch Gmünd. Seit 2012 ist er Professor für Designmanagement an der Hochschule Ostwestfalen-Lippe in Lemgo.

02

"My work philosophy is to aim for the best for and together with the customer, because every task is a new adventure."

„Meine Arbeitsphilosophie ist es, dass ich das Beste für und mit dem Kunden erreichen will, denn jede Aufgabe ist ein neues Abenteuer."

What are the main challenges in a designer's everyday life?
The movement between inspiration and stringency, order and creative chaos. Those who only follow a routine dry out and stop growing.

What impressed you most during the Red Dot judging process?
The concentrated competence of jurors and our interaction, the professionalism of the Red Dot team and the courage of the large number of companies that take part in the global competition.

What do you take special notice of when you are assessing a product as a jury member?
That the product is suitable for daily use, its ergonomic quality, and that there are no unnecessary frills. I also check if the design matches the brand, and I look for the product that stands out in a positive way.

What significance does winning an award in a design competition have for the manufacturer?
The role design plays in a company is strengthened and it is "officially" recognised as a successful tool.

Was sind große Herausforderungen im Alltag eines Designers?
Die Bewegung zwischen Inspiration und Stringenz, Ordnung und schöpferischem Chaos. Wer nur Routine macht, trocknet aus und wächst nicht mehr.

Was hat Sie bei der Red Dot-Jurierung am meisten beeindruckt?
Die geballte Kompetenz der Juroren und unser Zusammenspiel, die Professionalität des Red Dot-Teams und der Mut der vielen Unternehmen, die sich dem globalen Wettbewerb stellen.

Worauf achten Sie besonders, wenn Sie ein Produkt als Juror bewerten?
Auf Alltagstauglichkeit, Ergonomie, gestalterische Ordnung und den Verzicht auf Schnickschnack. Ich schaue auch, ob die Gestaltung zur Marke passt, und suche das Produkt, das sich positiv abhebt.

Welche Bedeutung hat die Auszeichnung in einem Designwettbewerb für den Hersteller?
Die Position von Design im Unternehmen wird gestärkt und „offiziell" als erfolgreiches Instrument anerkannt.

01

Dr Luisa Bocchietto
Italy
Italien

Luisa Bocchietto, architect and designer, graduated from the Milan Polytechnic. She has worked as a free-lancer undertaking projects for local development, building renovations and urban planning. As a visiting professor she teaches at universities and design schools, she takes part in design conferences and international juries, publishes articles and organises exhibitions on architecture and design. Over the years, her numerous projects aimed at supporting the spread of design quality. She was a member of the Italian Design Council at the Ministry for Cultural Heritage, the Polidesign Consortium of the Milan Polytechnic, the CIDIC Italo-Chinese Council for Design and Innovation and the CNAC National Anti-Counterfeiting Council at the Ministry of Economic Development. From 2008 until 2014 she was National President of the ADI (Association for Industrial Design). Currently, she is a member of the Icsid Board, and since January 2015 she has been Editorial Director of PLATFORM, a new magazine about Architecture and Design.

Luisa Bocchietto, Architektin und Designerin, graduierte am Polytechnikum Mailand. Sie arbeitet freiberuflich und führt Projekte für die lokale Entwicklung, Gebäudeumbauten und Stadtplanung durch. Als Gastprofessorin lehrt sie an Universitäten und Designschulen, sie nimmt an Designkonferenzen und internationalen Jurys teil, veröffentlicht Artikel und betreut Ausstellungen über Architektur und Design. Ihre zahlreichen Projekte über die Jahre hinweg verfolgten das Ziel, die Verbreitung von Designqualität zu unterstützen. Sie war Mitglied des italienischen Rates für Formgebung am Kulturministerium, der Polidesign-Vereinigung des Polytechnikums Mailand, der CIDIC, der italienisch-chinesischen Vereinigung für Design und Innovation, und der CNAC, der Nationalen Vereinigung gegen Fälschungen am Ministerium für wirtschaftliche Entwicklung. Von 2008 bis 2014 war sie nationale Präsidentin der ADI, des Verbandes für Industriedesign. Aktuell ist sie Gremiumsmitglied des Icsid und seit Januar 2015 Chefredakteurin von PLATFORM, einem neuen Magazin über Architektur und Design.

02

"My work philosophy is to simplify, initiate sensible projects, generate quality and share joy."
„Meine Arbeitsphilosophie besteht darin, Dinge und Vorgänge zu vereinfachen, sinnvolle Projekte anzustoßen, Qualität zu schaffen und Freude zu teilen."

You are the editorial director of the new magazine "PLATFORM". What fascinates you about this job?
It allows me to explore the design aspects that I am more passionate about while at the same time keeping in touch with all the extraordinary people I have met during my ADI presidency.

The Red Dot Award has been uncovering the best designs for 60 years now. Which product innovation would you like to see in the next 60 years?
Products that are meant to improve people's lives – not only in an aesthetic and functional sense, but also in a social and ethical one.

What is the advantage of commissioning an external designer compared to having an in-house design team?
Working with an internal technical team means optimising in a linear way, while engaging with external designers entails the opportunity of a "lateral deviation", which often leads to unexpected and great results.

Sie sind die Chefredakteurin des neuen Magazins „PLATFORM". Was fasziniert Sie an dieser Rolle?
Sie erlaubt mir, die Gestaltungsaspekte, die mich besonders begeistern, näher zu untersuchen und zugleich mit all den außergewöhnlichen Menschen, die ich während meiner ADI-Präsidentschaft kennengelernt habe, in Kontakt zu bleiben.

Der Red Dot Award ermittelt seit 60 Jahren die besten Gestaltungen. Welche Produktinnovation würden Sie sich für die nächsten 60 Jahre wünschen?
Produkte, die darauf abzielen, das Leben zu verbessern – nicht nur in einem ästhetischen und funktionalen Sinn, sondern auch in einem sozialen und ethischen.

Worin liegt der Vorteil, einen externen Designer zu beauftragen, im Vergleich zu einem Inhouse-Designteam?
Mit einem internen technischen Team zu arbeiten, bedeutet, Dinge auf lineare Weise zu optimieren, wohingegen die Zusammenarbeit mit externen Designern die Gelegenheit für eine „laterale Abweichung" bietet, die oft zu unerwarteten und großartigen Ergebnissen führt.

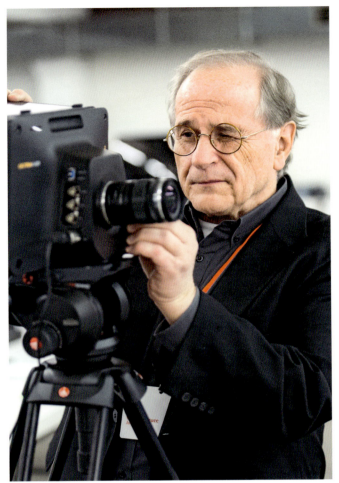

01

Gordon Bruce
USA
USA

Gordon Bruce is the owner of Gordon Bruce Design LLC and has been a design consultant for 40 years working with many multinational corporations in Europe, Asia and the USA. He has worked on a very wide range of products, interiors and vehicles – from aeroplanes to computers to medical equipment to furniture. From 1991 to 1994, Gordon Bruce was a consulting vice president for the Art Center College of Design's Kyoto programme and, from 1995 to 1999, chairman of Product Design for the Innovative Design Lab of Samsung (IDS) in Seoul, Korea. In 2003, he played a crucial role in helping to establish Porsche Design's North American office. For many years, he served as head design consultant for Lenovo's Innovative Design Center (IDC) in Beijing and he is presently working with Bühler in Switzerland and Huawei Technologies Co., Ltd. in China. Gordon Bruce is a visiting professor at several universities in the USA and in China and also acts as an author and design publicist. He recently received Art Center College of Design's "Lifetime Achievement Award".

Gordon Bruce ist Inhaber der Gordon Bruce Design LLC und seit mittlerweile 40 Jahren als Designberater für zahlreiche multinationale Unternehmen in Europa, Asien und den USA tätig. Er arbeitete bereits an einer Reihe von Produkten, Inneneinrichtungen und Fahrzeugen – von Flugzeugen über Computer bis hin zu medizinischem Equipment und Möbeln. Von 1991 bis 1994 war Gordon Bruce beratender Vizepräsident des Kioto-Programms am Art Center College of Design sowie von 1995 bis 1999 Vorsitzender für Produktdesign beim Innovative Design Lab of Samsung (IDS) in Seoul, Korea. Im Jahr 2003 war er wesentlich daran beteiligt, das Büro von Porsche Design in Nordamerika zu errichten. Über viele Jahre war er leitender Designberater für Lenovos Innovative Design Center (IDC) in Beijing. Aktuell arbeitet er für Bühler in der Schweiz und Huawei Technologies Co., Ltd. in China. Gordon Bruce ist Gastprofessor an zahlreichen Universitäten in den USA und in China und als Buchautor sowie Publizist tätig. Kürzlich erhielt er vom Art Center College of Design den Lifetime Achievement Award.

02

"My work philosophy is to
maintain a sense of curiosity.
Then, I conceptualise an idea
from my inquisitiveness, test it
and start the process over again."
„Meine Arbeitsphilosophie besteht
darin, Neugierde zu bewahren.
In einem zweiten Schritt konzipiere
ich eine Idee auf Basis meiner
Wissbegierde, teste sie und beginne
den Prozess von vorne."

**Which IT project that you were involved in are
you most proud of?**
I was one of three designers who helped design the
first "Massively Parallel Processor", a super computer
produced by the Thinking Machines Company.
The unique design, using a series of black translucent
boxes while accommodating many issues based on
improved usability, resulted in a paradigm shift for
large computers.

**The Red Dot Award has been uncovering the
best designs for 60 years now. Which product
innovation would you like to see in the next
60 years, and why?**
I would like to see a common theme in design move
more and more towards Mother Nature's design.

**What significance does winning an award in
a design competition have for the designer?**
Winning a design award from a very well-recognised
design competition gives designers affirmation that
they are making the right choices in their design
decision processes.

**Auf welches IT-Projekt, an dem Sie beteiligt waren,
sind Sie besonders stolz?**
Ich war einer von drei Designern, die dabei halfen,
den ersten „Massively Parallel Processor", einen Super-
Computer der Thinking Machines Company, zu
gestalten. Das einzigartige Design, das aus einer Reihe
lichtdurchlässiger Kästen besteht und gleichzeitig
viele Aspekte verbesserter Benutzerfreundlichkeit
beherbergt, begründete einen Paradigmenwechsel für
große Computer.

**Der Red Dot Award ermittelt seit 60 Jahren die
besten Gestaltungen. Welche Produktinnovation
würden Sie sich für die nächsten 60 Jahre
wünschen, und warum?**
Ich würde gerne einen einheitlichen Trend im Designbereich sehen, der sich mehr und mehr zu einer Gestaltung im Stil von Mutter Natur hinbewegt.

**Welche Bedeutung hat die Auszeichnung in einem
Designwettbewerb für den Designer?**
Eine Auszeichnung in einem allgemein anerkannten
Designwettbewerb zu gewinnen, gibt Designern die
Bestätigung, dass sie die richtigen Entscheidungen
im Designprozess getroffen haben.

01

Rüdiger Bucher
Germany
Deutschland

Rüdiger Bucher, born in 1967, graduated in political science from the Philipps-Universität Marburg and subsequently completed the postgraduate study course "Interdisciplinary studies on France" in Freiburg, Germany. While still at school he wrote for daily newspapers and magazines, before joining publishing house Verlagsgruppe Ebner Ulm in 1995, where he was in charge of "Scriptum. Die Zeitschrift für Schreibkultur" (Scriptum. The magazine for writing culture) for five years. In 1999 he became Chief Editor of Chronos, the leading German-language special interest magazine for wrist watches, with the same publishing house. During his time as Chief Editor, since 2005, Chronos has positioned itself internationally with subsidiary magazines and licensed editions in China, Korea, Japan and Poland. At the same time, Rüdiger Bucher established a successful corporate publishing department for Chronos. Since 2014, he has been Editorial Director of the business area "Watches" at the Ebner publishing house and besides Chronos he has also been in charge of the sister magazines concerning watches and classic watches as well as the New York-based "WatchTime".

Rüdiger Bucher, geboren 1967, studierte Politikwissenschaft an der Philipps-Universität Marburg und schloss daran den Aufbaustudiengang „Interdisziplinäre Frankreich-Studien" in Freiburg an. Schon als Schüler schrieb er für verschiedene Tageszeitungen und Zeitschriften, bevor er 1995 zum Ebner Verlag Ulm kam und dort fünf Jahre lang „Scriptum. Die Zeitschrift für Schreibkultur" betreute. Im selben Verlag wurde er 1999 Redaktionsleiter von „Chronos", dem führenden deutschsprachigen Special-Interest-Magazin für Armbanduhren. Während seiner Amtszeit als Chefredakteur, seit 2005, hat sich Chronos mit Tochtermagazinen und Lizenzausgaben in China, Korea, Japan und Polen international aufgestellt. Gleichzeitig baute Rüdiger Bucher für Chronos einen erfolgreichen Corporate-Publishing-Bereich auf. Seit 2014 verantwortet er als Redaktionsdirektor des Ebner Verlags im Geschäftsbereich „Uhren" neben Chronos auch die Schwestermagazine „Uhren-Magazin", „Klassik Uhren" sowie die in New York beheimatete „WatchTime".

01 **Chronos Special Uhrendesign**
Published once a year each
September since 2013.
Erscheint seit 2013 einmal
jährlich im September.

02 Chronos is present around the
globe with issues of differing
themes plus special issues.
Mit verschiedenen Ausgaben
und Sonderheften ist Chronos
rund um den Globus vertreten.

02

"My work philosophy is to inform our readers competently, while at the same time entertaining and occasionally surprising them with something new and unexpected."
„Meine Arbeitsphilosophie ist es, den Leser kompetent zu informieren, ihn dabei zu unterhalten und ihn gelegentlich durch Neues, Unerwartetes zu überraschen."

What fascinates you about your job as Chief Editor of Chronos?
It is a very multi-faceted position. We distribute our information using different channels and at the same time we are closely in touch with what's happening in a highly interesting industry, which is characterised by a fascinating tension between past and future.

When did you first think consciously about design?
As a child I already preferred using specific glasses and a specific cutlery set. Apart from the look, it was above all the well-designed haptic qualities that appealed to me.

The Red Dot Award has been uncovering the best designs for 60 years now. Which product innovations would you like to see in the next 60 years, and why?
A technology that does away with cables and at the same time is non-hazardous to health.

What significance does winning an award in a design competition have for the manufacturer?
An objective quality seal increases credibility, generates attention and enhances its profile.

Was fasziniert Sie an Ihrem Beruf als Chefredakteur von Chronos?
Er ist sehr vielseitig: Wir verbreiten unsere Informationen auf unterschiedlichsten Kanälen und sind zugleich nah dran an einer hochinteressanten Branche, die sich in Technik und Design in einem faszinierenden Spannungsfeld zwischen Vergangenheit und Zukunft bewegt.

Wann haben Sie das erste Mal bewusst über Design nachgedacht?
Schon als Kind benutzte ich bestimmte Gläser und bestimmtes Besteck lieber als andere. Neben der Optik sprach mich vor allem eine gelungene Haptik an.

Der Red Dot Award ermittelt seit 60 Jahren die besten Gestaltungen. Welche Produktinnovation würden Sie sich für die nächsten 60 Jahre wünschen, und warum?
Eine Technik, die es erlaubt, auf Kabel zu verzichten, und dabei gesundheitlich unbedenklich ist.

Welche Bedeutung hat die Auszeichnung in einem Designwettbewerb für den Hersteller?
Ein objektives Gütesiegel erhöht die Glaubwürdigkeit, schafft Aufmerksamkeit und schärft sein Profil.

01

Wen-Long Chen
Taiwan
Taiwan

Wen-Long Chen has been CEO of the Taiwan Design Center since 2013 and has been designated as "Taiwan's Top Boss in Design". He has accumulated over 25 years of practical experience in design management and product design. In 1988, he founded Nova Design with support from Chinfon Trading Group, and has since led and completed hundreds of design projects. During his time at Nova Design, Wen-Long Chen led the company to tremendous growth with visionary design thinking. In response to clients' needs, he utilised the "Design System Competitiveness" as the core value of development, established a KMO (Knowledge Management Officer) system to oversee six branch offices in Asia, North America and Europe, and invested millions of dollars in the most advanced facilities. Today, Nova Design has over 200 employees worldwide and has won over 100 international design awards since 2006. Wen-Long Chen continues to propel the development of Taiwan's design industry using "Design x Knowledge Management", striving to shape Taiwan into a powerful international design force.

Wen-Long Chen ist seit 2013 CEO des Taiwan Design Centers und ist zu „Taiwan's Top Boss in Design" ernannt worden. Er verfügt über mehr als 25 Jahre praktische Erfahrung im Designmanagement und Produktdesign. 1988 gründete er mit Unterstützung der Chinfon Trading Group Nova Design und hat seitdem Hunderte Designprojekte durchgeführt. Während dieser Zeit führte und leitete er Nova Design mit visionärer Denkweise zu enormem Erfolg. Als Antwort auf die Bedürfnisse der Kunden nutzte er „Design System Competitiveness" als zentralen Entwicklungswert, etablierte das System „KMO" (Knowledge Management Officer), um sechs Zweigbüros in Asien, Nordamerika und Europa zu betreuen, und investierte mehrere Millionen Dollar in die fortgeschrittensten Einrichtungen. Heute hat Nova Design mehr als 200 Angestellte weltweit und erhielt seit 2006 über 100 internationale Designauszeichnungen. Wen-Long Chen treibt weiterhin die Entwicklung der taiwanesischen Designindustrie voran, indem er „Design x Knowledge Management" nutzt und danach strebt, Taiwan zu einer weltweit starken Designmacht zu formen.

01 SYM Fighter 4V 150

02 Transformed Crane
Commissioned by SANY Heavy Industry Co., Ltd. and exhibited at "bauma China 2012", the International Trade Fair for Construction Machinery, Building Material Machines, Construction Vehicles and Equipment.
Transformierter Kran
In Auftrag gegeben von SANY Heavy Industry Co., Ltd. und ausgestellt auf der „bauma China 2012", der internationalen Messe für Baumaschinen, Baumaterialmaschinen, Konstruktion, Baufahrzeuge und Baugeräte.

02

> "My work philosophy is: do the right thing right, which will eventually lead you to the right people and the right resources."
> „Meine Arbeitsphilosophie lautet: Mache die richtige Sache richtig, denn das wird dich letztlich zu den richtigen Menschen und den richtigen Quellen führen."

Which product area will Taiwan have most success with in the next ten years?
Our efficient design industry will achieve success as a whole. There are both manufacturers with strong technology expertise and high adaptivity as well as a rich amount of creative talent in various fields.

Which country do you consider to be a pioneer in product design?
I would still consider Germany the pioneer in product design, due to the value they place on craftsmanship. Even though nowadays most breakthrough innovation comes from US-based start-ups, I believe in the end, it is hard to outperform Germany, both in terms of aesthetics and practicality.

What significance does winning an award in a design competition have for the designer?
While winning the award is proof of their career progress, the most beneficial effect comes from designers having the means to position themselves in the industry and determine their status.

In welchen Produktbereichen wird Taiwan in den nächsten zehn Jahren den größten Erfolg haben?
Unsere effiziente Designindustrie wird als Ganzes erfolgreich sein. Es gibt sowohl Hersteller mit großem Technologiewissen und hoher Adaptionsfähigkeit als auch eine Vielzahl an kreativen Talenten in den verschiedensten Bereichen.

Welches Land halten Sie für einen Pionier im Bereich Produktgestaltung?
Ich halte Deutschland nach wie vor für den Pionier im Produktdesign, weil es großen Wert auf handwerkliches Können legt. Obwohl die meisten bahnbrechenden Innovationen heutzutage von Start-up-Unternehmen aus den USA kommen, glaube ich, dass es letztlich schwer ist, Deutschland zu übertreffen, sowohl in puncto Ästhetik als auch in Praktikabilität.

Welche Bedeutung hat die Auszeichnung in einem Designwettbewerb für den Designer?
Während die Auszeichnung eine Bestätigung des Karrierefortschritts der Designer ist, besteht ihr hilfreichster Effekt darin, dass sie ihnen ermöglicht, sich in der Branche zu positionieren und ihren eigenen Status zu ermitteln.

01

Vivian Wai Kwan Cheng
Hong Kong
Hongkong

On leaving Hong Kong Design Institute after 19 years of educational service, Vivian Cheng founded "Vivian Design" in 2014 to provide consultancy services and promote her own art in jewellery and glass. She graduated with a BA in industrial design from the Hong Kong Polytechnic University and was awarded a special prize in the Young Designers of the Year Award hosted by the Federation of Hong Kong Industries in 1987, and the Governor's Award for Industry: Consumer Product Design in 1989, after joining Lambda Industrial Limited as the head of the Product Design team. In 1995 she finished her Master degree and joined the Vocational Training Council teaching product design and later became responsible for, among others, establishing an international network with design-related organisations and schools. Vivian Cheng was the International Liaison Manager at the Hong Kong Design Institute (HKDI) and member of the Chartered Society of Designers Hong Kong, member of the Board of Directors of the Hong Kong Design Centre (HKDC) and board member of the Icsid from 2013 to 2015. Furthermore, she has been a panel member for the government and various NGOs.

Nach 19 Jahren im Lehrbetrieb verließ Vivian Cheng 2014 das Hong Kong Design Institute und gründete „Vivian Design", um Beratungsdienste anzubieten und ihre eigene Schmuck- und Glaskunst weiterzuentwickeln. 1987 machte sie ihren BA in Industriedesign an der Hong Kong Polytechnic University. Im selben Jahr erhielt sie einen Sonderpreis im Wettbewerb „Young Designers of the Year", veranstaltet von der Federation of Hong Kong Industries, sowie 1989 den Governor's Award for Industry: Consumer Product Design, nachdem sie bei Lambda Industrial Limited als Leiterin des Produktdesign-Teams angefangen hatte. 1995 beendete sie ihren Master-Studiengang und wechselte zum Vocational Training Council, wo sie Produktdesign unterrichtete und später u. a. für den Aufbau eines internationalen Netzwerks mit Organisationen und Schulen im Designbereich verantwortlich war. Vivian Cheng war International Liaison Manager am Hong Kong Design Institute (HKDI), Mitglied der Chartered Society of Designers Hong Kong, Vorstandsmitglied des Hong Kong Design Centre (HKDC) und Gremiumsmitglied des Icsid. Außerdem war sie Mitglied verschiedener Bewertungsgremien der Regierung und vieler Nichtregierungsorganisationen.

02

"My work philosophy is: things
I do today will be a bit better than
yesterday's. I look into tomorrow,
while enjoying today, and articulate
what I experienced yesterday."

„Meine Arbeitsphilosophie ist:
Dinge, die ich heute mache, werden
ein bisschen besser als die gestrigen
sein. Ich schaue auf das Morgen,
genieße das Heute und spreche aus,
was ich gestern erfahren habe."

When did you first think consciously about design?
I always wanted to dress only in the way I like since
I was a small child. And I was told that I started to
draw on walls at the age of three. So I believe I was
born to appreciate art and design.

**What do you take special notice of when you are
assessing a product as a jury member?**
I pay special attention to whether the design combines well with the choice of materials and the
required technology, the craftsmanship, its positioning
in the market, and that the user's emotional needs
have been taken into account.

**What is the advantage of commissioning an
external designer compared to having an in-house
design team?**
An external designer can bring an outsider's view and
provide the opportunity for a second party to challenge the product's functions and the way it operates.
It might also help with the extension of product
functions as well as the application of new materials
and technology in different ways.

**Wann haben Sie das erste Mal bewusst über
Design nachgedacht?**
Schon als Kleinkind wollte ich nur die Kleidung tragen,
die mir gefiel. Und ich weiß von Erzählungen, dass ich
bereits im Alter von drei Jahren begann, Wände zu
bemalen. Daher glaube ich, dass ich dazu geboren bin,
Kunst und Design zu würdigen.

**Worauf achten Sie besonders, wenn Sie ein
Produkt als Jurorin bewerten?**
Ich achte besonders darauf, dass die Gestaltung gut
zur Wahl der Materialien und Technologien, der
Verarbeitung und der Marktpositionierung passt und
dass die emotionalen Bedürfnisse des Benutzers
berücksichtigt wurden.

**Worin liegt der Vorteil, einen externen Designer
zu beauftragen, im Vergleich zu einem Inhouse-
Designteam?**
Ein externer Designer kann eine Sicht von außen ein-
bringen und Raum für eine zweite Seite schaffen, die
die Funktionen und die Handhabung der Produkte
hinterfragt. Zudem kann es dabei helfen, die Produkt-
funktionen zu erweitern und neue Materialien und
Technologien auf unterschiedliche Arten einzubeziehen.

443

01

Datuk Prof. Jimmy Choo OBE
Malaysia /
Great Britain
Malaysia /
Großbritannien

Datuk Professor Jimmy Choo OBE is descended from a family of Malaysian shoemakers and learned the craft from his father. He studied at Cordwainers College, which is today part of the London College of Fashion. After graduating in 1983, he founded his own couture label and opened a shoe shop in London's East End whose regular customers included the late Diana, Princess of Wales. In 1996, Choo launched his ready-to-wear line with Tom Yeardye. Choo sold his share in the business in November 2001 to Equinox Luxury Holdings Ltd, charging them with the ongoing use of the label on the luxury goods market, while he continued to run his couture line. Choo now spends his time promoting design education. He is an ambassador for footwear education at the London College of Fashion, a spokesperson for the British Council in their promotion of British Education to foreign students and also spends time working with the non-profit programme, Teach For Malaysia. In 2003, Jimmy Choo was honoured for his contribution to fashion by Queen Elizabeth II who appointed him "Officer of the Order of the British Empire".

Datuk Professor Jimmy Choo OBE, der einer malaysischen Schuhmacher-Familie entstammt und das Handwerk von seinem Vater lernte, studierte am Cordwainers College, heute Teil des London College of Fashion. Nach seinem Abschluss 1983 gründete er sein eigenes Couture-Label und eröffnete ein Schuhgeschäft im Londoner East End, zu dessen Stammkundschaft auch Lady Diana gehörte. 1996 führte Choo gemeinsam mit Tom Yeardye seine Konfektionslinie ein und verkaufte seine Anteile an dem Unternehmen im November 2001 an die Equinox Luxury Holdings Ltd. Diese beauftragte er damit, das Label auf dem Markt für Luxusgüter fortzuführen, während er sich weiter um seine Couture-Linie kümmerte. Heute fördert Jimmy Choo die Designlehre. Er ist Botschafter für Footwear Education am London College of Fashion sowie Sprecher des British Council für die Förderung der Ausbildung ausländischer Studenten in Großbritannien und arbeitet für das gemeinnützige Programm „Teach for Malaysia". Für seine Verdienste in der Mode verlieh ihm Königin Elisabeth II. 2003 den Titel „Officer of the Order of the British Empire".

01 **Red Sandal**
In modern curves, made with
silk and a leather strap.
Rote Sandale
In modernen Kurven, hergestellt
aus Seide und einem Lederband.

02 **Light Brown Sandal**
Made with traditional woven
cotton from East Malaysia
(Pua Kumbu) with leather strap.
Leichte braune Sandale
Hergestellt aus traditioneller,
gewobener Baumwolle aus
Ostmalaysia (Pua Kumbu) mit
Lederband.

02

"My work philosophy is my
philosophy of life: always
move forward. Work hard.
Believe in yourself."
„Meine Arbeitsphilosophie ist
meine Lebensphilosophie:
Gehe immer vorwärts. Arbeite
hart. Glaube an dich selbst."

How does it feel to be a brand and a person of the same name?
I am very proud of what I have achieved with Jimmy Choo Couture. I no longer hold shares in Jimmy Choo London, but I am proud of co-founding a company that has become a global phenomenon.

Which country do you consider to be a pioneer in product design?
England has a rich history of producing leading design talent – and that continues to this day.

What impressed you most during the Red Dot judging process?
That professionals from around the world gather in one place, which gives us the opportunity to choose the most outstanding talent among the best.

What do you take special notice of when you are assessing a product as a jury member?
I always look for something that I haven't seen before – e.g. a new take on a well-known product. To catch my eye, the product also has to be of the highest quality, functional and durable.

Wie ist es, zugleich eine Marke und eine Person mit demselben Namen zu sein?
Ich bin sehr stolz auf das, was ich mit Jimmy Choo Couture erreicht habe. Ich besitze keine Anteile mehr an Jimmy Choo London, aber ich bin stolz darauf, ein Unternehmen mitgegründet zu haben, das sich zu einem globalen Phänomen entwickelt hat.

Welche Nation ist für Sie Vorreiter im Produktdesign?
England hat in der Vergangenheit eine Vielzahl führender Designtalente hervorgebracht – und das bis heute.

Was hat Sie bei der Red Dot-Jurierung am meisten beeindruckt?
Dass sich Profis aus aller Welt an einem Ort versammeln, was uns die Gelegenheit gibt, die hervorstechendsten Talente aus den Besten auszuwählen.

Worauf achten Sie besonders, wenn Sie ein Produkt als Juror bewerten?
Ich suche immer nach etwas, das ich vorher noch nicht gesehen habe – etwa eine neue Interpretation eines bekannten Produktes. Um meine Aufmerksamkeit zu erregen, muss es sowohl die höchsten Qualitätsstandards erfüllen als auch funktional und langlebig sein.

01

Vincent Créance
France
Frankreich

After graduating from the Ecole Supérieure de Design Industriel, Vincent Créance began his career in 1985 at the Plan Créatif Agency where he became design director and developed, among other things, numerous products for hi-tech and consumer markets, for France Télécom and RATP (Paris metro). In 1996 he joined Alcatel as Design Director for all phone activities on an international level. In 1999, he became Vice President Brand in charge of product design and user experience as well as all communications for the Mobile Phones BU. During the launch of the Franco-Chinese TCL and Alcatel Mobile Phones joint venture in 2004, Vincent Créance advanced to the position of Design and Corporate Communications Director. In 2006, he became President and CEO of MBD Design, one of the major design agencies in France, providing design solutions in transport design and product design. Créance is a member of the APCI (Agency for the Promotion of Industrial Creation), on the board of directors of ENSCI (National College of Industrial Creation), and a member of the Strategic Advisory Board for Strate College.

Vincent Créance begann seine Laufbahn nach seinem Abschluss an der Ecole Supérieure de Design Industriel 1985 bei Plan Créatif Agency. Hier stieg er 1990 zum Design Director auf und entwickelte u. a. zahlreiche Produkte für den Hightech- und Verbrauchermarkt, für die France Télécom oder die RATP (Pariser Metro). 1996 ging er als Design Director für sämtliche Telefonaktivitäten auf internationaler Ebene zu Alcatel und wurde 1999 Vice President Brand, zuständig für Produktdesign und User Experience sowie die gesamte Kommunikation für den Geschäftsbereich „Mobile Phones". Während des Zusammenschlusses des französisch-chinesischen TCL und Alcatel Mobile Phones 2004 avancierte Vincent Créance zum Design and Corporate Communications Director. 2006 wurde er Präsident und CEO von MBD Design, einer der wichtigsten Designagenturen in Frankreich, und entwickelte Designlösungen für Transport- und Produktdesign. Créance ist Mitglied von APCI (Agency for the Promotion of Industrial Creation), Vorstand des ENSCI (National College of Industrial Design) und Mitglied im wissenschaftlichen Beirat des Strate College.

02/03

"My work philosophy is: when risks appear, it becomes interesting!"

„Meine Arbeitsphilosophie lautet: Wenn Risiken auftauchen, wird es interessant!"

What are the main challenges in a designer's everyday life?
For a young one: to acquire experience in order to avoid big mistakes. For a senior one: to forget his or her experience in order to avoid big mistakes!

The Red Dot Award has been uncovering the best designs for 60 years now. Which product innovations would you like to see in the next 60 years, and why?
If everything we read about global warming etc. is true, it is obvious that product innovations have to find answers to these issues.

What significance does winning an award in a design competition have for the designer?
It's a recognition from his peers, always pleasant and helps to improve his image.

What significance does winning an award in a design competition have for the manufacturer?
It's a very good means to check the "non-rational" performance of their products. Because people also want pleasure, even from professional goods.

Was sind große Herausforderungen im Alltag eines Designers?
Für einen jungen Designer: Erfahrung sammeln, um große Fehler zu vermeiden. Für einen erfahrenen Designer: seine Erfahrung vergessen, um große Fehler zu vermeiden!

Der Red Dot Award ermittelt seit 60 Jahren die besten Gestaltungen. Welche Produktinnovationen würden Sie sich für die nächsten 60 Jahre wünschen, und warum?
Falls alles, was wir über die globale Erwärmung etc. lesen, wahr ist, ist offensichtlich, dass Produktinnovationen für diese Probleme Lösungen finden müssen.

Welche Bedeutung hat die Auszeichnung in einem Designwettbewerb für den Designer?
Sie ist eine Anerkennung seiner Partner, immer erfreulich und hilfreich, um sein Image zu verbessern.

Welche Bedeutung hat die Auszeichnung in einem Designwettbewerb für den Hersteller?
Es ist ein gutes Mittel, um die nicht-rationale Leistung seiner Produkte zu überprüfen. Denn Menschen wollen Wohlgefallen, selbst bei Fachprodukten.

01

Martin Darbyshire
Great Britain
Großbritannien

Martin Darbyshire is a founder and CEO of the internationally renowned design consultancy tangerine, working for clients such as Asiana, Azul, B/E Aerospace, Huawei, Nikon, The Royal Mint, Snoozebox.com and Virgin Australia. Before founding tangerine in 1989, he worked for Moggridge Associates and then in San Francisco at ID TWO (now IDEO). Martin Darbyshire is responsible for tangerine's commercial management, leading design projects and creating new business opportunities. Most notably, he led the multidisciplinary team that created both generations of the "Club World" business-class aircraft seating for British Airways – the world's first fully flat bed in business class which, since its launch in 2000, has remained the profit engine of the airline and transformed the industry. Besides, Martin Darbyshire has been a UK Trade and Investment ambassador for the UK Creative Industries sector, he is a recognised industry spokesperson, an advisor on design and innovation and was a board member of the Icsid.

Martin Darbyshire ist Gründer und CEO des international renommierten Designbüros tangerine, das für Kunden wie Asiana, Azul, B/E Aerospace, Huawei, Nikon, The Royal Mint, Snoozebox.com und Virgin Australia tätig ist. Bevor er tangerine 1989 gründete, arbeitete er für Moggridge Associates und danach in San Francisco bei ID TWO (heute IDEO). Martin Darbyshire verantwortet das kaufmännische Management von tangerine, wozu die Leitung von Designprojekten und die Entwicklung neuer Geschäftsmöglichkeiten gehört. Unter seiner Leitung stand insbesondere das multidisziplinäre Team, das beide Generationen der Business-Class-Sitze „Club World" für British Airways gestaltet hat – das weltweit erste komplett flache Bett in einer Business Class, das der Airline seit seiner Markteinführung im Jahr 2000 enorme Umsatzzahlen beschert und die Branche nachhaltig verändert hat. Darüber hinaus ist Martin Darbyshire für das Ministerium für Handel und Investition des Vereinigten Königreichs Botschafter für den Bereich der Kreativindustrie, ein anerkannter Sprecher der Branche sowie Berater für Design und Innovation und er war Gremiumsmitglied des Icsid.

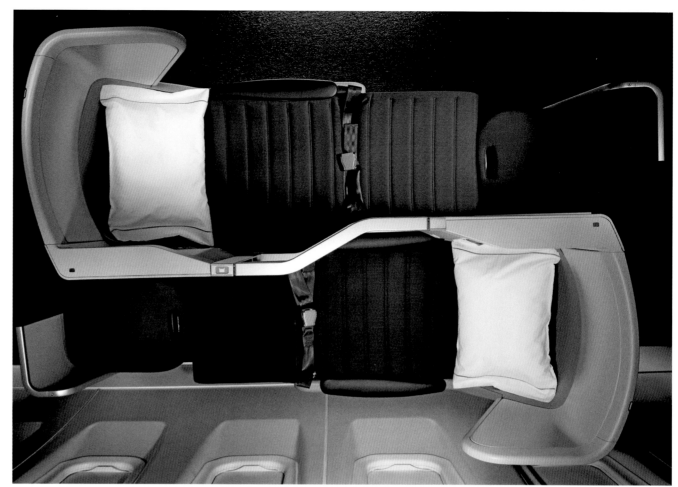

02

"My work philosophy is about challenging preconceptions and creating breakthrough change, improving lives and generating wealth."

„Meine Arbeitsphilosophie besteht darin, vorgefasste Meinungen zu hinterfragen und bahnbrechende Veränderungen zu erzielen, die das Leben verbessern und Wohlstand erzeugen."

The Red Dot Award has been uncovering the best designs for 60 years now. Which product innovations would you like to see in the next 60 years, and why?
I would like to see more applications of modern technologies and a greater focus on using modern technological advances to develop cheaper medical products, for instance creating low-cost portable equipment for developing countries.

What do you take special notice of when you are assessing a product as a jury member?
When assessing a new car entry for instance, I look at the interior first, taking great interest in the detailing and finishes before looking at the exterior. For me it is important that both have been given equal importance and attention by the design team.

What significance does winning an award in a design competition have for the designer?
It is a good feeling to know that your peers have judged you worthy of it, because a Red Dot carries such prestige within the design and commercial community.

Der Red Dot Award ermittelt seit 60 Jahren die besten Gestaltungen. Welche Produktinnovationen würden Sie sich für die nächsten 60 Jahre wünschen, und warum?
Ich würde gerne mehr Anwendungen moderner Technologien und ein größeres Augenmerk auf moderne technologische Fortschritte sehen, um günstigere Medizinprodukte zu entwickeln, wie zum Beispiel bei der Herstellung von kostengünstigen, tragbaren Geräten für Entwicklungsländer.

Worauf achten Sie besonders, wenn Sie ein Produkt als Juror bewerten?
Wenn ich etwa ein Auto bewerte, schaue ich zuerst auf die Innenausstattung, wobei mich die Details und die Verarbeitung besonders interessieren. Erst dann inspiziere ich das Äußere. Für mich ist es wichtig, dass beide die gleiche Aufmerksamkeit vom Designteam erhalten haben.

Welche Bedeutung hat die Auszeichnung in einem Designwettbewerb für den Designer?
Es ist ein gutes Gefühl zu wissen, dass dich deine Fachkollegen als dessen würdig erachten, da ein Red Dot ein so großes Ansehen in der Design- und Geschäftswelt genießt.

01

Robin Edman
Sweden
Schweden

Robin Edman has been the chief executive of SVID, the Swedish Industrial Design Foundation, since 2001. After studying industrial design at Rhode Island School of Design he joined AB Electrolux Global Design in 1981 and parallel to this started his own design consultancy. In 1989, Robin Edman joined Electrolux North America as vice president of Industrial Design for Frigidaire and in 1997, moved back to Stockholm as vice president of Electrolux Global Design. Throughout his entire career he has worked towards integrating a better understanding of users, their needs and the importance of design in society at large. His engagement in design related activities is reflected in the numerous international jury appointments, speaking engagements, advisory council and board positions he has held. Robin Edman served on the board of Icsid from 2003 to 2007, the last term as treasurer. Since June 2015, he is the president of BEDA (Bureau of European Design Associations).

Robin Edman ist seit 2001 Firmenchef der SVID, der Swedish Industrial Design Foundation. Nach seinem Industriedesign-Studium an der Rhode Island School of Design kam er 1981 zu AB Electrolux Global Design. Zeitgleich startete er seine eigene Unternehmensberatung für Design. 1989 wechselte Edman zu Electrolux North America als Vizepräsident für Industrial Design für Frigidaire und kehrte 1997 als Vizepräsident von Electrolux Global Design nach Stockholm zurück. Während seiner gesamten Karriere hat er daran gearbeitet, ein besseres Verständnis für Nutzer zu entwickeln, für deren Bedürfnisse und die Wichtigkeit von Design in der Gesellschaft insgesamt. Sein Engagement in designbezogenen Aktivitäten weist sich durch zahlreiche Jurierungsberufungen aus sowie durch Rednerverpflichtungen und Positionen in Gremien sowie Beratungsausschüssen. Von 2003 bis 2007 war Robin Edman Mitglied im Vorstand des Icsid, in der letzten Amtsperiode als Schatzmeister. Seit Juni 2015 ist er Präsident von BEDA (Bureau of European Design Associations).

02

"My work philosophy is to create
a better place to live in for as
many people as possible ...
and to have fun while doing so."
„Meine Arbeitsphilosophie besteht
darin, für so viele Menschen wie
möglich einen besseren Lebensraum zu schaffen ...
und dabei Spaß zu haben."

**Which product area will Sweden have most success
with in the next ten years?**
Assuming the definition of "product" follows the now
common view of both goods and services, I would like
to stress the enormous potential in the public sector.
The areas ranging from the development of the health
care sector to regional and local innovation will include an array of design-focused competences that
will transform our future society.

**The Red Dot Award has been uncovering the
best designs for 60 years now. Which product
innovations would you like to see in the next
60 years?**
A combination of solutions that will secure the
continuation of humankind on earth and decrease
our footprint on earth.

**What significance does winning an award in a
design competition have for the manufacturer?**
It means recognition and is a great marketing tool
that gives a seal of credibility and drives internal
innovation.

**In welchem Produktbereich wird Schweden in den
nächsten zehn Jahren den größten Erfolg haben?**
Voraussetzend, dass sich die Definition des Begriffs
„Produkt" auf dessen zurzeit gängige Ansicht als Güter
und Dienstleistungen bezieht, möchte ich das enorme
Potenzial im öffentlichen Sektor hervorheben. Die Bereiche von der Entwicklung der Gesundheitsversorgung
bis hin zu regionalen und lokalen Innovationen werden
ein breites Spektrum designbezogener Kompetenzen
beinhalten, das unsere Gesellschaft zukünftig
verändern wird.

**Der Red Dot Award ermittelt seit 60 Jahren die
besten Gestaltungen. Welche Produktinnovationen
würden Sie sich für die nächsten 60 Jahre
wünschen?**
Eine Kombination aus Lösungen, die das Fortbestehen
der Menschheit auf der Erde sichern und unseren
ökologischen Fußabdruck auf der Erde verringern.

**Welche Bedeutung hat die Auszeichnung in einem
Designwettbewerb für den Hersteller?**
Sie bedeutet Anerkennung und ist ein großartiges
Marketingwerkzeug, das ein Siegel der Glaubwür-
digkeit darstellt und die interne Motivation antreibt.

01

02

Hans Ehrich
Sweden
Schweden

Hans Ehrich, born 1942 in Helsinki, Finland, has lived and been educated in Finland, Sweden, Germany, Switzerland, Spain and Italy. From 1962 to 1967, he studied metalwork and industrial design at the University College of Arts, Crafts and Design (Konstfackskolan), Stockholm. In 1965, his studies as a designer took him to Turin and Milan. With Tom Ahlström, he co-founded and became a director of A&E Design AB in 1968, a company which he still heads as director. From 1982 to 2002, he was managing director of Interdesign AB, Stockholm. Hans Ehrich has designed for, among others, Alessi, Anza, ASEA, Cederroth, Colgate, Fagerhults, Jordan, RFSU, Siemens, Turn-O-Matic and Yamagiwa. His work has been exhibited at many international exhibitions and collections and he has received numerous awards.

Hans Ehrich, 1942 in Helsinki, Finnland, geboren, lebte und lernte in Finnland, Schweden, Deutschland, der Schweiz, Spanien und Italien. Von 1962 bis 1967 studierte er Metall- und Industriedesign am University College of Arts, Crafts and Design (Konstfackskolan) in Stockholm. 1965 führten ihn Studien als Designer nach Turin und Mailand. 1968 war er zusammen mit Tom Ahlström Gründungsdirektor von A&E Design AB, Stockholm, das er heute immer noch als Direktor leitet, und von 1982 bis 2002 war er geschäftsführender Direktor von Interdesign AB, Stockholm. Hans Ehrich gestaltete u. a. für Alessi, Anza, ASEA, Cederroth, Colgate, Fagerhults, Jordan, RFSU, Siemens, Turn-O-Matic und Yamagiwa. Seine Arbeiten sind in zahlreichen internationalen Ausstellungen und Sammlungen vertreten und er wurde vielfach ausgezeichnet.

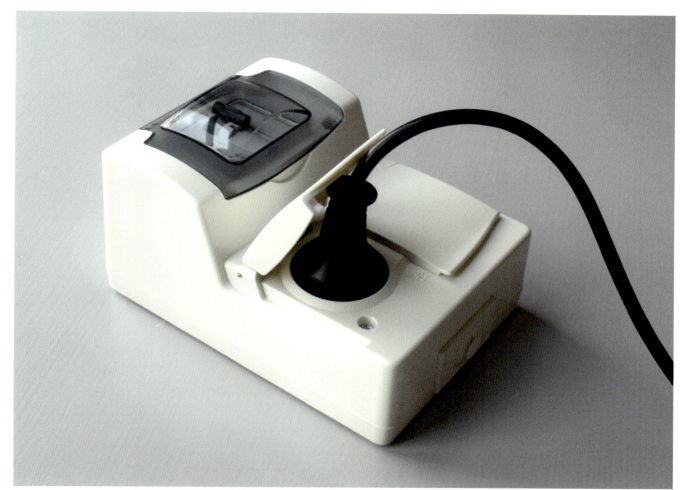

03

"My work philosophy has always been: create for tomorrow with a foothold on yesterday."
„Meine Arbeitsphilosophie war stets: Gestalte für morgen mit einem Halt in der Vergangenheit."

This year you have donated the complete collection from your 50 years of work to the Swedish National Museum. Which motto do you think would be most fitting for this collection?
"Handmade models and prototypes, approximately 400 objects, which represent half a century of successful Swedish product design."

Which of the projects in your lifetime are you particularly proud of?
I am particularly proud of the "Stockholm II" folding stool for museums, which is in use worldwide.

What impressed you most during the Red Dot judging process?
The high design standards, the good functionality and the appealing material quality of many of the submitted products.

What is the advantage of commissioning an external designer compared to having an in-house design team?
An external designer often has a broader spectrum of professional expertise with more influences than an in-house-designer.

Sie haben in diesem Jahr die gesamte Sammlung Ihrer 50 Berufsjahre dem schwedischen National-museum überlassen. Unter welches Motto möchten Sie diese Sammlung am liebsten stellen?
„Handgefertigte Modelle und Prototypen, etwa 400 Objekte, die ein halbes Jahrhundert erfolgreiches schwedisches Produktdesign repräsentieren."

Auf welches Projekt in Ihrem Leben sind Sie besonders stolz?
Auf den weltweit verbreiteten Museumsklapphocker „Stockholm II".

Was hat Sie bei der Red Dot-Jurierung am meisten beeindruckt?
Das hohe Gestaltungsniveau, die gute Funktionalität und die ansprechende Materialqualität einer Vielzahl der eingereichten Produkte.

Worin liegt der Vorteil, einen externen Designer zu beauftragen, im Vergleich zu einem Inhouse-Designteam?
Ein externer Designer verfügt meistens über eine breiter gefächerte Berufskompetenz mit mehr Einfalls-winkeln als ein Inhouse-Designer.

01

Joachim H. Faust
Germany
Deutschland

Joachim H. Faust, born in 1954, studied architecture at the Technical University of Berlin, the Technical University of Aachen, as well as at Texas A&M University (with Professor E. J. Romieniec), where he received his Master of Architecture in 1981. He worked as a concept designer in the design department of Skidmore, Owings & Merrill in Houston, Texas and as a project manager in the architectural firm Faust Consult GmbH in Mainz. From 1984 to 1986, he worked for KPF Kohn, Pedersen, Fox/Eggers Group in New York and as a project manager at the New York office of Skidmore, Owings & Merrill. In 1987, Joachim H. Faust took over the management of the HPP office in Frankfurt am Main. Since 1997, he has been managing partner of the HPP Hentrich-Petschnigg & Partner GmbH + Co. KG in Düsseldorf. He also writes articles and gives lectures on architecture and interior design.

Joachim H. Faust, 1954 geboren, studierte Architektur an der TU Berlin und der RWTH Aachen sowie – bei Professor E. J. Romieniec – an der Texas A&M University, wo er sein Studium 1981 mit dem Master of Architecture abschloss. Er war Entwurfsarchitekt im Design Department des Büros Skidmore, Owings & Merrill, Houston, Texas, sowie Projektleiter im Architekturbüro der Faust Consult GmbH in Mainz. Anschließend arbeitete er im Büro KPF Kohn, Pedersen, Fox/Eggers Group in New York und war Projektleiter im Büro Skidmore, Owings & Merrill in New York. 1987 übernahm Joachim H. Faust die Leitung des HPP-Büros in Frankfurt am Main und ist seit 1997 geschäftsführender Gesellschafter der HPP Hentrich-Petschnigg & Partner GmbH + Co. KG in Düsseldorf. Er ist zudem als Autor tätig und hält Vorträge zu Fachthemen der Architektur und Innenarchitektur.

02

"My work philosophy is:
to your own self be true."

„Meine Arbeitsphilosophie lautet:
Bleibe dir selbst treu."

What piece of advice has been useful in your youth or at the beginning of your professional career?
To face big things with composure, and small ones with close attention.

What will a single-family home look like in the year 2050?
Modular, prefabricated, interactive, adaptive, but hopefully also personal with human-centred standards and materials.

The Red Dot Award has been uncovering the best designs for 60 years now. Which product innovation would you like to see in the next 60 years, and why?
Intrinsic value is one of the highest goals in the design of a product or building. This means continuity in the best sense of a "classic".

What has impressed you most during the Red Dot judging process?
The wide range of products and the sensitivity for material conformity have increased every year. And with that my enthusiasm to take part.

Welcher Ratschlag hat Sie in Ihrer Jugend oder frühen Berufskarriere weitergebracht?
Großen Dingen begegnet man mit Gelassenheit, kleinen mit besonderer Aufmerksamkeit.

Wie wird ein Einfamilienhaus im Jahr 2050 aussehen?
Modular, vorfabriziert, interaktiv, adaptiv, aber hoffentlich auch privat mit menschlich angemessenem Maßstab und Material.

Der Red Dot Award ermittelt seit 60 Jahren die besten Gestaltungen. Welche Produktinnovation würden Sie sich für die nächsten 60 Jahre wünschen, und warum?
Werthaltigkeit ist eines der höchsten Ziele in der Gestaltung eines Produkts bzw. Bauwerks. Das bedeutet Kontinuität im besten Sinne des „Klassikers".

Was hat Sie bei der Red Dot-Jurierung am meisten beeindruckt?
Die Vielfalt der Produkte und die Sensibilität für Materialkonformität sind in jedem Jahr gestiegen. Damit natürlich auch meine Begeisterung, mitzuwirken.

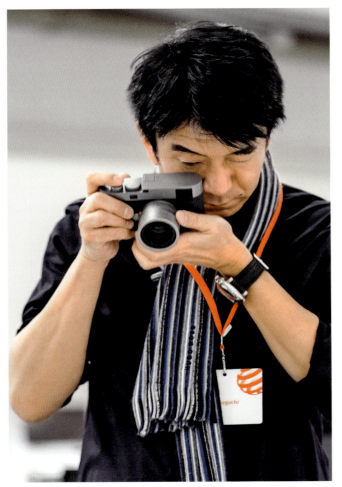

01

Hideshi Hamaguchi
USA / Japan
USA / Japan

Hideshi Hamaguchi graduated with a Bachelor of Science in chemical engineering from Kyoto University. Starting his career with Panasonic in Japan, Hamaguchi later became director of the New Business Planning Group at Panasonic Electric Works, Ltd. and then executive vice president of Panasonic Electric Works Laboratory of America, Inc. In 1993, he developed Japan's first corporate Intranet and also led the concept development for the first USB flash drive. Hideshi Hamaguchi has over 15 years of experience in defining strategies and decision-making, as well as in concept development for various industries and businesses. As Executive Fellow at Ziba Design and CEO at monogoto, he is today considered a leading mind in creative concept and strategy development on both sides of the Pacific and is involved in almost every project this renowned business consultancy takes on. For clients such as FedEx, Polycom and M-System he has led the development of several award-winning products.

Hideshi Hamaguchi graduierte als Bachelor of Science in Chemical Engineering an der Kyoto University. Seine Karriere begann er bei Panasonic in Japan, wo er später zum Direktor der New Business Planning Group von Panasonic Electric Works, Ltd. und zum Executive Vice President von Panasonic Electric Works Laboratory of America, Inc. aufstieg. 1993 entwickelte er Japans erstes Firmen-Intranet und übernahm zudem die Leitung der Konzeptentwicklung des ersten USB-Laufwerks. Hideshi Hamaguchi verfügt über mehr als 15 Jahre Erfahrung in der Konzeptentwicklung sowie Strategie- und Entscheidungsfindung in unterschiedlichen Industrien und Unternehmen. Als Executive Fellow bei Ziba Design und CEO bei monogoto wird er heute als führender Kopf in der kreativen Konzept- und Strategieentwicklung auf beiden Seiten des Pazifiks angesehen und ist in nahezu jedes Projekt der renommierten Unternehmensberatung involviert. Für Kunden wie FedEx, Polycom und M-System leitete er etliche ausgezeichnete Projekte.

01 **Cintiq 24HD**
    for Wacom, 2012
    für Wacom, 2012

02 **FedEx World Service
    Centre, 1999**
    The concept that changed
    the way FedEx understands
    and treats its customers.
    Das Konzept, das FedEx ein
    besseres Verständnis für
    seine Kunden und deren
    Behandlung vermittelte.

02

"My work philosophy is:
all I need is less."

„Meine Arbeitsphilosophie lautet:
Alles, was ich brauche, ist weniger."

**What impressed you most during the Red Dot
judging process?**
This year still felt like the end of a transitional
period – a time to resolve some of the critical ten-
sions that have emerged out of the massive changes
of technology, consumer experience, and business
models in the past ten years.

**What challenges do you see for the future
in design?**
The challenge is finding the sweet spot between
what resonates with the consumer and what is true
to the brand.

**Do you see a correlation between the design
quality of a company's products and the economic
success of this company?**
I see a strong correlation between them. If a company
has a good design in all three phases of consumer
interaction – attract, engage and extend – it should
directly impact its success.

**Was hat Sie bei der Red Dot-Jurierung am meisten
beeindruckt?**
Dieses Jahr fühlte sich immer noch wie das Ende einer
Übergangszeit an – eine Zeit, um einen Teil der kritischen Spannungen, die aus den massiven Veränderungen der Technologie, Konsumentenerfahrung und
Geschäftsmodelle während der letzten zehn Jahre
resultieren, aufzulösen.

**Welche zukünftigen Herausforderungen sehen
Sie im Designbereich?**
Die Herausforderung besteht darin, das richtige Verhältnis zwischen dem, was beim Konsumenten Anklang
findet, und dem, was der Wahrheit der Marke entspricht, zu finden.

**Sehen Sie einen Zusammenhang zwischen der
Designqualität, die sich in den Produkten eines
Unternehmens äußert, und dem wirtschaftlichen
Erfolg dieses Unternehmens?**
Ich sehe eine starke Korrelation zwischen beiden.
Wenn ein Unternehmen gute Gestaltung für alle drei
Phasen der Interaktion mit dem Konsumenten –
auffallen, einnehmen, ausbauen – bietet, dann sollte
sich das unmittelbar auf seinen Erfolg auswirken.

01

Prof. Renke He
China
China

Professor Renke He, born in 1958, studied civil engineering and architecture at Hunan University in China. From 1987 to 1988, he was a visiting scholar at the Industrial Design Department of the Royal Danish Academy of Fine Arts in Copenhagen and, from 1998 to 1999, at North Carolina State University's School of Design. Renke He is dean and professor of the School of Design at Hunan University and is also director of the Chinese Industrial Design Education Committee. Currently, he holds the position of vice chair of the China Industrial Design Association.

Professor Renke He wurde 1958 geboren und studierte an der Hunan University in China Bauingenieurwesen und Architektur. Von 1987 bis 1988 war er als Gastprofessor für Industrial Design an der Royal Danish Academy of Fine Arts in Kopenhagen tätig, und von 1998 bis 1999 hatte er eine Gastprofessur an der School of Design der North Carolina State University inne. Renke He ist Dekan und Professor an der Hunan University, School of Design, sowie Direktor des Chinese Industrial Design Education Committee. Er ist derzeit zudem stellvertretender Vorsitzender der China Industrial Design Association.

02

"My work philosophy is: happy life, happy design."

„Meine Arbeitsphilosophie ist: Ein glückliches Leben führt zu glücklicher Gestaltung."

Which country do you consider to be a pioneer in product design?
The USA, because companies like Apple combine technologies with service design, business model and interaction design in order to create brand new products for the global markets.

What are the main challenges in a designer's everyday life?
Finding a balance between business and social responsibility in design.

What significance does winning an award in a design competition have for the designer?
It is the ultimate recognition of a designer's professional skills and reputation in this competitive world.

What significance does winning an award in a design competition have for the manufacturer?
A design award is a wonderful ticket to success in the marketplace.

Welche Nation ist für Sie Vorreiter im Produktdesign?
Die USA. Unternehmen wie Apple vereinen Technologien mit Servicedesign, Geschäftsmodell und Interaction Design, um so brandneue Produkte für den globalen Markt zu entwerfen.

Was sind große Herausforderungen im Alltag eines Designers?
Das Gleichgewicht zwischen Geschäft und sozialer Verantwortung im Design zu finden.

Welche Bedeutung hat die Auszeichnung in einem Designwettbewerb für den Designer?
Sie ist die beste Bestätigung für die Fertigkeiten und den Ruf eines Designers in unserer wettbewerbsorientierten Welt.

Welche Bedeutung hat die Auszeichnung in einem Designwettbewerb für den Hersteller?
Eine Auszeichnung in einem Designwettbewerb ist eine wunderbare Fahrkarte zum Markterfolg.

01

Prof. Herman Hermsen
Netherlands
Niederlande

Professor Herman Hermsen, born in 1953 in Nijmegen, Netherlands, studied at the ArtEZ Institute of the Arts in Arnhem from 1974 to 1979. Following an assistant professorship, he began his career in teaching in 1985. Since 1979, he is an independent jewellery and product designer. Until 1990, he taught product design at the Utrecht School of the Arts (HKU), after which time he returned to Arnhem as lecturer at the Academy. Hermsen has been professor of product and jewellery design at the University of Applied Sciences in Düsseldorf since 1992. He gives guest lectures at universities and colleges throughout Europe, the United States and Japan, and began regularly organising specialist symposia in 1998. He has also served as juror for various competitions. Herman Hermsen has received numerous international awards for his work in product and jewellery design, which is shown worldwide in solo and group exhibitions and held in the collections of renowned museums, such as the Cooper-Hewitt Museum, New York; the Pinakothek der Moderne, Munich; and the Museum of Arts and Crafts, Kyoto.

Professor Herman Hermsen, 1953 in Nijmegen in den Niederlanden geboren, studierte von 1974 bis 1979 am ArtEZ Institute of the Arts in Arnheim und ging nach einer Assistenzzeit ab 1985 in die Lehre. Seit 1979 ist er unabhängiger Schmuck- und Produktdesigner. Bis 1990 unterrichtete er Produktdesign an der Utrecht School of the Arts (HKU) und kehrte anschließend nach Arnheim zurück, um an der dortigen Hochschule als Dozent zu arbeiten. Seit 1991 ist Hermsen Professor für Produkt- und Schmuckdesign an der Fachhochschule Düsseldorf; er hält Gastvorlesungen an Hochschulen in ganz Europa, den USA und Japan, organisiert seit 1988 regelmäßig Fachsymposien und ist Juror in verschiedenen Wettbewerbsgremien. Für seine Arbeiten im Produkt- und Schmuckdesign, die weltweit in Einzel- und Gruppenausstellungen präsentiert werden und sich in den Sammlungen großer renommierter Museen befinden – z. B. Cooper-Hewitt Museum, New York, Pinakothek der Moderne, München, und Museum of Arts and Crafts, Kyoto –, erhielt Herman Hermsen zahlreiche internationale Auszeichnungen.

02

"My work philosophy is: if 'less or more' is wanted, it should at least contribute to the poetry."

„Meine Arbeitsphilosophie ist: Wenn ‚weniger oder mehr' gewünscht wird, sollte es wenigstens zur Poesie beitragen."

What can laypeople learn, when they look at award-winning products in design museums?
They can learn that design is not a direct translation of emotions, but that a lot of thought has gone into which design language can aesthetically communicate the perception of the product's function, and how this perception has developed in different cultures and eras.

Which of the projects in your lifetime are you particularly proud of?
I have created designs which I believe provide a poetic answer; they communicate the essence of a concept, such as my lamp "Charis" for Classicon.

Which nation to you consider to be a pioneer in product design?
I have come to increasingly appreciate Scandinavian design, because the designers have developed a successful combination of design language, function, zeitgeist, innovation and high-quality manufacturing.

Was können Laien lernen, wenn sie in Design-museen hervorragend gestaltete Produkte betrachten?
Dass Gestaltung keine direkte Umsetzung von Emotionen ist, dass aber genau überlegt wurde, welche Formensprache die Sichtweise auf die Produktfunktion ästhetisch kommunizieren kann. Und wie sich diese Sichtweisen in den unterschiedlichen Kulturen und Epochen entwickelten.

Auf welches Projekt in Ihrem Leben sind Sie besonders stolz?
Ich habe natürlich Entwürfe, von denen ich denke, mit der gestalterischen Umsetzung eine poetische Antwort gefunden zu haben, die die Essenz des Konzeptes kommuniziert; z. B. meine Lampe „Charis" für Classicon.

Welche Nation ist für Sie Vorreiter im Produktdesign?
Immer mehr schätze ich das skandinavische Design, denn die Designer entwickelten ein gelungenes Zusammenspiel aus Formensprache, Funktion, Zeitgeist, Innovation und qualitativ sehr guter Herstellung.

01

Prof. Carlos Hinrichsen
Chile
Chile

Professor Carlos Hinrichsen, born in 1957, graduated as an industrial designer in Chile in 1982 and earned his master's degree in engineering in Japan in 1991. Currently, he is dean of the Faculty of Engineering and Business at the Gabriela Mistral University in Santiago. From 2007 to 2009 he was president of Icsid and currently serves as senator within the organisation. He has since been heading research projects in the areas of innovation, design and education, and in 2010 was honoured with the distinction "Commander of the Order of the Lion of Finland". From 1992 to 2010 he was director of the School of Design Duoc UC, Chile and from 2011 to 2014 director of the Duoc UC International Affairs. He has led initiatives that integrate trade, engineering, design, innovation, management and technology in Asia, Africa and Europe and is currently design director for the Latin American Region of Design Innovation. Since 2002, Carlos Hinrichsen has been an honorary member of the Chilean Association of Design. Furthermore, he has been giving lectures at various conferences and universities around the world.

Professor Carlos Hinrichsen, 1957 geboren, erlangte 1982 seinen Abschluss in Industriedesign in Chile und erhielt 1991 seinen Master der Ingenieurwissenschaft in Japan. Aktuell ist er Dekan der Fakultät „Engineering and Business" an der Universität Gabriela Mistral in Santiago. Von 2007 bis 2009 war er Icsid-Präsident und ist heute Senator innerhalb der Organisation. Seither leitet er Forschungsprojekte in den Bereichen Innovation, Design sowie Erziehung und wurde 2010 mit der Auszeichnung „Commander of the Order of the Lion of Finland" geehrt. Von 1992 bis 2010 war er Direktor der School of Design Duoc UC in Chile und von 2011 bis 2014 Direktor der Duoc UC International Affairs. In Asien, Afrika und Europa leitete Carlos Hinrichsen Initiativen, die Handel, Ingenieurwesen, Design, Innovation, Management und Technologie integrieren, und ist aktuell Designdirektor der Latin American Region of Design Innovation. Seit 2002 ist er Ehrenmitglied der Chilean Association of Design. Außerdem hält er Vorträge auf Konferenzen und in Hochschulen weltweit.

02

"My work philosophy is well
described by the following quote:
'If you can imagine it, you can
create it. If you can dream it, you
can become it.'"

„Meine Arbeitsphilosophie lässt sich
gut mit folgendem Zitat beschreiben:
‚Wenn du es dir vorstellen kannst,
kannst du es kreieren. Wenn du
davon träumen kannst, kannst du
es werden.'"

When did you first think consciously about design?
I realised as a child that design contributes to human
happiness, and I have seen this insight confirmed over
the years.

**What impressed you most during the Red Dot
judging process?**
This time I saw how products offer realisations of the
desires and dreams of the users, as well as those of
the designers and producers. In product categories,
design quality and innovation are playing a key role
in turning technological innovations into good and
useful solutions. For this the Red Dot Award is like
a mirror, always reflecting what is going on in the
current design industry and market. It reveals the
prevalent trends and enables us to foresee other
potential trends, all of which have opened an unpre-
cedented field of knowledge and expectations.

**Wann haben Sie das erste Mal bewusst über
Design nachgedacht?**
Als Kind habe ich erkannt, dass Design dazu beiträgt,
dass sich Menschen glücklich fühlen. Und diese
Erkenntnis hat sich über die Jahre hinweg bestätigt.

**Was hat Sie bei der Red Dot-Jurierung am meisten
beeindruckt?**
Dieses Mal habe ich gesehen, wie Produkte Umset-
zungen der Wünsche und Träume sowohl der
Nutzer als auch der Designer und Hersteller bieten.
In den Produktkategorien spielen Designqualität
und -innovation eine zentrale Rolle dabei, technische
Neuerungen in gute und nützliche Lösungen zu überführen. Daher ist der Red Dot Award wie ein Spiegel,
der wiedergibt, was momentan in der Designbranche
und auf dem Markt passiert. Er offenbart die vor-
herrschenden Trends und ermöglicht uns, andere po-
tenzielle Trends vorherzusehen, die alle zusammen
einen neuartigen Bereich an Wissen und Erwartungen
eröffnet haben.

01

Tapani Hyvönen
Finland
Finnland

Tapani Hyvönen graduated in 1974 as an industrial designer from the present Aalto University School of Arts, Design and Architecture. He founded the design agency "Destem Ltd." in 1976 and was co-founder of ED-Design Ltd. in 1990, one of Scandinavia's largest design agencies. He has served as CEO and president of both agencies until 2013. Since then, he has been a visiting professor at, among others, Guangdong University of Technology in Guangzhou and Donghua University in Shanghai, China. His many award-winning designs are part of the collections of the Design Museum Helsinki and the Cooper-Hewitt Museum, New York. Tapani Hyvönen was an advisory board member of the Design Leadership Programme at the University of Art and Design Helsinki from 1989 to 2000, and a board member of the Design Forum Finland from 1998 to 2002, as well as the Icsid from 1999 to 2003 and again from 2009 to 2013. He has been a jury member in many international design competitions, a member of the Finnish-Swedish Design Academy since 2003 and a board member of the Finnish Design Museum since 2011.

Tapani Hyvönen graduierte 1974 an der heutigen Aalto University School of Arts, Design and Architecture zum Industriedesigner. 1976 gründete er die Design-agentur „Destem Ltd." und war 1990 Mitbegründer der ED-Design Ltd., einer der größten Designagenturen Skandinaviens, die er beide bis 2013 als CEO und Präsident leitete. Seitdem lehrt er als Gastprofessor u. a. an der Guangdong University of Technology in Guangzhou und der Donghua University in Shanghai, China. Seine vielfach ausgezeichneten Arbeiten sind in den Sammlungen des Design Museum Helsinki und des Cooper-Hewitt Museum, New York, vertreten. Tapani Hyvönen war von 1989 bis 2000 in der Bera-tungskommission des Design Leadership Programme der University of Art and Design Helsinki, von 1998 bis 2002 Vorstandsmitglied des Design Forum Finland sowie von 1999 bis 2003 und von 2009 bis 2013 des Icsid. Er ist international als Juror tätig, seit 2003 Mitglied der Finnish-Swedish Design Academy und seit 2011 Vorstandsmitglied des Finnish Design Museum.

02

"My work philosophy is to be
open to new ideas."

„Meine Arbeitsphilosophie ist es,
für neue Ideen offen zu sein."

**What advice did you find helpful in your younger
years or in the early days of your career?**
Be curious, look and study different aspects seriously
and don't fall in love with the first idea that comes.

**Are there any designers who are role models for
you?**
A designer who puts good design, function, aesthetics and structure in balance. Alvar Aalto's beautiful, simple and functional designs are something
I appreciate.

**The Red Dot Award has been uncovering the
best designs for 60 years now. Which product
innovation would you like to see in the next
60 years?**
Technology will become invisible. We will have
products similar to the ones from now but they will
feature a new kind of intelligence.

**What significance does winning an award in a
design competition have for the manufacturer?**
It challenges the company to invest in design.

**Welcher Ratschlag hat Sie in Ihrer Jugend oder
frühen Berufskarriere weitergebracht?**
Neugierig zu sein, sich ernsthaft verschiedene Aspekte
anzuschauen und zu studieren und sich nicht in die
erstbeste Idee, die daherkommt, zu verlieben.

Gibt es Designer, die Ihnen als Vorbilder dienen?
Designer, die gute Gestaltung, Funktion, Ästhetik und
Konstruktion in Einklang bringen. Alvar Aaltos schöne,
schlichte und funktionale Gestaltungen schätze ich
durchaus.

**Der Red Dot Award ermittelt seit 60 Jahren die
besten Gestaltungen. Welche Produktinnovation
würden Sie sich für die nächsten 60 Jahre
wünschen?**
Technologie wird unsichtbar werden. Wir werden
ähnliche Produkte wie die heutigen haben, aber sie
werden eine neue Art von Intelligenz aufweisen.

**Welche Bedeutung hat die Auszeichnung in einem
Designwettbewerb für den Hersteller?**
Es fordert das Unternehmen dazu heraus, in
Gestaltung zu investieren.

01

Guto Indio da Costa
Brazil
Brasilien

Guto Indio da Costa, born in 1969 in Rio de Janeiro, studied product design and graduated from the Art Center College of Design in Switzerland in 1993. He is design director of Indio da Costa A.U.D.T., a consultancy based in Rio de Janeiro, which develops architectural, urban planning, design and transportation projects. It works with a multidisciplinary strategic-creative group of designers, architects and urban planners, supported by a variety of other specialists. Guto Indio da Costa is a member of the Design Council of the State of Rio de Janeiro, former Vice President of the Brazilian Design Association (Abedesign) and founder of CBDI (Brazilian Industrial Design Council). He has been active as a lecturer and a contributing writer to different design magazines and has been a jury member of many design competitions in Brazil and abroad.

Guto Indio da Costa, geboren 1969 in Rio de Janeiro, studierte Produktdesign und machte 1993 seinen Abschluss am Art Center College of Design in der Schweiz. Er ist Gestaltungsdirektor von Indio da Costa A.U.D.T., einem in Rio de Janeiro ansässigen Beratungsunternehmen, das Projekte in Architektur, Stadtplanung, Design- und Transportwesen entwickelt und mit einem multidisziplinären, strategisch-kreativen Team aus Designern, Architekten und Stadtplanern sowie mit der Unterstützung weiterer Spezialisten operiert. Guto Indio da Costa ist Mitglied des Design Councils des Bundesstaates Rio de Janeiro, ehemaliger Vize-Präsident der brasilianischen Designvereinigung (Abedesign) und Gründer des CBDI (Industrial Design Council Brasilien). Er ist als Lehrbeauftragter aktiv, schreibt für verschiedene Designmagazine und ist als Jurymitglied zahlreicher Designwettbewerbe in und außerhalb Brasiliens tätig.

01 "Fabrimar Lucca"
Line of faucets, winner of
the Iconic Awards 2014 by
the German Design Council
Armaturserie, Gewinner
des Iconic Awards 2014
des German Design Council

02 São Paulo Bus Shelter
This bus shelter, of which thou-
sands are already built, is the
winner of an international tender
realised by the Government of
the City of São Paulo won by the
Brazilian company Ótima S/A.
Diese Bushaltestelle, von der
bereits Tausende gebaut wurden,
ging als Sieger aus einem internationalen Wettbewerb hervor, den
die Regierung der Stadt São Paulo
durchgeführt und den das brasi-
lianische Unternehmen Ótima S/A
gewonnen hat.

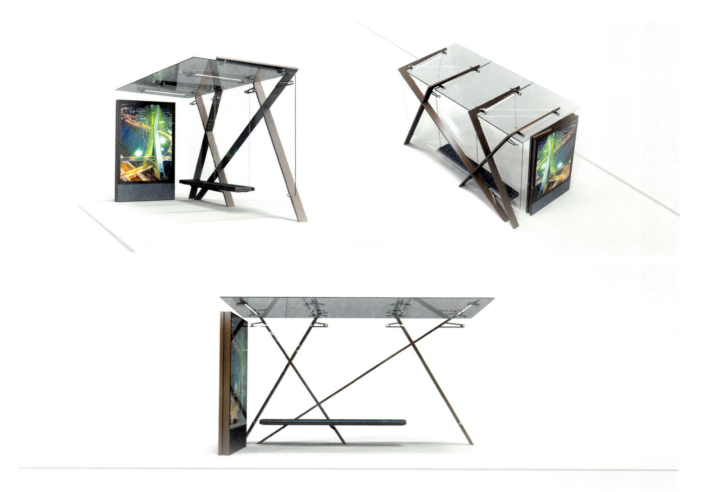

02

"My work philosophy focuses
on how to please and cleverly
surprise the user not only
through aesthetics but also
through functional and technical
innovations."

„Meine Arbeitsphilosophie konzentriert sich darauf, wie man Benutzer
erfreuen und geschickt überraschen
kann, nicht nur durch Ästhetik,
sondern auch durch funktionelle
und technische Innovationen."

Which product design area will Brazil have most success with in the next ten years?
Considering the world's urgent need for a more eco-friendly industrial production and that Brazil's vast and diversified natural resources could lead to the research and development of eco-friendly materials, Brazilian design has the opportunity to play a leading role in this new eco-production revolution.

What are your sources of inspiration?
People. Observing the way people behave, the way people work, live or enjoy life.

The Red Dot Award has been uncovering the best designs for 60 years now. Which product innovations would you like to see in the next 60 years, and why?
I would love to see innovations that lead to zero-footprint production, where waste is easily transformed into resources. Designers can play a leading role in this transformation.

In welchen Produktbereichen wird Brasilien in den nächsten zehn Jahren den größten Erfolg haben?
In Anbetracht dessen, dass die Welt ein dringendes Bedürfnis nach einer umweltfreundlicheren industriellen Herstellung hat und dass Brasiliens riesige Vorkommen an verschiedenen Rohstoffen zur Erforschung und Entwicklung umweltfreundlicher Materialien dienen können, hat brasilianisches Design die Gelegenheit, eine führende Rolle in dieser Revolution hin zu einer neuen, umweltfreundlichen Produktion zu spielen.

Was sind Ihre Inspirationsquellen?
Menschen. Zu beobachten, wie sie sich verhalten, wie sie arbeiten, leben oder das Leben genießen.

Der Red Dot Award ermittelt seit 60 Jahren die besten Gestaltungen. Welche Produktinnovationen würden Sie sich für die nächsten 60 Jahre wünschen, und warum?
Innovationen hin zu einer umweltneutralen Produktion, in der Abfälle einfach wieder in Rohstoffe umgewandelt werden. Designer können bei dieser Umwandlung eine führende Rolle spielen.

01

Prof. Cheng-Neng Kuan
Taiwan
Taiwan

In 1980, Professor Cheng-Neng Kuan earned a Master's degree in Industrial Design (MID) from the Pratt Institute in New York. He is currently a full professor and the vice president of Shih-Chien University, Taipei, Taiwan. With the aim of developing a more advanced design curriculum in Taiwan, he founded the Department of Industrial Design, in 1992. He served as department chair until 1999. Moreover, Professor Kuan founded the School of Design in 1997 and had served as the dean from 1997 to 2004 and as the founding director of the Graduate Institute of Industrial Design from 1998 to 2007. He had also held the position of the 16th chairman of the Board of China Industrial Designers Association (CIDA), Taiwan. His fields of expertise include design strategy and management as well as design theory and creation. Having published various books on design and over 180 research papers and articles, he is an active member of design juries in his home country and internationally. He is a consultant to major enterprises on product development and design strategy.

1980 erwarb Professor Cheng-Neng Kuan einen Master-Abschluss in Industriedesign (MID) am Pratt Institute in New York. Derzeit ist er ordentlicher Professor und Vizepräsident der Shih-Chien University in Taipeh, Taiwan. 1992 gründete er mit dem Ziel, einen erweiterten Designlehrplan zu entwickeln, das Department of Industrial Design in Taiwan. Bis 1999 war Professor Kuan Vorsitzender des Instituts. Darüber hinaus gründete er 1997 die School of Design, deren Dekan er von 1997 bis 2004 war. Von 1998 bis 2007 war er Gründungsdirektor des Graduate Institute of Industrial Design. Zudem war er der 16. Vorstandsvorsitzende der China Industrial Designers Association (CIDA) in Taiwan. Seine Fachgebiete umfassen Designstrategie, -management, -theorie und -kreation. Neben der Veröffentlichung verschiedener Bücher über Design und mehr als 180 Forschungsarbeiten und Artikel ist er aktives Mitglied von Designjurys in seiner Heimat sowie auf internationaler Ebene. Zudem ist er als Berater für Großunternehmen im Bereich Produktentwicklung und Designstrategie tätig.

02

"My work philosophy is to bridge the known and unknown, familiarity and strangeness and to make new and good things happen continuously."

„Meine Arbeitsphilosophie besteht darin, eine Brücke zwischen Bekanntem und Unbekanntem, Familiärem und Fremdem zu schlagen und ständig neue und gute Dinge zu verwirklichen."

What challenges will designers face in the future?
In response to the interaction of technology (Internet, IoT, Big Data, 3D printing), entrepreneurship (crowdfunding) and micro lifestyles (cross-culture, cross-age, cross-region), designers will not only have to discover new ways of thinking and working, but also face creative challenges from individuals without a design background.

What do you take special notice of when you are assessing a product as a jury member?
As a language, design has to integrate all criteria; however, what concerns me most in terms of the degree of integration is the expressive uniqueness and the exquisite qualities of a product.

What is the advantage of commissioning an external designer compared to having an in-house design team?
It can broaden the vision for the design with regards to creativity and through the differentiation of designs improves the discovery and interpretation of a brand's DNA from different perspectives.

Was sind Herausforderungen für Designer in der Zukunft?
Als Reaktion auf die Interaktion zwischen Technologie (Internet, Internet der Dinge, Big Data, 3D-Drucken), Unternehmergeist (Crowdfunding) und Mikro-Lebensstilen (kultur-, alters- und regionenübergreifend) werden Designer nicht nur neue Arten des Denkens und Arbeitens entdecken müssen, sondern sich auch kreativen Herausforderungen von Individuen ohne Gestaltungshintergrund stellen müssen.

Worauf achten Sie besonders, wenn Sie ein Produkt als Juror bewerten?
Als eine Sprache muss Design alle Kriterien integrieren. Was mir jedoch am wichtigsten bei dem Grad der Integration ist, sind die Einzigartigkeit des Ausdrucks und die hervorragenden Qualitäten eines Produktes.

Worin liegt der Vorteil, einen externen Designer zu beauftragen, im Vergleich zu einem Inhouse-Designteam?
Es kann die Vision des Designs kreativ erweitern und verbessert durch Differenzierung der Entwürfe die Entdeckung und Interpretation der Marken-DNA aus verschiedenen Perspektiven.

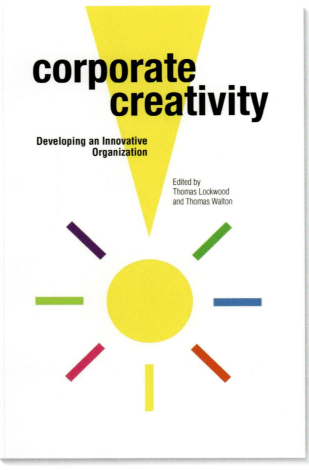

01

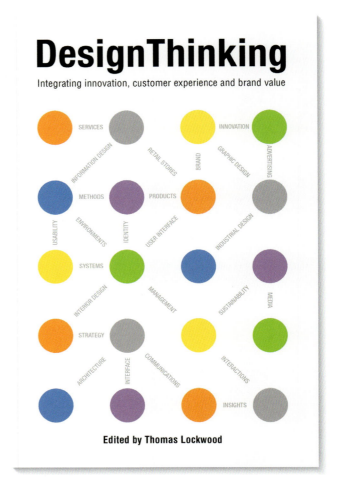

02

Dr Thomas Lockwood
USA
USA

Thomas Lockwood is the author of books on design thinking, design management, corporate creativity and design strategy. He has a PhD, works in design management and is regarded as a pioneer in the integration of design and innovation practices with business as well as for founding international design and user-experience organisations. He was guest professor at Pratt Institute in New York City, is a Fellow at the Royal Society of the Arts in London and a frequent design award judge. Lockwood is a founding partner of Lockwood Resource, an international recruiting firm specialising in design leadership recruiting. From 2005 to 2011, he was president of DMI, the Design Management Institute, a non-profit research association in Boston. From 1995 to 2005, he was design director at Sun Microsystems and StorageTek, and creative director at several design and advertising firms for a number of years. In addition, he manages a blog about design leadership at lockwoodresource.com.

Thomas Lockwood ist Autor von Büchern zu den Themen Design Thinking, Designmanagement, Corporate Creativity und Designstrategie und als Doktor der Philosophie im Designmanagement tätig. Er gilt als Vordenker für die Integration der Design- und Innovationspraxis ins Geschäftsleben sowie für die Gründung internationaler Design- und User-Experience-Organisationen. Er war Gastprofessor am Pratt Institute in New York City, ist Mitglied der Royal Society of the Arts in London und regelmäßig Juror bei Designwettbewerben. Lockwood ist Gründerpartner von Lockwood Resource, einer internationalen Rekrutierungsfirma, die sich auf die Anwerbung von Führungspersonal für den Designbereich spezialisiert hat. Von 2005 bis 2011 war er Präsident des DMI (Design Management Institute), einer gemeinnützigen Forschungsorganisation in Boston. Von 1995 bis 2005 war er Designdirektor bei Sun Microsystems und StorageTek und über mehrere Jahre hinweg Kreativdirektor in verschiedenen Design- und Werbefirmen. Darüber hinaus führt er den Blog lockwoodresource.com über Mitarbeiterführung im Designsektor.

01 Corporate Creativity (2008) explores
how to develop more creativity and
innovation power at the individual,
team and organisational levels.
Corporate Creativity (2008) untersucht,
wie sich Kreativität und Innovationskraft
auf individueller, Team- und Organisati-
onsebene steigern lassen.

02 Design Thinking (2010) presents best
practice cases of design thinking
methods applied to innovation, brand
building, service design and customer
experience.
Design Thinking (2010) präsentiert
Best-Practice-Fallstudien gestalterischer
Denkansätze für Innovation, Marken-
bildung, Service-Design und Kunden-
erfahrung.

"My work philosophy is to learn, grow, and change. When we learn, we grow. When we grow, we change. I choose to embrace this process."

„Meine Arbeitsphilosophie ist zu lernen, zu wachsen und zu verän- dern. Wenn wir lernen, wachsen wir. Wenn wir wachsen, verändern wir uns. Ich habe entschieden, mir diesen Prozess zu eigen zu machen."

In your opinion, which decade brought the greatest progress in the design sector?
I think the years from 2006 to 2011, the Great Recession, were fabulous for design. Because many CEOs realised that recovery would not just mean going back to business as usual. These CEOs gradually realised that design is the connection between their innovation ideas and new customers.

The Red Dot Award has been uncovering the best designs for 60 years now. Which product innovations would you like to see in the next 60 years?
I would love to see innovations in food distribution, in energy development, in health, and in equality of wealth.

What impressed you most during the Red Dot judging process?
I love the process: to review all the entries by myself first, and then to review them again with the team of judges for discussions, which are very insightful.

Welches Jahrzehnt hat Ihrer Meinung nach den größten Fortschritt im Designbereich gebracht?
Ich glaube, die Jahre von 2006 bis 2011, die große Rezession, waren fabelhaft für Design. Vielen CEOs wurde bewusst, dass ein Aufschwung nicht einfach bedeuten würde, dass alles wie gehabt weitergeht. Diese CEOs haben nach und nach gemerkt, dass Design die Verbindung zwischen ihren Innovationsideen und neuen Kunden darstellt.

Der Red Dot Award ermittelt seit 60 Jahren die besten Gestaltungen. Welche Produktinnovationen würden Sie sich für die nächsten 60 Jahre wünschen?
Ich würde gerne mehr Innovationen in der Nahrungs- mittelverteilung, der Energieentwicklung, im Gesund- heitsbereich und im Angleichen des Wohlstands sehen.

Was hat Sie bei der Red Dot-Jurierung am meisten beeindruckt?
Ich liebe den Prozess: alle Einreichungen zuerst allein zu begutachten, um sie dann erneut zusammen mit dem Jurorenteam anzuschauen und sehr erkenntnis- reiche Diskussionen zu führen.

01

Lam Leslie Lu
Hong Kong
Hongkong

Lam Leslie Lu received a Master of Architecture from Yale University in Connecticut, USA in 1977, and was the recipient of the Monbusho Scholarship of the Japanese Ministry of Culture in 1983, where he conducted research in design and urban theory in Tokyo. He is currently the principal of the Hong Kong Design Institute and academic director of the Hong Kong Institute of Vocational Education. Prior to this, he was head of the Department of Architecture at the University of Hong Kong. Lam Leslie Lu has worked with, among others, Cesar Pelli and Associates, Hardy Holzman Pfeiffer and Associates, Kohn Pedersen Fox and Associates and Shinohara Kazuo on the design of the Centennial Hall of the Institute for Technology Tokyo. Moreover, he was visiting professor at Yale University and the Delft University of Technology as well as assistant lecturer for the Eero Saarinen Chair at Yale University. He also lectured and served as design critic at major international universities such as Columbia, Cambridge, Delft, Princeton, Yale, Shenzhen, Tongji, Tsinghua and the Chinese University Hong Kong.

Lam Leslie Lu erwarb 1977 einen Master of Architecture an der Yale University in Connecticut, USA, und war 1983 Monbusho-Stipendiat des japanischen Kulturministeriums, an dem er die Forschung in Design und Stadttheorie in Tokio leitete. Derzeit ist er Direktor des Hong Kong Design Institutes und akademischer Direktor des Hong Kong Institute of Vocational Education. Zuvor war er Leiter des Architektur-Instituts an der Universität Hongkong. Lam Leslie Lu hat u. a. mit Cesar Pelli and Associates, Hardy Holzman Pfeiffer and Associates, Kohn Pedersen Fox and Associates und Shinohara Kazuo am Design der Centennial Hall des Instituts für Technologie Tokio zusammengearbeitet, war Gastprofessor an der Yale University und der Technischen Universität Delft sowie Assistenz-Dozent für den Eero-Saarinen-Lehrstuhl in Yale. Er hielt zudem Vorträge und war Designkritiker an großen internationalen Universitäten wie Columbia, Cambridge, Delft, Princeton, Yale, Shenzhen, Tongji, Tsinghua und der chinesischen Universität Hongkong.

01 Centennial Hall of
   the Tokyo Institute of
   Technology
   Die Centennial Hall
   des Instituts für
   Technologie Tokio

"My work philosophy is to advance knowledge and talent through calculated strategies and research while drawing in fresh thinking."

„Meine Arbeitsphilosophie besteht darin, Wissen und Talent durch wohl kalkulierte Strategien und Forschung voranzutreiben und gleichzeitig frische Denkansätze einzubeziehen."

Which of the projects in your lifetime are you particularly proud of?
I spent a period of time working with Shinohara Kazuo on the Centennial Hall building. The depth of thought and the level of experimentation in space, form and structure was the furthest I ever got in the design of a building. The interactive experience of designing as a duo set the standard for all my designs to follow.

What significance does winning an award in a design competition have for the designer?
The need to be recognised by one's peers is paramount and maybe more important than public acclaim. We need it not for our ego but for our soul.

What is the advantage of commissioning an external designer compared to having an in-house design team?
Change and new conceptions can only come about with the ebb and flow of talent, either through internal deployment or external injection.

Auf welches Ihrer bisherigen Projekte sind Sie besonders stolz?
Ich habe eine Zeit lang mit Shinohara Kazuo am Centennial-Hall-Gebäude zusammengearbeitet. Die Tiefe der Gedanken und das Niveau der Experimente in Bezug auf Raum, Form und Struktur waren die extremste Arbeit, die ich je bei der Gestaltung eines Gebäudes geleistet habe. Die interaktive Erfahrung des Gestaltens zu zweit hat Maßstäbe für alle meine nachfolgenden Entwürfe gesetzt.

Welche Bedeutung hat die Auszeichnung in einem Designwettbewerb für den Designer?
Die Notwendigkeit, von seinen Partnern anerkannt zu werden, ist entscheidend und womöglich wichtiger als öffentliche Anerkennung. Wir brauchen sie nicht für unser Ego, sondern für unsere Seele.

Worin liegt der Vorteil, einen externen Designer zu beauftragen, im Vergleich zu einem Inhouse-Designteam?
Veränderung und neue Ideen können nur mit der Ebbe und Flut an Talenten entstehen, entweder durch internen Personaleinsatz oder durch Zuführung von außen.

01

Wolfgang K. Meyer–Hayoz
Switzerland
Schweiz

Wolfgang K. Meyer–Hayoz studied mechanical engineering, visual communication and industrial design and graduated from the Stuttgart State Academy of Art and Design. The professors Klaus Lehmann, Kurt Weidemann and Max Bense have had formative influence on his design philosophy. In 1985, he founded the Meyer-Hayoz Design Engineering Group with offices in Winterthur/Switzerland and Constance/Germany. The design studio offers consultancy services for national as well as international companies in the five areas of design competence: design strategy, industrial design, user interface design, temporary architecture and communication design, and has received numerous international awards. From 1987 to 1993, Wolfgang K. Meyer-Hayoz was president of the Swiss Design Association (SDA). He is a member of the Association of German Industrial Designers (VDID), Swiss Marketing and the Swiss Management Society (SMG). Wolfgang K. Meyer-Hayoz also serves as juror on international design panels and supervises Change Management and Turnaround projects in the field of design strategy.

Wolfgang K. Meyer-Hayoz absolvierte Studien in Maschinenbau, Visueller Kommunikation sowie Industrial Design mit Abschluss an der Staatlichen Akademie der Bildenden Künste in Stuttgart. Seine Gestaltungsphilosophie prägten die Professoren Klaus Lehmann, Kurt Weidemann und Max Bense. 1985 gründete er die Meyer-Hayoz Design Engineering Group mit Büros in Winterthur/Schweiz und Konstanz/Deutschland. Das Designstudio bietet Beratungsdienste für nationale wie internationale Unternehmen in den fünf Designkompetenzen Designstrategie, Industrial Design, User Interface Design, temporäre Architektur und Kommunikationsdesign und wurde bereits vielfach ausgezeichnet. Von 1987 bis 1993 war Wolfgang K. Meyer-Hayoz Präsident der Swiss Design Association (SDA); er ist Mitglied im Verband Deutscher Industrie Designer (VDID), von Swiss Marketing und der Schweizerischen Management Gesellschaft (SMG). Wolfgang K. Meyer-Hayoz engagiert sich auch als Juror internationaler Designgremien und moderiert Change-Management- und Turnaround-Projekte im designstrategischen Bereich.

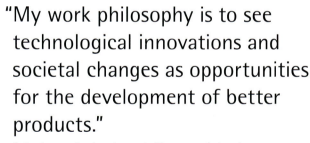

01 Bike Loft
Automated bicycle parking system with a closed parking box for each bicycle, with parking reservation through GPS and payment via smart-phone. For Bike Loft GmbH, Switzerland.
Automatisiertes Bikeparking-System mit einer geschlos-senen Parking-Box pro Rad, mit Parkreservierung per GPS und der Zahlung mittels Smartphone. Für die Bike Loft GmbH, Schweiz.

02 Steam to Power–STP
Energy-efficiency plant for Talbot New Energy AG
Energieeffizienzanlage für die Talbot New Energy AG

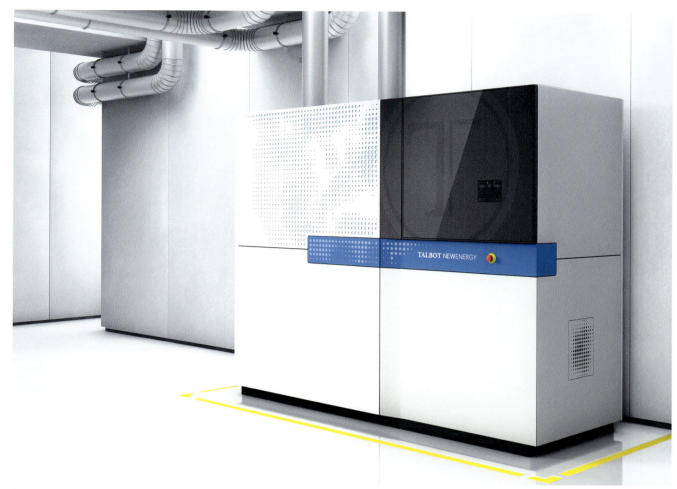

02

"My work philosophy is to see technological innovations and societal changes as opportunities for the development of better products."

„Meine Arbeitsphilosophie lautet, die technologischen Innovationen und gesellschaftlichen Veränderun-gen als Chancen für die Entwicklung besserer Produkte zu begreifen."

You have been a Red Dot juror for many years. What significance does being appointed to the jury have for you?
I see it is an expression of the appreciation for my work as a designer, and I appreciate the exchange with international colleagues from all areas of design.

What advice has been helpful in your youth or at the beginning of your professional career?
The advice to always be open to new developments, be they technical or societal, and to fathom what they mean for new design approaches.

What do you take special notice of when you are assessing a product as a juror?
In the field of medical technology products in particular, the aspects of safety, clarity of function, manufacturing processes, hygienic properties and stability of value are important criteria.

What significance does winning an award in a design competition have for the designer?
A competition gives designers the opportunity to compete internationally and thus get to know their skills better.

Sie sind langjähriger Red Dot–Juror. Welchen Wert hat eine Berufung in diese Jury für Sie?
Sie ist für mich Ausdruck der Wertschätzung meiner gestalterischen Arbeit, und ich selbst schätze den Austausch mit den internationalen Kollegen aus allen Sparten des Designs.

Welche Ratschläge haben Sie in Ihrer Jugend oder frühen Berufskarriere weitergebracht?
Immer offen zu sein für neue Entwicklungen, seien es technische oder gesellschaftliche, und auszuloten, was dies für neue Designansätze bedeutet.

Worauf achten Sie besonders, wenn Sie ein Produkt als Juror bewerten?
Speziell in den Produktsortimenten der Medizintechnik sind für mich die Aspekte Sicherheit, Sinnfälligkeit, Herstellungsprozesse, Hygieneeigenschaften und Wertbeständigkeit wichtige Kriterien.

Welche Bedeutung hat die Auszeichnung in einem Designwettbewerb für den Designer?
Ein Wettbewerb gibt Designern die Gelegenheit, sich international zu messen und so ihre Qualifikation besser kennenzulernen.

01

Prof. Jure Miklavc
Slovenia
Slowenien

Professor Jure Miklavc graduated in industrial design from the Academy of Fine Arts in Ljubljana, Slovenia, and has nearly 20 years of experience in the field of design. Miklavc started his career working as a freelance designer, before founding his own design consultancy, Studio Miklavc. Studio Miklavc works in the fields of product design, visual communications and brand development and is a consultancy for a variety of clients from the industries of light design, electronic goods, user interfaces, transport design and medical equipment. Sports equipment designed by the studio has gained worldwide recognition. From 2013 onwards, the team has been working for the prestigious Italian motorbike manufacturer Bimota. Designs by Studio Miklavc have received many international awards and have been displayed in numerous exhibitions. Jure Miklavc has been involved in design education since 2005 and is currently a lecturer and head of industrial design at the Academy of Fine Arts and Design in Ljubljana.

Professor Jure Miklavc machte seinen Abschluss in Industrial Design an der Academy of Fine Arts and Design in Ljubljana, Slowenien, und verfügt über nahezu 20 Jahre Erfahrung im Designbereich. Er arbeitete zunächst als freiberuflicher Designer, bevor er sein eigenes Design-Beratungsunternehmen „Studio Miklavc" gründete. Studio Miklavc ist in den Bereichen Produktdesign, visuelle Kommunikation und Markenentwicklung sowie in der Beratung zahlreicher Kunden der Branchen Lichtdesign, elektronische Güter, Benutzeroberflächen, Transport-Design und medizinisches Equipment tätig. Die von dem Studio gestalteten Sportausrüstungen erfahren weltweit Anerkennung. Seit 2013 arbeitet das Team für den angesehenen italienischen Motorradhersteller Bimota. Studio Miklavc erhielt bereits zahlreiche Auszeichnungen sowie Präsentationen in Ausstellungen. Seit 2005 ist Jure Miklavc in der Designlehre tätig und aktuell Dozent und Head of Industrial Design an der Academy of Fine Arts and Design in Ljubljana.

02

"My work philosophy is: quality over quantity."

„Meine Arbeitsphilosophie lautet: Qualität über Quantität."

You are involved in design education. What fascinates you about this job?
The exchange between the practical knowledge that I gain in my studio and the theoretical knowledge that is based on the Academy. Mixing those influences enriches both parts of my professional life.

When did you first think consciously about design?
When I was in primary school, I spent all the time enhancing, modifying and redesigning my toys. This sounds like just playing, but I added electric motors, new metal parts or just enhanced their finish.

What impressed you most during the Red Dot judging process?
All jury members took the task very seriously and displayed a high level of professionalism. It is a privilege and remarkably easy to work with so many renowned experts. The work is quite intense and focused, but enjoyable.

Sie sind in der Designlehre tätig. Was fasziniert Sie an dieser Aufgabe?
Der Austausch zwischen praktischem Wissen, das ich in meinem Designstudio erwerbe, und dem theoretischen Wissen, das auf der Akademie vermittelt wird. Diese Einflüsse zu vermischen, bereichert beide Bereiche meines beruflichen Lebens.

Wann haben Sie das erste Mal bewusst über Design nachgedacht?
Als ich in der Grundschule war, verbrachte ich meine gesamte Zeit damit, mein Spielzeug zu verbessern, zu verändern und umzugestalten. Das klingt nach bloßem Spielen, aber ich baute auch elektrische Motoren ein, neue Teile aus Metall oder veränderte einfach die Lackierung.

Was hat Sie bei der Red Dot-Jurierung am meisten beeindruckt?
Alle Mitglieder der Jury nahmen die Aufgabe sehr ernst und zeigten sich sehr professionell. Es ist ein Privileg und bemerkenswert einfach, mit so vielen berühmten Experten zusammenzuarbeiten. Die Arbeit ist ziemlich intensiv und konzentriert, aber unterhaltsam.

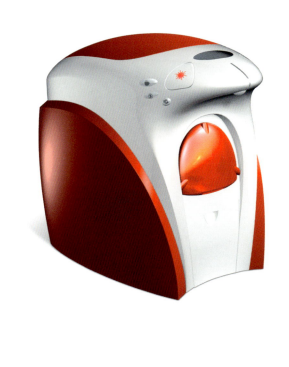

01

Prof. Ron A. Nabarro
Israel

Israel

Professor Ron A. Nabarro is an industrial designer, strategist, entrepreneur, researcher and educator. He has been a professional designer since 1970 and has designed more than 750 products to date in a wide range of industries. He has played a leading role in the emergence of age-friendly design and age-friendly design education. From 1992 to 2009, he was a professor of industrial design at the Technion Israel Institute of Technology, where he founded and was the head of the graduate programme in advanced design studies and design management. Currently, Nabarro teaches design management and design thinking at DeTao Masters Academy in Shanghai, China. From 1999 to 2003, he was an executive board member of Icsid and now acts as a regional advisor. He is a frequent keynote speaker at conferences, has presented TEDx events, has lectured and led design workshops in over 20 countries and consulted to a wide variety of organisations. Furthermore, he is co-founder and CEO of Senior-touch Ltd. and design4all. The principle areas of his research and interest are design thinking, age-friendly design and design management.

Professor Ron A. Nabarro ist Industriedesigner, Stratege, Unternehmer, Forscher und Lehrender. Seit 1970 ist er praktizierender Designer, gestaltete bisher mehr als 750 Produkte für ein breites Branchenspektrum und spielt eine führende Rolle im Bereich des altersfreundlichen Designs und dessen Lehre. Von 1992 bis 2009 war er Professor für Industriedesign am Technologie-Institut Technion Israel, an dem er das Graduierten-programm für fortgeschrittene Designstudien und Designmanagement einführte und leitete. Aktuell unterrichtet Nabarro Designmanagement und Design Thinking an der DeTao Masters Academy in Shanghai, China. Von 1999 bis 2003 war er Vorstandsmitglied des Icsid, für den er aktuell als regionaler Berater tätig ist. Er ist ein gefragter Redner auf Konferenzen, hat bei TEDx-Veranstaltungen präsentiert, hielt Vorträge und Workshops in mehr als 20 Ländern und beriet eine Vielzahl von Organisationen. Zudem ist er Mitbegründer und Geschäftsführer von Senior-touch Ltd. und design4all. Die Hauptbereiche seiner Forschung und seines Interesses sind Design Thinking, altersfreundliches Design und Designmanagement.

02

"My work philosophy is to improve the lives of ageing people by gathering leading companies, entrepreneurs, designers and technologists in order to educate and innovate together."

„Meine Arbeitsphilosophie ist die Verbesserung des Lebens älterer Menschen durch Innovation und Erziehung in Zusammenarbeit mit führenden Firmen, Unternehmern, Designern und Technologen."

What impressed you most during the Red Dot judging process?
The most important aspect of the jury process was the way jury members treat the process, in particular the respect for each work presented and the commitments to the high standards of adjudication.

Do you see a correlation between the design quality of a company's products and the economic success of this company?
In most cases this "formula" could work, still it is important to acknowledge the importance of marketing and not put everything on the designers' shoulders. We also can see designs that have been ahead of their time and totally failed, although the design was brilliant.

Which project would you like to realise one day?
At this stage of my life I find more interest in sharing my professional experience and life experience with young designers and design students.

Was hat Sie bei der Red Dot-Jurierung am meisten beeindruckt?
Der wichtigste Aspekt während der Jurierung war die Art, wie die Jurymitglieder dem Prozess begegnet sind, insbesondere der Respekt für jede einzelne Arbeit und die Hingabe an die hohen Jurierungsstandards.

Sehen Sie einen Zusammenhang zwischen der Designqualität, die sich in den Produkten eines Unternehmens äußert, und dem wirtschaftlichen Erfolg dieses Unternehmens?
In den meisten Fällen funktioniert diese „Formel", dennoch ist es wichtig, auch die Rolle des Marketings zu erkennen und nicht alles auf die Schultern der Gestalter zu laden. Wir kennen auch Gestaltungen, die ihrer Zeit voraus waren, aber komplett gescheitert sind, obgleich die Gestaltung brillant war.

Welches Projekt würden Sie gerne einmal realisieren?
In dieser Phase meines Lebens interessiert es mich mehr, meine professionelle Erfahrung und Lebenserfahrung mit jungen Designern und Designstudenten zu teilen.

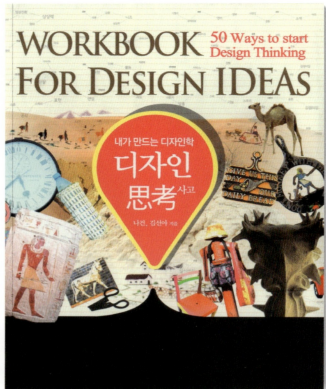

01

Prof. Dr. Ken Nah
Korea

Korea

Professor Dr Ken Nah graduated with a Bachelor of Arts in industrial engineering from Hanyang University, South Korea, in 1983. He deepened his interest in human factors/ergonomics by earning a master's degree from Korea Advanced Institute for Science and Technology (KAIST) in 1985 and he gained a PhD from Tufts University in 1996. In addition, Ken Nah is also a USA Certified Professional Ergonomist (CPE). He was the dean of the International Design School for Advanced Studies (IDAS) and is currently professor of design management as well as director of the Human Experience and Emotion Research (HE.ER) Lab at IDAS, Hongik University, Seoul. From 2002 he was the director of the International Design Trend Center (IDTC). Ken Nah was the director general of "World Design Capital Seoul 2010". Alongside his work as a lecturer he is also the vice president of the Korea Association of Industrial Designers (KAID), the Ergonomics Society of Korea (ESK) and the Korea Institute of Design Management (MIDM), as well as the chairman of the Design and Brand Committee of the Korea Consulting Association (KCA).

Professor Dr. Ken Nah graduierte 1983 an der Hanyang University in Südkorea als Bachelor of Arts in Industrial Engineering. Sein Interesse an Human Factors/Ergonomie vertiefte er 1985 mit einem Master-Abschluss am Korea Advanced Institute for Science and Technology (KAIST) und promovierte 1996 an der Tufts University. Darüber hinaus ist Ken Nah ein in den USA zertifizierter Ergonom (CPE). Er war Dekan der International Design School for Advanced Studies (IDAS) und ist aktuell Professor für Design Management sowie Direktor des „Human Experience and Emotion Research (HE.ER)"-Labors an der IDAS, Hongik University, Seoul. Von 2002 an war er Leiter des International Design Trend Center (IDTC). Ken Nah war Generaldirektor der „World Design Capital Seoul 2010". Neben seiner Lehrtätigkeit ist er Vizepräsident der Korea Association of Industrial Designers (KAID), der Ergonomics Society of Korea (ESK) und des Korea Institute of Design Management (MIDM) sowie Vorsitzender des „Design and Brand"-Komitees der Korea Consulting Association (KCA).

01 **Workbook for Design Ideas**
The Korean and Chinese
version on 50 ways to start
design thinking
Die koreanische und chinesi-
sche Version über 50 Wege, mit
Design Thinking zu beginnen

"My work philosophy is to do my best in all areas and every moment of every day, since time never stops and opportunity never waits."

„Meine Arbeitsphilosophie besteht darin, in allen Bereichen und in jedem Moment mein Bestes zu geben, da die Zeit ständig voran-schreitet und sich Gelegenheiten nicht zweimal bieten."

What motivates you to get up in the morning?
My question every morning before getting up is: "What if today were my last day?" This question motivates me to get back to work and use my time and energy in the best possible way.

When did you first think consciously about design?
It was in the winter of 1987, when I had to decide what majors to choose at university. Reading books and articles, I instantly fell in love with the words "Human Factors", defined as designing for people and optimising living and working conditions. Not only physical and physiological characteristics, but also psychological ones are important in "design". Since then, design has been everything to me!

What do you take special notice of when you are assessing a product as a jury member?
I pay attention to the "balance" between form and function. The product should also be well balanced between logic and emotion.

Was motiviert Sie, morgens aufzustehen?
Die Frage, die ich mir jeden Morgen vor dem Aufstehen stelle, ist: „Was wäre, wenn heute mein letzter Tag wäre?" Sie motiviert mich, wieder an die Arbeit zu gehen und meine Zeit und Energie optimal zu nutzen.

Wann haben Sie das erste Mal bewusst über Design nachgedacht?
Es war im Winter 1987, als ich an der Universität meine Hauptfächer wählen musste. Beim Lesen vieler Bücher und Artikel hatte ich mich sofort in die Worte „Human Factors" verliebt. Human Factors bedeutet, etwas für Menschen zu gestalten und die Lebens- und Arbeitsverhältnisse zu optimieren. Nicht nur physische und physiologische, sondern auch psychologische Eigenschaften sind wichtig in der Gestaltung. Seither ist Design mein Ein und Alles!

Worauf achten Sie besonders, wenn Sie ein Produkt als Juror bewerten?
Ich achte auf das Gleichgewicht zwischen Form und Funktion. Zudem sollte das Produkt auch zwischen Logik und Emotion gut ausgewogen sein.

01

Prof. Dr. Yuri Nazarov
Russia
Russland

Professor Dr Yuri Nazarov, born in 1948 in Moscow, teaches at the National Design Institute in Moscow where he is also provost. As an actively involved design expert, he serves on numerous boards, for example as president of the Association of Designers of Russia, as a corresponding member of Russian Academy of Arts, and as a member of the Russian Design Academy. Yuri Nazarov has received a wide range of accolades for his achievements: he is a laureate of the State Award of the Russian Federation in Literature and Art as well as of the Moscow Administration' Award, and he also has received a badge of honour for "Merits in Development of Design".

Professor Dr. Yuri Nazarov, 1948 in Moskau geboren, lehrt am National Design Institute in Moskau, dessen Rektor er auch ist. Als engagierter Designexperte ist er in zahlreichen Gremien des Landes tätig, zum Beispiel als Präsident der Russischen Designervereinigung, als korrespondierendes Mitglied der Russischen Kunstakademie sowie als Mitglied der Russischen Designakademie. Für seine Verdienste wurde Yuri Nazarov mit einer Vielzahl an Auszeichnungen geehrt. So ist er Preisträger des Staatspreises der Russischen Föderation in Literatur und Kunst sowie des Moskauer Regierungspreises und besitzt zudem das Ehrenabzeichen für „Verdienste in der Designentwicklung".

01 **Roly-Poly**
Mascot for the Russian
Federation's pavilion of the
World Expo Milano 2015 real-
ised with Vitaly Stavitcky.
Maskottchen für den rus-
sischen Pavillon auf der
Weltausstellung Expo Milano
2015, umgesetzt mit Vitaly
Stavitcky.

"My work philosophy ranges from functional designs for social regional projects and low-income individuals to engage in joint work with young designers and to integrate Russian design on an international scale."

„Meine Arbeitsphilosophie reicht von betont funktionalen Gestaltungen für soziale regionale Projekte wie für Geringverdienende bis zur Anbindung junger Designer sowie des russischen Designs an internationales Niveau."

What is, in your opinion, the significance of design quality in the product categories you evaluated?
The significance of design quality lies in confirming the usability and safety of the products.

What are the important criteria for you as a juror in the assessment of a product?
Creative ideas and the quality of their realisation.

What impressed you most during the Red Dot judging process?
The most outstanding for me was the mutual understanding of and similarities between our viewpoints.

Do you see a correlation between the design quality of a company's products and the economic success of this company?
It depends on how we interpret economic success. If we talk about profit it may simply be due to a hot commodity. But real design quality always implies taking into account consumer preferences.

Wie schätzen Sie den Stellenwert der Designqualität in den von Ihnen beurteilten Produktkategorien ein?
Der Stellenwert der Designqualität liegt darin, die Bedienbarkeit und Sicherheit der Produkte zu bekräftigen.

Worauf achten Sie als Juror, wenn Sie ein Produkt bewerten?
Auf kreative Ideen und die Qualität ihrer Umsetzung.

Was hat Sie bei der Red Dot-Jurierung am meisten beeindruckt?
Das Hervorstechendste für mich waren das gegenseitige Verständnis und die Ähnlichkeiten unserer Standpunkte.

Sehen Sie einen Zusammenhang zwischen der Designqualität, die sich in den Produkten eines Unternehmens äußert, und dem wirtschaftlichen Erfolg dieses Unternehmens?
Das hängt davon ab, wie man wirtschaftlichen Erfolg interpretiert. Sprechen wir von Profit, mag das schlicht an einem „heißen" Produkt liegen. Aber echte Designqualität impliziert immer die Berücksichtigung der Vorlieben der Verbraucher.

01

Ken Okuyama
Japan
Japan

Ken Kiyoyuki Okuyama, industrial designer and CEO of KEN OKUYAMA DESIGN, was born 1959 in Yamagata, Japan, and studied automobile design at the Art Center College of Design in Pasadena, California. He has worked as a chief designer for General Motors, as a senior designer for Porsche AG, and as design director for Pininfarina S.p.A., being responsible for the design of Ferrari Enzo, Maserati Quattroporte and many other automobiles. He is also known for many different product designs such as motorcycles, furniture, robots and architecture. KEN OKUYAMA DESIGN was founded in 2007 and provides business consultancy services to numerous corporations. Ken Okuyama also produces cars, eyewear and interior products under his original brand. He is currently a visiting professor at several universities and also frequently publishes books.

Ken Kiyoyuki Okuyama, Industriedesigner und CEO von KEN OKUYAMA DESIGN, wurde 1959 in Yamagata, Japan, geboren und studierte Automobildesign am Art Center College of Design in Pasadena, Kalifornien. Er war als Chief Designer bei General Motors, als Senior Designer bei der Porsche AG und als Design Director bei Pininfarina S.p.A. tätig und zeichnete verantwortlich für den Ferrari Enzo, den Maserati Quattroporte und viele weitere Automobile. Zudem ist er für viele unterschiedliche Produktgestaltungen wie Motorräder, Möbel, Roboter und Architektur bekannt. KEN OKUYAMA DESIGN wurde 2007 als Beratungsunternehmen gegründet und arbeitet für zahlreiche Unternehmen. Ken Okuyama produziert unter seiner originären Marke auch Autos, Brillen und Inneneinrichtungsgegenstände. Derzeit lehrt er als Gastprofessor an verschiedenen Universitäten und publiziert zudem Bücher.

02

"My design philosophy reads:
modern, simple, timeless."

„Meine Designphilosophie lautet:
Modern, schlicht, zeitlos."

**What trends have you noticed in the field of
"Vehicles" in recent years?**
Driving performance is no longer a sales point.
Transport design should propose not only mobility but
also a new lifestyle. A car's design reflects its owner's
character and lifestyle more than ever.

**Do you see a correlation between the design
quality of a company's products and the economic
success of this company?**
The correlation is the result of a clear vision and the
teamwork that made it happen, plus the personalities
of individual team members.

**What are the important criteria for you as a juror
in the assessment of a product?**
A juror has to determine a product's value to society
and the market. Therefore, an objective view and wide
ranging knowledge of technology, materials, manufacturing, etc. are necessary.

**Welche Trends konnten Sie im Bereich „Fahrzeuge"
in den letzten Jahren ausmachen?**
Fahr-Performance ist kein Verkaufsargument mehr. Im
Segment „Transport" sollte Gestaltung nicht nur auf
Mobilität abzielen, sondern auch auf einen neuen
Lebensstil. Das Design eines Autos spiegelt mehr denn
je den Charakter und Lebensstil seines Besitzers wider.

**Sehen Sie einen Zusammenhang zwischen der
Designqualität, die sich in den Produkten eines
Unternehmens äußert, und dem wirtschaftlichen
Erfolg dieses Unternehmens?**
Diese Wechselwirkung ist das Ergebnis einer klaren
Vision und der ihr zugrunde liegenden Teamarbeit –
plus der Persönlichkeiten der einzelnen Teammitglieder.

**Worauf achten Sie als Juror, wenn Sie ein Produkt
bewerten?**
Ein Juror muss den Wert bestimmen, den ein Produkt
für die Gesellschaft und den Markt hat. Daher sind
eine objektive Sichtweise und eine große Bandbreite
an Wissen über Technik, Werkstoffe, Herstellung etc.
notwendig.

01

Simon Ong
Singapore
Singapur

Simon Ong, born in Singapore in 1953, graduated with a master's degree in design from the University of New South Wales and an MBA from the University of South Australia. He is the group managing director and co-founder of Kingsmen Creatives Ltd., a leading communication design and production group with 18 offices across the Asia Pacific region and the Middle East. Kingsmen has won several awards, such as the President's Design Award, SRA Best Retail Concept Award, SFIA Hall of Fame, Promising Brand Award, A.R.E. Retail Design Award and RDI International Store Design Award USA. Simon Ong is actively involved in the creative industry as chairman of the design group of Manpower, the Skills & Training Council of Singapore Workforce Development Agency. Moreover, he is a member of the advisory board of the Singapore Furniture Industries Council and School of Design & Environment at the National University of Singapore, Design Business Chamber Singapore and Interior Design Confederation of Singapore.

Simon Ong, geboren 1953 in Singapur, erhielt einen Master in Design der University of New South Wales und einen Master of Business Administration der University of South Australia. Er ist Vorstandsvorsitzender und Mitbegründer von Kingsmen Creatives Ltd., eines führenden Unternehmens für Kommunikationsdesign und Produktion mit 18 Geschäftsstellen im asiatisch-pazifischen Raum sowie im Mittleren Osten. Kingsmen wurde vielfach ausgezeichnet, u. a. mit dem President's Design Award, SRA Best Retail Concept Award, SFIA Hall of Fame, Promising Brand Award, A.R.E. Retail Design Award und RDI International Store Design Award USA. Simon Ong ist als Vorsitzender der Designgruppe von Manpower, der „Skills & Training Council of Singapore Workforce Development Agency", aktiv in die Kreativindustrie involviert, ist unter anderem Mitglied des Beirats des Singapore Furniture Industries Council, der School of Design & Environment an der National University of Singapore, des Design Business Chamber Singapore und der Interior Design Confederation of Singapore.

02

"My work philosophy is 'less is more' in everything we do. Having less 'quantity' or details enables us to have more time to focus on quality and what matters."

„Meine Arbeitsphilosophie ist ‚Weniger ist mehr', in allem, was wir tun. Die Reduktion von Quantität oder Details gibt uns mehr Zeit, uns auf die Qualität und das Wesentliche zu konzentrieren."

What motivates you to get up in the morning?
Humour or having a good laugh. In our rapidly developing world, we are often so caught up in our work that we tend to forget the finest thing in life – that is to laugh.

In your opinion, which decade brought the greatest progress in the design sector?
Affordable computer software in the 1990s opened up vast opportunities to go beyond traditional processes in design thinking. Designs could be "tested" through walk-through imaging – saving time and resources for prototyping.

What challenges will designers face in the future?
Designers will have to think ahead and look at what lies beyond sustainable design.

Was motiviert Sie, morgens aufzustehen?
Etwas zum Lachen zu haben. In unserer schnelllebigen Welt sind wir oft so stark in unsere Arbeit verstrickt, dass wir dazu neigen, das Beste im Leben zu vergessen – und zwar zu lachen.

Welches Jahrzehnt hat Ihrer Meinung nach den größten Fortschritt im Designbereich gebracht?
Erschwingliche Computer-Software in den 1990er Jahren eröffnete riesige Möglichkeiten, über die Grenzen traditioneller Prozesse im Design Thinking hinauszugehen. Entwürfe konnten durch Computer-simulationen „getestet" werden, um so Zeit und Ressourcen bei der Herstellung von Prototypen zu sparen.

Vor welchen Herausforderungen werden Designer in der Zukunft stehen?
Designer werden vorausdenken und ausmachen müssen, was nach nachhaltiger Gestaltung kommt.

01

Prof. Martin Pärn
Estonia
Estland

Professor Martin Pärn, born in Tallinn in 1971, studied industrial design at the University of Industrial Arts Helsinki (UIAH). After working in the Finnish furniture industry he moved back to Estonia and undertook the role of the ambassadorial leader of design promotion and development in his native country. He was actively involved in the establishment of the Estonian Design Centre and continues directing the organisation as chair of the board. Martin Pärn founded the multidisciplinary design office "iseasi", which creates designs ranging from office furniture to larger instruments and from small architecture to interior designs for the public sector. Having received many awards, Pärn begun in 1995 with the establishment and development of design training in Estonia and is currently head of the Design and Engineering's master's programme, a joint initiative of the Tallinn University of Technology and the Estonian Academy of Arts, which aims, among other things, to create synergies between engineers and designers.

Professor Martin Pärn, geboren 1971 in Tallinn, studierte Industriedesign an der University of Industrial Arts Helsinki (UIAH). Nachdem er in der finnischen Möbelindustrie gearbeitet hatte, ging er zurück nach Estland und übernahm die Funktion des leitenden Botschafters für die Designförderung und -entwicklung des Landes. Er war aktiv am Aufbau des Estonian Design Centres beteiligt und leitet seither die Organisation als Vorstandsvorsitzender. Martin Pärn gründete das multidisziplinäre Designbüro „iseasi", das ebenso Büromöblierung wie größere Instrumente, „kleine Architektur" oder Interior Designs im öffentlichen Sektor gestaltet. Vielfach ausgezeichnet, startete Pärn 1995 mit der Entwicklung und dem Ausbau der Designlehre in Estland und ist heute Leiter des Masterprogramms Design und Engineering, einer gemeinsamen Initiative der Tallinn University of Technology und der Estonian Academy of Arts, u. a. mit dem Ziel, durch den Zusammenschluss Synergien von Ingenieuren und Designern zu erreichen.

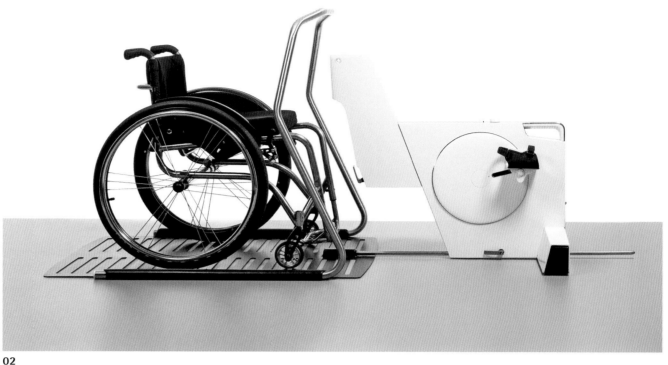

02

"My work philosophy is to search
for something that is obvious,
but yet unnoticed."
„Meine Arbeitsphilosophie besteht
darin, nach etwas zu suchen,
das offensichtlich, aber bisher
unbemerkt geblieben ist."

What motivates you to get up in the morning?
Sun or, in short, life itself. It is full of new and unseen
miracles I do not want to miss.

What challenges will designers face in the future?
The challenge of shifting the focus from fast consumer success towards long-lasting effectiveness and
sustainability.

**What do you take special notice of when you are
assessing a product as a jury member?**
I am looking more for the new "Why?" than the new
"How?".

**What significance does winning an award in
a design competition have for the designer?**
It means they have gained the respect of their
colleagues, and thus it is a matter of honour.
The real credits have to be earned in the field.

Was motiviert Sie, morgens aufzustehen?
Die Sonne oder, kurz gesagt, das Leben an sich. Es ist
voller neuer und unbemerkter Wunder, die ich nicht
verpassen möchte.

**Vor welchen Herausforderungen werden Designer
in der Zukunft stehen?**
Vor der Herausforderung, das Augenmerk vom
schnellen Markterfolg auf langlebige Leistungsfähigkeit und Nachhaltigkeit zu verschieben.

**Worauf achten Sie besonders, wenn Sie ein
Produkt als Juror bewerten?**
Ich suche mehr nach einem neuen „Warum?" als nach
einem neuen „Wie?".

**Welche Bedeutung hat die Auszeichnung in einem
Designwettbewerb für den Designer?**
Es bedeutet, dass der Designer den Respekt seiner
Kollegen gewonnen hat, was daher eine Frage der
Ehre ist. Die echten Auszeichnungen müssen in der
Praxis verdient werden.

01

02/03

Dr Sascha Peters
Germany
Deutschland

Dr Sascha Peters is founder and owner of the agency for material and technology HAUTE INNOVATION in Berlin. He studied mechanical engineering at the RWTH Aachen, Germany, and product design at the ABK Maastricht, Netherlands. He wrote his doctoral thesis at the University of Duisburg-Essen, Germany, on the complex of problems in communication between engineering and design. From 1997 to 2003, he led research projects and product developments at the Fraunhofer Institute for Production Technology IPT in Aachen and subsequently became head of the Design Zentrum Bremen. Sascha Peters is author of various specialised books on sustainable raw materials, smart materials, innovative production techniques and energetic technologies. He is a leading material expert and trend scout for new technologies. Since 2014, he has been an advisory board member of the funding initiative "Zwanzig20 – Partnerschaft für Innovation" (2020 – Partnership for innovation) by order of the German Federal Ministry of Education and Research.

Dr. Sascha Peters ist Gründer und Inhaber der Material- und Technologieagentur HAUTE INNOVATION in Berlin. Er studierte Maschinenbau an der RWTH Aachen und Produktdesign an der ABK Maastricht. Seine Doktorarbeit schrieb er an der Universität Duisburg-Essen über die Kommunikationsproblematik zwischen Engineering und Design. Von 1997 bis 2003 leitete er Forschungsprojekte und Produktentwicklungen am Fraunhofer-Institut für Produktionstechnologie IPT in Aachen und war anschließend bis 2008 stellvertretender Leiter des Design Zentrums Bremen. Sascha Peters ist Autor zahlreicher Fachbücher zu nachhaltigen Werkstoffen, smarten Materialien, innovativen Fertigungsverfahren und energetischen Technologien und zählt zu den führenden Materialexperten und Trendscouts für neue Technologien. Seit 2014 ist er Mitglied im Beirat der Förderinitiative „Zwanzig20 – Partnerschaft für Innovation" im Auftrag des Bundesministeriums für Bildung und Forschung.

04

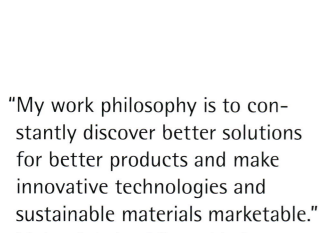

"My work philosophy is to constantly discover better solutions for better products and make innovative technologies and sustainable materials marketable."

„Meine Arbeitsphilosophie ist, beständig bessere Lösungen für bessere Produkte zu finden und innovative Technologien und nachhaltige Materialien marktfähig zu machen."

What impressed you most during the Red Dot judging process?
The jury with its experts from so many different countries. This network of opinions emerging from different cultural backgrounds is the basis for global acceptance and the success of the Red Dot Award.

What properties must a new material have to convince you of its outstanding quality?
With regards to our resources becoming ever scarcer, the issues we are faced with concerning waste disposal and the challenges resulting from a growing world population, I judge the development of a material as outstanding when it opens up the possibility for sustainable use and leaves a particularly small ecological footprint.

Which material development from the last hundred years has had the biggest influence on today's world?
Plastics. They enable the creation of almost limitless properties. However, most of them are not bio-degradable.

Was hat Sie bei der Red Dot-Jurierung am meisten beeindruckt?
Die Herkunft der Jury mit Experten aus den unterschiedlichsten Ländern. Dieses Netzwerk aus Meinungen verschiedener kultureller Einflüsse ist die Grundlage für die globale Akzeptanz und den Erfolg des Red Dot Awards.

Welche Eigenschaften muss ein neues Material vorweisen, um Sie von seiner herausragenden Qualität zu überzeugen?
Mit Blick auf die knapper werdenden Ressourcen, die Probleme, die wir mit der Abfallentsorgung haben, und die Herausforderungen, die sich durch die wachsende Weltbevölkerung ergeben, bewerte ich Materialentwicklungen als herausragend, wenn sie eine Möglichkeit zu einer nachhaltigen Nutzung aufzeigen und einen besonders geringen ökologischen Fußabdruck offenbaren.

Welche Materialentwicklung der letzten hundert Jahre hat den größten Einfluss auf die heutige Zeit?
Kunststoffe. Ihre Qualitäten lassen sich nahezu beliebig einstellen. Ihr Großteil ist jedoch nicht biologisch abbaubar.

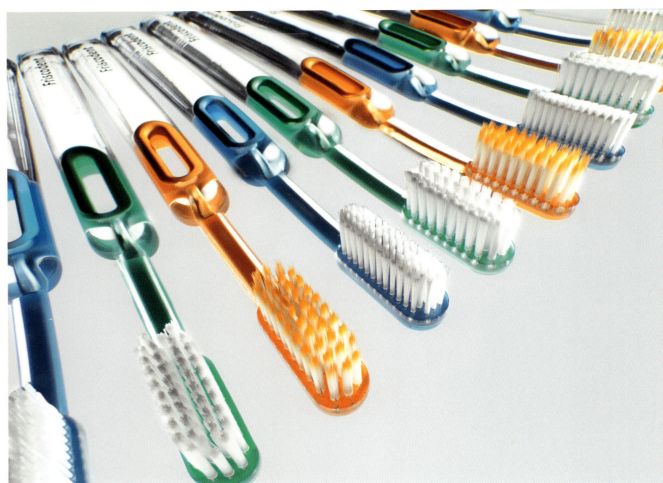

01

Oliver Stotz
Germany
Deutschland

Oliver Stotz, born in 1961 in Stuttgart, Germany, studied industrial design at the University Essen, Germany and at the Royal College of Art in London. As the founder of his own studio "stotz-design.com" in Wuppertal, Germany, he has more than 20 years of experience in the fields of industrial and corporate design. His studio has created established brands such as Proseat in the automotive sector and Blomus in the field of glass, porcelain and ceramic. With his eight-strong team, Stotz advises companies in various industries regarding the implementation and realisation of new design concepts. Since 2010, Oliver Stotz has been a board member of the foundation "Mia Seeger Stiftung", which promotes young designers after graduation with the "Mia Seeger Preis" award. In addition, he is a lecturer at the design department of the University Wuppertal. His design achievements have received several awards and have been on display in numerous exhibitions.

Oliver Stotz, 1961 in Stuttgart geboren, studierte Industriedesign an der Universität Essen und am Royal College of Art in London. Als Gründer des Studios „stotz-design.com" mit Sitz in Wuppertal kann er inzwischen auf mehr als 20 Jahre Berufserfahrung in den Bereichen Industrial und Corporate Design zurückgreifen. In seinem Studio entstanden bereits etablierte Marken wie Proseat im Automotive-Sektor oder auch Blomus im Bereich Glas, Porzellan, Keramik. Mit seinem achtköpfigen Team berät Stotz national wie international Unternehmen aus unterschiedlichsten Branchen bei der Implementierung und Umsetzung von neuen Designkonzepten. Seit 2010 ist Oliver Stotz Vorstandsmitglied der „Mia Seeger Stiftung", die junge Designerinnen und Designer nach ihrem Studienabschluss mit dem „Mia Seeger Preis" fördert. Darüber hinaus ist er als Dozent an der Universität Wuppertal im Fachbereich Design tätig. Seine Designleistungen wurden vielfach ausgezeichnet und in zahlreichen Ausstellungen präsentiert.

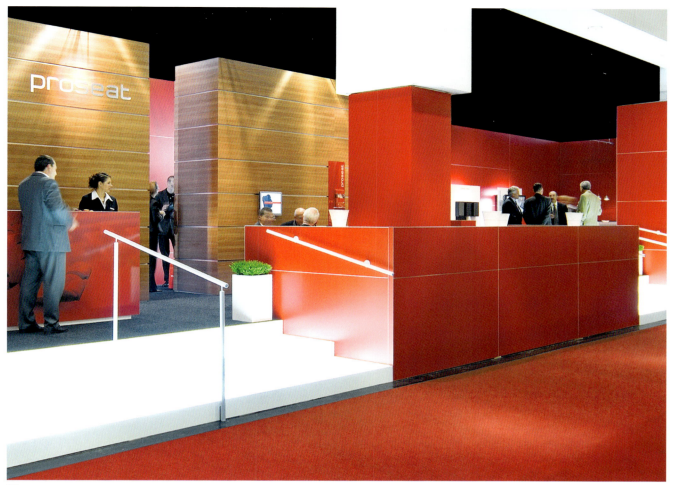

02

"My work philosophy is:
a good designer is always
ahead of his time."

„Meine Arbeitsphilosophie ist:
Ein guter Designer ist immer
seiner Zeit voraus."

In which product industry will Germany be most successful in the next ten years?
Germany is currently technology leader in the field of transportation and will achieve more successes in this field. However, networked thinking of the global players is a prerequisite for this to happen.

What are the main challenges in a designer's everyday life?
Keeping it simple is always the main challenge.

The Red Dot Award has been uncovering the best designs for 60 years now. Which product innovation would you like to see in the next 60 years, and why?
Driven by the thought that in future we will have to deal with the digitisation of many processes, as a result of which more and more things will become virtual, I wish to see materiality as product innovation, that haptic qualities become an enrichment of things.

What do you take special notice of when you are assessing a product as a jury member?
Whether it is an original design.

Mit welcher Produktbranche wird Deutschland in den nächsten zehn Jahren am erfolgreichsten sein?
Im Bereich Transportation ist Deutschland Technologieführer und wird Erfolge erzielen können. Voraussetzung ist allerdings, dass die Global Player vernetzt denken.

Was sind große Herausforderungen im Alltag eines Designers?
Keep it simple! Das ist immer wieder die große Aufgabe.

Der Red Dot Award ermittelt seit 60 Jahren die besten Gestaltungen. Welche Produktinnovation würden Sie sich für die nächsten 60 Jahre wünschen, und warum?
Von dem Gedanken getrieben, sich zukünftig mit der Digitalisierung vieler Prozesse auseinandersetzen zu müssen, in deren Folge immer mehr Dinge virtuell werden, wünsche ich mir bei den Produkten als Innovation die Materialität, das Haptische als konkret erlebbare Bereicherung der Dinge.

Worauf achten Sie besonders, wenn Sie ein Produkt als Juror bewerten?
Darauf, ob die Eigenständigkeit gewährleistet ist.

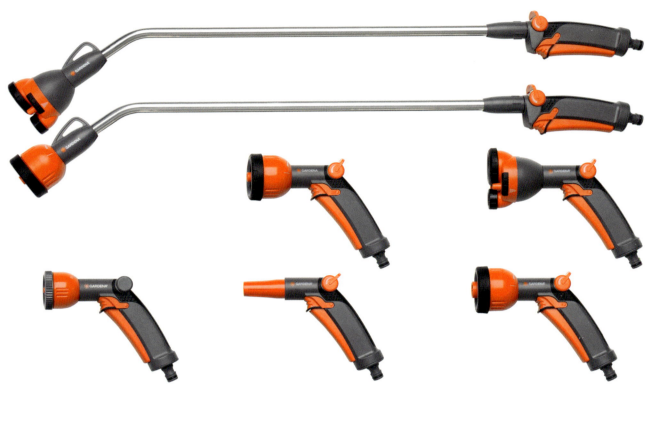

01

Aleks Tatic
Germany / Italy
Deutschland / Italien

Aleks Tatic, born 1969 in Cologne, Germany, is product designer and founder of Tatic Designstudio in Milan, Italy. After his studies at the Art Center College of Design in the USA and Switzerland, he specialised in his focal areas, sports and lifestyle products, in various international agencies in London and Milan. Afterwards, he guided the multiple award-winning Italian design studio Attivo Creative Resource to international success, leading the agency for 12 years. Together with his multicultural team of designers and product specialists, he today designs and develops – amongst others – sailing yachts, sporting goods, power tools, FMCGs and consumer electronics for European and Asian premium brands. Aleks Tatic lectures practice-oriented industrial design and innovation management at various European universities and seminars.

Aleks Tatic, geboren 1969 in Köln, ist Produktdesigner und Gründer der Agentur Tatic Designstudio in Mailand. Nach seinem Studium am Art Center College of Design in den USA und der Schweiz hat er sich zunächst in verschiedenen internationalen Büros in London und Mailand auf sein Schwerpunktgebiet Sport- und Lifestyleprodukte spezialisiert. Danach führte er zwölf Jahre lang das mehrfach ausgezeichnete italienische Designbüro Attivo Creative Resource zu internationalem Erfolg. Heute gestaltet und entwickelt er mit seinem multikulturellen Team von Designern und Produktspezialisten u. a. Segeljachten, Sportgeräte, Hobby- und Profiwerkzeuge, FMCGs und Unterhaltungselektronik für europäische und asiatische Premiummarken. Aleks Tatic unterrichtet an verschiedenen europäischen Hochschulen und Seminaren praxisorientiertes Industriedesign und Innovationsmanagement.

02

"My work philosophy is:
it has to be fun (for my
colleagues and me)!"

„Meine Arbeitsphilosophie ist:
Es muss uns (meinen Kollegen
und mir) Spaß machen!"

Which of the projects in your lifetime are you particularly proud of?
That would be our first sailing yacht, which we designed and then promptly won the "Boat of the Year" award for, even though we were new to the industry. This is the perfect proof that you can design any product even without specialising in the industry.

The Red Dot Award has been uncovering the best designs for 60 years now. Which product innovation would you like to see in the next 60 years, and why?
I would love to have someone launch Scotty's beamer in the market. This would cut travel times to our clients significantly.

What significance does winning an award in a design competition have for the manufacturer?
Many of our clients want to enter the Red Dot Award, because it is often hard for their clients – consumers or buying decision makers – to judge the design quality. Awards such as the Red Dot, as internationally recognised quality seals, are excellent indicators for this purpose.

Auf welches Projekt in Ihrem Leben sind Sie besonders stolz?
Auf unsere erste Segeljacht, die wir als komplett Branchenfremde gestaltet und dann prompt die Auszeichnung „Boat of the Year" gewonnen haben – der perfekte Beweis dafür, dass man jedes Produkt auch ohne Branchenspezialisierung gestalten kann.

Der Red Dot Award ermittelt seit 60 Jahren die besten Gestaltungen. Welche Produktinnovation würden Sie sich für die nächsten 60 Jahre wünschen, und warum?
Ich würde mich sehr freuen, wenn endlich jemand Scottys „Beamer" auf den Markt bringen würde. Die Reisezeiten zu unseren Kunden würden sich endlich deutlich verringern.

Welche Bedeutung hat die Auszeichnung in einem Designwettbewerb für den Hersteller?
Viele unserer Kunden möchten gerne beim Red Dot Award mitmachen. Denn für ihre Kunden – Verbraucher oder Kaufentscheider – ist die Designqualität oft schwierig zu beurteilen. Auszeichnungen wie der Red Dot als international anerkanntes Qualitätssiegel bilden hierfür herausragende Gradmesser.

01

Nils Toft
Denmark
Dänemark

Nils Toft, born in Copenhagen in 1957, graduated as an architect and designer from the Royal Danish Academy of Fine Arts in Copenhagen in 1985. He also holds a Master's degree in Industrial Design and Business Development. Starting his career as an industrial designer, Nils Toft joined the former Christian Bjørn Design in 1987, an internationally active design studio in Copenhagen with branches in Beijing and Ho Chi Minh City. Within a few years, he became a partner of CBD and, as managing director, ran the business. Today, Nils Toft is the founder and managing director of Designidea. With offices in Copenhagen and Beijing, Designidea works in the following key fields: communication, consumer electronics, computing, agriculture, medicine, and graphic arts, as well as projects in design strategy, graphic and exhibition design.

Nils Toft, geboren 1957 in Kopenhagen, machte seinen Abschluss als Architekt und Designer 1985 an der Royal Danish Academy of Fine Arts in Kopenhagen. Er verfügt zudem über einen Master im Bereich Industrial Design und Business Development. Zu Beginn seiner Karriere als Industriedesigner trat Nils Toft 1987 bei dem damaligen Christian Bjørn Design ein, einem international operierenden Designstudio in Kopenhagen, das mit Niederlassungen in Beijing und Ho-Chi-Minh-Stadt vertreten ist. Innerhalb weniger Jahre wurde er Partner bei CBD und leitete das Unternehmen als Managing Director. Heute ist Nils Toft Gründer und Managing Director von Designidea. Mit Büros in Kopenhagen und Beijing operiert Designidea in verschiedenen Hauptbereichen: Kommunikation, Unterhaltungselektronik, Computer, Landwirtschaft, Medizin und Grafikdesign sowie Projekte in den Bereichen Designstrategie, Grafik- und Ausstellungsdesign.

02

"My work philosophy is: design is a language that carries your brand and tells the story of how good your products are."

„Meine Arbeitsphilosophie lautet: Design ist eine Sprache, die deine Marke transportiert und erzählt, wie gut deine Produkte sind."

When did you first think consciously about design?
I was around seven or eight years old when my parents introduced me to arts and design. I remember how I, in my mind, redesigned the interior of my playmates' homes, when I saw how different they looked compared to mine.

What motivates you to get up in the morning?
I have always looked at life as a big apple and I can't wait to get up in the morning to take the next bite.

What impressed you most during the Red Dot judging process?
The vast number of exceptional entries and the fantastic discussions with the other jury members.

What significance does winning an award in a design competition have for the designer?
It is a tribute to the uniqueness of their talent and a reminder not to take talent for granted.

Wann haben Sie das erste Mal bewusst über Design nachgedacht?
Ich war ungefähr sieben oder acht Jahre alt, als mich meine Eltern an Kunst und Design heranführten. Ich erinnere mich daran, wie ich in meinem Kopf die Inneneinrichtung der Häuser meiner Spielkameraden umgestaltete, als ich sah, wie stark sie sich von meiner unterschied.

Was motiviert Sie, morgens aufzustehen?
Ich habe das Leben schon immer als einen großen Apfel betrachtet und kann es kaum erwarten, morgens aufzustehen und den nächsten Bissen zu nehmen.

Was hat Sie bei der Red Dot-Jurierung am meisten beeindruckt?
Die riesige Anzahl bemerkenswerter Einreichungen und die fantastischen Diskussionen mit den anderen Jurymitgliedern.

Welche Bedeutung hat die Auszeichnung in einem Designwettbewerb für den Designer?
Sie zollt der Einzigartigkeit seines Talentes Tribut und erinnert daran, Talent nicht als selbstverständlich anzusehen.

01

Prof. Danny Venlet
Belgium
Belgien

Professor Danny Venlet was born in 1958 in Victoria, Australia and studied interior design at Sint-Lukas, the Institute for Architecture and Arts in Brussels. Back in Australia in 1991, Venlet started to attract international attention with large-scale interior projects such as the Burdekin hotel in Sydney, and Q-bar, an Australian chain of nightclubs. His design projects range from private mansions, lofts, bars and restaurants all the way to showrooms and offices of large companies. The interior projects and the furniture designs of Danny Venlet are characterised by their contemporary international style. He says that the objects arise from an interaction between art, sculpture and function. These objects give a new description to the space in which they are placed – with respect, but also with relative humour. Today, Danny Venlet teaches his knowledge to students at the Royal College of the Arts in Ghent.

Professor Danny Venlet wurde 1958 in Victoria, Australien, geboren und studierte Interior Design am Sint-Lukas Institut für Architektur und Kunst in Brüssel. Nachdem er 1991 wieder nach Australien zurückgekehrt war, begann er, mit der Innenausstattung großer Projekte wie dem Burdekin Hotel in Sydney und der Q-Bar, einer australischen Nachtclub-Kette, internationale Aufmerksamkeit zu erregen. Seine Designprojekte reichen von privaten Wohnhäusern über Lofts, Bars und Restaurants bis hin zu Ausstellungsräumen und Büros großer Unternehmen. Die Innenausstattungen und Möbeldesigns von Danny Venlet sind durch einen zeitgenössischen, internationalen Stil ausgezeichnet und entspringen, wie er sagt, der Interaktion zwischen Kunst, Skulptur und Funktion. Seine Objekte geben den Räumen, in denen sie sich befinden, eine neue Identität – mit Respekt, aber auch mit einer Portion Humor. Heute vermittelt Danny Venlet sein Wissen als Professor an Studenten des Royal College of the Arts in Gent.

02

"My work philosophy is to create objects that, while embodying the theory of relativity, leave us no other choice than to act differently."

„Meine Arbeitsphilosophie besteht darin, Objekte zu kreieren, die – obwohl sie die Relativitätstheorie verkörpern – uns keine andere Wahl lassen, als anders zu handeln."

Which design project was the greatest intellectual challenge for you?
The Easyrider for Bulo, because my briefing consisted of the two opposing concepts of "relaxation/office", which gave birth to, if I may say so, a great innovation.

Which country do you consider to be a pioneer in product design?
Of course Belgium, due to its excellent training, its history and companies which are willing to take the design path.

What is the advantage of commissioning an external designer compared to having an in-house design team?
The biggest advantage is that an external designer is free-spirited and not conditioned by the constraints that an in-house designer might have encountered working in a company. In order to innovate you need to be able to look beyond!

Welches Ihrer bisherigen Designprojekte war die größte Herausforderung für Sie?
Der Easyrider für Bulo, da mein Briefing aus den zwei gegensätzlichen Konzepten „Entspannung/Büro" bestand, aus denen eine, wenn ich das einmal so sagen darf, großartige Innovation hervorging.

Welche Nation ist für Sie Vorreiter im Produktdesign?
Natürlich Belgien, aufgrund seiner hervorragenden Ausbildungsmöglichkeiten, seiner geschichtlichen Vergangenheit und seiner Unternehmen, die gewillt sind, auf Design zu setzen.

Worin liegt der Vorteil, einen externen Designer zu beauftragen, im Vergleich zu einem Inhouse-Designteam?
Der größte Vorteil ist, dass ein externer Designer über mehr geistige Freiheit verfügt und nicht von den Beschränkungen konditioniert ist, denen ein Inhouse-Designer in einer Firma unterworfen ist. Um Innovationen zu kreieren, muss man die Möglichkeit haben, über den Tellerrand hinauszuschauen!

01

Günter Wermekes
Germany
Deutschland

Günter Wermekes, born in Kierspe, Germany in 1955, is a goldsmith and designer. After many years of practice as an assistant and head of the studio of Professor F. Becker in Düsseldorf, he founded his own studio in 1990. He attracted great attention with his jewellery collection "Stainless steel and brilliant", which he has been presenting at national and international fairs since 1990. His designs are also appreciated by renowned manufacturers such as BMW, Rodenstock, Niessing or Tecnolumen, for which he designed, among other things, accessories, glasses, watches and door openers. In exhibitions and lectures around the world he has illustrated his personal design philosophy, which is: "Minimalism means reducing things more and more to their actual essence, and thus making it visible." He has also implemented this motto in his design of the Red Dot Trophy. In 2014, Günter Wermekes was one of ten finalists in the design competition for the medal of the Tang Prize, the "Asian Nobel Prize". His works have won numerous prizes and are part of renowned collections.

Günter Wermekes, 1955 geboren in Kierspe, ist Goldschmied und Designer. Nach langjähriger Tätigkeit als Assistent und Werkstattleiter von Professor F. Becker in Düsseldorf gründete er 1990 sein eigenes Studio. Große Aufmerksamkeit erregte er mit seiner Schmuckkollektion „Edelstahl und Brillant", die er ab 1990 auf nationalen und internationalen Messen präsentierte. Sein Design schätzen auch namhafte Hersteller wie BMW, Rodenstock, Niessing oder Tecnolumen, für die er u. a. Accessoires, Brillen, Uhren und Türdrücker entwarf. In Ausstellungen und Vorträgen weltweit verdeutlicht er seine persönliche Gestaltungsphilosophie, dass „Minimalismus bedeutet, Dinge so auf ihr Wesen zu reduzieren, dass es dadurch sichtbar wird". Diesen Leitspruch hat er auch in seinem Entwurf der Red Dot Trophy umgesetzt. 2014 war Günter Wermekes einer von zehn Finalisten des Gestaltungswettbewerbs für die Preismedaille des „Tang Prize", des „Asiatischen Nobelpreises". Seine Arbeiten wurden mehrfach prämiert und befinden sich in bedeutenden Sammlungen.

01

01 Der Halbkaräter (The Half-Carat-Piece)
Ring model R0011 made from stainless steel and a 0.5 ct tw/vsi brilliant-cut diamond for the "Stainless steel and brilliant" collection
Ringmodell R0011 aus Edelstahl und einem Brillanten von 0,5 ct tw/vsi für die Kollektion „Edelstahl und Brillant"

02 Red Dot Trophy
Trophy made from stainless steel and acrylic glass for the Red Dot Design Award, Essen, Germany, 2013
Preistrophäe aus Edelstahl und Plexiglas für den Red Dot Design Award, Essen, 2013

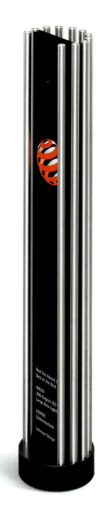

02

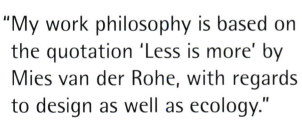
"My work philosophy is based on the quotation 'Less is more' by Mies van der Rohe, with regards to design as well as ecology."
„Meine Arbeitsphilosophie fußt auf dem Zitat von Mies van der Rohe ‚Weniger ist mehr', und zwar in gestalterischer Hinsicht wie auch im Hinblick auf Ökologie."

When did you first consciously think about design?
Around 1970/72, when my art teacher at grammar school gave me the following assignment: Design and produce a wooden chair. The chair has never been completed, but my interest in designing objects of everyday life was awakened.

What are your sources of inspiration?
Contrasts, borders, irritations. Actually, anything you can discover outside the box. And the music of Johann Sebastian Bach as a common thread.

The Red Dot Award has been uncovering the best designs for 60 years now. Which product innovations would you like to see in the next 60 years?
I would like to see innovations that are dealing responsibly with our living environment, planet Earth.

What impressed you most during the Red Dot judging process?
The fact that an awareness of good and useful design is attracting increasing attention.

Wann haben Sie das erste Mal bewusst über Design nachgedacht?
So um 1970/72, als mein Kunstlehrer am Gymnasium mir die Aufgabe stellte: Entwirf und fertige einen Holzstuhl. Der Stuhl ist zwar nie fertig geworden, aber das Interesse für die Gestaltung von Alltagsgegenständen war geweckt.

Was sind Ihre Inspirationsquellen?
Gegensätze, Grenzbereiche, Irritationen. Eigentlich alles, was man jenseits des „Tellerrands" entdecken kann. Und als „Roter Faden" die Musik von Johann Sebastian Bach.

Der Red Dot Award ermittelt seit 60 Jahren die besten Gestaltungen. Welche Produktinnovationen würden Sie sich für die nächsten 60 Jahre wünschen?
Ich wünsche mir Innovationen, die verantwortungsvoll mit unserem Lebensraum Erde umgehen.

Was hat Sie bei der Red Dot-Jurierung am meisten beeindruckt?
Dass das Bewusstsein für gutes und sinnvolles Design immer größere Kreise zieht.

Alphabetical index manufacturers and distributors
Alphabetisches Hersteller- und Vertriebs-Register

Alphabetical index manufacturers and distributors
Alphabetisches Hersteller- und Vertriebs-Register

Alphabetical index designers
Alphabetisches Designer-Register

Alphabetical index designers
Alphabetisches Designer-Register

Alphabetical index designers
Alphabetisches Designer-Register

Alphabetical index designers
Alphabetisches Designer-Register

Alphabetical index designers
Alphabetisches Designer-Register

Alphabetical index designers
Alphabetisches Designer-Register

reddot edition

Editor | Herausgeber
Peter Zec

Project management | Projektleitung
Jennifer Bürling

Project assistance | Projektassistenz
Sophie Angerer
Tatjana Axt
Theresa Falkenberg
Constanze Halsband
Danièle Huberty
Estelle Limbah
Judith Lindner
Melanie Masino
Anamaria Sumic

Editorial work | Redaktion
Bettina Derksen, Simmern, Germany
Eva Hembach, Vienna, Austria
Catharina Hesse
Burkhard Jacob
Karin Kirch, Essen, Germany
Karoline Laarmann, Dortmund, Germany
Sarah Latussek, Cologne, Germany
Bettina Laustroer, Wuppertal, Germany
Kirsten Müller, Essen, Germany
Astrid Ruta, Essen, Germany
Martina Stein, Otterberg, Germany
Achim Zolke

Proofreading | Lektorat
Klaus Dimmler (supervision), Essen, Germany
Mareike Ahlborn, Essen, Germany
Jörg Arnke, Essen, Germany
Wolfgang Astelbauer, Vienna, Austria
Sabine Beeres, Leverkusen, Germany
Dawn Michelle d'Atri, Kirchhundem, Germany
Annette Gillich-Beltz, Essen, Germany
Eva Hembach, Vienna, Austria
Karin Kirch, Essen, Germany
Norbert Knyhala, Castrop-Rauxel, Germany
Laura Lothian, Vienna, Austria
Regina Schier, Essen, Germany
Anja Schrade, Stuttgart, Germany

Translation | Übersetzung
Heike Bors-Eberlein, Tokyo, Japan
Patrick Conroy, Lanarca, Cyprus
Stanislaw Eberlein, Tokyo, Japan
William Kings, Wuppertal, Germany
Cathleen Poehler, Montreal, Canada
Tara Russell, Dublin, Ireland
Jan Stachel-Williamson, Christchurch, New Zealand
Philippa Watts, Exeter, Great Britain
Andreas Zantop, Berlin, Germany
Christiane Zschunke, Frankfurt am Main, Germany

Layout | Gestaltung
Lockstoff Design GmbH, Grevenbroich, Germany
Susanne Coenen
Katja Kleefeld
Judith Maasmann
Stephanie Marniok
Lena Overkamp
Nicole Slink

Photographs | Fotos
Dragan Arrigler ("Alpina ESK Pro", juror Jure Miklavc)
Kaido Haagen ("Fleximoover", juror Martin Pärn)
Jäger & Jäger ("Lodelei", juror Martin Pärn)
Kompan marketing department ("elements", juror Nils Toft)
MBD Design ("Optifuel Trailer", juror Vincent Créance)
Masaki Ogawa ("Black+white", juror Herman Hermsen)
Christophe Recoura ("Alstom – Paris Tram T7"/
"Bombardier Regio 2N", juror Vincent Créance)
studio aisslinger ("Upcycling & tuning"/"Luminaire",
juror Werner Aisslinger)
Roberto Turci ("Helmet for Bimota SA", juror Jure Miklavc)
Wagner Ziegelmeyer ("Fabrimar Lucca", juror
Guto Indio da Costa)

Page | Seite
472
Name | Name
Tokyo_Institute_of_Technology_Centennial_Hall_2009
Copyright | Urheber
Wiiii
Source | Quelle
http://commons.wikimedia.org/wiki/File:Tokyo_
Institute_of_Technology_Centennial_Hall_2009.jpg

Distribution of this photograph is protected by copyright
law. A license is required for distribution. More information
is available at
Diese Fotografie unterliegt hinsichtlich der Verbreitung
dem Urheberrechtsschutz. Die Verbreitung bedarf einer
Lizenz. Nähere Angaben dazu finden Sie unter
http://creativecommons.org/licenses/by-sa/3.0/deed.en

Jury photographs | Jurorenfotos
Simon Bierwald, Dortmund, Germany

In-company photos | Werkfotos der Firmen

Production and litography |
Produktion und Lithografie
tarcom GmbH, Gelsenkirchen, Germany
Gregor Baals
Jonas Mühlenweg
Bernd Reinkens

Printing | Druck
Dr. Cantz'sche Druckerei Medien GmbH,
Ostfildern, Germany

Bookbindery | Buchbinderei
BELTZ Bad Langensalza GmbH
Bad Langensalza, Germany

Red Dot Design Yearbook 2015/2016
Living: 978-3-89939-174-9
Doing: 978-3-89939-175-6
Working: 978-3-89939-176-3
Set (Living, Doing & Working): 978-3-89939-173-2

© 2015 Red Dot GmbH & Co. KG, Essen, Germany

The Red Dot Award: Product Design
competition is the continuation of the
Design Innovations competition.
Der Wettbewerb „Red Dot Award: Product Design"
gilt als Fortsetzung des Wettbewerbs
„Design Innovationen".

All rights reserved, especially those of translation.
Alle Rechte vorbehalten, besonders die der Übersetzung
in fremde Sprachen.

No liability is accepted for the completeness
of the information in the appendix.
Für die Vollständigkeit der Angaben im Anhang
wird keine Gewähr übernommen.

Publisher + worldwide distribution |
Verlag + Vertrieb weltweit
Red Dot Edition
Design Publisher | Fachverlag für Design
Contact | Kontakt
Sabine Wöll
Gelsenkirchener Str. 181
45309 Essen, Germany
Phone +49 201 81418-22
Fax +49 201 81418-10
E-mail edition@red-dot.de
www.red-dot-edition.com
www.red-dot-shop.com
Book publisher ID no. | Verkehrsnummer
13674 (Börsenverein Frankfurt)

**Bibliographic information published
by the Deutsche Nationalbibliothek**
The Deutsche Nationalbibliothek
lists this publication in the Deutsche
Nationalbibliografie; detailed bibliographic
data are available on the Internet at
http://dnb.ddb.de
Bibliografische Information
der Deutschen Nationalbibliothek
Die Deutsche Nationalbibliothek verzeichnet
diese Publikation in der Deutschen
Nationalbibliografie; detaillierte
bibliografische Daten sind im Internet über
http://dnb.ddb.de abrufbar.